Understanding Human Communication

Understanding Human Communication

THIRTEENTH EDITION

Ronald B. Adler

SANTA BARBARA CITY COLLEGE

George Rodman

BROOKLYN COLLEGE
CITY UNIVERSITY OF NEW YORK

Athena du Pré

UNIVERSITY OF WEST FLORIDA

OXFORD   NEW YORK
OXFORD UNIVERSITY PRESS

Oxford University Press is a department of the University of Oxford.
It furthers the University's objective of excellence in research,
scholarship, and education by publishing worldwide. Oxford is
a registered trade mark of Oxford University Press in the UK and
certain other countries.

Published in the United States of America by Oxford University Press
198 Madison Avenue, New York, NY 10016, United States of America.

Library of Congress Cataloging-in-Publication Data

Names: Adler, Ronald B. (Ronald Brian), 1946– author. | Rodman, George R.,
    1948– author. | DuPré, Athena, author.
Title: Understanding human communication / Ronald B. Adler, Santa Barbara
    City College; George Rodman, Brooklyn College, City University of
    New York ; Athena duPré, University of West Florida.
Description: Thirteenth edition. | New York: Oxford University Press, [2016]
Identifiers: LCCN 2016033665 | ISBN 9780190297084
Subjects: LCSH: Communication. | Interpersonal communication.
Classification: LCC P90 .A32 2016 | DDC 302.2—dc23 LC record available at
    https://lccn.loc.gov/2016033665

9 8 7 6 5 4 3 2 1

Printed by LSC Communications, United States of America

Brief Contents

Contents

CHAPTER 3 Communication and Culture 67

PART TWO COMMUNICATION ELEMENTS

CHAPTER 4 Language 95

CHAPTER ⑤ **Listening 123**

CHAPTER ⑥ **Nonverbal Communication 153**

PART THREE INTERPERSONAL COMMUNICATION

CHAPTER 7 Understanding Interpersonal Communication 181

CHAPTER 8 Managing Conflict in Interpersonal Relationships 213

PART FIVE PUBLIC COMMUNICATION

Preface

If you want to push most communication professors' buttons, claim that the principles they study and teach are "just common sense."

The truth is that communication, like many things in life, may *look* straightforward. But beneath the veneer of simplicity, it's fraught with challenges and questions. For example:

- Why do others misunderstand you? (And why do they accuse you of not understanding them?)
- How can you know when another person is telling the truth?
- When it comes to communicating, are men from Mars and women from Venus, or are we more or less the same?
- Why do we often get the most defensive when we know we're wrong?
- What makes some teams so effective and others disappointing?
- Why are so many well-informed speakers uninteresting and hard to understand?
- Why do people feel so nervous when speaking in public, and what can they do about it?

Understanding Human Communication answers questions such as these. It aims to provide an engaging, comprehensive, useful introduction to the academic study of human communication as it is practiced in the 21st century. To see how well this book succeeds, we invite you flip to any page and ask three questions: Is the content important? Is the explanation clear? Is it useful?

Approach

This 13th edition builds on the approach that has served well over half a million students and their professors. Rather than take sides in the theory-versus-skills debate that often rages in our discipline, *Understanding Human Communication* treats scholarship and skill development as mutually reinforcing. Its reader-friendly approach strives to present material clearly without being overly simplistic. Examples on virtually every page make concepts clear and interesting. A contemporary design makes the material inviting, as do amusing and instructive cartoons and photos that link concepts in the text to today's world.

New to This Edition

Beyond its user-friendly voice and engaging design, this edition reflects both the growth of scholarship and changing trends in the academic marketplace. Longtime users will discover not only a more contemporary look and feel but expanded coverage of key concepts and a wealth of new learning tools.

New Content

- **Expanded coverage of social media.** Throughout the book, readers will find new and updated coverage of topics including social media etiquette and the uses and gratifications of social media (Chapter 1), the risks of overusing mediated communication channels (Chapters 1 and 7), identity management on social media (Chapter 2), online social support (Chapter 5), the differences between mediated and in-person close relationships (Chapter 7), tips for meeting an online date for the first time (Chapter 7), working in virtual groups (Chapter 9), and cyberspace work environments (Chapter 10). New and updated "Understanding Communication Technology" boxes highlight the increasingly important role of technology in human communication.

- **Updated discussions of communication and culture.** Chapter 3 is devoted entirely to communication and culture. It focuses on cocultural factors that students are likely to encounter close to home every day: ethnicity, regional differences, gender/sexual orientation, religion, physical abilities, age/generation, and socioeconomic status. Updated material addresses cultural hegemony and intersectionality theory. The discussion of culture extends throughout the book, most visibly in "Understanding Diversity" boxes. New and updated boxes in this edition explore the challenge of managing personal pronouns in an era of increasing gender fluidity (Chapter 3), the effect of language on worldview (Chapter 4), and the advantages of multicultural teams (Chapter 10). In-text discussions and captioned photos address topics such as microaggressions (Chapter 2), cultural conflict styles (Chapter 8), and cultural leadership preferences (Chapter 9).

- **New tips for career success.** "@Work" boxes appear throughout the text, covering new and expanded topics such as the connection between communication skills and career success (Chapter 1), professional identity management and job interviewing (Chapter 2), the risks of humblebragging (Chapter 2), vocal cues and career success (Chapter 6), and new examples of effective presentations at work (Chapters 11–14). An online appendix, "Communicating for Career Success," is available at **www.oup.com/us/ adleruhc** and contains advice on how to communicate professionally in seeking employment and once on the job.

- **New chapter-opening profiles.** These stories highlight real-life communication challenges and are woven into the fabric of the chapter content. Profiles feature interesting and relevant personal stories, including those of Zappos founder Tony Hsieh, television personality and former Olympian Caitlyn Jenner, young publishing executive Erica Nicole, body language expert Amy Cuddy, relationship expert Brené Brown, startup whiz Matt Mullenweg, human rights activist Malala Yousafzai, students who have overcome disabilities, and others who have fought for equal rights. Questions at the end of the profiles prompt students to connect the material to their own lives.

- **New and expanded coverage of important topics in each chapter.** For example:

  o Chapter 1 discusses the social and physical benefits of effective communication. It includes a new "Understanding Communication Technology" box on controlling social media use and a new checklist on social media etiquette.

  o Chapter 2 now includes an expanded discussion of self-esteem, social influences on the self-concept, how stereotyping and scripts degrade the

quality of communication, and how empathy and frame switching provide better understanding of others.

o Chapter 3 includes a new "Understanding Diversity" box on gender
pronouns as well as expanded discussions of cultural dominance and
religion-based stereotyping.

o Chapter 4 contains a new discussion of the differences between connotative and denotative meaning. In addition, it includes a clearer, expanded
explanation of pragmatic rules and a more complete explanation of how
power relationships are expressed in language use. Chapter 4 also offers
a current view of similarities and differences between characteristically
male and female speech.

o Chapter 5 offers new evidence on the personal and career benefits of
effective listening, as well as gender differences in social support.

o Chapter 6 addresses contemporary speech mannerisms such as uptalk
and vocal fry.

o Chapter 7 now includes treatment of the role communication plays in
maintaining friendships, family connections, and romantic relationships. It also includes a clearer treatment of how dialectical tensions
shape communication in close relationships, and the roles of lies and
evasions in relational maintenance.

o Chapter 8 (now titled "Managing Conflict in Interpersonal Relationships") has been reorganized to present both familiar and new material
in a clearer and more useful way.

o Chapter 9 includes new material on transformational leadership.

o Chapter 10 includes three new checklists on teamwork and a new table
on decision-making methods.

o Chapters 11–14 have all new sample speeches, outlines, and analyses.
Their topics include many forms of diversity, including LGBTQ life, as
well as current controversies such as gun control. Chapter 13 provides
new examples of how to spark visual interest in a speech, how to use
vocal citations, and how to incorporate photos, videos, and audio files
into a presentation.

Learning Tools

- **Checklists** in every chapter, many of them new for this edition, provide
handy reference tools to help students build their skill sets and internalize
what they have learned. New checklists address how to use social media
courteously (Chapter 1); perception checking (Chapter 2); minimizing misunderstandings (Chapter 4); mindful listening, paraphrasing, and controlling defensiveness (Chapter 5); being a better friend (Chapter 7); creating
positive communication climates (Chapter 8); getting slackers to do their
share in groups and working with difficult bosses (Chapter 9); and dealing
with difficult team members (Chapter 10).

- **Self-assessments** invite students to evaluate and improve their communication skills and to consider their identities as communicators. These features
include quizzes to help students understand more about their listening
styles (Chapter 5), love languages (Chapter 7), interpersonal communication climates (Chapter 8), leadership and followership styles (Chapter 9),
team effectiveness (Chapter 10), and more.

- **Learning Objectives** now correspond to major headings in each chapter and coordinate with the end-of-chapter summary and review. They provide a clear map of what students need to learn and where to find that material.

- A new "**Making the Grade**" section at the end of each chapter helps students test and deepen their mastery of the material. Organized by learning objective, this section summarizes the key points from the text and includes related questions and prompts to promote understanding and application.

- "**Understanding Communication Technology**" boxes highlight the increasingly important role of technology in human communication.

- "**Understanding Diversity**" boxes provide a more in-depth treatment of intercultural communication topics.

- "**@Work**" boxes show students how key concepts from the text operate in the workplace.

- "**Ethical Challenge**" boxes engage students in debates such as whether honesty is always the best policy, the acceptability of presenting multiple identities, and how to deal effectively with difficult group members.

- **Key Terms** are boldfaced on first use and listed at the end of each chapter, and a new **Marginal Glossary** helps students learn new terms.

- **Activities** at the end of each chapter can be completed in class and help students apply the material to their everyday lives. Additional activities are available in the Instructor's Manual (*The Complete Guide to Teaching Communication*) at **https://arc2.oup-arc.com/**.

- **Ask Yourself** prompts in the margins invite students to apply the material to their own lives. These also provide a confidence-building opportunity to get students speaking in class before undertaking formal presentations.

- Marginal **cultural idioms** not only highlight the use of idioms in communication but also help nonnative English speakers appreciate the idiosyncratic expressions and colloquialisms they normally take for granted.

- An **enhanced support package** for every chapter (described in detail below) includes video links, pre- and post-reading quizzes, activities, discussion topics, examples, tools for recording and uploading student speeches for assessment, an online gradebook, and more.

Optional Chapter

Along with the topics included in the text itself, a custom chapter is available on **Mass Communication**. Ask your Oxford University Press representative for details, or see the *Understanding Human Communication* website at **www.oup.com/us/adleruhc**.

Ancillary Package

The 13th edition of *Understanding Human Communication* contains a robust package of ancillary materials that will make teaching more efficient and learning more effective. Instructors and students alike will be pleased to find a complete suite of supplements.

Online Learning

This edition of *Understanding Human Communication* offers a host of options for online learning:

- **Dashboard** delivers high-quality content, tools, and assessments to track student progress in an intuitive, web-based learning environment.

Dashboard gives instructors the ability to manage digital content from *Understanding Human Communication*, 13th edition, and its supplementary materials in order to make assignments, administer tests, and track student progress. Assessments are designed to accompany this text and are automatically graded so that instructors can easily check students' progress as they complete their assignments. The color-coded gradebook illustrates at a glance where students are succeeding and where they can improve. Dashboard is engineered to be simple, informative, and mobile. All Dashboard content is engineered to work on mobile devices, including iOS platforms.

With this edition's Dashboard, professors and students have access to a variety of interactive study and assessment tools designed to enhance their learning experience, including:

- o Multiple choice pre- and posttests to accompany each chapter
- o Interactive drag-and-drop questions in each chapter
- o Animations with assessment questions based on the 13th edition's figures, which help reinforce difficult and abstract concepts
- o Short video clips with assessment in each chapter to show communication in action and help students apply what they have learned
- o Examples of professional and student speeches, accompanied by review and analysis questions
- o Interactive flashcards to aid in self-study
- **Course cartridges** are an alternative to Dashboard. They are available for a variety of Learning Management Systems, including Blackboard Learn, Canvas, Moodle, D2L, and Angel. Course cartridges allow instructors to create their own course websites integrating student and instructor resources available on the Ancillary Resource Center and Companion Website. Contact your Oxford University Press representative for access or for more information about these supplements or customized options.

For Instructors

Ancillary Resource Center (ARC) at **https://arc2.oup-arc.com/**. This convenient, instructor-focused website provides access to all of the up-to-date teaching resources for this text—at any time—while guaranteeing the security of grade-significant resources. In addition, it allows OUP to keep instructors informed when new content becomes available. The following items are available on the ARC:

- The Complete Guide to Teaching Communication, written by co-author Athena du Pré, provides a complete syllabus, teaching tips, preparation checklists, grab-and-go lesson plans, high-impact activities, links to relevant video clips, and coordinating PowerPoint lecture slides and Prezi presentations.
- A comprehensive **Computerized Test Bank** includes 60 exam questions per chapter in multiple-choice, short-answer, and essay formats. The questions have been extensively revised for this edition, are labeled according to difficulty, and include the page reference and chapter section where the answers may be found.
- **PowerPoint and Prezi lecture slides** include key concepts, video clips, discussion questions, and other elements to engage students. They correspond to content in the lesson plans, making them ready to use and fully editable so that preparing for class is faster and easier than ever.

- *Now Playing*, **Instructor's Edition**, includes an introduction on how to incorporate film and television clips in class, as well as even more film examples, viewing guides and assignments, a complete set of sample responses to the discussion questions in the student edition, a full list of references, and an index by subject for ease of use. *Now Playing* also has an accompanying companion website at www.oup.com/us/nowplaying, which features descriptions of films from previous editions and selected film clips.

For Students

- *Now Playing*, **Student Edition**, available free in a package with a new copy of the book, explores contemporary films and television shows through the lens of communication principles. Updated yearly, it illustrates how communication concepts play out in a variety of situations, using mass media that are interactive, familiar, and easily accessible to students.

- The **Companion Website** is an open-access student website at **www.oup. com/us/adleruhc** that offers activities, audio tutorials, chapter outlines, review questions, worksheets, practice quizzes, flashcards, and other study tools. This companion site is perfect for students who are looking for a little extra study material online.

Acknowledgments

Anyone involved with creating a textbook knows that success isn't possible without the contributions of many people.

We owe a debt to our colleagues, whose reviews helped shape the edition you are holding. In particular, we wish to thank Anastacia Kurylo of St. Joseph's College for her insightful comments on Chapter 3, and Molly Steen of the University of California, Santa Barbara, for her guidance on communication strategies for job-seekers. Thanks yet again to Russ Proctor, University of Northern Kentucky, for sharing his work and insights. We also thank the following reviewers commissioned by Oxford University Press:

Theresa Albury	*Miami Dade College*
Mark Bergmooser	*Monroe County Community College*
Jaime Bochantin	*University of North Carolina, Charlotte*
Kelly Crue	*St. Cloud Technical and Community College*
Lisa Fitzgerald	*Austin Community College*
David Flatley	*Central Carolina Community College*
Sarah D. Fogle	*Embry-Riddle University*
Karley Goen	*Tarleton State University*
Donna L. Halper	*Lesley University*
Lysia Hand	*Phoenix College*
Milton Hunt	*Austin Community College*
Amy K. Lenoce	*Naugatuck Valley Community College*
Allyn Lueders	*East Texas Baptist University*
Kim P. Nyman	*Collin College*
Christopher Palmi	*Lewis University*
Evelyn Plummer	*Seton Hall University*

We also continue to be grateful to the many professors whose reviews of previous editions continue to bring value to this book:

Spence, Richland Community College; **Sarah Stout**, Kellogg Community College; **Don Taylor**, Blue Ridge Community College; **Cornelius Tyson**, Central Connecticut State University; **Curt VanGeison**, St. Charles Community College; **Robert W. Wawee**, The University of Houston–Downtown; **Kathy Wenell-Nesbit**, Chippewa Valley Technical College; **Shawnalee Whitney**, University of Alaska, Anchorage; **Princess Williams**, Suffolk County Community College; **Rebecca Wolniewicz**, Southwestern College; and **Jason Ziebart**, Central Carolina Community College.

Many thanks are due to colleagues who developed and refined elements of the ancillary package:

Mary Ann McHugh: Test Bank, Dashboard, Companion Website

Tanika Smith and Windolyn Yarberry: Dashboard

John James: PowerPoints/Prezis

The enhanced package that is the result of their efforts will help instructors teach more effectively and students succeed in mastering the material in this text.

As most instructors know, ancillaries are anything but secondary. The previous edition's ancillaries were reviewed for accuracy, ease of use, efficacy, relevancy, and rigor. It was this direct instructor feedback that we used to craft the current edition's program. We would like to thank the following reviewers of ancillary materials for their thoughtful insights:

Jessica Akey	*Champlain College*
Manuel G. Aviles-Santiago	*Arizona State University*
Jaime Bochantin	*DePaul University/University of North Carolina, Charlotte*
James Canney	*Naugatuck Valley Community College*
Kelly Crue	*St. Cloud Technical and Community College*
Stuart Doyle	*Embry-Riddle University*
Vance Elderkin	*Alamance Community College*
Rebecca A. Ellison	*Jefferson College*
Milton Hunt	*Austin Community College*
Audrey E. Kali	*Framingham State University*
James Keller	*Lone Star College CyFair*
Sarah Kercsmar	*University of Kentucky*
Randall E. King	*Indiana Wesleyan University*
Maria LeBerre	*Northern Virginia Community College*
Michelle M. Maresh-Fuehrer	*Texas A&M University, Corpus Christi*
James L. Redfield	*St. Cloud Technical and Community College*
Emily Richardson	*University of Pikeville*
Delwin E. Richey	*Tarleton State University*
Jacqueline A. Shirley	*Tarrant County College*
Kim G. Smith	*Bishop State Community College*
Linda H. Straubel	*Embry-Riddle University*
Charlotte Toguchi	*Kapi'olani Community College*
Archie Wortham	*Northeast Lakeview College*
Windolyn Yarberry	*Florida State College at Jacksonville*

In an age when publishing is becoming increasingly corporate, impersonal, and sales driven, we continue to be grateful for the privilege and pleasure of working with the professionals at the venerable Oxford University Press. They blend the best old-school practices with cutting-edge thinking.

Our Editor, Toni Magyar, has been a hardworking advocate for this book. We count our lucky stars for all the contributions of Development Editor Lauren Mine. She is an author's dream—smart, responsive, resourceful, and congenial. Associate Editor Paul Longo has been a valuable member of the *UHC* team, coordinating the ancillary package, figure updates, and countless other details. Senior Production Editor Barbara Mathieu's steady hand and Art Director Michele Laseau's design talents have transformed this project from a plain manuscript into the handsome book you are now reading. Marketing Manager David Jurman and the entire OUP sales team have gone the extra mile in bringing this book to users and supporting their teaching efforts. We are grateful to Sherri Adler and Sandy Cooke for their resourcefulness and the artistic sense they applied in choosing photos in these pages. We are also grateful to the eagle-eyed James Fraleigh for reviewing this manuscript and to Susan Monahan for her indexing talents.

Finally, as always, we thank our partners Sherri, Linda, and Grant for their good-natured understanding and support while we've worked on this edition for more than a year. When it comes to communication, they continue to be the best judges of whether we practice what we preach.

Ron Adler
George Rodman
Athena du Pré

About the Authors

Ronald B. Adler is Professor of Communication Emeritus, at Santa Barbara City College. He is coauthor of *Interplay: The Process of Interpersonal Communication; Essential Communication; Looking Out, Looking In;* and *Communicating at Work: Principles and Practices for Business and the Professions.*

George Rodman is Professor in the Department of Television and Radio at Brooklyn College, City University of New York, where he founded the graduate media studies program. He is the author of *Mass Media in a Changing World, Making Sense of Media,* and several books on public speaking.

Athena du Pré is Distinguished University Professor of Communication at the University of West Florida. She is the author of *Communicating About Health: Current Issues and Perspectives* and coauthor of *Essential Communication,* as well as other books, journal articles, and chapters on communicating effectively.

Understanding Human Communication

Communication: What and Why

1

CHAPTER OUTLINE

LEARNING OBJECTIVES

1.1

Apply the transactional communication model described on pages 9–10 to a specific incident, explaining how that exchange is part of a relational, symbolic process.

1.2

Identify the types of communication that operate in human interaction.

1.3

Compare and contrast face-to-face and mediated communication, including social media.

1.4

Describe the effects of changing communication technology on the study of human communication in recent decades.

1.5

Explain the key needs you and others attempt to satisfy by communicating.

1.6

Suggest ways of improving your communication competence in a specific situation.

1.7

Identify how misconceptions about communication can create problems, and suggest how changes in communication can lead to better outcomes.

You don't have to be CEO of a billion-dollar company to recognize the importance of effective communication.

?

Tony Hsieh uses communication to build relationships with colleagues and customers. What role does communication play in the quality of your relationships at school, at work, and in your personal life?

?

Describe the communication style of someone you admire. In what ways does that person display respect for others? How does he or she encourage mutual understanding?

?

What communication skills do you think will be most essential in the career you envision for yourself? Why?

ZAPPOS IS A BILLION-DOLLAR COMPANY. But if you go looking for the CEO, don't ask directions to the executive suite. Tony Hsieh (pronounced *Shay*) sits at a standard-issue desk in the midst of the company's busy call center. "I think I would probably get lonely in an office. I'd be away from all the action," explains the energetic visionary.[1]

Hsieh maintains a legendary fervor for open communication. By being in the same room as everyone else, he can easily share information and listen to employees' ideas in real time. It's an unconventional approach with unconventional results: Zappos skyrocketed to success as an online retail company in about 10 years. Behind that success is the philosophy that supportive relationships, enhanced by effective communication, matter more than anything else.

We'll return to Hsieh as an example throughout the chapter. His success makes a strong case for the importance of communication. But perhaps the strongest argument for studying this subject is its central role in our lives. The average person spends 7 out of every 10 waking hours actively communicating with family members, friends, coworkers, teachers, and even strangers.[2] With computers, phones, tablets, and all the rest, it's possible to carry on several conversations at one time. Of course, more communication isn't always better communication. In this chapter, we begin to explore how to make wise choices about what messages we share with others and how we share them.

Communication Defined

The term *communication* isn't as simple as it might seem. People use it in a variety of ways that are only vaguely related:

- A dog scratches at the back door to be let out.
- Data flows from one computer database to another in a cascade of electronic impulses.
- Strangers who live thousands of miles apart notice each other's social media postings, and they build a relationship via email, text messaging, and instant messaging.
- Locals offer directions to a group of confused-looking people who seem to be from out of town.
- A religious leader gives a sermon encouraging the congregation to get more involved in the community.

We need to narrow our focus before going on. A look at this book's table of contents shows that it does not deal with animals. Neither is it about Holy Communion, the bestowing of a material thing, or many of the other subjects mentioned in the *Oxford English Dictionary*'s 1,200-word definition of *communication*. What, then, are we talking about when we use the term?

Characteristics of Communication

As its title suggests, this is a book about understanding human communication—so we'll start by explaining what it means to study communication that is unique to members of our species. For our purposes we'll define human **communication** as *the process of creating meaning through symbolic interaction*. Examining this definition reveals some important insights.

> **communication** The process of creating meaning through symbolic interaction.

Communication Is a Process   We often think about communication as if it occurs in discrete, individual acts such as one person's utterance or a conversation. But in fact, communication is a continuous, ongoing process. There are probably people in your life who have changed your outlook through their words and actions. This change typically occurs over time, not instantly.

As a case in point, consider the CEO of Zappos. Tony Hsieh recognizes that creating a culture of happiness at Zappos requires more than a speech, a memo from HR, or slogans on the wall. Developing trust, creativity, and respect takes time. Hsieh made a commitment to listen to team members every day and take their ideas seriously. Open communication and effective listening have allowed Zappos to become one of the best places to work in the country.[3]

Consistency is key. Even what appears to be an isolated message is often part of a much larger process. Consider, for example, a friend's compliment about your appearance. Your interpretation of those words will depend on a long series of experiences stretching far back in time: How have others judged your appearance? How do you feel about your looks? How honest has your friend been in the past? How have you been feeling about each other recently? All this history will help shape your response to the friend's remark. In turn, the

Communication (without an "s") refers to the study of how people share messages. That's the primary focus of this book. *Communications* (with an "s") usually refers to the technologies that enable the exchange of information.

Can you give examples of the proper use of "communication" versus "communications"?

Communication is something we do *with*, not *to* others.

How well is your communication synchronized with others in important relationships?

symbol An arbitrary sign used to represent a thing, person, idea, event, or relationship in ways that make communication possible.

More than 150 years after the U.S. Civil War, the Confederate flag still evokes strong emotions. Some view it as a symbol of racism, while others see it as honoring the sacrifices of brave Confederate soldiers.

Which symbols communicate messages that draw you toward others? Which ones create discord?

words you speak and the way you say them will shape the way your friend behaves toward you and others—both in this situation and in the future.

This simple example shows that it's inaccurate to talk about "acts" of communication as if they occur in isolation. To put it differently, communication isn't a series of incidents pasted together like photographs in a scrapbook; instead, it is more like a motion picture in which the meaning comes from the unfolding of an interrelated series of images. The fact that communication is a process is reflected in the transactional model introduced later in this chapter.

Communication Is Relational, Not Individual   Communication isn't something we do *to* others; rather, it is something we do *with* them. Like many types of dancing, communication depends on the involvement of a partner. A great dancer who doesn't consider and adapt to the skill level of his or her partner can make both people look bad. In communication and in dancing, even two highly skilled partners must work at adaptation and coordination. Finally, relational communication—like dancing—is a unique creation that arises out of the way in which the partners interact: It varies with different partners.

Psychologist Kenneth Gergen captures the relational nature of communication well when he points out how our success depends on interaction with others. As he says, "one cannot be 'attractive' without others who are attracted, a 'leader' without others willing to follow, or a 'loving person' without others to affirm with appreciation."[4]

Because communication is relational, or transactional, it's often a mistake to suggest that just one person is responsible for a relationship. Although it's easy to blame each other for a disappointing outcome, that's often fruitless and counterproductive. It's usually far better to ask, "How did we handle this situation poorly, and what can we do to make it better?"

The transactional nature of communication shows up in school, where teachers and students influence one another's behavior. For example, teachers who regard some students negatively may treat them with subtle or overt disfavor. As a result, these students are likely to react to their teachers' behavior negatively, which reinforces the teachers' original attitudes and expectations.[5]

Communication Is Symbolic   Chapter 4 discusses the nature of symbols in more detail, but this idea is so important that it needs an introduction now. **Symbols** are used to represent things, processes, ideas, or events in ways that make communication possible.

One feature of symbols is their arbitrary nature. For example, there's no logical reason why the letters in the word *book* should stand for the object you're reading now. Speakers of Spanish call it a *libro*, and Germans call it a *Buch*. Even in English, another term would work just as well as long as everyone agreed to use it in the same way.

Conflicts can arise when people attach different meanings to a symbol. (See the Confederate flag photo on this page.) Is calling your friend a "gangsta" a joke or an insult? Are flowers offered after a fight an expression of apology or an attempt to avoid guilt? How people feel about each other depends a great deal on how they interpret one another's actions.

Animals don't use symbols in the varied and complex ways that we do. There's nothing symbolic about a dog scratching at the door to be let out; there is a natural connection between the door and the dog's goal. By contrast, the significance of a word or action is only arbitrarily related to the meaning we give it. Besides reflecting our identity, symbolic communication allows people to think or talk about the past, explain the present, and speculate about the future.

Modeling Communication

So far we have introduced a basic definition of communication and considered its characteristics. This information is useful, but it only begins to describe the process we will examine throughout this book. One way to deepen your understanding is to look at some models that describe what happens when two or more people interact. Over the years, scholars have developed an increasingly accurate and sophisticated view of this process.

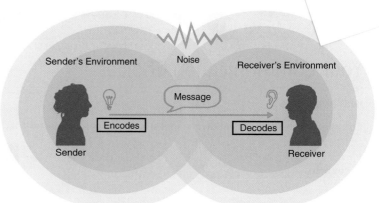

FIGURE 1-1 Linear Communication Model

A Linear Model Until about 50 years ago, researchers viewed communication as something that one person "does" to another.[6] In this **linear communication model**, communication is like giving an injection: A **sender encodes** (puts into symbols) ideas and feelings into some sort of **message** and then conveys them to a **receiver**, who **decodes** (attaches meaning to) them (Figure 1-1).

One important element of the linear model is the communication **channel**—the method by which a message is conveyed between people. Face-to-face contact is the most obvious channel. Writing is another channel. In addition to these long-used forms, **mediated communication** channels include telephone, email, instant messaging, faxes, voice mail, and video chats. (The word *mediated* reflects the fact that these messages are conveyed through some sort of communication medium.) The self-assessment on page 8 will help you appreciate how the channel you choose can help determine the success of your messages.

At first glance, the linear model suggests that communication is a straightforward matter: If you choose your words correctly, your message should get through without distortion. But even in the closest relationships, misunderstanding is common. In one study, researchers invited several pairs of people into their lab.[7] Some were married; others were strangers. The subjects invariably predicted that the married couples would understand each other better than strangers. In reality, the level of understanding was about the same. This finding highlights what the researchers called the "closeness-communication bias." Overestimating how well we understand others can result in potentially serious misunderstandings.

The channel you choose can make a big difference in the effect of a message. For example, if you want to say "I love you," a generic e-card probably wouldn't have the same effect as a handwritten note. Likewise, saying "I love you" for the first time via text message could make a very different statement than saying the words in person.

Why are misunderstandings—even in our closest relationships—so common? One factor is what scholars call **noise**—a broad category that includes any force that interferes with the accurate reception of a message. Noise can occur at every stage of the communication process. Three types of noise can disrupt communication—external, physiological, and psychological. *External noise* (also called "physical" noise) includes those factors outside the receiver that make hearing difficult, as well as many other kinds of distractions. For instance, a weak signal would make it hard for you to understand another person on the phone, and sitting in the rear of an auditorium might make hearing a speaker's remarks difficult. External noise can disrupt communication almost anywhere

linear communication model A characterization of communication as a one-way event in which a message flows from sender to receiver.

sender The originator of a message.

encode Put thoughts into symbols, most commonly words.

message A sender's planned and unplanned words and nonverbal behaviors.

receiver One who notices and attends to a message.

decode To attach meaning to a message.

channel The medium through which a message passes from sender to receiver.

mediated communication Communication sent via a medium other than face-to-face interaction.

noise External, physiological, and psychological distractions that interfere with the accurate transmission and reception of a message.

Your Communication Choices

INSTRUCTIONS:

Consider which communication channel(s) you would use in each situation described below.

Scenario	Your Communication Choice				
	Face-to-Face	Phone	Email	Text	Social Media
1. You have been concerned about a friend. The last time you were together you asked, "Is anything wrong?" Your friend replied, "I'm fine." Now it's been several weeks since you have heard from your friend, and you're worried. Which channel do you think is best for gauging your friend's true emotions?					
2. You're angry and frustrated with a professor and want to deal with this concern before the problem gets worse. Which communication choice offers you the best opportunity to address the problem?					
3. On Thursday your boss tells you it's okay to come in late Monday morning. You're worried he will forget that he gave you permission. What channel(s) should you use to make sure he remembers?					
4. You're applying for a job when a friend says, "You won't believe the photo of you that I'm going to post from the party last weekend!" How would you try to dissuade your friend from posting the photo?					
5. You just ended a long-time relationship. What's the best way to let your friends and family know?					
6. What is a communication challenge you currently face? Which channel(s) would be best suited to addressing it?					

EVALUATING YOUR RESPONSES

Explain the reasoning behind your choices. After reading this book, retake this assessment to see if your choices have changed.

---

in our model—in the sender, channel, message, or receiver. *Physiological noise* involves biological factors in the receiver or sender that interfere with accurate reception: illness, fatigue, and so on. *Psychological noise* refers to forces within a communicator that interfere with the ability to express or understand a message accurately. For instance, worrying about a recent conflict might make it hard to focus on work or school. In the same way, you might be so upset to learn you failed a test that you would be unable (perhaps unwilling) to understand clearly where you went wrong.

A linear model shows that communicators often occupy different **environments**—fields of experience that help them understand others' behavior. In communication terminology, *environment* refers not only to a physical location but also to the personal experiences and cultural backgrounds that participants bring to a conversation.

environment Both the physical setting in which communication occurs and the personal perspectives of the parties involved.

Consider just some of the factors that might contribute to different environments:

- A might belong to one ethnic group and **B** to another.
- A might be rich and **B** poor.
- A might be in a rush and **B** have nowhere to go.
- A might have lived a long, eventful life, and **B** might be young and inexperienced.
- A might be passionately concerned with the subject and **B** indifferent to it.

Notice how the model in Figure 1-1 (page 7) shows that the environments of the sender and receiver overlap. This area represents the background that the communicators must have in common. As the shared environment becomes smaller, communication becomes more difficult. Consider a few examples in which different perspectives can make understanding difficult:

- Bosses who have trouble understanding the perspective of their employees will be less effective managers, and workers who do not appreciate the challenges of being a boss are more likely to be uncooperative (and probably less suitable for advancement).
- Parents who have trouble recalling their youth are likely to clash with their children, who have never known and may not appreciate the responsibility that comes with parenting.
- Members of a dominant culture who have never experienced how it feels to be outside of it may not appreciate the concerns of people from nondominant cocultures, who may not have experienced how it feels to be in the majority.

Differing environments make understanding others challenging but certainly not impossible. Hard work and many of the skills described in this book provide ways to bridge the gap that separates all of us to a greater or lesser degree. For now, recognizing the challenge that comes from dissimilar environments is a good start. You can't solve a problem until you recognize that it exists.

A Transactional Model   Because of its simplicity, the linear model does not effectively represent the way most communication operates. The transactional communication model in Figure 1-2 (page 10) presents a more accurate picture in several respects.

Most notably, the **transactional model** shows that sending and receiving are simultaneous. Although some types of mass communication do flow in a one-way, linear manner, most types of personal communication are two-way exchanges. The roles of sender and receiver that seemed separate in the linear model are now superimposed and redefined as those of *communicators*. This new term reflects the fact that at a given moment we are capable of receiving, decoding, and responding to another person's behavior, while at the same time that other person is receiving and responding to ours.

Consider, for instance, the significance of a friend's yawn as you describe your romantic problems. Or imagine the blush you may see as you tell one of your raunchier jokes to a new acquaintance. Nonverbal behaviors like these show that most face-to-face communication is a two-way affair. The discernible response of a receiver to a sender's message is called **feedback**. Not all feedback is nonverbal, of course. Sometimes it is oral, as when you ask an instructor questions about an upcoming test or volunteer your opinion of a friend's new haircut. In other cases it is written, as when you answer the questions on a midterm exam or respond to a letter from a friend. Figure 1-2 makes the importance of feedback clear. It shows that most communication is, indeed, a two-way affair.

transactional communication model   A characterization of communication as the simultaneous sending and receiving of messages in an ongoing, irreversible process.

feedback   The discernible response of a receiver to a sender's message.

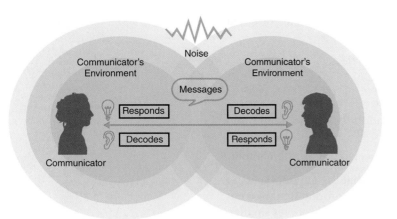

FIGURE 1-2 Transactional Communication Model

Some forms of mediated communication, such as email and text messaging, don't appear to be simultaneous. Even here, though, the process is more complicated than the linear model suggests. For example, if you've ever waited impatiently for the response to a text message or instant message (IM), you understand that even a nonresponse can have symbolic meaning. Is the unresponsive recipient busy? Thoughtful? Offended? Indifferent? Whether or not your interpretation is accurate, the silence is a form of communication.

Another weakness of the traditional linear model is the questionable assumption that all communication involves encoding. We certainly choose symbols to convey most verbal messages. But what about the many nonverbal cues that occur whether or not people speak: facial expressions, gestures, postures, vocal tones, and so on? Cues like these clearly do offer information about others, although they are often unconscious and thus don't involve encoding. For this reason, the transactional model replaces the term *encodes* with the broader term *responds*, because it describes both intentional and unintentional actions that can be observed and interpreted.

Types of Communication

Within the domain of human interaction, there are several types of communication. Each occurs in a different context. Despite the features they all share, each has its own characteristics.

Intrapersonal Communication

intrapersonal communication
Communication that occurs within a single person.

By definition, **intrapersonal communication** means "communicating with oneself."[8] One way that each of us communicates internally is by listening to the little voice in our mind. Take a moment and listen to what it is saying. Try it now, before reading on. Did you hear it? It may have been saying something like, "What little voice? I don't have any little voice!" This voice is the "sound" of your thinking.

Rather than listening to other people's definitions of success, Tony Hsieh tunes in to his own thoughts. "I made a list of the happiest periods in my life, and I realized that none of them involved money," he says.[9] Instead, relationships emerged as his prime source of satisfaction. "Connecting with a friend and talking through the entire night until the sun rose made me happy," says Hsieh. "Trick-or-treating in middle school with a group of my closest friends made me happy." As a result, he made interacting with friends a priority in his life.

The way we mentally process information influences our interaction with others. Even though intrapersonal communication doesn't include other people directly and may not be apparent, it does affect almost every type of interaction. You can understand the role of intrapersonal communication by imagining your thoughts in each of the following situations:

- You are planning to approach a stranger whom you would like to get to know better.

- You pause a minute and look at the audience before beginning a 10-minute speech.

- The boss yawns while you are asking for a raise.

- A friend seems irritated lately, and you're not sure whether you are responsible.

The way you handle all of these situations would depend on the intrapersonal communication that precedes or accompanies your overt behavior. Much of Chapter 2 deals with the perception process in everyday situations, and part of Chapter 13 focuses on the intrapersonal communication that can minimize anxiety when you deliver a speech.

Dyadic/Interpersonal Communication

Social scientists call two persons interacting a **dyad**, and they often use the term **dyadic communication** to describe this type of communication. Dyadic communication can occur in person or via mediated channels that include telephone, email, text messaging, instant messaging, and social networking websites.

Dyadic is the most common type of personal communication. It is also one of the most powerful predictors of relationship quality. Researchers in one study found that they could reliably predict whether family members were satisfied with each other by studying how they joked around, shared news of their day, and discussed their relationships together.[10] Even communication within larger groups (think of classrooms, parties, and work environments as examples) often consists of multiple, shifting dyadic encounters.

Dyadic interaction is sometimes considered identical to **interpersonal communication**, but as Chapter 7 explains, not all two-person interaction can be considered interpersonal in the fullest sense of the word. In fact, you will learn that the qualities that characterize interpersonal communication aren't limited to twosomes. They can be present in threesomes or even in small groups.

Dyadic relationships are as important in business as in personal life. At Zappos, staffers are not evaluated on how many calls they field or how much merchandise they sell, but on the quality of the one-on-one relationships they build with customers. They are encouraged to get to know callers and to spend time exceeding their expectations, even when the result is not a direct sale. The effect is that the company's customer service reputation is exceptionally high, increasing the likelihood of future sales.

Dyadic communication is arguably the context in which most close relationships operate.

How are your dyadic relationships—with friends, romantic partners, and even strangers—maintained through communication? What channels do you use?

Small Group Communication

Small groups are a fixture of everyday life. Your family is an example. So are an athletic team, a team of coworkers in several time zones who communicate online, and a team working on a class project. In **small group communication**, every person can participate actively with the other members.

Whether small groups meet in person or via mediated channels, they possess characteristics that are not present in a dyad. For instance, in a group, the majority of members can put pressure on those in the minority to conform, either consciously or unconsciously, but in a dyad no such majority pressure exists. Conformity pressures can also be comforting, leading group members to take risks that they would not dare to take if they were alone or in a dyad. With their greater size, groups also have the ability to be more creative than dyads. Finally, communication in groups is affected strongly by the type of leader who is in a position of authority. Groups are such an important communication setting that Chapters 9 and 10 focus exclusively on them.

Organizational Communication

Larger, more permanent collections of people engage in **organizational communication** when they work collectively to achieve goals. Organizations operate for a variety of reasons: commercial (e.g., corporations), nonprofit (e.g.,

dyad A two-person unit.

dyadic communication Two-person communication.

interpersonal communication Communication in which the parties consider one another as unique individuals rather than as objects. It is characterized by minimal use of stereotyped labels; unique, idiosyncratic social rules; and a high degree of information exchange.

small group communication Communication within a group of a size such that every member can participate actively with the other members.

organizational communication Communication that occurs among a structured collection of people in order to meet a need or pursue a goal.

Communication Skills and Career Success

Regardless of the job, people spend most of their working lives communicating.[11] Consider emails: The average worker receives nearly 12,000 every year and spends the equivalent of 111 workdays responding to them.[12] Combine that with telephone and face-to-face conversations, instant messaging, team meetings, videoconferences, presentations, and many other types of interaction, and you'll see that communication is at the heart of the workplace.[13]

On-the-job communication isn't just frequent; it's essential for success. Most adults recognize this fact. In one survey, more than 3,000 U.S. adults were asked what skills they believe are most important "to get ahead in the world today." Across the board, communication skills topped the list—ahead of math, writing, logic, and scientific ability.[14]

Communication skills are more important today than ever. The only jobs that have shown consistent wage growth over the last two decades are those that require social skills, all of which involve communication.[15] Traditional middle-skill jobs, such as clerical or factory work, have been largely replaced or made scarce by technology. High-wage, high-demand jobs combine technical and interpersonal expertise: Think physical therapy, general contracting, computer programming (usually a team endeavor), and medicine.[16]

Employers also recognize that communication skills are indispensable. In an annual survey, representatives from a wide range of industries ranked the "ability to verbally communicate with persons inside and outside the organization"

as the most essential skill for career success. In fact, communication skill was rated as more important than "technical knowledge related to the job."[17] Other research reinforces the value of communication. An analysis of almost 15 million job advertisements from across all occupations revealed that the ability to speak and write effectively were the most requested skills, identified twice as often as any other quality.[18]

Evidence like this makes it clear that communication skills can make the difference between a successful and a disappointing career. For more on increasing your efficiency and productivity at work, see *Communicating for Career Success* at www.oup.com/us/adleruhc.

charities and religious groups), political (e.g., government or political action groups), health-related (e.g., hospitals and doctor's offices), and even recreational (e.g., sports leagues).

Zappos has built a work environment designed to nurture teamwork, happiness, and individuality. The open floor plan encourages ongoing communication, and Hsieh supports a "no job title, no hierarchy" approach in which associates work together to foster continual improvement and innovation based on their own ideas, not managers'.[19]

Regardless of the context, the unique qualities of organizational communication make it worth studying. For example, it involves specific roles (e.g., sales associate, general manager, corporate trainer) that shape what people communicate about and their relationships with one another. It also involves complex and fascinating communication networks. As you'll read in Chapter 3, each organization develops its own culture, and analyzing the traditions and customs of organizations is a useful field of study.

Public Communication

Public communication occurs when a group becomes too large for all members to contribute. One characteristic of public communication is an unequal amount of speaking. One or more people are likely to deliver their remarks to the remaining members, who act as an audience. This leads to a second characteristic of public settings: limited verbal feedback. The audiences aren't able to talk back in a two-way conversation the way they might in a dyadic or small group setting. This doesn't mean that speakers <u>operate in a vacuum</u> when delivering their remarks. Audiences often have a chance to ask questions and offer brief comments, and their nonverbal reactions offer a wide range of clues about their reception of the speaker's remarks.

Public speakers usually have a greater chance to plan and structure their remarks than do communicators in smaller settings. For this reason, several chapters of this book describe the steps you can take to prepare and deliver an effective speech.

Mass Communication

Mass communication consists of messages that are transmitted to large, widespread audiences via electronic and print media: newspapers, magazines, television, radio, blogs, websites, and so on. As you can see in the Mass Communication section of this book's companion website (www.oup.com/us/adleruhc), this type of communication differs from the interpersonal, small group, organizational, and public varieties in several ways.

- First, most mass messages are aimed at a large audience without any personal contact between sender and receivers.
- Second, many of the messages sent via mass communication channels are developed, or at least financed, by large organizations. In this sense, mass communication is far less personal and more of a product than the other types of communication we have examined so far.
- Finally, although blogs have given ordinary people the chance to reach enormous audiences, the bulk of mass messages are still controlled by corporate, media, and governmental sources that determine what messages will be delivered to consumers, how they will be constructed, and when they will be delivered.

Communication in a Changing World

Over the past several decades, the nature of communication has changed dramatically. Today we are equipped with a range of communication technologies that, even two decades ago, would have been the stuff of fantasy and science fiction.

Yet along with the technological opportunities in today's world, communication challenges abound. How do we use the newest communication tools in ways that make life richer rather than simply busier? How can we deal with people whose communication practices differ dramatically from our own? This section will provide some tools to help answer these questions.

Lizzie Velasquez was once taunted online as "the world's ugliest woman," and online trolls urged her to kill herself. Instead, she has become a successful motivational speaker. Her speech "How Do You Define Yourself?" has been viewed more than 10 million times on YouTube.

What would you talk about if you had the microphone and an attentive group of listeners?

public communication Communication that occurs when a group becomes too large for all members to contribute. It is characterized by an unequal amount of speaking and by limited verbal feedback.

mass communication The transmission of messages to large, usually widespread audiences via broadcast, print, multimedia, and other forms of media, such as recordings and movies.

cultural idiom
operate in a vacuum: operate independently of outside influences

⊘ ASK YOURSELF

How have techno-
logical changes in your
lifetime affected the
way you communicate
with people?

Changing Technology

Figure 1-3 shows that communication technology is changing more rapidly than ever before. For most of human history, face-to-face speech was the primary form of communication. Writing developed approximately 5,000 years ago, but until the last few centuries, the vast majority of people were illiterate. In most societies only a small elite class mastered the arts of reading and writing. Books were scarce, and the amount of information available was small. Speaking and listening were the predominant communication "technologies."

By the mid-18th century, literacy grew in industrial societies, giving ordinary people access to ideas that had been available only to the most privileged. By the end of the 19th century, affordable rail travel increased mobility, and the telegraph made possible transmission of both news and personal messages over vast distances.

The first half of the 20th century introduced a burst of communication technology. The invention of the telephone extended the reach of both personal and business relationships. Radio and, later, television gave mass audiences a taste of the wider world. Information was no longer a privilege of the elite class.

By the dawn of the 21st century, cellular technologies and the Internet broadened the ability to communicate even further, beyond the dreams of earlier generations. Pocket-sized telephones made it cheap and easy to talk, send data, and exchange images with people around the globe. Now, new fiber-optic technology allows for more than 150 million phone calls every second.[20] Videoconferencing is another channel for remote connection, allowing us to see one another's facial cues, body movements, and gestures almost as if we were face-to-face.

The accelerating pace of innovations in communication technology is astonishing: It took 38 years for radio to reach 50 million listeners. It took television only 13 years to capture the same number of viewers. It took less than 4 years for the Internet to attract 50 million users. Facebook added 100 million users in less than 9 months.[21]

FIGURE 1-3 The Accelerating Pace of Communication Technology

Face-to-Face Communication
For most of human history, face-to-face speech was the primary form of communication. Speaking and listening were the predominant communication "technologies."

Literate Populations
By the mid-eighteenth century, literacy had developed in industrial societies, giving ordinary people access to ideas that had been available only to the most privileged.

| 3000+ BC | 1700s | 1800s |

Writing
Writing developed about 5,000 years ago but until the past few centuries, the vast majority of people were illiterate and books were scarce.

Trains, Telegraphs, Mail Service
By the end of the nineteenth century, affordable rail travel increased mobility and the telegraph enabled the transmission of news and personal messages across vast distances.

1620: Sign language (Spain)
1605: First newspaper (Germany)
C. 1450: First metal movable-type printing press (Germany)
868: First surviving book printed (China)
105: Paper invented (China)

AD
BC

300: World's first library (Egypt)
776: Carrier pigeons (Greece)
900: First postal service (China)
3000+: Writing invented

Changing Discipline

The study of communication has evolved to reflect the changing world. The first systematic analysis of how to communicate effectively was Aristotle's *Rhetoric*, written about 2,500 years ago.[22] The ancient philosopher set forth specific criteria for effective speaking (called the "Canons of Rhetoric"), which still can be used to judge effective public communication. In various forms, rhetoric has been part of a classical liberal arts education since Aristotle's era. Today it is commonly taught in public speaking courses that are offered in most colleges and universities.

In the early 20th century, the study of communication expanded from the liberal arts, where it had been housed for more than 2,000 years, and began to capture the attention of social scientists. As persuasive messages began to reach large numbers of people via print, film, and broadcasting, scholars began to study how mediated messages shaped attitudes and behaviors. During and after World War II, the effectiveness of government propaganda was an important focus of research.[23] Since then, media effects has become one of the most widely studied areas in the field of communication.

In the 1950s, researchers began asking questions about human relationships in family and work settings, marking the beginning of research on small group communication.[24] The analysis of decision making and other small group communication processes emerged as a major area of study in the field and continues today.

In the 1960s, social scientists expanded their focus to study how communication operates in personal relationships.[25] Since then, scholars have studied a wide range of phenomena, including how relationships develop, the nature of social support, the role of emotions, how honesty and deception operate, and how new technologies affect interpersonal relationships. Other branches of the discipline examine how health care providers and clients interact, the influence of gender on interaction, and how people from different backgrounds communicate with

The 1930s cartoon detective Dick Tracy communicated with a two-way wristwatch radio. Such a device seemed like science fiction to earlier generations, but even more sophisticated technologies are commonplace today.

Think about the communication technologies you use. How would your life be different without them?

Phones, Radio, and Television
Telephones reduced distance, expanding the reach of both personal relationships and commerce. Radio and later, television gave mass audiences a taste of the wider world. Information was no a privilege of the elite.

1900s

1963: First communication satellite
1936: Regular TV broadcasts (London)
1920: First radio broadcast
1876: First telephone demonstrated
1843: Long-distance telegraph

Digital Age
By the dawn of the 21st century, communication technology had expanded beyond the dreams of earlier generations.

2000s

2015: Wearable technologies become ubiquitous
2010: Tablets enhance mobile computing
2009: Smartphone sales top 170 million
2006: Twitter launched
2005: YouTube.com appears online
1997: First social network (SixDegrees.com)
1996: Instant messaging developed
1994: Personal blogging begins
1992: First text message sent
1991: World Wide Web begins
1981: IBM markets first personal computer
1975: First microcomputer, the Altair 8800
1973: First cell phone call
1972: First email with "@" in address
1969: ARPANET (forerunner to Internet)

Into the Future
Analysis predicts that we may one day communicate via lifelike holograms and have embedded implants that allow us to exchange messages via brainwaves.

"Anyone following me on Twitter already knows what I did this past summer."

Source: Alex Gregory The New Yorker Collection/The Cartoon Bank

Web 2.0 The Internet's evolution from a one-way medium into a "masspersonal" phenomenon.

social media Digital communication channels used primarily for personal reasons, often to reach small groups of receivers.

one another, to name just a few areas. The scope of the field has expanded far beyond its rhetorical roots.

Understanding Social Media

Given that Tony Hsieh has 3 million Twitter followers and is CEO of an online company, you might assume that he loves social media. You'd only be partially correct, however. The Zappos leader says he loves what social media can accomplish under the right conditions, but he is adamant that it's only a tool, and sometimes not the most effective one. "As unsexy and low-tech as it may sound, our belief is that the telephone is one of the best branding devices out there," Hsieh says.[26] He emphasizes what he calls a personal emotional connection, which may happen online but is more common in real-time encounters.

In this section, we consider how to use social media effectively and when to set it aside in favor of other communication channels. Until recently, when people heard the word *media*, they most likely thought of television, radio, and other forms of mass communication. But today, not all media are aimed at mass audiences. The term **Web 2.0** is often used to describe how the Internet has evolved from a one-way medium into what one scholar called a *masspersonal* phenomenon.[27] Individual users now interact in a host of ways that include social networking sites, video- and photo-sharing services, and blogs. If you blog, tweet, post photos on a website such as Tumblr, or maintain a page on Facebook or some other social networking site, you have experience with Web 2.0. You're not only a consumer of mediated messages but a creator of them.

As the name suggests, people use **social media** for personal reasons, often to reach small groups of receivers. You're using social media when you exchange text messages, emails, and instant messages, and when you use social networking websites such as Facebook and Google+. As mentioned earlier, the number of social media technologies has exploded in the past few decades, giving communicators today an array of choices that would have amazed someone from a previous era.

Social media are different from the mass variety in some important ways. Most obvious is the *variable size of the target audience*. Whereas the mass media are aimed at large audiences, the intended audience in social media can vary. On one hand, you typically address emails, text messages, and IMs to a single receiver, or maybe a few. In fact, you'd probably be embarrassed to have some of your personal messages circulate more widely. On the other hand, blogs, tweets, and other postings are often aimed at much larger groups of receivers.

Unlike mass media, social media are *interactive*: The recipients of your messages can—and usually do—talk back. About two-thirds of teens say they have made new friends online, and 9 in 10 say they keep in touch with established friends via technology.[28] The photo-sharing service Snapchat is a good example. It allows users to spontaneously send in-the-moment experiences with friends and followers, who can easily respond with photos of their own.[29] This sort of back-and-forth sharing reflects the difference between traditional print media, in which communication is essentially one way, and far more interactive web-based social media.

Unlike traditional forms of mass communication, social media are also distinguished by *user-generated content*. You decide what goes on your page and what topics are covered. There aren't any market researchers to tell you what the audience wants. No staff writers, editors, designers, or marketers craft your message. It's all you.

Despite these characteristics, the boundary between mass and interpersonal communication isn't as clear as it might first seem. Consider, for example, YouTube and other streaming video websites. They provide a way for individuals to publish their own content (e.g., your graduation or a baby's first birthday party) for a limited number of interested viewers. On the other hand, some videos go "viral," receiving thousands or even millions of hits. For example, YouTube videos by skateboard enthusiast Andrew Shrock have been viewed 160 million times.[30]

Twitter is another example of the fuzzy boundary between personal and mass media. Many people broadcast updates to a rather small group of interested parties. ("I'm at the concert—Great seats!") On the other hand, millions of fans follow the tweets of favorite celebrities. Twitter offers an interesting blend of messages from real friends and celebrities, "strangely intimate and at the same time celebrity-obsessed," as one observer put it. "You glance at your Twitter feed over that first cup of coffee, and in a few seconds you find out that your nephew got into med school and Shaquille O'Neal just finished a cardio workout in Phoenix."[31]

Blogs also straddle the categories of mass and social media. Some are highly personal: You can set one up and share your opinions with anybody who cares to read them. Others (such as the Huffington Post, the Daily Beast, and Bloomberg) are much closer to traditional mass media, published regularly and reaching audiences numbering in the hundreds of thousands.

People may use social media to get information, connect with others, and be entertained.

What are the most common ways you use social media? How well do these satisfy your personal and practical needs?

richness A term used to describe the abundance of nonverbal cues that add clarity to a verbal message.

Mediated Versus Face-to-Face Communication

As Figure 1-4 shows, both face-to-face and mediated channels are important ways to communicate. What does in-person communication have in common with the mediated options? How is it different?

In some ways, mediated and face-to-face communication are quite similar. Both include the elements described on pages 7–10: senders, receivers, channels, feedback, and so on. Both are used to satisfy physical, identity, social, and practical needs, as we discuss in the next section. Despite these similarities, the two forms of communication differ in some important ways.

Message Richness Social scientists use the term **richness** to describe the abundance of nonverbal cues that add clarity to a verbal message. As Chapter 6 explains in detail, face-to-face communication is rich because it abounds with nonverbal messages that give communicators cues about the meanings of one another's words and offer hints about their feelings.[32] By comparison, most mediated communication is a much leaner channel for conveying information.

% of all teens who spend time with friends

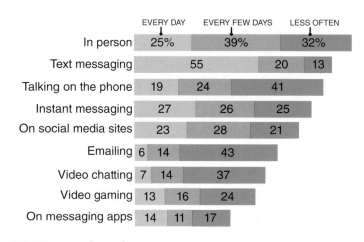

	EVERY DAY	EVERY FEW DAYS	LESS OFTEN
In person	25%	39%	32%
Text messaging	55	20	13
Talking on the phone	19	24	41
Instant messaging	27	26	25
On social media sites	23	28	21
Emailing	6	14	43
Video chatting	7	14	37
Video gaming	13	16	24
On messaging apps	14	11	17

FIGURE 1-4 Media Used to Keep in Touch with Friends

synchronous communication Communication that occurs in real time.

asynchronous communication Communication that occurs when there's a lag between receiving and responding to messages.

Synchronicity Communication that occurs in real time, such as through in-person or phone conversations, is **synchronous**. By contrast, **asynchronous communication** occurs when there's a lag between receiving and responding to messages. Voice mail messages are asynchronous. So are "snail mail" letters, emails, and Twitter postings. When you respond to asynchronous messages, you have more time to carefully consider your wording or to ask others for advice about what to say. You might even choose not to respond at all. You can ignore most problematic text messages without much fallout. But that isn't a good option if the person who wants an answer gets you on the phone or confronts you in person.

Permanence What happens in a face-to-face conversation is transitory. By contrast, the text and video you send via hard copy or mediated channels can be stored indefinitely and forwarded to others. Sometimes permanence is useful. For example, you might want a record documenting your boss's permission for time off work. In other cases, though, permanence can work against you. You probably wish you could delete an email sent in anger or embarrassing photos posted online.

How People Use Social Media

In the mid-20th century, researchers began to study the question, "What do media do to people?" They sought answers by measuring the effects of print and broadcast media on users. Did programming influence viewers' use of physical violence? Did it affect academic success? What about family communication patterns?

In the following decades, researchers began to explore a different question: "What do people do *with* media?"[33] This branch of study became known as *uses and gratifications* theory. In the digital age, researchers continue to explore how we use both social media and face-to-face communication.

The uses listed there fall into four broad categories:[34]

1. **Information:** Asking questions such as: What do people think of a new film or musical group? Can anybody trade work hours this weekend? Is there a good Honda mechanic nearby? Can your network provide leads on getting your dream job?

2. **Personal relationships:** Seeing what your friends are up to, tracking down old classmates, announcing changes in your life to the people in your personal networks, and finding a romantic partner.

3. **Personal identity:** Observing others as models to help you become more effective, getting insights about yourself from trusted others, and asserting your personal values and getting feedback from others.

4. **Entertainment:** Gaming online with a friend, sharing your music playlists with others, joining the fan base of your favorite star, and finding interest or activity groups to join.

(?) ASK YOURSELF

How does the amount and nature of your mediated communication affect your in-person relationships?

Functions of Communication

So why do we speak, listen, read, write, and text so much? There's a good reason: Communication satisfies many of our needs.

Physical Needs

Communication is so important that it is necessary for physical health. Lonely people typically experience abnormally high levels of pain, depression, and fatigue.[35] An absence of satisfying communication can even jeopardize life itself. People who feel socially isolated tend to become ill and die at younger ages than their than more socially engaged peers.[36] The reasons for this are numerous. Evidence suggests that loneliness not only affects people at a psychological level; the stress can actually cause changes in their body chemistry and compromise their immune systems.[37]

Identity Needs

Communication does more than enable us to survive. It is how we learn who we are. As you'll read in Chapter 2, our sense of identity comes from the way we interact with other people.

Deprived of communication with others, we would have no sense of identity. This fact is illustrated by the case of the famous "Wild Boy of Aveyron," who spent his early childhood without any apparent human contact. The boy was discovered in 1800 while digging for vegetables in a French village garden.[38] He showed no behaviors one would expect in a social human. The boy could not speak but uttered only weird cries. More significant than this absence of social skills was his lack of any identity as a human being. As author Roger Shattuck put it, "the boy had no human sense of being in the world. He had no sense of himself as a person related to other persons."[39] Only after the influence of a loving "mother" did the boy begin to behave—and, we can imagine, think of himself as a human. Contemporary stories support the essential role that communication plays in shaping identity. In 1970, authorities discovered a 12-year-old girl (whom they called Genie) who had spent virtually all her life in an otherwise empty, darkened bedroom with almost no human contact. The child could not speak and had no sense of herself as a person until she was removed from her family and "nourished" by a team of caregivers.[40]

Like Genie and the boy of Aveyron, each of us enters the world with little or no sense of identity. We gain an idea of who we are from the ways others define us. As Chapter 3 explains, the messages we receive in early childhood are the strongest, but the influence of others continues throughout life. Chapter 3 also explains how we use communication to manage the way others view us.

Social Needs

Besides helping define who we are, communication provides a vital link with others. Researchers and theorists have identified a range of social needs we satisfy by communicating: pleasure (e.g., "because it's fun," "to have a good time"); affection (e.g., "to help others," "to let others know I care"); inclusion (e.g., "because I need someone to talk to or be with," "because it makes me less lonely"); escape (e.g., "to put off doing something I should be doing"); relaxation (e.g., "because it allows me to unwind"); and control (e.g., "because I want someone to do something for me," "to get something I don't have").[41]

Tony Hsieh, who is just shy of billionaire status, gave up his condo a few years ago in favor of a simpler, more communication-oriented lifestyle. He and his pet alpaca now live in a luxury trailer in a Las Vegas mobile home park.[42] Living in homes with around 200 square feet of living space each isn't a hardship for Hsieh or his neighbors, he says, because they spend most of their "at home" time together at the park's campfire area and community kitchen.

As you look at this list of social needs for communicating, imagine how empty your life would be if these needs weren't satisfied. Then notice that it would be impossible to fulfill them without communicating with others. Because relationships with others are so vital, some theorists have gone as far as to argue that communication is the primary goal of human existence. Anthropologist Walter Goldschmidt terms the drive for meeting social needs as the "human career."[43]

Practical Needs

We shouldn't overlook the everyday, important functions that communication serves. Communication is the tool that lets us tell the hair stylist to take just a

Whether it's explaining where it hurts, giving instructions, offering an explanation, or conveying a host of other messages, communication is essential to getting your needs met and gaining others' cooperation.

How could you meet practical needs better by improving your communication skills?

little off the sides, direct the doctor to where it hurts, and inform the plumber that the broken pipe needs attention *now*!

Beyond these obvious needs, a wealth of research demonstrates that communication is an important key to effectiveness in a variety of everyday settings. For example, at least 4 of the top 10 qualities employers look for in job candidates involve communication. These include the ability to work well with team members, verbally communicate with people, write well, and influence others.[44]

Communication is just as important outside of work. People are typically happiest with friends they consider to be good communicators[45] and with romantic partners who are good listeners and who share personal feelings and information.[46] Communication comes in many forms. A hug, gesture, or meaningful look can be as powerful as words. A college student interviewed in one study describes what happened during a memorable argument with her sister: "This one day when she brought up one of her concerns I just ignored her. To my surprise she stopped nagging me and then began to cry. . . . Because she cried, I realized how important the issue was. It changed my perspective and now I always try to listen."[47]

Communication Competence: What Makes an Effective Communicator?

It's easy to recognize good communicators, and even easier to spot poor ones. But what are the characteristics that distinguish effective communicators from their less successful counterparts? Answering this question has been one of the leading challenges for communication scholars. Although all the answers aren't yet in, research has identified a great deal of important and useful information about communication competence.

Communication Competence Defined

> **communication competence** The ability to maintain a relationship on terms acceptable to all parties.

Although scholars are still working to clarify the nature of **communication competence**, most would agree that effective communication involves achieving one's goals in a manner that, ideally, maintains or enhances the relationship in which it occurs.[48] This definition suggests several important characteristics of communication competence. To take a closer look at the ways you communicate well and how you might improve, fill out the "Strengths and Goals" self-assessment on page 22 and/or ask a friend to rate you on the communication dimensions listed. As you think about the results, notice how they reflect the following principles.

There Is No "Ideal" Way to Communicate A variety of communication styles can be effective. Keep in mind that certain types of communication may succeed in one situation yet fail in another. The joking insults you routinely trade with a friend might be insensitive and discouraging if he or she had just suffered a personal setback. Similarly, the language you use with your peers might offend a family member. For this reason, being a competent communicator requires flexibility in understanding what approach is likely to work best in a given situation.

Competence Is Situational Because competent behavior varies so much from one situation and person to another, it's a mistake to think that communication competence is a trait that a person either possesses or lacks. It's more accurate to talk about *degrees* or *areas* of competence. You and the people you know are probably quite competent in some areas and less so in others. You might deal quite skillfully with peers, for example, but feel clumsy interacting with people much older or younger, wealthier or poorer, or more or less attractive than yourself. In fact, your competence with one person may vary from one situation to another.

This means that it's an overgeneralization to say, in a moment of distress, "I'm a terrible communicator!" It would be more accurate to say, "I didn't handle this situation very well, even though I'm better in others."

Competence Is Relational　Because communication is transactional, something we do with others rather than to them, behavior that is competent in one relationship isn't necessarily competent in others.

An early study on relational satisfaction illustrated that what constitutes satisfying communication varies from one relationship to another.[49] Researchers Brant Burleson and Wendy Sampter hypothesized that people with sophisticated communication skills (such as managing conflict well, giving ego support to others, and providing comfort to relational partners) would be better at maintaining friendships than would be less skilled communicators. To their surprise, the results did not support this hypothesis. In fact, friendships were most satisfying when partners possessed matching skill levels. Apparently, relational satisfaction arises in part when our style matches those of the people with whom we interact.

The same principle holds true in the case of jealousy. People deal with jealousy in a variety of ways, including keeping close tabs on the partner, acting indifferent, decreasing affection, talking the matter over, and acting angry. Researchers have found that no type of behavior is effective or ineffective in every relationship.[50] Approaches that work well with some people are hurtful to others. Findings like these demonstrate that competence arises out of developing ways of interacting that work for you and for the other people involved.

Competence Can Be Learned　To some degree, biology is destiny when it comes to communication style.[51] Studies of identical and fraternal twins suggest that traits including sociability, anger, and relaxation seem to be partially a function of our genetic makeup. Fortunately, biology isn't the only factor that shapes how we communicate: Communication is a set of skills that anyone can learn. Even one of the most universally feared communication challenges, public speaking, becomes easier with training. College students who practice giving speeches and receive helpful feedback are significantly less likely than their peers to feel anxious about public speaking in the future.[52] Even without systematic training, it's possible to develop communication skills through the processes of trial and error and observation. We learn from our own successes and failures, as well as from observing other models—both positive and negative.

Characteristics of Competent Communicators

Although competent communication varies from one situation to another, scholars have identified several common denominators that characterize effective communication in most contexts.

A Wide Range of Behaviors　Effective communicators are able to choose their actions from a wide range of behaviors. To understand the importance of having a large communication repertoire, imagine that someone you know repeatedly tells jokes—perhaps discriminatory ones—that you find offensive. You could respond to these jokes in a number of ways. You could:

- Say nothing, figuring that the risks of bringing the subject up would be greater than the benefits.
- Ask a third party to say something to the joke teller about the offensiveness of the jokes.
- Hint at your discomfort, hoping that your friend would get the point.
- Joke about your friend's insensitivity, counting on humor to soften the blow of your criticism.

The late Steve Jobs, cofounder of Apple Computer, was legendary for both conveying his visions and abusing subordinates. This dual nature illustrates that communication competence is situational.

In what situations are you most successful as a communicator? In what situations would you like to improve?

? ASK YOURSELF

What lessons you have learned through experience about communicating effectively?

cultural idioms
common denominators: features common to several instances
counting on: depending on
soften the blow: ease the effect

Your Communication Strengths and Goals

INSTRUCTIONS:

You can rate yourself on each of the items below and/or invite people who know you well to rate you.

Description	Your Communication Strengths and Goals			
	Rarely	**Sometimes**	**Often**	**Almost Always**
1. Well informed and prepared for meetings				
2. Clear and confident when expressing ideas				
3. Impatient with others				
4. Confident speaking before an audience				
5. Good at helping people understand complex information				
6. Apt to spend more time talking than listening				
7. Known to text or talk on the phone during class, meetings, or personal conversations				
8. Fascinated by different customs and worldviews				
9. Well organized and good at meeting deadlines				
10. Instrumental in helping others reach agreement				
11. In contact with a wide range of people				
12. Inclined to say things in the heat of the moment and then regret it later				
13. Open to new ideas and ways of thinking				
14. Attentive to implied meanings and what people convey through body language and tone of voice				
15. Intended meaning frequently misunderstood by others				
16. Inclined to avoid discussing matters that involve conflict or sensitive issues				
17. Apt to interrupt others to challenge what they are saying				

EVALUATING YOUR RESPONSES

Most of us are better at some forms of communication than others. Evaluate your scores in light of the information below to see where you are already strong and where you might strengthen your communication skills.

Listening

Three of the most common barriers to listening are impatience, inattentiveness, and eagerness to defend one's point of view. If you scored "rarely" on items 3, 6, 7, and 17, give yourself a pat on the back. If not, consider how you might build your listening skills.

Interpersonal Communication

Interpersonal communication involves a complicated array of skills. If you scored "rarely" on items 3, 15, and 17, you are probably effective at expressing yourself and taking time to negotiate meaning with others. Likewise, if you scored "often" or "almost always" on item 14, you are probably tuned in to messages that are implied but not spoken aloud, concepts we'll discuss in Chapters 6 and 7. If you scored "rarely" on item 2, you probably have a good grip on your emotional reactions. However, if you scored "often" or "almost always" on item 16, you may be bottling up your emotions rather than engaging in open communication.

Diversity Awareness

If you scored "often" or "almost always" on items 8, 11, and 13, you have what it takes to be interculturally competent. Curiosity and open-mindedness are assets.

Group and Team Skills
If you scored high ("often" or "almost always") on items 1, 2, 9, and 10, you have many of the qualities valued in team members—preparedness, confidence, patience, and diplomacy. But if you scored high on items 3 or 6, you may come off as intimidating or indifferent at times, which can hamper effective teamwork.

Public Speaking Skills
If you scored "often" or "almost always" on items 2, 4, and 5, you have an edge when it comes to public speaking. The measure of a great speaker is not an absence of nerves but the ability to process information and summon the confidence to present it clearly and powerfully to others.

- Express your discomfort in a straightforward way, asking your friend to stop telling the offensive jokes, at least around you.
- Simply demand that your friend stop.

With this choice of responses at your disposal (and you can probably think of others as well), you could pick the one that had the best chance of success. But if you were able to use only one or two of these responses when raising a delicate issue—always keeping quiet or always hinting, for example—your chances of success would be much smaller. Indeed, many poor communicators are easy to spot by their limited range of responses. Some are chronic jokers. Others are always belligerent. Still others are quiet in almost every situation. Like a piano player who knows only one tune or a chef who can prepare only a few dishes, these people are forced to rely on a small range of responses again and again, whether or not they are successful.

Ability to Choose the Most Appropriate Behavior   Simply possessing a large array of communication skills isn't a guarantee of effectiveness. It's also necessary to know which of these skills will work best in a particular situation. Choosing the best way to send a message is rather like choosing a gift: What is appropriate for one person won't be appropriate for another one at all. This ability to choose the best approach is essential because a response that works well in one setting would flop miserably in another one.

Although it's impossible to say precisely how to act in every situation, there are at least three factors to consider when you are deciding which response to choose: the context, your goal, and the other person.

Skill at Performing Behaviors   After you have chosen the most appropriate way to communicate, it's still necessary to perform the required skills effectively. There is a big difference between knowing about a skill and being able to put it into practice. Simply being aware of alternatives isn't much help, unless you can skillfully put these alternatives to work.

Just reading about communication skills in the following chapters won't guarantee that you can start using them flawlessly. As with any other skills—playing a musical instrument or learning a sport, for example—the road to competence in communication is not a short one. You can expect that your first efforts at communicating differently will be awkward. After some practice you will become more skillful, although you will still have to think about the new way of speaking or listening. Finally, after repeating the new skill again and again, you will find you can perform it without conscious thought.

Empathy/Perspective Taking   People have the best chance of developing an effective message when they understand the other person's point of view. And because others aren't always good at expressing their thoughts and feelings clearly, the ability to imagine how an issue might look from the other's point of view is an important skill. The value of taking the other's perspective suggests one reason why listening is so important. Not only does it help us understand others, it also gives us information to develop strategies about how to best influence them.

? ASK YOURSELF

How do you typically respond when someone says something that makes you feel uncomfortable, such as telling an off-color joke?

Because empathy is such an important element of communicative competence, much of Chapters 2 and 5 are devoted to this topic.

Cognitive Complexity   Cognitive complexity is the ability to construct a variety of frameworks for viewing an issue. Cognitive complexity is an ingredient of communication competence because it allows us to make sense of people using a variety of perspectives. For example, if a friend tells you about something that is bothering her, it might be most helpful if you just listen quietly. Or she might prefer that you help her analyze and solve the problem. Or perhaps she just wants to hear that someone understands how she feels. People who are able to choose and embody a range of listening responses based on the speaker's needs are highly prized as good listeners.[53]

Self-Monitoring   Psychologists use the term *self-monitoring* to describe the process of paying close attention to one's behavior and using these observations to shape the way one behaves. Unlike Calvin in the cartoon above, high self-monitors are able to separate a part of their consciousness and observe their behavior from a detached viewpoint, making observations such as:

> *"I'm making a fool out of myself."*
>
> *"I'd better speak up now."*
>
> *"This approach is working well. I'll keep it up."*

Chapter 2 explains how too much self-monitoring can be problematic. Still, people who are aware of their behavior and the impression it makes are more skillful communicators than people who are low self-monitors. They tend to adapt more readily to new work[54] and school environments. By contrast, low self-monitors are often unaware that their communication skills are lacking.[55] (Calvin, in the nearby cartoon, does a nice job of illustrating this problem.)

Commitment to the Relationship   One feature that distinguishes effective communication in almost any context is commitment. For example, romantic couples who frequently express their commitment to each other tend to stay together longer and rate themselves as more satisfied than couples who are in committed relationships but don't talk about it much.[56]

Communicating Competently with Social Media

Perhaps you've found yourself in situations like these:

- You want to bring up a delicate issue with a friend or family member or at work. You aren't sure whether to do so in person, on the phone, or via some mediated channel such as texting or email.

- You're enjoying a film at the theater—until another moviegoer starts a cell phone conversation.

Dear Social Media, I Need a Little Space

Something interesting happened when tech writer Eric Griffith left social media: nothing. "I didn't miss out on anything world-shaking, the globe kept on spinning," Griffith says. "Eventually, it felt utterly normal to not be on Twitter all the time."[57] He did, however, experience a personal rise in productivity and uninterrupted time with friends. No one is suggesting a total break from the Twitter feed. Well, some people are. But they usually announce that via—you guessed it—social media.

Here are the most common reasons people give for limiting time on social media:

1. **To Reclaim Time and Focus**
   Taken to extremes, social media drains effort that could be spent on other activities. On average, college students spend about 3 times more time on their cell phones than they do on school and 4 times more than they spend on work-related activities.[58,59]

2. **To Live in the Moment**
   The lure of online activities can detract from the here and now. "How many of you have sat down for dinner with friends or family at a restaurant, you take out your phone to check Facebook or Twitter?" asks reformed social media junkie Chris Mullen.[60] To participate in the online chatter, "you take a pic of everyone at dinner. Then you tweet: 'at din w/ fam,'" Mullen predicts. All the while, the people closest to you (physically speaking) are distanced from interacting with you.

3. **To Give Your Ego a Break**
   "One of the main problems with social media is you are often bombarded by others' accomplishments," observes a writer for *Elite Daily*.[61] It's a distorted reality in that few people post unflattering photos of themselves or broadcast their failures. A steady diet of Facebook, Twitter, and Instagram can make life can seem dull and inadequate by comparison.

4. **To Break the Approval Craving**
   Social media posts are often designed to get a validating response from others. That can foster an unhealthy dependency. "If you are waiting for people to like your status or your photos, you might be seeking and spending a lot of time waiting on others to approve of the online version of yourself," warns pop culture writer Christen Grumstrip.[62]

   Our relationship with social media is best described as "it's complicated." One online story urging people to cut back on social media use garnered more than 2 million Facebook likes,[63] which suggests more than a little ambivalence about that advice. And no wonder. A little space doesn't mean a permanent breakup. Eric Griffith, who had quit Twitter, is now back on it. But the break gave him a healthy perspective. As with any relationship, he says, "when [social media] takes over your life, it's not a good thing."[64]

- A friend posts a picture of you online that you would rather others not see.
- Someone you care about is spending too much time online, crowding out face-to-face interactions.
- Reading a text message while you walk to an appointment, you bump into another pedestrian.
- Even though you know it's a bad idea, you read a text message while driving and nearly hit a cyclist.
- You are copied on an email or text message that was obviously not meant for your eyes.
- You receive so many email messages that you have trouble meeting school and work deadlines.

None of these situations would have existed a generation ago. They highlight the need for a set of social agreements that go beyond the general rules of communicative competence outlined in this chapter. The following pages offer some guidelines that have evolved in recent years. Although they won't cover every situation involving mediated communication, they can help you avoid some problems and deal most effectively with others that are bound to arise.

Choose the Best Medium Sometimes the choice of a medium is a no-brainer. If a friend says, "Call me while I'm on the road," you know what to do. If your boss or professor only responds to emails, then it would be foolish to use any other approach. But in many other situations, you have several options available. Table 1-1 outlines the advantages and drawbacks of the most common ones.

Choosing the right medium is just as important in personal relationships. Anyone who has been dumped via text message knows that it only adds insult to injury. Just because there is an option that allows you to avoid a difficult conversation doesn't mean you should take the easy way out. Many difficult conversations are better when conducted face-to-face. These types of conversations include, but aren't limited to, sharing really bad news, ending a relationship, and trying to resolve a conflict.

In situations like these, a useful guideline is what's been called the "platinum rule": treating others as *they* would like to be treated. Ask yourself how the recipient of your message would prefer to receive it, and act accordingly, even though it's difficult.

Be Careful What You Post You may be forever haunted by mistakes in the form of text and photos posted online. As a cautionary tale, consider the case of Jean-Sun Hannah Ahn. A few days after being crowned Miss Seattle in 2012, she tweeted to her friends back in Arizona: "Ew, I'm seriously hating Seattle right now . . . Take me back to az! Ugh can't stand cold rainy Seattle and the annoying people." The message made news in the town where she had just been honored. After she publicly apologized, Ahn vowed that, in the future, "I will call my friends or text them if I'm feeling down or want to complain about something."[65] Her experience is a reminder that, once sent, electronic messages often have a life of their own.

Some incautious posts can go beyond being simply amusing. Consider the phenomenon of "sexting." One survey revealed that 10% of young adults between

---

TABLE 1-1

Choosing a Communication Channel

Choosing the best communication channel can make the difference between success and failure on the job. This table offers guidelines for choosing the channel that is best suited for a particular situation.

	SYNCHRONICITY	RICHNESS OF INFORMATION CONVEYED	SENDER'S CONTROL OVER MESSAGE	CONTROL OVER RECEIVER'S ATTENTION	EFFECTIVENESS FOR DETAILED MESSAGES
FACE-TO-FACE	Synchronous	High	Moderate	Highest	Weak
TELEPHONE, TELE-CONFERENCING, AND VIDEOCONFERENCING	Synchronous	Moderate	Moderate	Moderate	Weak
VOICE MAIL	Asynchronous	Moderate	High	Low	Weak
EMAIL	Asynchronous	Low	High	Low	High
INSTANT MESSAGING	Almost synchronous	Low	High	Varies	Weak
TEXT MESSAGING AND TWITTER	Varies	Low	High (given briefness of message)	Low	Good for brief messages
HARD COPY (E.G., HANDWRITTEN OR TYPED MESSAGE)	Asynchronous	Low	High	Low	High

Source: Adapted from R. B. Adler, J. M. Elmhorst, & K. Lucas. (2003). *Communicating at work: Strategies for success in business and the professions* (11th ed.). New York: McGraw-Hill, p. 14.

the ages of 14 and 24 have texted or emailed a nude or partially nude image of themselves to someone else, and 15% have received such pictures or videos of someone else they know.[66] Perhaps even more disturbing, 8% reported that they had been forwarded nude or partially nude images of someone they knew.[67]

When minors are involved, authorities can make arrests for manufacturing, disseminating, or even possessing child pornography. Far worse, some teens have committed suicide when explicit photos were posted online.[68] Even without such dire consequences, it's not hard to imagine the unpleasant consequences of a private photo or text going public.

Be Considerate Mediated communication calls for its own rules of etiquette. Here are a few. (The checklist on page 30 provides a handy reminder of these.)

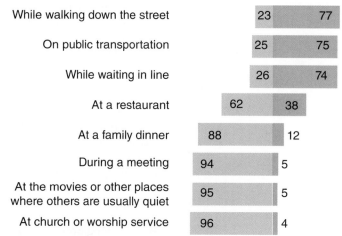

	Generally not OK	Generally OK
While walking down the street	23	77
On public transportation	25	75
While waiting in line	26	74
At a restaurant	62	38
At a family dinner	88	12
During a meeting	94	5
At the movies or other places where others are usually quiet	95	5
At church or worship service	96	4

FIGURE 1-5 When Is It Acceptable to Use a Cell Phone?

Respect Others' Need for Undivided Attention It might be hard to realize that some people are insulted when you divide your attention between your in-person conversational partner and distant contacts. As one observer put it, "While a quick log-on may seem, to the user, a harmless break, others in the room receive it as a silent dismissal. It announces: 'I'm not interested.'"[69]

As Figure 1-5 shows, most people understand that using mobile devices in social settings is distracting and annoying. Chapter 5 has plenty to say about the challenges of listening effectively when you are multitasking. Even if you think you can understand others while dealing with communication media, it's important to realize that they may perceive you as being rude.

Keep Your Tone Civil If you've ever shot back a nasty reply to a text or instant message, posted a mean comment on a blog, or forwarded an embarrassing email, you know that it's easier to behave badly when the recipient of your message isn't right in front of you.

The tendency to transmit messages without considering their consequences is called **disinhibition**, and research shows it is more likely in mediated channels than in face-to-face contact.[70] Sometimes communicators take disinhibition to the extreme, blasting off angry—even vicious—emails, text messages, and website postings. The common term for these outbursts is **flaming**. Flames are problematic because of their emotional and irreversible nature. Even once you've calmed down, the aggressive message can continue to cause pain.

Flaming isn't the only type of mediated harassment. Ongoing "cyberbullying" has become a widespread phenomenon, often with dire consequences. More than 4 in 10 teens report being the target of online harassment. Recipients of cyberbullying often feel helpless and scared, to such a degree that one report found they are 8 times more likely to carry a weapon to school than other students. There are at least 41 reported cases in which victims of cyberbullying committed suicide in the United States, Canada, Great Britain, and Australia,[71] a sobering statistic in light of reports that 81% of cyberbullies admit their only reason for bullying is because "it's funny."[72]

One way to behave better in asynchronous situations is to ask yourself a simple question before you send, post, or broadcast: Would you deliver the same message to the recipient in person? If your answer is no, then you might want to think before hitting the "send" key.

disinhibition The tendency to transmit messages without considering their consequences.

flaming Sending angry and/or insulting emails, text messages, and website postings.

Respect Privacy Boundaries Sooner or later you're bound to run across information about others that you suspect they would find embarrassing. If your relationship is close enough, you might consider sending out a for-your-information alert. In other cases, you may intentionally run a search about someone. Even if you uncover plenty of interesting information, it can be smart to avoid mentioning what you've discovered to the object of your searching.

Be Mindful of Bystanders If you spend even a little time in most public spaces, you're likely to encounter communicators whose use of technology interferes with others: restaurant patrons whose phone alerts intrude on your conversation, pedestrians who are more focused on their handheld device than on avoiding others, or people in line who are trying to pay the cashier and use their cell phone at the same time. If you aren't bothered by this sort of behavior, it can be hard to feel sympathetic with others who are offended by it. Nonetheless, this is another situation in which the platinum rule applies: Consider treating others the way *they* would like to be treated.

Balance Mediated and Face Time It's easy to make a case that many relationships are better because of social media. And as you've already read in this chapter, some research supports this position. But even with all the benefits of communication technology, it's worth asking whether there is such a thing as *too much* mediated socializing. (Read about some people's efforts to scale back their social media use in the Understanding Communication Technology box on page 25.)

There is a link between heavy reliance on mediated communication and conditions including depression, loneliness, and social anxiety.[73] People who spend excessive time online may begin to experience problems at school or work and withdraw further from their offline relationships.[74] Many people who pursue exclusively online social contacts do so because they have social anxiety or low social skills to begin with. For these people, retreating further from offline relationships may diminish their already low social skills.

Researchers have been especially interested in determining when cyber communication crosses the bridge from normal to excessive.[75] How many of the following points describe you?

- I find it difficult to avoid the urge to go online (e.g., to check Facebook on my phone).
- I spend more and more time online.
- I often spend more time online than I anticipated or intended.
- I have tried to reduce my social media use, but I have not been successful at cutting back.
- My time online causes me to do poorly at work, home, or school.
- I engage in social or recreational activities less than I should because of my time online.

Although experts disagree about whether Internet addiction disorder (IAD) is a certifiable addiction or just a symptom of another issue, they suggest several strategies for reining in excessive use of digital media. Unlike other addictions, such as those to drugs and alcohol, treatment for Internet addiction focuses on moderation and controlled use of the Internet rather than abstinence. If you are worried about your Internet use:

- Keep track of the amount of time you spend online so you can accurately assess whether it's too much.
- Plan a limited amount of online time in your daily schedule and see if you can stick to your plan.
- Make a list of problems in your life that may have occurred because of your time spent online.

- If you do not feel able to change your behavior on your own, seek the help of a counselor or therapist.

Be Safe   Many people fail to realize the hazards of posting certain information in public forums, and other people don't realize that what they are posting is public. You may post your "on vacation" status in Facebook, assuming that only your friends can see your message. But if a friend uses a public computer or lets another friend see your page, unintended recipients are viewing your information.[76]

As a rule, don't disclose information in a public-access medium that you would not tell a stranger on the street. Even personal emails present a problem: They can be forwarded, and accounts can be hacked. The safest bet is to assume that mediated messages can be seen by unintended recipients, some of whom you may not know or trust.

Careless use of social media can damage more than your reputation. Using a cell phone while driving is just as dangerous as driving under the influence of alcohol or drugs. Cell phone use while driving (hand-held or hands-free) lengthens a driver's reaction time as much as having a blood alcohol concentration at the legal limit of .08%.[77] In the United States alone, drivers distracted by cell phones cause about 5,474 deaths and 448,000 injuries every year.[78] Even a hands-free device doesn't eliminate the risks. Drivers carrying on phone conversations are 18% slower to react to brake lights. They also take 17% longer to regain the speed they lost when they braked.[79]

"I need a more interactive you."

Source: Mick Stevens The New Yorker Collection/
The Cartoon Bank

Clarifying Misconceptions About Communication

Having spent time talking about what communication is, we also ought to identify some things it is not. Recognizing some misconceptions is important, not only because they ought to be avoided by anyone knowledgeable about the subject, but also because following them can get you into trouble.

Communication Does Not Always Require Complete Understanding

Most people operate on the implicit but flawed assumption that the goal of all communication is to maximize understanding between communicators. Although some understanding is necessary for us to comprehend one another's thoughts, there are some types of communication in which understanding as we usually conceive it isn't the primary goal.[80] Consider, for example, the following:

- *Social rituals.* "How's it going?" you ask. "Great," the other person replies. The primary goal in exchanges like these is mutual acknowledgment: There's obviously no serious attempt to exchange information.

- *Many attempts to influence others.* A quick analysis of most television commercials shows that they are aimed at persuading viewers to buy products, not to understand the content of the commercial. In the same way, many of our attempts at persuading another to act as we want don't involve a desire to get the other person to understand what we want—just to comply with our wishes.

coordination Interaction in which participants interact smoothly, with a high degree of satisfaction but without necessarily understanding one another well.

- *Deliberate ambiguity and deception.* When you decline an unwanted invitation by saying, "I can't make it," you probably want to create the impression that the decision is really beyond your control. (If your goal was to be perfectly clear, you might say, "I don't want to get together. In fact, I'd rather do almost anything than accept your invitation.") As Chapters 4 and 7 explain in detail, we often equivocate precisely because we want to obscure our true thoughts and feelings.

- *Coordinated action.* Examples are conversations in which satisfaction doesn't depend on full understanding. The term **coordination** has been used to describe situations in which participants interact smoothly, with a high degree of satisfaction but without necessarily understanding one another well.[81] Coordination without understanding can be satisfying in far more important situations. Consider the words "I love you." This is a phrase that can have many meanings: Among other things, it can mean "I admire you," "I feel great affection for you," "I desire you," "I am grateful to you," "I feel guilty," "I want you to be faithful to me," or even "I hope *you* love *me*."[82] It's not hard to picture a situation in which partners gain great satisfaction—even over a lifetime—without completely understanding that the mutual love they profess actually is quite different for each of them. "You mean you mostly love me because I've been there for you? Hey, a *dog* is there for you!"

Communication Will Not Solve All Problems

"If I could just communicate better . . ." is the sad refrain of many unhappy people who believe that if they could just express themselves better, their relationships would improve. Though this is sometimes true, it's an exaggeration to say that communicating—even communicating clearly—is a guaranteed panacea.

Communication Isn't Always a Good Thing

In truth, communication is neither good nor bad in itself. Rather, its value comes from the way it is used. Communication can be a tool for expressing warm feelings and useful facts, but under different circumstances the same words and actions can cause both physical and emotional pain.

Meanings Rest in People, Not Words

It's a mistake to think that, just because you use a word in one way, others will do so, too. Sometimes differing interpretations of symbols are easily caught, as when we might first take the statement "He's loaded" to mean the subject has had too much to drink, only to find out that he is quite wealthy. In other cases, however, the ambiguity of words and nonverbal behaviors isn't so apparent and thus has more far-reaching consequences. Remember, for instance, a time when someone said to you, "I'll be honest," and only later did you learn that those words hid precisely the opposite fact. In Chapter 4 you'll read a great deal more about the problems that come from mistakenly assuming that meanings rest in words.

Communication Is Not Simple

Most people assume that communication is an aptitude that people develop without the need for training—rather like breathing. After all, we've been swapping ideas with one another since early childhood, and there are lots of people who communicate pretty well without ever having had a class on the subject. Though this picture of communication as a natural ability seems accurate, it's actually a gross oversimplification.

Many people do learn to communicate skillfully because they have been exposed to models of such behavior by those around them. This principle of modeling explains why children who grow up in homes with stable relationships between family members have a greater chance of developing such relationships themselves. But even the best communicators aren't perfect. They often suffer the frustration of being unable to get a message across effectively, and they frequently misunderstand others. Furthermore, even the most successful people you know probably can identify ways in which their relationships could profit from better communication. These facts show that communication skills are rather like athletic ability: Even the most inept of us can learn to be more effective with training and practice, and those who are talented can always become better.

More Communication Isn't Always Better

Although it's certainly true that not communicating enough is a mistake, there are also situations when *too much* communication is ill advised. Sometimes excessive communication simply is unproductive, as when we "talk a problem to death," going over the same ground again and again without making any headway. And there are times when communicating too much can actually aggravate a problem. We've all had the experience of "talking ourselves into a hole"—making a bad situation worse by pursuing it too far. As two noted communication scholars put it, "more and more negative communication merely leads to more and more negative results."[83]

There are even times when *no* communication is the best course. Any good salesperson will tell you that it's often best to stop talking and let the customer think about the product. And when two people are angry and hurt, they may say things they don't mean and will later regret. At times like these it's probably best to spend a little time cooling off, thinking about what to say and how to say it.

Along with their benefits, the technologies that keep us connected have a downside. When your boss, colleagues, and customers can reach you at any time, you can become too distracted to be productive.

Communication researchers discovered that remote workers developed two strategies to reduce contact and thereby increase their efficiency.[84] The first simply involved *disconnecting* from time to time: logging off the computer, forwarding the phone call to voice mail, or simply ignoring incoming messages. The researchers labeled the second strategy *dissimulation.* Teleworkers disguise their activities to discourage contact: changing their instant message status to "in a meeting" or posting a fake "out of the office" message online.

It's important to note that these strategies were typically used not to avoid work but to get more done. These findings show that too much connectivity is like many other parts of life: More isn't always better. In closing, let's revisit the example set by Tony Hsieh. His ideals of open communication and kindness are reflected in his company's practices. Zappos associates are authorized to listen to and help solve customers' problems and even help them find products on competitors' websites if Zappos is out of stock. "You create thousands of stories, and those stories spread the word about Zappos," Hsieh says. "We're not trying to maximize every transaction. We're trying to build a lifelong relationship with each of our customers, one call at a time."[85] Ultimately, a powerful connection with other people is what communication is all about.

? ASK YOURSELF

Describe a time when a misunderstanding led to a mistake or a comical situation. How did interpretations of the initial message(s) differ?

ETHICAL CHALLENGE

To Communicate or Not to Communicate?

Think of an occasion (real or hypothetical) in which another person is urging you to keep the channels of communication open. You know communicating more with this person will cause the situation to deteriorate, yet you don't want to appear uncooperative. What should you do?

MAKING THE GRADE

For more resources to help you understand and apply the information in this chapter, visit the *Understanding Human Communication* website at www.oup.com/us/adleruhc.

OBJECTIVE 1.1   Apply the transactional communication model described on pages 9–10 to a specific incident, explaining how that exchange is part of a relational, symbolic process.

- *Communication*, as it is examined in this book, is the process of creating meaning through symbolic interaction.
- The transactional model described on pages 9–10 is superior to the more simplistic linear model in representing the process-oriented nature of human interaction.
  - > Describe a real-life example that illustrates how communication is symbolic, transactional, and process oriented.
  - > Have you ever overreacted to something someone said, only to realize later that you misinterpreted that person's intent? How can the knowledge that communication is a process help you put things in perspective?
  - > How can your understanding of the transactional model help you communicate more effectively in an important relationship?

OBJECTIVE 1.2   Identify the types of communication that operate in human interaction.

- Communication operates in six contexts: intrapersonal, dyadic/interpersonal, small group, organizational, public, and mass.
  - > Provide examples of communication in three of the contexts listed above.
  - > In what way would you most like to improve your communication relevant to interpersonal communication? To small group communication? To public speaking?
  - > What goals would you most like to set for yourself as a communicator? Which contexts will your goals involve?

OBJECTIVE 1.3   Compare and contrast face-to-face and mediated communication, including social media.

- Emerging technology makes communication easier in some ways but more difficult in others.
  - > What percentage of your communication is face-to-face? What percentage involves social media or other mediated forms?

  - > How satisfied are you with the amount of time you spend on social media and the quality of your mediated interactions with people?
  - > In what ways does social media enhance your relationships? In what ways does social media take away from them?

OBJECTIVE 1.4   Describe the effects of changing communication technology on the study of human communication in recent decades.

- After millennia with little change, the pace of innovation in communication technologies has accelerated in the last 150 years.
- The study of communication has also evolved, taking a more expansive approach to human interaction.
  - > How do the communication technologies you use differ from those of your parents' generation? Your grandparents'?
  - > How is your life enriched by using today's communication technologies? Do these technologies diminish the quality of your relationships in any ways?

OBJECTIVE 1.5   Explain the key needs you and others attempt to satisfy by communicating.

- Communication satisfies several types of needs: physical, identity, social, and practical.
  - > Using examples from your own life, describe each type of need that communication meets.

OBJECTIVE 1.6   Suggest ways of improving your communication competence in a specific situation.

- Communication competence is situational and relational in nature, and it can be learned.
- Competent communicators are able to choose and perform appropriately from a wide range of behaviors, as well as being cognitively complex self-monitors who can take the perspective of others and who are committed to important relationships.
  - > Give examples from your own life to illustrate the situational and relational nature of communication competence.
  - > Based on the criteria on pages 21–24, how would you rate your own communicative competence?
  - > How can you improve your communicative competence by applying the material in this chapter? Give specific examples.

jbode@mid-america.edu midnight
Due Monday April 16 @ Email.
Responses to be 1-2pgs

OBJECTIVE 1.7 Identify how misconceptions about communication can create problems, and suggest how changes in communication can lead to better outcomes.

- Communication doesn't always require complete understanding.
- Communication is not always a good thing that will solve every problem.
- More communication is not always better.
- Meanings are in people, not in words.
- Communication is neither simple nor easy.
  - > Give examples of how each misconception described in this chapter can lead to problems.
  - > Which misconception(s) have been most problematic in your life? Give examples of each.
  - > How can you avoid succumbing to the misconceptions listed in this chapter? Give examples of how you could communicate more effectively.

KEY TERMS

asynchronous communication p. 18

channel p. 7

communication p. 5

communication competence p. 20

coordination p. 30

decode p. 7

disinhibition p. 27

dyad p. 11

dyadic communication p. 11

encode p. 7

environment p. 8

feedback p. 9

flaming p. 27

interpersonal communication p. 11

intrapersonal communication p. 10

linear communication model p. 7

mass communication p. 13

mediated communication p. 7

message p. 7

noise p. 7

organizational communication p. 11

public communication p. 13

receiver p. 7

richness p. 17

sender p. 7

small group communication p. 11

social media p. 16

symbol p. 6

synchronous communication p. 18

transactional communication model p. 9

Web 2.0 p. 16

ACTIVITIES

1. **Analyzing Your Communication Behavior** Prove for yourself that communication is both frequent and important by observing your interactions for a one-day period. Record every occasion in which you are involved in some sort of human communication as it is defined in this chapter. Based on your findings, answer the following questions:

   a. What percentage of your waking day is involved in communication?

   b. What percentage of time do you spend communicating in the following contexts: intrapersonal, dyadic, small group, and public?

   c. What percentage of your communication is devoted to satisfying each of the following types of needs: physical, identity, social, and practical? (Note that you might try to satisfy more than one type at a time.)

   Based on your analysis, describe 5 to 10 ways you would like to communicate more effectively. For each item on your list of goals, describe who is involved (e.g., "my boss," "people I meet at parties") and how you would like to communicate differently (e.g., "act less defensively when criticized," "speak up more instead of waiting for them to approach me"). Use this list to focus your studies as you read the remainder of this book.

2. **Social Media Analysis** Construct a diary of the ways you use social media in a three-day period. For each instance when you use social media (email, a social networking website, phone, Twitter, etc.), describe:

   a. The kind(s) of social media you use

   b. The nature of the communication (e.g., "Wrote on friend's Facebook wall," "Reminded roommate to pick up dinner on the way home")

   c. The type of need you are trying to satisfy (information, relational, identity, entertainment)

   Based on your observations, describe the types of media you use most often and the importance of social media in satisfying your communication needs.

3. **Medium and Message Effectiveness** Send the same message to four friends, but use a different medium for each person. For example, ask the question, "How's it going?" Use the following media:

   - Social media message
   - Email
   - Text message
   - Telephone

   Notice how each response differs and what that may say about the nature of the medium.

The Self, Perception, and Communication

2

CHAPTER OUTLINE

Communication and the Self 36

Perceiving Others 42

Communication and Identity Management 55

LEARNING OBJECTIVES

2.1

Identify the communicative influences that shape the self-concept.

2.2

Examine the perceptual tendencies and situational factors that influence our perceptions of other people.

2.3

Describe how the process of identity management operates, in both face-to-face and online communication.

As you reflect on Caitlyn Jenner's story, think about the following questions:

?

What messages have you received from others and the media about the kind of person you should be? What do you like about others' expectations of you? What expectations do you wish were different?

?

What communication strategies have you used to influence how other people see you?

?

How do your perceptions of key people in your life shape the nature of your relationships with them?

"[I] LOOKED IN the mirror and I felt, 'Oh my God,'" says Caitlyn Jenner, reflecting on her groundbreaking photo shoot as a woman. It was the first time, reflects the reality TV star and Olympic champion, that her public image matched her self-concept. "Seeing that image was powerful to me," she recalls.[1]

Bruce Jenner was in the limelight for nearly 40 years, first as a record-breaking athlete and later as patriarch of the Jenner–Kardashian family. He was aware that the world viewed him as the quintessential "macho male."[2] Few people suspected that, on the inside, he felt more like a woman than a man. Failing to address the discrepancy felt like a lie, Jenner admits. Nonetheless he felt compelled to keep up the façade for 65 years because, as he puts it, "I don't want to disappoint people."[3]

Jenner's saga is a dramatic illustration of how our identities are shaped by messages from others. Like Jenner, we gain a sense of who we are partly in response to what others tell us. When we accept others' definition of us, we act in ways to reinforce it. But if we reject the identity others impose on us, we create messages aimed at convincing them to view us differently. Either way, communication is at the heart of defining our self-concept and social identity, as we will see in this chapter.

Communication and the Self

Nothing is more fundamental to understanding how we communicate than our sense of self. For that reason, the following pages introduce the notion of self-concept and explain how the way we view ourselves shapes our interaction with others.

Self-Concept Defined

The **self-concept** is a set of relatively stable perceptions that each of us holds about ourselves. It includes our conception of what is unique about us and what makes us both similar to and different from others. You might imagine the self-concept as a mental mirror that reflects how we view ourselves. This picture includes not only our physical features but also emotional states, talents, likes and dislikes, values, and roles.

We will have more to say about the nature of the self-concept shortly, but first, consider how this theoretical construct applies to you. Who are you? What is your gender identity? What is your age? Your religion? Occupation?

There are many ways of identifying yourself. Take a few more minutes and list as many ways as you can to identify who you are. You'll need this list later in this chapter, so be sure to complete it now. Try to include all the characteristics that describe you, including:

- Your moods or feelings
- Your appearance and physical condition
- Your social traits
- Talents you possess or lack
- Your intellectual capacity
- Your belief systems (religion, philosophy)
- Your strong beliefs
- Your social roles

Even a list of 20 or 30 terms would be only a partial description. To make this written self-portrait even close to complete, your list would have to be hundreds—or even thousands—of words long.

Of course, not all items on such a list would be equally important. For example, you might define yourself primarily by your social roles (parent, veteran), culture (Mexican American, Chinese), or beliefs (libertarian, feminist). Others might define themselves more in terms of physical qualities (tall, Deaf), or accomplishments and skills (athletic, scholar).

An important element of the self-concept is **self-esteem**: our evaluations of self-worth. One person's self-concept might include being religious, tall, or athletic. That person's self-esteem is shaped by how he or she feels about these qualities. Self-esteem is established early in life. By the age of 5, most children have developed a sense of self-worth.[4] They can offer accounts of how good they are at concrete skills like running and spelling.

Self-esteem influences the way we communicate and how people view us. People with high self-esteem are typically more confident in their perceptions and less likely to be swayed by peer pressure than are people with low self-esteem.[5] They are also more likely to approach conflict in a collaborative manner[6] and to exhibit an air of confidence and belonging in the workplace.[7] Not surprisingly, people with healthy self-esteem tend to be viewed positively by their peers, reinforcing their sense of being appreciated and valued.[8]

The opposite is true of low self-esteem. People who think poorly of themselves often communicate in ways that cause others to doubt them.[9] For example, a person who lacks confidence in social situations may avoid parties and study groups. As a result, she may come off as disinterested and unfriendly, which can add to her sense of being a social outcast, making her even less confident about reaching out. Take the self-assessment on page **39** to reflect on your own level of self-esteem.

Despite its obvious benefits, self-esteem doesn't guarantee success in personal and professional relationships.[10] People with an exaggerated sense of self-worth may *think* they make a great impression, but objective tests and the reactions of others don't always match this belief. It's easy to see how people with an inflated sense of self-worth could irritate others by coming across as condescending know-it-alls, especially when their self-worth is challenged.[11]

self-concept The relatively stable set of perceptions each individual holds of himself or herself.

self-esteem The part of the self-concept that involves evaluations of self-worth.

ASK YOURSELF

How would you describe your level of self-esteem? How is it affected by the messages you receive in face-to-face exchanges and via social media?

For years, civil rights leader Rachel Dolezal identified and presented herself as African American. When outed as white by her estranged biological parents, she argued that heredity doesn't equal racial identity.

What factors shape and affect your identity?

personality The set of enduring characteristics that define a person's temperament, thought processes, and social behavior.

reflected appraisal The influence of others on one's self-concept.

significant other A person whose opinion is important enough to affect one's self-concept strongly.

Biology, Personality, and the Self

Take another look at the list of terms you used to describe yourself on page 37. You'll almost certainly find some that reflect your **personality**—characteristic ways that you think and behave across a variety of situations. Personality tends to be stable throughout life: Research suggests that our temperament at age 3 is highly predictive of how we will behave as adults.[12]

People around the world embody many of the same basic personality traits. For example, languages as diverse as Hungarian, Dutch, English, Italian, and Korean have words to describe people who are extroverted, patient, emotional, honest, humble, and imaginative.[13]

Personality is determined partly by our genes. Researchers estimate that we inherit about 40% of our personality traits.[14] If your parents or grandparents are shy, for example, you may have a genetic tendency to be reserved around strangers. Or you might have a natural tendency to be novelty seeking, emotionally expressive,[15] or assertive,[16] or to exhibit many other traits. To some degree, we come programmed to communicate in particular ways.

Even though you may have a disposition toward some personality traits, you can do a great deal to control how you actually communicate.[17] Even shy people can learn how to reach out to others, and those with aggressive tendencies can learn to communicate in more sociable ways. One author put it this way: "Experiences can silence genes or activate them. Even shyness is like Silly Putty once life gets hold of it."[18] Throughout this book you will learn about communication skills that, with practice, you can build into your repertoire.

Although we inherit a portion of our personality, that isn't the only factor that influences our communication, as you will see next. So far we've talked about what the self-concept is, but at this point you may be asking what it has to do with the study of human communication. We can begin to answer this question by looking at how you came to possess your own self-concept.

External Influence on the Self-Concept

Along with heredity, our identity is shaped by communication with others. The term **reflected appraisal** describes how we develop an image of ourselves from the way we think others view us. We begin to grasp other people's impressions of us—both positive and negative—early on. As this happens, we are likely to internalize many of the messages we receive. This process continues throughout life, especially when messages come from **significant others**—people whose opinions we especially value. A teacher from long ago, a special friend or relative, or perhaps a barely known acquaintance can all leave an imprint on how you view yourself. To see the importance of significant others, ask yourself how you arrived at your opinion of you as a student, as a potential romantic partner, as a competent worker, and so on, and you will see that these self-evaluations were probably influenced by the way others regarded you.

Caitlyn Jenner is a good example. Even though she felt like a female on the inside, she felt obligated to fulfill society's expectations of male behavior for most of her life.[19] She says she understood what was at stake: Those who defy social expectations risk being judged, rejected, and even physically harmed. Transgender individuals in the United States are nearly twice as likely as others to be violently attacked.[20] Although Jenner has the benefits of bodyguards and wealth, her fame makes her subject to public opinion. Some critics have called her "sick and delusional," and others refuse to adopt feminine pronouns in reference to her.[21]

Even if the discrepancy between how you feel on the inside and how others see you is not as great as Jenner's, you probably feel the pressure to act in socially approved ways. For example, you may try to hold back tears in a stressful business meeting, lower your voice in church, or adopt the latest fashion or hairstyle.

When Kate Winslet agreed to model for L'Oréal cosmetics, she insisted on a "no retouching" clause in her contract. The actor says she doesn't want to contribute to unrealistic standards of attractiveness that make women feel bad about their appearance.

How do idealized images in the media affect your self-esteem?

SELF-ASSESSMENT

Communication and Your Self-Esteem

	Mostly True	Mostly False
1. People enjoy talking to me.		
2. If someone criticizes my work, I feel horrible.		
3. When I face a difficult communication challenge, I know I can succeed if I work at it.		
4. When people tell me they love me, I have a hard time believing it.		
5. I am comfortable admitting when I am wrong.		
6. People would like me more if I were better looking or more successful.		
7. I feel confident making big decisions about my relationships.		
8. I frequently let people down.		
9. It is more important that I am comfortable with myself than that others like me.		
10. I am frequently afraid of saying the wrong thing or looking stupid.		

Evaluating Your Responses

Give yourself 1 point for every odd-numbered statement you answered Mostly True and 1 point for every even-numbered question you answered Mostly False. Add them all together and look for your score below. Then ask yourself how your self-esteem shapes the way you communicate.

Scores

8 to 10 —You have very high self-esteem. This can be a bonus in your personal and professional life. However, be careful not to present yourself as infallible or superior to others.

5 to 7 —You have moderately high self-esteem. You are mostly self-assured and confident. However, you may feel insecure in some situations and relationships. When you find your confidence lagging, consider why. Perhaps developing new communication strategies or changing your internal dialogue will help.

3 to 6 —You have somewhat low self-esteem. Feeling bad about yourself can create a harmful pattern in which you doubt the positive things people say about you and accept poor treatment. Instead, try to surround yourself with people who value and honor you.

0 to 2 —You have very low self-esteem. Perhaps the messages you have received from loved ones and the media have caused you to doubt your self-worth. Look back at the list of characteristics you wrote about yourself (p. 37). Which of these make you strong and unique? When you start to focus on your positive qualities, others are more likely to take note of them as well.

Along with messages from others, our own assessments shape our self-concept. This occurs through the process of **social comparison**: evaluating ourselves in terms of how we compare with others.

Are you attractive or ugly? A success or failure? Intelligent or stupid? It depends on those you measure yourself against. For instance, young women who regularly compare themselves with ultra-thin media models tend to develop negative appraisals of their own bodies.[22] Men, too, who compare themselves to media-idealized male physiques, evaluate their bodies negatively.[23] These patterns tend to persist even when people know the images they are viewing have been retouched.[24] This has led some models and advertisers to insist on photos that are realistic rather than computer-enhanced. (See the photo on page 38.)

social comparison Evaluating ourselves in terms of how we compare with others.

Is Honesty Always the Best Policy?

Even with the best of intentions, there are cases when an honest message is likely to reduce another person's self-esteem. Consider a few examples:

- Your friend, an aspiring artist, asks, "What do you think of my latest painting?" You think it's terrible.

- After a long, hard week, you are looking forward to spending the evening at home. A somewhat insecure friend who just broke off a long romantic relationship calls to ask if you want to get together. You don't.

- A good friend asks to use your name as a reference for a potential employer. You can't honestly tell the employer that your friend is qualified for the job.

In situations like these, how do you avoid diminishing another person's self-esteem while being honest? Based on your conclusions, is it possible to always be both honest and supportive?

Social media are a powerful source of comparison. For example, researchers in one study found that people who spent significant time viewing photos of Facebook friends later reported reduced self-esteem.[25]

As we grow older, we tend to develop a more defined and enduring sense of who we are, and we are slightly less affected by other people's opinions of us.[26] We might find that constructive criticism no longer damages our self-confidence very much. Yet if we have negative feelings about ourselves, we may cling to those. If you see yourself as a terrible student, for example, you might respond to a high grade by thinking, "I was just lucky" or "The professor must be an easy grader."

You might argue that not every part of one's self-concept is shaped by others. You are right that some traits are objectively recognizable through self-observation. Nobody needs to tell you that you are taller than others, speak with an accent, can run quickly, and so on. However, the *significance* we attach to such attributes depends greatly on our social environment. If two friends, one Jamaican and one American, watch a woman cross the street, they may have different impressions of her. The American may conclude, "Look how slender that woman is. She must have a healthy diet." However, the Jamaican friend may reflect, "I wonder what sort of stress has caused her to be so thin. She is clearly unwell."[27]

Culture and the Self-Concept

Cultures affect our self-concept in both obvious and subtle ways. As you'll read in Chapter 3, most Western cultures are highly individualistic, meaning they value individuality and independence. By contrast, some cultures—most Asian ones, for example—are traditionally much more collective, meaning they value group identity more than personal autonomy. This cultural dynamic has powerful implications for communication. In job interviews, for example, people from the Americas, Canada, Australia, and Europe are likely to describe and even exaggerate their accomplishments. However, people in many Asian cultures value modesty instead. In Chinese written language, for example, the pronoun *I* looks very similar to the word for *selfish*.[28] They are likely to downplay their accomplishments, with the shared understanding that the interviewer will see that they are both highly accomplished and admirably humble about it.[29]

The Self-Concept and Communication with Others

So far we've focused on how the self-concept has been shaped by our interpretations of messages from our cultural environment and influential others. Now we will explore how the self-concept shapes the way we communicate *with* other people.

Figure 2-1 illustrates the relationship between the self-concept and behavior. It illustrates how the self-concept both shapes and is shaped by much of our communication behavior.

We can begin to examine the process by considering the self-concept you bring to an event. Suppose, for example, that one element of your self-concept is "nervous with authority figures." That image probably comes from the evaluations of significant others in the past—perhaps teachers or former employers. If you view yourself as nervous with authority figures like these, you will probably behave in nervous ways when you encounter them in the future—in a teacher–student conference or a job interview. That nervous behavior is likely to influence how others view your personality, which in turn will shape how they respond to you—probably in ways that reinforce the self-concept you brought to the event. Finally, the responses of others will affect the way you interpret future events: other job interviews, meetings with professors, and so on. This cycle illustrates how the

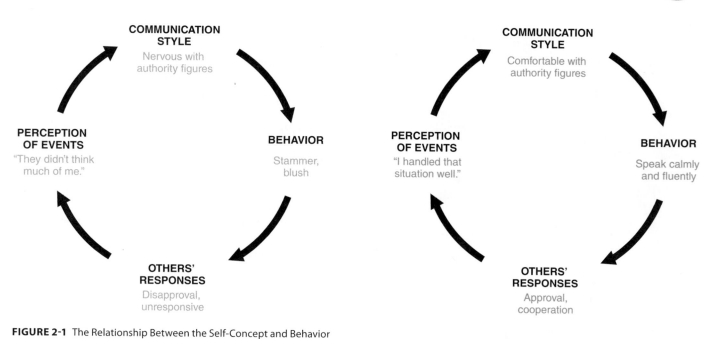

FIGURE 2-1 The Relationship Between the Self-Concept and Behavior

chicken-and-egg nature of the self-concept, which is shaped by significant others in the past, helps to govern your present behavior and influences the way others view you.

The Self-Fulfilling Prophecy and Communication

The self-concept is such a powerful force that it can influence our future behavior and that of others. A **self-fulfilling prophecy** occurs when a person's expectation of an outcome and subsequent behavior make the outcome more likely to occur. This happens all the time. For example, think of the following instances, which you may have experienced:

- You expected to become nervous during a job interview, and your anxiety caused you to answer questions poorly.

- You anticipated having a good (or terrible) time at a party, and your expectations led you to act in ways that shaped the outcome to fit your prediction.

- A teacher or boss explained a new task to you, saying that you probably wouldn't do well at first. You took these comments to heart, and as a result you didn't do well.

- A friend described someone you were about to meet, saying that you wouldn't like the person. You then looked for—and found—reasons to dislike the new acquaintance.

In each of these cases, the outcome happened at least in part because of the expectation that it would happen.

There are two types of self-fulfilling prophecies. The first type occurs when your expectations influence your behavior. In sports you have probably <u>psyched yourself</u> into playing either better or worse than usual. The same principle operates for public speakers: Those who feel anxious often create self-fulfilling prophecies that

self-fulfilling prophecy A prediction or expectation of an event that makes the outcome more likely to occur than would otherwise have been the case.

cultural idiom
psyched yourself: boosted confidence by thinking positively

"I don't sing because I am happy. I am happy because I sing."

Source: Edward Frascino The New Yorker Collection/ The Cartoon Bank

ASK YOURSELF

What self-fulfilling prophecies have you imposed or have others imposed on you?

cause them to perform less effectively.[30] (Chapter 11 offers advice on overcoming this kind of communication apprehension.)

Research has demonstrated the power of this first type of self-fulfilling prophecy. In one study, communicators who predicted that conflict episodes would be intense were likely to be highly emotional during them and to engage in personal attacks.[31] On the bright side, people who expect that others will be friendly and accepting usually act friendlier and more outgoing themselves.[32] As you might expect, they find a warmer reception than people who are fearful of rejection. As the cartoon on this page suggests, self-fulfilling prophecies can be physiologically induced. Researchers have found that putting a smile on your face, even if you're not in a good mood, can lead to a more positive disposition.[33]

A second type of self-fulfilling prophecy occurs when one person's expectations govern another's actions.[34] This principle was demonstrated in a classic experiment.[35] Researchers told teachers that 20% of the children in a certain elementary school showed unusual potential for intellectual growth. The names of the 20% were drawn randomly. Eight months later these unusual or "magic" children showed significantly greater gains in IQ than did the remaining children, who had not been singled out for the teachers' attention. The change in the teachers' behavior toward these allegedly "special" children led to changes in the children's performance. What the teachers communicated to the students—the message "I think you're bright"—affected the students' self-concepts, which ultimately affected their outcomes.

This type of self-fulfilling prophecy has been shown to help shape the self-concept and thus behavior in a wide range of settings. In medicine, for example, patients who unknowingly receive placebos—substances that have no curative value—may respond favorably, as if they had received an effective drug. The self-fulfilling prophecy operates in families as well. If parents disparage their children, the children's self-concepts may soon incorporate this idea, and they may fail at many of the tasks they attempt. On the other hand, if children are told they are capable, or lovable, or kind, there is a much greater chance they will behave in those ways.[36]

The self-fulfilling prophecy is an important force in communication, but it doesn't explain all behavior. There are certainly times when the expectation of an event's outcome won't bring about that outcome. Believing you'll do well in a job interview when you're clearly not qualified for the position is unrealistic. Similarly, there will probably be people you don't like and occasions you won't enjoy, no matter what your attitude. To connect the self-fulfilling prophecy with the "power of positive thinking" is an oversimplification.

As we keep these qualifications in mind, it's important to recognize the tremendous influence that self-fulfilling prophecies play in our lives. To a great extent we are what we believe we are. In this sense, we and those around us constantly create our self-concepts and thus ourselves.

Perceiving Others

The first part of this chapter explored how our self-perceptions affect the way we communicate. The following pages examine how the ways we perceive others shape our interactions with them.

Steps in the Perception Process

In 1890 the psychologist William James described an infant's world as "one great blooming, buzzing confusion."[37] Babies—all humans, in fact—need some mechanisms to sort out the avalanche of stimuli that bombard us every moment. As you will read in the following pages, many of these stimuli involve others' behavior, and how we deal with those stimuli shapes our communication. We sort out and make sense of others' behavior in three steps: selection, organization, and interpretation.

Selection   Because we're exposed to more input than we can possibly manage, the first step in perception is the **selection** of which data we will attend to. Some external factors help shape what we notice about others. For example, stimuli that are *intense* often attract our attention. Something that is louder, larger, or brighter than its surroundings stands out.

We also pay attention to *contrasts* or *changes* in stimulation. Put differently, unchanging people or things become less noticeable. This principle explains why we take consistently wonderful people for granted when we interact with them frequently. It's only when they stop being so wonderful or go away that we appreciate them.

Along with external factors like intensity and contrast, *internal factors* shape how we make sense of others. For example, our *motives* often determine how we perceive people. Someone on the lookout for a romantic adventure will be especially aware of attractive potential partners, whereas the same person in an emergency might be oblivious to anyone but police or medical personnel.

Our *emotional state* also shapes what we select. To some extent, your mood reflects your self-talk—what you tell yourself when you are sad or upset. People whose internal dialogue focuses on better days ahead typically experience better moods and more happiness than people who focus on negative events in the past.[38] Once started, focusing on the positive can create a spiral. If you're happy about your relationship, you will be more likely to interpret your partner's behavior in a charitable way. This, in turn, can lead to greater happiness. Of course, the same process can work in the opposite direction. One study revealed that, when spouses who felt uncertain about the status of their marriage saw their partners conversing with strangers, they were likely to perceive the conversations as relational threats, even when the conversations seemed quite ordinary to outsiders.[39]

Organization   After selecting information from the environment, we must mentally arrange it in some meaningful way to make sense of the world. We call this stage **organization**.

The raw sense data we perceive can be organized in more than one way, as the photo on page 44 illustrates. If you look at it long enough, you'll realize that it can be viewed in two ways: as a frontal image or as a profile. The same principle applies to communication behavior. Is your friend's "frown" a sign that she is in unhappy or just confused? Is an apology sincere or phony? In a larger sense, does an unreturned text message reflect disinterest or distraction?

We organize our perceptions of other people using *perceptual schemata*, cognitive frameworks that allow us to give order to the information we have selected.[40] Four types of schema help us classify others:

1. *Physical constructs* classify people according to their appearance, such as beautiful or ugly, fat or thin, young or old.

2. *Role constructs* use social position, such as student, attorney, wife.

3. *Interaction constructs* focus on social behavior, such as friendly, helpful, aloof, sarcastic.

4. *Psychological constructs* refer to internal states of mind and dispositions, such as confident, insecure, happy, neurotic.

Our perceptions of others are based on a variety of factors that are usually unconscious.

What stimuli shape the way you perceive people in your everyday life?

selection The perceptual act of attending to some stimuli in the environment and ignoring others.

organization The perceptual process of organizing stimuli into patterns.

Profile or frontal view? It depends on how you look at it.

Can you shift your evaluation of messages you receive of others? What does this say about the "truth"?

interpretation The perceptual process of attaching meaning to stimuli that have previously been selected and organized.

The kinds of constructs we use strongly affect the way we perceive others. For example, college students who watched advertisements for a violent mixed martial arts event typically preferred ads that showed men, rather than women, fighting. This was especially true among viewers who felt strongly that women should be protected from harm and should not engage in combat, even during sports.[41]

In online environments in which few nonverbal cues exist, it's harder to categorize people we haven't met in person. In text-based situations, strangers often rely on words to form impressions of others. In one study, for example, students developed opinions of strangers they couldn't see based on their screen names.[42] Experimenters asked college students to form impressions of fictional characters with names like "packerfan4" and "stinkybug." Using just the screen names, most of the respondents assigned attributes including biological sex, ethnicity, and age to the supposed owners of these names.

Once we have an organizing schema to classify people, we use it to make generalizations about members of the groups who fit our categories. For example, if religion plays an important part in your life, you might think of members of your faith differently than you do others. If ethnicity is an important issue for you, you probably tune in to the differences between members of various ethnic groups. There's nothing wrong with generalizations about groups as long as they are accurate. In fact, it would be impossible to get through life without them. But faulty overgeneralizations can lead to problems of stereotyping, which we address on page 49.

Interpretation   Once we have selected and organized our perceptions, we interpret them in a way that makes some sort of sense. **Interpretation** plays a role in virtually every type of communication. Is the person who smiles at you across a crowded room interested in romance or simply being polite? Is a friend's kidding a sign of affection or irritation? Should you take an invitation to "drop by anytime" literally or not? Table 2-1 summarizes several factors that cause us to interpret a person's behavior in one way or another.

Although we have talked about selection, organization, and interpretation separately, the three phases of perception can occur in differing sequences. For example, a parent's or babysitter's past interpretation (such as "Jason is a trouble-maker") can influence future selections (his behavior becomes especially noticeable) and the organization of events (when there's a fight, the assumption is that Jason started it). As with all communication, perception is an ongoing process in which it is hard to pin down beginnings and endings.

Influences on Perception

A variety of factors, from physiological to cultural and social, influence how we select, organize, and interpret data about others.

Physiological Influences   Some communication problems come from ignoring physiological (bodily) differences in how each of us experiences the world.[43]

TABLE 2-1
Factors That Affect Interpretation of Behavior

FACTOR	EXAMPLE
Degree of familiarity	You may perceive people who are familiar to you as more extraverted, stable, friendly, and conscientious than people you regard as strangers.[44]
Relational satisfaction	If a friend or romantic partner hurts your feelings, you are more likely to forgive and forget if you are generally happy in the relationship than if you are dissatisfied with it.[45]
Personal experience	If landlords have unfairly charged you in the past, you might be skeptical about an apartment manager's assurances of refunding your cleaning deposit.
Assumptions about human behavior	A boss who assumes that people are lazy and avoid responsibility may misjudge an employee's contributions.
Expectations	If you go into a conversation expecting a hostile attitude, you're likely to hear a negative tone in the other person's voice—even if that tone isn't intended by the speaker.
Knowledge of others	If you know a friend has just been rejected by a lover or fired from a job, you'll interpret his or her aloof behavior differently than if you were unaware of what happened.

For example, *age* can shape perceptions. Young children often seem egocentric, selfish, and uncooperative. A parent's exasperated "Can't you see I'm too tired to play?" might not make sense to a 4-year-old full of energy, who imagines that everyone else must feel the same.

Health and nutrition also influence the way we interact. When you're ill or have been working long hours or studying late for an exam, the world can seem quite different than when you are well rested. People who are sleep deprived perceive time intervals as longer than they really are. So the 5 minutes you spend waiting for a friend to show up may seem longer, leaving you feeling more impatient than you otherwise would be.[46] Your own experience probably confirms that nutrition also shapes communication. Being hungry (and getting grumpy) or having overeaten (and getting tired) affects how we interact with others. One study found that teenagers who reported that their families did not have enough to eat were almost twice as likely as other teens to report difficulty getting along with others.[47]

Biological cycles also affect perception and communication. Each of us has a daily cycle in which all sorts of changes constantly occur, including variations in body temperature, sexual drive, alertness, tolerance to stress, and mood.[48] These cycles can affect the way we relate to one another. Researchers in one study found that men had more negative views of their partners after a night in which they didn't sleep well. For women, it was the other way around: Their sleep patterns were disrupted when they perceived problems in the relationship. The researchers observe that sleep, mood, and interpersonal communication are mutually influential.[49]

Some differences in perception are rooted in *neurology*, the functioning of the nervous system. For instance, people with ADHD (attention deficit hyperactivity disorder) are easily distracted from tasks and have difficulty delaying gratification.[50] Those with ADHD might find a long lecture boring and tedious, while other audience members are fascinated by it.[51] In addition, young women with ADHD are more likely to have conflicts with their mothers and have fewer romantic relationships than those without ADHD.[52] People with bipolar disorder experience significant mood swings in which their perceptions of events, friends, and even family members shift dramatically. The National Institute of Mental Health estimates that between 5 million and 7 million Americans are affected by these

Even though they're not intentionally malicious, some casual remarks known as microaggressions can be perceived as hurtful, especially when they're directed at members of marginalized groups. This comment might have been intended as a compliment, but it felt like a slight to some people.

Have you ever experienced or committed microaggressions?

sex A biological category such as male, female, or intersex.

gender Socially constructed roles, behaviors, activities, and attributes that a society considers appropriate for men and/or women.

androgynous Combining both masculine and feminine traits.

two disorders alone[53]—and there are many other psychological conditions that influence people's perceptions.

Cultural and Social Influences   As we will discuss in depth in Chapter 3, culture provides a perceptual filter that influences the way we interpret even the simplest events. This filter extends to sex and gender roles, occupational roles, and relational roles, to name a few.

Sex and Gender Roles   Social expectations about masculine and feminine behavior have a powerful influence on our identity and how we communicate. Consider a few examples:

- A mother says to her daughter, "Don't act that way. It isn't ladylike."
- A same-sex couple is frustrated when people assume that one of them plays a more masculine role and one a more feminine role in the relationship.
- You overhear a colleague call your female supervisor "a bitch." You realize that term is never used to describe men.

For most people, gendered expectations began the moment they were born—or even before, with the selection of clothing, blankets, toys, and nursery décor. Yet we don't often question the assumptions that underlie cultural expectations about gender.

Let's start with a few misconceptions. One is that people are either male or female. Actually, hormone levels and other factors make biological sex a more complicated formula than you might think. The feelings and actions of both men and women are shaped by a mixture of estrogen and testosterone, which influence how the body and mind develop. Because these hormones are present in both men and women in varying degrees, the difference between male and female is more of a continuum than a question of either–or.[54] For example, people who are intersex may have hormonal profiles and physical characteristics of both sexes.

A second misconception is that "sex" and "gender" are the same. In fact, **sex** is a biological category (male, female, intersex, and so on), whereas **gender** is a socially constructed set of expectations about what it means to be "masculine" or "feminine." A statement about a man's long hair looking girlish reveals a particular cultural assumption. In many cultures, long hair is considered masculine. Thus, it is possible—even necessary—for a woman to embody some attributes associated with masculine gender, and vice versa for men. Indeed, as you will soon see, there are advantages to embodying both "masculine" and "feminine" qualities.

A third misconception is that masculine and feminine behaviors occur at opposite poles on a single continuum. Early theorizing suggested that stereotypical masculine and feminine behaviors are not opposites but rather two separate sets of behavior.[55] One alternative to the masculine–feminine dichotomy is the idea of four basic gender identities, including masculine, feminine, **androgynous** (combining masculine and feminine traits), and undifferentiated (neither masculine nor feminine). Even this model barely scratches the surface. As society has become more accepting of nuanced gender identities, people have begun to identify themselves using a wide array of terms, such as pansexual, asexual, queer, and questioning. We'll talk more about these in Chapter 3.

Although the differences between men's and women's behavior are not as great as many people think, evidence points to some differences at the extremes. For example, especially masculine males tend to see their interpersonal relationships as opportunities for competitive interaction, as opportunities to win something. Especially feminine females typically see their interpersonal relationships as opportunities to be nurturing, to express their feelings and emotions.[56] Androgynous individuals tend to see their relationships as opportunities to behave in a variety of ways, depending on the nature of the relationship, the context in which it takes place, and the myriad other variables affecting what might constitute appropriate behavior. These variables are often ignored by the sex-typed masculine males and feminine females, who have a smaller repertoire of behaviors.

Women and men often judge the same behaviors quite differently. Men tend to rate verbal and nonverbal behaviors as more flirtatious and seductive than do women.[57] By contrast, women are more likely than men to consider sexually suggestive behavior in the workplace to be destructive, unfair, and illegal.[58] Younger women are more likely to perceive harassment than older ones, who, presumably, have a different set of expectations about what kinds of communication are and aren't appropriate. Also, attitudes play a role. People who disapprove of socializing and dating between coworkers are more likely to perceive harassment than those who accept this sort of relationship.[59]

"How is it gendered?"

Source: Edward Koren The New Yorker Collection/
The Cartoon Bank

Occupational Roles　The kind of work we do also governs our view of the world.

Perhaps the most dramatic illustration of how occupational roles shape perception is a famous study from the early 1970s. Stanford psychologist Philip Zimbardo recruited a group of well-educated middle-class young men.[60] He randomly chose 11 to serve as "guards" in a mock prison set up in the basement of Stanford's psychology building. He issued the guards uniforms, handcuffs, whistles, and billy clubs. The remaining 10 participants became "prisoners" and were placed in rooms with metal bars, bucket toilets, and cots.

Zimbardo let the "guards" establish their own rules for the experiment. The rules were tough: no talking during meals and rest periods and after lights-out. Troublemakers received short rations.

Faced with these conditions, the "prisoners" began to resist. The guards reacted to the rebellion by punishing protesters, and what had been an experiment suddenly became very real. Some guards turned sadistic, physically and verbally abusing the prisoners. The experiment was scheduled to go on for 2 weeks, but after 6 days Zimbardo had to call it off. It seems that society's designation of our occupational roles has a profound impact on how we behave.

Relational Roles　Think back to the "Who am I?" list you made earlier in this chapter. It's likely your list included roles you play in relation to others: daughter, roommate, spouse, friend, and so on. Roles like these don't just define who you are—they also affect your perception.

Take, for example, the role of parent. As most new mothers and fathers will attest, having a child alters the way they see the world. They might perceive their crying baby as a helpless soul in need of comfort, whereas nearby strangers have a less charitable appraisal. As the child grows, parents often pay more attention to the messages in the child's environment. One father we know said he never noticed how much football fans curse and swear until he took his 6-year-old to a game with him. In other words, his role as father affected what he heard and how he interpreted it.

Sworn to "protect and serve," some police have been accused of racial profiling.

Have you ever been unfairly stereotyped by others? Have you ever been guilty of unfairly stereotyping?

The roles involved in romantic love can also dramatically affect perception. These roles have many labels: partner, spouse, boyfriend/girlfriend, sweetheart, and so on. There are times when your affinity biases the way you perceive the object of your affection. You may see your sweetheart as more attractive than other people do, and you might overlook some faults that others notice.[61] Your romantic role can also change the way you view others. One study found that when people are in love, they view other romantic candidates as less attractive than they normally would.[62]

Perhaps the most telltale sign of the effect of "love goggles" is when they come off. Many people look at former romantic partners are wonder, "What did I ever see in that person?" The answer—at least in part—is that you saw what your relational role led you to see.

Narratives, Perception, and Communication

We all have our own story of the world, and often our story is quite different from those of others. A family member or roommate might think your sense of humor is inappropriate, whereas you think you're quite clever. You might blame an unsatisfying class on the professor, who you think is a long-winded bore. On the other hand, the professor might characterize the students as superficial and lazy and blame the class environment on them. (The discussion of emotive language in Chapter 4 will touch on the sort of name-calling embedded in the previous sentences.)

Social scientists call these created personal stories **narratives**.[63] In a few pages we will look at how a tool called "perception checking" can help bridge the gap between different narratives. For now, though, the important point is that differing narratives can lead to problematic communication.

After they take hold, narratives offer a framework for explaining behavior and shaping future communication. For example, the stories that children in blended families tell about their stepparents both shape and reflect how they feel about them.[64] *Sudden* narratives reflect a sense that the adults married quickly and didn't communicate effectively with their children before involving them in a radically new family structure. By contrast, *idealized* narratives portray stepparents as heroic figures who help children feel like they are part of "real families." Once formed, narratives shape communication. If you view a stepparent as unwanted, you'll probably be wary, or even hostile. On the other hand, regarding the stepparent as a welcome addition—or even a savior—will no doubt lead to more positive behavior.

A classic study on long-term happy marriages demonstrates that shared narratives don't have to be accurate to be powerful.[65] Couples who report being happily married after 50 or more years seem to collude in a relational narrative that doesn't always reflect the facts. They often say agree that they rarely have conflict, although objective analysis reveals that they have had their share of disagreements and challenges. Without overtly agreeing to do so, they choose to blame outside forces or unusual circumstances for problems instead of attributing responsibility to each other. They offer the most charitable interpretations of each other's behavior, believing that the spouse acts with good intentions when things don't go well. They seem willing to forgive, or even forget, transgressions. Examining this research, one scholar concludes:

> Should we conclude that happy couples have a poor grip on reality? Perhaps they do, but is the reality of one's marriage better known by outside onlookers than by the players themselves? The conclusion is evident. One key to a long happy marriage is to tell yourself and others that you have one and then to behave as though you do![66]

narrative The stories people create and use to make sense of their personal worlds.

(?) ASK YOURSELF

Describe one narrative about yourself that you have constructed online. How conscious were you in crafting this narrative? How accurate is it?

cultural idiom

long-winded: speaking for a long time

Research like this demonstrates that the reality of what we experience isn't "out there"; rather, we create it with others through intrapersonal and interpersonal communication.

Common Perceptual Tendencies

Shared narratives may be desirable, but they can be hard to achieve. Some of the biggest obstacles to understanding and agreement arise from errors in interpretation, or what psychologists call **attribution**—the process of attaching meaning to behavior. We attribute meaning to both our own actions and the actions of others, but we often use different yardsticks. Researchers have identified several perceptual errors that can lead to inaccurate attributions—and to troublesome communication. By becoming aware of these errors, we can guard against them and avoid unnecessary conflicts.

We Make Snap Judgments   Our ancestors often had to make quick judgments about whether strangers were likely to be dangerous, and there are still times when this ability can be a survival skill.[67] But there are many cases when judging others without enough knowledge or information can get us into trouble. If you've ever been written off by a potential employer in the first few minutes of an interview or have been unfairly rebuffed by someone you just met, then you know the feeling. Snap judgments become particularly problematic when they are based on **stereotyping**—exaggerated beliefs associated with a categorizing system. Stereotypes based on "primitive categories" such as race, sex, and age may be founded on a kernel of truth, but they go beyond the facts at hand and make claims that usually have no valid basis.[68]

Three characteristics distinguish stereotypes from reasonable generalizations:

1. *Categorizing others on the basis of easily recognized but not necessarily significant characteristics.* For example, perhaps the first thing you notice about a person is his or her skin color—but that may not be nearly as significant as the person's intelligence or achievements.

2. *Ascribing a set of characteristics to most or all members of a group.* For example, you might unfairly assume that all older people are doddering or that all men are insensitive to women's concerns.

3. *Applying the generalization to a particular person.* Once you believe all old people are geezers and all men are jerks, it's a short step to considering a particular senior citizen as senile, or a particular man as a chauvinist pig.

By adulthood, we tend to engage in stereotyping frequently, effortlessly, and often unconsciously.[69] This process saves mental energy, but it can lead to mistaken assumptions about other people based on limited information. For example, after researchers gave college students bogus personality test results indicating that they were either "blue" or "yellow" personality types, they were more distrustful of people in the other color group.[70] If distrust can spring up that quickly, it's no surprise that it becomes entrenched in daily life, where a variety of factors may send the misguided message of "us" versus "them." For example, video games disproportionally depict minority male characters as excessively violent thugs and athletes, suggesting that they are distinctly different and less trustworthy than other characters.[71]

Once we create and hold stereotypes, we tend to seek out isolated behaviors that support our inaccurate beliefs. Even when the stereotype seems positive, it can be dehumanizing and unfair. People who frequently see Asian Americans depicted on television as brainy and polite are more likely than others to think that they are all alike and that success comes easily to them.[72] The reality, of course, is that Asian Americans are as diverse as any other group, and viewing them through such a narrow lens diminishes their potential as individuals.

attribution The process of attaching meaning.

stereotyping The perceptual process of applying exaggerated beliefs associated with a categorizing system.

Latina Americans have made the Twitter feed "Hispanic Girls United" a popular forum for calling out ethnic stereotypes, such as media photos in which Hispanic models are either depicted as hypersexual and wild or are airbrushed to appear "whiter" than they are.

Do you feel the media misrepresents social groups with which you identify?

cultural idiom

yardsticks: standards of comparison

One way to avoid the kinds of communication problems that come from excessive stereotyping is to "decategorize" others. Through this practice, you give yourself a chance to treat people as individuals instead of assuming that they possess the same characteristics as every other member of the group to which you assign them. The best antidote to stereotypes is willingness to admit how faulty they are. Participants in one study initially rated white managers to be more competent, achievement oriented, and manipulative than black managers, and black managers to be more interpersonally skilled and less polished than their white counterparts.[73] However, those judgments largely disappeared when the participants were informed that all of the managers were successful.

We Often Judge Ourselves More Charitably Than We Judge Others　In an attempt to convince ourselves and others that the positive face we show to the world is true, we tend to judge ourselves in the most generous terms possible. Social scientists call this tendency fundamental attribution error, or simply the **self-serving bias**.[74] When others suffer, we often blame the problem on their personal qualities. On the other hand, when we suffer, we find explanations outside ourselves. Consider a few examples:

- When someone else botches a job, we might think that person wasn't listening well or trying hard enough; when we botch a job, the problem was unclear directions or not enough time.
- When he lashes out angrily, we say he's being moody or too sensitive; when we blow off steam, it's because of the pressure we've been under.
- When others don't reply to your text or email, you might think they are inconsiderate, disrespectful, and unprofessional. However, when we don't reply, we may say we were too busy, we didn't see the message, or it didn't seem necessary to reply.[75]

Evidence like this suggests how uncharitable attitudes toward others can affect communication. Your harsh opinions of others can lead to judgmental messages. At the same time, you may be blind to ways that you can improve and defensive when people question your behavior.

We Pay More Attention to Negative Impressions Than Positive Ones　Research shows that when people are aware of both the positive and negative traits of another, they tend to be more influenced by the negative traits. For example, employers are often turned off when they find negative images of a job candidate on social media, even if the candidate is otherwise highly qualified.[76]

Sometimes a negativity bias makes sense. If an unappealing quality clearly outweighs any positive ones, you'd be foolish to ignore it. A surgeon with shaky hands and a teacher who hates children, for example, would be unsuitable for their jobs whatever their other virtues. But much of the time it's a bad idea to pay excessive attention to negative qualities and overlook positive ones. This is the mistake some people make when screening potential friends or dates. Of course, it's important to find people you truly enjoy, but expecting perfection can lead to much unnecessary loneliness.

We Are Influenced by What Is Most Obvious　Every time we encounter another person, we are bombarded with more information than we can possibly manage. You can appreciate this by spending 2 or 3 minutes just reporting on what you can observe about another person through your five senses. You will find that every time you seem to near the end, a new observation presents itself.

To make sense of others, we need to select from within an overwhelming amount of sense data. Several factors cause us to notice some messages and ignore others. We pay attention to stimuli that are *intense* (loud music, brightly dressed

self-serving bias The tendency to interpret and explain information in a way that casts the perceiver in the most favorable manner.

It's understandable to notice the most obvious features of others, but those aren't necessarily the most important ones.

What would others miss about you if they only attended to your most obvious features? How can you avoid this perceptual error when encountering strangers?

cultural idioms

lashes out: attacks verbally

blow off steam: release excess energy or anger

people), *repetitious* (dripping faucets, persistent people), or *contrastive* (a normally happy person who acts grumpy or vice versa). *Motives* also determine what information we select from our environment. If you're anxious about being late for a date, you'll notice clocks. If you're hungry, you'll become aware of restaurants, markets, and billboards advertising food. Motives also determine how we perceive people.

If intense, repetitious, or contrastive information were the most important thing to know about others, there would be no problem. But the most noticeable behavior of others isn't always the most important. For example:

- When two children (or adults, for that matter) fight, it may be a mistake to blame the one who lashes out first. Perhaps the other one was at least equally responsible, by teasing or refusing to cooperate.

- You might complain about an acquaintance whose malicious gossiping or arguing has become a bother, forgetting that, by previously tolerating that kind of behavior, you have been at least partially responsible.

- You might blame an unhappy work situation on the boss, overlooking other factors beyond her control, such as a change in the economy, the policy of higher management, or demands of customers or other workers.

We Cling to First Impressions, Even If Wrong   Labeling people according to our first impressions is an inevitable part of the perception process. These labels are a way of making interpretations. "She seems cheerful." "He seems sincere." "They sound awfully conceited."

Problems arise when the labels we attach are inaccurate, because after we form an opinion of someone, we tend to hang on to it and ignore any conflicting information. Suppose, for instance, you mention the name of your new neighbor to a friend. "Oh, I know him," your friend replies. "He seems nice at first, but it's all an act." Perhaps this appraisal is off base. Whether or not the judgment is accurate, after you accept your friend's evaluation, it will probably influence the way you respond to the neighbor. Your response may in turn influence your neighbor's behavior, creating a self-fulfilling prophecy.

The power of first impressions is important in personal relationships. A study of college roommates found that those who had positive initial impressions of each other were likely to have positive subsequent interactions, manage their conflicts constructively, and continue living together.[77] The converse was also true: Roommates who got off to a bad start tended to spiral negatively—hence the old adage, "You never get a second chance to make a first impression." Given the almost unavoidable tendency to form first impressions, the best advice we can offer is to keep an open mind. Be willing to change your opinion as events prove that your first impressions were mistaken.

We Tend to Assume That Others Are Similar to Us   We commonly imagine that others possess the same attitudes and motives that we do. This tendency often leads to mistaken assumptions, as in the following examples:

- You've heard an off-color joke that you found funny. You might assume that it won't offend a friend. It does.

- You've been bothered by an instructor's digressions during lectures. You decide that your instructor will probably be grateful for some constructive criticism, as you would be. Unfortunately, you're wrong.

- You lost your temper with a friend a week ago and said some things you regret. In fact, if someone said those things to you, you would consider the relationship finished. Imagining that your friend feels the same way, you avoid making contact. In fact, your friend feels that he was partly responsible and has avoided you because he thinks you're the one who wants to end things.

Examples like these show that others don't always think or feel the way we do and that an assumption of similarities can lead to problems. For example, men are more likely than women to think that flirting indicates an interest in having sex.[78]

How can you find out the other person's real position? Sometimes by asking directly, sometimes by checking with others, and sometimes by making an educated guess after you've thought the matter out. All these alternatives are better than simply assuming that everyone would react the way you do.

Don't misunderstand: We don't always commit the kind of perceptual errors described in this section. Sometimes, for instance, people are responsible for their misfortunes, and sometimes our problems are not our fault. Likewise, the most obvious interpretation of a situation may be the correct one. Nonetheless, research proves again and again that our perceptions of others are often distorted in the ways listed here. The moral, then, is clear: Don't assume that your first judgment of a person is accurate.

Perception in Mediated Communication

Mediated communication offers fewer cues about others than you get in face-to-face interaction. The ability to edit a message before posting makes it easy to craft an identity that doesn't match what you would encounter in person.

Despite the apparent barriers to accurate perception in mediated channels, people can form accurate impressions of one another in cyberspace—though not in the same ways as when they communicate in person.[79] Early in relationships, people who meet in person have more accurate impressions of one another than those who meet online.[80] But over time, people can get to know each other well online and may even share information about themselves that they would be reluctant to disclose in person.[81]

In text-based channels, receivers depend more on writing style to get a sense of the sender. This means that typos and grammatical mistakes can ruin your reputation on the job. Researchers in one study found that 74% of people notice spelling and grammar errors on corporate websites, and most of them avoid doing business with those companies as a result. It follows that employees with poor writing skills are a liability rather than an asset. One analyst compares sloppy writing to the unprofessionalism of wearing an egg-stained shirt or pajamas to work.[82]

Empathy, Perception, and Communication

By now it's clear that differing perceptions present a major challenge to communicators. One solution is to increase the ability to empathize. **Empathy** is the ability to re-create another person's perspective, to experience the world from the other's point of view.

empathy The ability to project oneself into another person's point of view, so as to experience the other's thoughts and feelings.

Dimensions of Empathy As we'll use the term here, empathy has three dimensions:[83]

1. On one level, empathy involves *perspective taking*—the ability to take on the viewpoint of another person. This understanding requires you to set aside your own opinions and see things as another person does.

2. Besides cognitive understanding, empathy also has an *emotional* dimension that allows you to experience what others are feeling. You might share their

fear, joy, sadness, and so on. When you combine the perspective-taking and emotional dimensions, you may see that empathizing allows you to experience the other's perception—in effect, to become that person temporarily.

3. Empathy requires a genuine *concern* for the welfare of the other person. When you empathize with others, you go beyond just thinking and feeling as they do; you genuinely care about their well-being.

It's easy to confuse empathy with **sympathy**, compassion for another's situation, but the concepts are different in two important ways. First, empathy means you have a personal sense of another's experience, whereas if you are sympathetic you do not necessarily share another person's experience. Second, we only sympathize when we accept the reasons for another's pain as valid, whereas it's possible to empathize without feeling sympathy. Perhaps you have witnessed a frustrated customer berating a cashier on a busy shopping day. You have felt similar frustration before, so you empathize. But this customer's behavior is so over the top that you have little sympathy for him. In the same way, you can empathize with a difficult relative or even a criminal without feeling much sympathy for that person. Empathizing allows you to understand another person's motives without requiring you to agree with them. After empathizing, you will almost certainly understand a person better, but sympathy won't always follow.

The ability to empathize seems to exist in a rudimentary form in even the youngest children. Virtually from birth, infants are attentive to the sound of other babies crying, and they are likely to cry along. Although toddlers are notorious for wanting toys all to themselves, when a playmate is hurt or upset, their desire to comfort may win out. Young children often stop playing to offer a hug, soothing words, or a toy to friends in distress, suggesting that empathy is a strong motivating force even in early life.[84]

People vary in their ability to understand how others feel. This is partly because of the abilities they are born with. Some people seem to have a hereditary capacity for greater empathy than do others.[85] But most people can develop their empathic capacity by being aware and attentive. Parents and teachers can help by role-modeling empathic behaviors[86] and by pointing out to children how their actions affect others ("Look how sad Jessica is because you took her toy. Wouldn't you be sad if someone took away your toys?"). Children who are encouraged to understand how others feel and who are rewarded for being considerate tend to demonstrate more empathy and prosocial behavior than their peers do.[87]

There is no consistent evidence that the ability to empathize is greater for one sex or another. On average, teenage girls tend to develop adult levels of empathy sooner than boys do, but the boys usually catch up.[88] Women are commonly considered to be more empathic toward children than men are, in part because of traditional parenting roles. But the latest research shows that people of any sex who identify with certain characteristics, such as being affectionate, gentle, and nurturing, are typically more responsive to babies' nonverbal cues than other adults are.[89]

Total empathy is impossible to achieve. Completely understanding another person's point of view is simply too difficult a task for humans with different backgrounds and limited communication skills. Nonetheless, it is possible to

"How would you feel if the mouse did that to you?"

Source: William Steig The New Yorker Collection/ The Cartoon Bank

sympathy Compassion for another's situation.

get a strong sense of what the world looks like through another person's eyes.

Perception Checking   Good intentions and a strong effort to empathize are one way to understand others. Along with a positive attitude, a simple tool can help you interpret the behavior of others more accurately. To see how this tool operates, consider how often others jump to mistaken conclusions about your thoughts, feelings, and motives:

> *"Why are you mad at me?"* (Who said you were?)
>
> *"What's the matter with you?"* (Who said anything was the matter?)
>
> *"Come on now. Tell the truth."* (Who said you were lying?)

As you'll learn in Chapter 7, even if your interpretation is correct, a dogmatic, mind-reading statement is likely to generate defensiveness. The skill of **perception checking** provides a better way to handle your interpretations. A complete perception check has three parts:

1. A description of the behavior you noticed
2. At least two possible interpretations of the behavior
3. A request for clarification about how to interpret the behavior

Perception checks for the preceding three examples would look like this:

> *"When you stomped out of the room and slammed the door* [behavior], *I wasn't sure whether you were mad at me* [first interpretation] *or just in a hurry* [second interpretation]. *How did you feel* [request for clarification]?"
>
> *"You haven't laughed much in the last couple of days* [behavior]. *I wonder whether something's bothering you* [first interpretation] *or whether you're just feeling quiet* [second interpretation]. *What's up* [request for clarification]?"
>
> *"You said you really liked the job I did* [behavior], *but there was something about your voice that made me think you may not like it* [first interpretation]. *Maybe it's just my imagination, though* [second interpretation]. *How do you really feel* [request for clarification]?"

Perception checking is a tool for helping us understand others accurately instead of assuming that our first interpretation is correct. Because its goal is mutual understanding, perception checking is a cooperative approach to communication. Besides leading to more accurate perceptions, it minimizes defensiveness. Instead of saying in effect, "I know what you're thinking . . . ," a perception check takes the more respectful approach that states or implies, "I know I'm not qualified to judge you without some help."

Sometimes a perception check is effective without all of the parts listed earlier:

> *"You haven't* <u>dropped by</u> *lately. Is anything the matter* [single interpretation combined with request for clarification]?"
>
> *"I can't tell whether you're kidding me about being cheap or if you're serious* [behavior combined with interpretations]. *Are you mad at me?"*
>
> *"Are you sure you don't mind driving? I can use a ride if it's no trouble, but I don't want to take you out of your way* [no need to describe behavior]."

perception checking   A three-part method for verifying the accuracy of interpretations, including a description of the sense data, two possible interpretations, and a request for confirmation of the interpretations.

CHECKLIST ✓
Check Your Perceptions Before Responding

The next time you feel puzzled or upset about someone's behavior, don't jump to conclusions. Instead, follow these steps to find out more.

☐ Describe the behavior. ("You didn't say anything in response to my ideas.")

☐ Suggest at least two interpretations of the behavior. ("Maybe you needed some time to think about it. Or perhaps my ideas weren't what you had in mind.")

☐ Request clarification about how to interpret the behavior. ("I value your opinion. What are the pros and cons of my ideas, as you see them?")

cultural idiom
dropped by: made an unplanned visit

Of course, a perception check can succeed only if your nonverbal behavior reflects the open-mindedness of your words. An accusing tone of voice or a hostile glare will contradict the sincerely worded request for clarification, suggesting that you have already made up your mind about the other person's intentions.

Communication and Identity Management

So far we have described how communication shapes the way communicators view themselves and others. In the remainder of this chapter we <u>turn the tables</u> and focus on **identity management**—the communication strategies people use to influence how others view them. In the following pages you will see that many of our messages aim at creating desired impressions.

Public and Private Selves

To understand why identity management exists, we have to discuss the notion of self in more detail. So far we have referred to the "self" as if each of us had only one identity. In truth, each of us possesses several selves, some private and others public. Often these selves are quite different.

The **perceived self** is a reflection of the self-concept. Your perceived self is the person you believe yourself to be in moments of honest self-examination. We can call the perceived self "private" because you are unlikely to reveal all of it to another person. You can verify the private nature of the perceived self by reviewing the self-concept list you developed while reading **page 37**. You'll probably find some elements of yourself there that you would not disclose to many people, and some that you would not share with anyone. You might, for example, be reluctant to share some feelings about your appearance ("I think I'm rather unattractive"), your intelligence ("I'm not as smart as I wish I were"), your goals ("The most important thing to me is becoming rich"), or your motives ("I care more about myself than about others").

In contrast to the perceived self, the **presenting self** is a public image—the way we want to appear to others. In most cases the presenting self we seek to create is a socially approved image: diligent student, loving partner, conscientious worker, loyal friend, and so on. Social norms often create a gap between the perceived and presenting selves.

Sociologist Erving Goffman used the word **face** to describe the presenting self, and he coined the term **facework** to describe the verbal and nonverbal ways we act to maintain our own presenting image and the images of others.[90] He argued that each of us can be viewed as a kind of playwright who creates roles that we want others to believe, as well as the performer who acts out those roles.

Facework involves two tasks: managing our own identity and communicating in ways that reinforce the identities that others are trying to present.[91] You can see how these two goals operate by recalling a time when you have used self-deprecating humor to defuse a potentially unpleasant situation. Suppose, for example, that a friend gave you confusing directions to a party that caused you to be late. "Sorry I got lost," you might have said. "I'm a terrible navigator." This sort of mild self-put-down accomplishes two things at once. It preserves the other person's face by implicitly saying, "It's not your fault." At the same time, your mild self-debasement shows that you're a nice person who doesn't find faults in others or make a big issue out of small problems.[92]

In most situations, we have a choice about how much information to reveal. The challenge is how to present an accurate image of our true selves without revealing information we might later regret.

How congruent are your public and private selves? Are you over- or undersharing?

identity management Strategies used by communicators to influence the way others view them.

perceived self The person we believe ourselves to be in moments of candor. It may be identical to or different from the presenting and ideal selves.

presenting self The image a person presents to others. It may be identical to or different from the perceived and ideal selves. See also **face**.

face The socially approved identity that a communicator tries to present.

facework Verbal and nonverbal behavior designed to create and maintain a communicator's face and the face of others.

cultural idiom
  turn the tables: reverse an existing situation

Characteristics of Identity Management

Now that you have a sense of what identity management is, we can look at some characteristics of this process.

We Have Multiple Identities   In the course of even a single day, most people play a variety of roles: respectful student, joking friend, friendly neighbor, and helpful worker, to suggest just a few. We even play a variety of roles with the same person. As a teenager, you almost certainly changed characters as you interacted with your parents. In one context you acted as the responsible adult ("You can trust me with the car!"), and in another context you were the helpless child ("I can't find my socks!"). At some times—perhaps on birthdays or holidays—you were a dedicated family member, and at other times you may have played the role of rebel. Likewise, in romantic relationships we switch among many ways of behaving, depending on the context: friend, lover, business partner, scolding critic, apologetic child, and so on.

None of us has a single self. We present ourselves differently, depending on the circumstances.

What multiple selves do you show the world? How authentic and appropriate are each of them?

The ability to construct multiple identities is one element of communication competence. For example, the style of speaking or even the language itself can reflect a choice about how to construct one's identity. We recall an African American colleague who was also minister of a Southern Baptist congregation consisting mostly of black members. On campus his manner of speaking was typically professorial, but a visit to hear him preach one Sunday revealed a speaker whose style was much more animated and theatrical, reflecting his identity in that context. Likewise, one scholar pointed out that bilingual Latinos in the United States often choose whether to use English or Spanish depending on the kind of identity they are seeking in a given conversation.[93]

Identity Management Is Collaborative   As we perform like actors trying to create a persona (character), our "audience" is made up of other actors who are trying to create their own. Identity-related communication is a kind of theater in which we collaborate with other actors to improvise scenes.

You can appreciate the collaborative nature of identity management by thinking about how you might handle a minor complaint about a friend's behavior:

You: By the way, Jenny said she texted you last week to invite us to her party. If you let me know, I guess I missed it.

Friend: Oh, sorry. I meant to tell you in person, but our schedules have been so crazy this week that I haven't seen you.

You: *(in friendly tone of voice)* That's okay. Maybe next time you can just forward the text to me.

Friend: No problem.

In this upbeat conversation, both you and your friend accept each other's bids for identity as basically thoughtful people. As a result, the conversation runs smoothly. Imagine, though, how different the outcome would be if your friend didn't accept your role as "nice person":

You: By the way, Jenny said she texted you last week to invite us to her party. If you let me know, I guess I missed it.

Friend: *(defensively)* Okay, so I forgot. It's not that big a deal. You're not perfect either!

At this point you have the choice of persisting in trying to play the original role: "Hey, I'm not mad at you, and I know I'm not perfect!" Or, you might switch to the new role of "unjustly accused person," responding with aggravation, "I never said I was perfect. But we're not talking about me here."

As this example illustrates, *collaboration* doesn't mean the same thing as *agreement*. The small issue of the forgotten text might mushroom into a fight. The point here is that virtually all conversations provide an arena in which communicators construct their identities in response to the behavior of others. As you read in Chapter 1, communication isn't made up of discrete events that can be separated from one another. Instead, what happens at one moment is influenced by what each party brings to the interaction and by what happened in their relationship up to that point.

Identity Management Can Be Conscious or Unconscious   At this point you might object to the notion of strategic identity management, claiming that most of your communication is spontaneous and not a deliberate attempt to present yourself in a certain way. However, you might acknowledge that some of your communication involves a conscious attempt to manage impressions.

There's no doubt that sometimes we are highly aware of managing our identities. You try hard to make a good impression during job interviews and first dates. You act differently around friends than at work or in the classroom. **Frame switching** involves adopting different perspectives based on the cultures and situations in which we find ourselves.[94] You may know people who aren't very good at this. They tend to say things that are unfitting for the situation and offend others by being more aggressive, casual, or standoffish than people expect them to be. Identity management can be especially hard work for people who operate in two or more cultures. A Filipino American man describes the duality of his work and family life this way: "At my first job I learned I had to be very competitive and fighting with my other coworkers for raises all the time, which I was not ready for—being brought up as nice and quiet in my family."[95]

In other cases, we engage in identity management without thinking much about it. You probably don't give a lot of conscious thought to whether you should smile or laugh, but much of the time this *is* a decision. Consider the gap between how we express humor online and in person: People don't actually LOL as much as their texts say they do. In reality, people are far less likely to laugh or smile when alone than with other people. Smiles and laughter are social cues as much (or more) as they are expressions of emotion. We tend to trust people who smile (as long as it seems genuine) more than those who don't,[96] and we generally perceive people who smile and laugh to be more attractive than stern people with similar features.[97]

We also develop **scripts** over time—habitual, reflexive ways of behaving. When you meet a new person or greet a customer at work, for example, you may say and do fairly consistent things. Public speakers often develop fairly elaborate scripts to talk about familiar topics. In many cases, scripts are effective and save us mental energy. They can have downsides, however, when we don't think clearly about the demands of the moment or how others may interpret our words and actions.

People Differ in Their Degree of Identity Management   Some people are much more aware of their identity management behavior than others. These high

frame switching Adopting the perspectives of different cultures.

script Habitual, reflexive way of behaving.

UNDERSTANDING DIVERSITY

Managing Identity and Coming Out

I grew up in a mid-size town in Mexico. My siblings and I were taught to never do anything to "dishonor" the name of our family.

From a very early age I sensed I was not a boy who liked girls. My father was very involved in Mexican rodeo, and he expected me to ride a horse, rope, and deal with cattle. I quit when I was twelve. My dad never understood why.

Middle school was hard. I coped with bullies by being funny and making fun of others before they could make fun of me. I had a girlfriend for a year. It was all pretend, but it helped ease the harassment. By high school I knew was gay and wanted to live openly. But my religious upbringing, peer group, and family's reputation were all in the way of me coming out.

I started living a double life during college. Guilt became the main emotion I felt. I became very cold and distant with everyone in my family. To cover up my identity, I had a girlfriend for three years. After college I met the first openly gay men who were comfortable and happy. I fell in love with one of them, and he helped me understand that I needed to have the courage to accept who I am and to stop hiding this from my family.

It took three more years for me to come out to my siblings. Those were the hardest conversations I have ever endured. My siblings urged me to not tell my parents, who they said would be upset and saddened. Unfortunately I followed their suggestion, and my mom passed without me ever sharing this part of myself. Still, I am sure she knew the truth. I have never told my dad directly that I am gay. When we talk, he always asks me how I am doing and how John (my partner) is doing. That is his way of letting me know he knows.

I can sleep much better at night since I came out sixteen years ago. Since then, I have stopped worrying what others think about me. There is no guilt and I feel complete. I know I am happy.

J. C. Rivas

self-monitors have the ability to pay attention to their own behavior and others' reactions and adjust their communication to create the desired impression. By contrast, low self-monitors express what they are thinking and feeling without much attention to the impression their behavior creates.[98]

There are advantages to being a high self-monitor. People who pay attention to themselves are generally good actors who can create the impression they want, acting interested when bored, or friendly when they really feel quite the opposite. This allows them to handle social situations smoothly, often putting others at ease. They are also good "people-readers" who can adjust their behavior to get the desired reaction from others. Along with these advantages, there are potential disadvantages to being an extremely high self-monitor. The analytical nature of high self-monitors may prevent them from experiencing events completely because a portion of their attention is devoted to viewing the situation from a detached position. High self-monitors' ability to act means that it is difficult to tell how they are really feeling. In fact, because high self-monitors change roles often, they may have a hard time knowing themselves how they really feel.

People who score low on the self-monitoring scale live life quite differently from their more self-conscious counterparts. They have a simpler, more focused idea of who they are and who they want to be. Low self-monitors are likely to have a narrower repertoire of behaviors, so that they tend to act in more or less the same way regardless of the situation. This means that low self-monitors are easy to read. "What you see is what you get" might be their motto. Although this lack of flexibility may make their social interaction less smooth in many situations, low self-monitors can be counted on to be straightforward communicators.

By now it should be clear that neither extremely high nor low self-monitoring is ideal. In some situations paying attention to yourself and adapting your behavior can be useful, but in other situations reacting without considering the effect on others is a better approach. This need for a range of behaviors demonstrates again the notion of communicative competence outlined in Chapter 1: Flexibility is the key to successful relationships.

Identity Management in the Workplace

Some advisors encourage workers to "just be yourself" on the job. But there are times when disclosing certain information about your personal life can damage your chances for success.[99] This is especially true for people with "invisible stigmas"—traits that, if revealed, may cause them to be viewed unfavorably.[100]

Many parts of a worker's identity have the potential to be invisible stigmas: religion (evangelical Christian, Muslim), sexual orientation (gay, lesbian, bisexual), and health (bipolar, HIV positive). What counts as a stigma to some people (politically progressive, conservative) might be favored in another organization.[101]

As you consider how to manage your identity at work, consider the following:

1. **Proceed with caution.** In an ideal world, everyone would be free to reveal themselves without hesitation. But in real life, total candor can have consequences, so it is best to move slowly.

2. **Assess the organization's culture.** If people in your workplace seem supportive of differences—and especially if they appear to welcome people like you—then revealing more of yourself may be safe.

3. **Consider the consequences of not opening up.** Keeping an important part of your identity secret can also take an emotional toll.[102] If keeping quiet is truly necessary, you may be better off finding a more welcoming place to work.

4. **Test the waters.** If you have a trusted colleague or manager, think about revealing yourself to that person and asking advice about whether and how to go further. But realize that even close secrets can leak, so be sure the person you approach can keep confidences.

Another aspect of identity management at work involves letting other people know about your achievements. See the "@Work" box on page 60 for more about striking a balance between self-promoting and bragging.

Why Manage Identities?

Why bother trying to shape others' opinions? Sometimes we create and maintain a front to follow social rules. As children we learn to act polite, even when bored. Likewise, part of growing up consists of developing a set of manners for various occasions: meeting strangers, attending school, going to religious services, and so on. Young children who haven't learned all the do's and don'ts of polite society often embarrass their parents by behaving inappropriately ("Mommy, why is that man so fat?"). But by the time they enter school, behavior that might have been excusable or even amusing just isn't acceptable. Good manners are often aimed at making others more comfortable. For example, able-bodied people may mask their discomfort on encountering someone with a disability by acting nonchalant or stressing similarities between themselves and the other person.

Social rules govern our behavior in a variety of settings. It would be impossible to keep a job, for example, without meeting certain expectations. Salespeople are obliged to treat customers with courtesy. Employees should appear reasonably respectful when talking to the boss. Some forms of clothing would be considered outrageous at work. By agreeing to take on a job, you sign an unwritten contract that you will present a certain face at work, whether or not that face reflects the way you feel at a particular moment.

Even when social roles don't dictate the proper way to behave, we often manage identities for a second reason: to accomplish personal goals. You might,

Humblebragging in Job Interviews

Life is hard for humblebraggers. Every golden moment inspires an ostensibly modest complaint about a dazzling accomplishment. Can't you just feel the envy—and anguish—of their social media followers?

> why does the Mercedes dealership always have fresh baked hot cookies?! don't they understand how mean that is?

> Why can't I look cool when I meet @TomHanks & he hands me his Emmy? Instead I get so excited & look like a goober.

> I'm wearing a ponytail, rolled out of bed from a nap, at the bar w/ my guy and guys r still hitting on me. Like really?

After reading even a few humblebrags, you probably agree that they are . . . well, annoying. A Harvard study confirms that trying to pass off a brag as a complaint usually doesn't fool anyone.[103] After reading humblebraggers' online posts, participants in the study gave them low marks in terms of likability, sincerity, and competence.

Despite the blatant faux modesty of humblebrags, sometimes it's necessary to self-promote, especially in the world of work. People were humblebragging in job interviews long before the term was even invented: "My greatest weakness? It's probably that I'm a perfectionist." Self-serving comments like this sound phony. But are they worse than honest confessions such as, "I'm not very organized" or "I'm not a team player"? How can you share your accomplishments without being self-defeating on one hand or boastful on the other?

In the second part of their humblebragging study, the Harvard team put those questions to the test. The researchers asked college students to respond in writing to the classic job interview question: "What is your

biggest weakness?" Trained evaluators then assessed how likely they would be to hire the students based on their responses. In the end, humblebraggers ("I find myself doing a lot of favors for others") were less likely to be hired than those who revealed honest weaknesses ("I sometimes tend to procrastinate").

Honesty carries the day, say the researchers: "While people do not love braggers or complainers, they at least see them as more sincere than humblebraggers."[104]

for example, dress up for a visit to traffic court in the hope that your front (responsible citizen) will convince the judge to treat you sympathetically. You might act sociable to your neighbors so they will agree to your request that they keep their dog off your lawn. We also try to create a desired impression to achieve one or more of the social needs described in Chapter 1: affection, inclusion, control, and so on. For instance, you might act more friendly and lively than you feel on meeting a new person so that you will appear likable. You could sigh and roll your eyes when arguing politics with a classmate to gain an advantage in an argument. You might smile and preen to show the attractive stranger at a party that you would like to get better acquainted. In situations like these you aren't being deceptive as much as putting your best foot forward.

All these examples show that it is difficult—even impossible—not to create impressions. After all, you have to send some sort of message. If you don't act friendly when meeting a stranger, you have to act aloof, indifferent, hostile, or in some other manner. If you don't act businesslike, you have to behave in an

cultural idiom

putting your best foot forward:
making the best appearance possible

alternative way: casual, goofy, or whatever. Often the question isn't whether or not to present a face to others; the question is only which face to present.

Identity Management in Mediated Communication

At first glance, computer-mediated communication (CMC) seems to have limited potential for identity management. As you read in Chapter 1, text-based messages lack the "richness" of other channels. They are less vivid and detailed than, say, face-to-face communication and video chats. Text messages don't convey the postures, gestures, or facial expressions that impart important cues about what people really mean. They even lack the vocal information available in telephone messages. These limitations might seem to make it harder to create and manage an identity at a distance.

Recently, though, communication scholars have begun to recognize that what is missing in CMC can actually be an advantage for communicators who want to manage the impressions they make. For example, research shows that teens tend to strategically post photos that make them appear attractive and socially engaged with others and omit those that are less flattering.[105] Email authors can edit their messages to include just the right amount of clarity, ambiguity, humor, or logic to convey the desired impression.[106] Unlike face-to-face communication, electronic correspondence allows a sender to say difficult things without forcing the receiver to respond immediately, and it permits the receiver to ignore a message rather than give an unpleasant response. With CMC you don't have to worry about stammering or blushing, apparel or appearance, or any other unseen factor that might detract from the impression you want to create. Options like these show that CMC can serve as a tool for impression management.

Taking impression management even further, strangers online can change their age, history, personality, appearance, and other matters that would be impossible to hide in person.[107] A quarter of teens have pretended to be a different person online, and a third confess to having given false information about themselves while emailing and instant messaging. A survey of one online dating site's participants found that 86% felt others misrepresented their physical appearance in their posted descriptions.[108] For more about the pros and cons of virtual identity management, see the "Understanding Communication Technology" box on the next page.

"On the Internet, nobody knows you're a dog."

Source: Peter Steiner The New Yorker Collection/
The Cartoon Bank

Identity Management and Honesty

After reading this far, you might think that identity management sounds like an academic label for manipulation or phoniness. If the perceived self is the "real" you, it might seem that any behavior that contradicts it would be dishonest.

There certainly are situations in which identity management is dishonest. A manipulative date who pretends to be affectionate in order to gain sexual favors is clearly unethical and deceitful. So are job applicants who lie about their academic records to get hired and salespeople who pretend to be dedicated to customer service when their real goal is to make a quick buck. But managing identities doesn't necessarily make you a liar. In fact, it is almost impossible to imagine how we could communicate effectively without making decisions about which front to present in one situation or another. It would be ludicrous for you to act the same way with strangers as you do with close friends, and nobody would show the same face to a 2-year-old as to an adult.

cultural idiom

make a quick buck: earn money
easily

UNDERSTANDING COMMUNICATION TECHNOLOGY

Identity Management in Social Media

Go ahead and roll your eyes. The video-game version of Kim Kardashian is clearly a "kustomized" (her word) version of the real woman. Virtual Kim lives in a world filled with parties and photo shoots where she looks her best 24 hours a day. No real human could pull this off.

Now turn the lens on yourself. Is your online image an accurate depiction of who you are in the flesh? The discrepancy may not approach Kardashian proportions, but for most of us, the answer is no.

Social media is a powerful tool for identity management. People typically experience a boost in self-esteem when they believe people they care about are viewing positive images of them on Facebook.[109]

There are downsides to online identities, however, as celebrities have long known. For one, it's easy to wonder if people like the real you or only the stylized, online version. "My ideal self . . . lives in a perpetually clean house and . . . always takes the time to put on makeup before she leaves the house," observes freelance writer Kelsey Sunstrum.[110] Online, she largely lives up to that image. But in person, she doesn't smile all the time or always look her best. In fact, she experiences frequent bouts of depression that her online followers don't see.

Second, your self-esteem can take a hit if you compare yourself to other people's online identities. Young adults (both male and female) who frequently view and comment on peers' Facebook images are more likely than others to feel dissatisfied with their own appearance.[111]

Another downside is that online identities are subject to sabotage. Anyone with a smart phone can snap and post unflattering photos of you for the world to see. At the same time, it's easier than ever for people to publicly comment on your appearance and behavior.

How can you stay emotionally healthy while managing your online identity? Experts offer the following advice:

☐ **Don't overdo it.** Too much screen time can make the virtual world feel more real than actual face time. Scale back to keep a healthy perspective.[112]

☐ **Avoid adding to the negativity.** As we have seen all too often (*ahem, Nicki Minaj and Katy Perry*), one negative post often leads to others. Resist the impulse to post something negative, even in response to a critical comment aimed at you. People who know the real you know won't be fooled by unfair criticism.

☐ **Strive for authenticity.** As tempting as it is to post glamour shots all the time, the ultimate selfie is one that shows you as you really are.

As one cultural observer reminds us, even Kim Kardashian "puts her Spanx on one leg at a time. Being human is, by definition, ordinary."[113]

Each of us has a repertoire of faces—a cast of characters—and part of being a competent communicator is choosing the best role for the situation. Consider a few examples:

- You have been communicating online for several weeks with someone you just met, and the relationship is starting to turn romantic. You have a physical trait that you haven't mentioned yet.
- You offer to teach a friend a new skill: playing the guitar, operating a computer program, or sharpening a tennis backhand. Your friend is making slow progress with the skill, and you find yourself growing impatient.
- At a party with a companion, you meet someone you find very attractive, and you are pretty sure the feeling is mutual. You feel an obligation to

spend most of your time with the person with whom you came, but the opportunity here is very appealing.

- At work you face a belligerent customer. You don't believe that anyone has the right to treat you this way.

- A friend or family member makes a joke about your appearance that hurts your feelings. You aren't sure whether to make an issue of the remark or pretend that it doesn't bother you.

In each of these situations—and in countless others every day—you have a choice about how to act. It is an oversimplification to say that there is only one honest way to behave in each circumstance and that every other response would be insincere and dishonest. Instead, impression management involves deciding which face—which part of yourself—to reveal. For example, when teaching a new skill, you can choose to display the patient instead of the impatient side of yourself. In the same way, at work you have the option of acting hostile or nondefensive in difficult situations. With strangers, friends, or family you can choose whether or not to disclose your feelings. Which face to show to others is an important decision, but in any case you are sharing a real part of yourself. You may not be revealing everything—but, as you will learn in Chapter 7, complete self-disclosure is rarely appropriate.

It's also worth noting that not all misrepresentations are intentional. Researchers have used the term *foggy mirror* to describe the gap between participants' self-perceptions and a more objective assessment. An online dater who describes himself or herself as being "average" in weight might be engaging in wishful thinking rather than telling an outright lie.[114]

We began the chapter with Caitlyn Jenner, who says that although the gender transition has been difficult, it feels good that her social identity now matches her self-perception. As she puts it, there's "nothing better in life [than] to wake up in the morning, look at yourself in the mirror and feel comfortable with yourself and who you are."[115]

ETHICAL CHALLENGE
Honesty and Multiple Identities

Explore the ethics of multiple identities by identifying two situations from your life:

1. A time when you presented a public identity that didn't match your private self in a manner that wasn't unethical.

2. A situation (real or hypothetical) in which you have presented or could present a dishonest identity.

Based on the situations you and your classmates present, develop a code of ethics that identifies the boundary between ethical and unethical identity management.

MAKING THE GRADE

For more resources to help you understand and apply the information in this chapter, visit the *Understanding Human Communication* website at www.oup.com/us/adleruhc.

OBJECTIVE 2.1   Identify the communicative influences that shape the self-concept.

- The self-concept is a set of relatively stable perceptions that each of us holds about ourselves.

- Although we are born with some innate personality characteristics, communication and cultural/social factors also help shape our self-concept.

- Once established, the self-concept can lead us to create self-fulfilling prophecies that determine how we behave in response to our own and other people's perceptions of us.

> Name at least four factors that influence a person's self-concept. Explain the role of communication relevant to each of these factors.

> Describe a relationship or event that had a powerful impact on your self-concept. What role did communication play?

> Recognizing the power of self-fulfilling prophecies, present a scenario in which you might use communication to help someone believe in him- or herself and consequently achieve an important goal.

OBJECTIVE **2.2**   **Examine the perceptual tendencies and situational factors that influence our perceptions of other people.**

- Perception is a multistage process that includes selection, organization, and interpretation.

- Our perceptions of other people are influenced by a range of factors including our emotions, how we feel physiologically, cultural expectations, and social roles.

- We often incorporate our perceptions into personal narratives that not only tell a story but suggest a particular interpretation that others may accept or challenge.

- Perceptual tendencies and errors can affect the way we view and communicate with others.

- Empathy is a valuable tool for increasing understanding of others and hence communicating more effectively with them, both in person and online.

- Perception checking is one tool for increasing the accuracy of perceptions and for increasing empathy.

  > List and explain at least five perceptual tendencies that shape how individuals make judgments about other people.

  > Describe an experience in which you met someone new. How did your impressions of that person evolve in the context of the perceptual phases of selection, organization, and interpretation?

  > How could greater empathy have changed an interpersonal conflict you have experienced?

OBJECTIVE **2.3**   **Describe how the process of identity management operates, in both face-to-face and online communication.**

- Identity management consists of strategic communication designed to influence others' perceptions of an individual.

- Identity management is usually collaborative.

- Identity management occurs for two reasons: (1) to follow social rules and conventions and (2) to achieve a variety of content and relational goals.

- Communicators engage in creating impressions by managing their manner, appearance, online posts, and the settings in which they interact with others.

- Although identity management might seem manipulative, it can be an authentic form of communication. Because each person has a variety of faces that he or she can present, choosing which one to present is not necessarily being dishonest.

  > Explain how people use facework and frame switching to manage their private and public identities.

> Describe at least three different identities you present. Compare and contrast the communication strategies you use to support each.

> Have you ever been judged on the basis of your sex or gender? In what ways have you judged other people? What perceptual tendencies contribute to such judgments? What advice do you have for avoiding hurtful and unfair judgments?

KEY TERMS

androgynous p. 46

attribution p. 49

empathy p. 52

face p. 55

facework p. 55

frame switching p. 57

gender p. 46

identity management p. 55

interpretation p. 44

narrative p. 48

organization p. 43

perceived self p. 55

perception checking p. 54

personality p. 38

presenting self p. 55

reflected appraisal p. 38

script p. 57

selection p. 43

self-concept p. 37

self-esteem p. 37

self-fulfilling prophecy p. 41

self-serving bias p. 50

sex p. 46

significant other p. 38

social comparison p. 39

stereotyping p. 49

sympathy p. 53

ACTIVITIES

1. **Exploring Narratives** Think of a situation in which you and others have achieved relational harmony by sharing a narrative. Then think of a situation in which you and another person used different narratives to describe a set

of circumstances. What were the consequences of having different narratives?

2. **Identifying Your Identities** Keep a 1-day log listing the identities you create in different situations: at school and at work, and with strangers, various family members, and different friends. For each identity:

a. Describe the persona you are trying to project (e.g., "responsible son or daughter," "laid-back friend," "attentive student").

b. Explain how you communicate to promote this identity. What kinds of things do you say (or not say)? How do you act?

Communication and Culture

CHAPTER OUTLINE

LEARNING OBJECTIVES

3.1

Analyze the influence of cultures and cocultures on intergroup communication, explaining the concepts of salience and generalizations.

3.2

Distinguish among the following cultural norms and values, and explain how they influence communication: individualism and collectivism, high and low context, uncertainty avoidance, power distance, talk and silence, competition and cooperation.

3.3

Evaluate the influence of factors such as race and ethnicity, sexual identity, religion, age, and socioeconomic status on communication.

3.4

Use the criteria in this chapter to assess your intercultural and cocultural communication competence, and describe how you could communicate more competently.

Robin Luo's story shows that even a simple statement such as "How are you?" can be interpreted many different ways. Take a moment to think about your own experiences involving culture and communication.

?

Have you ever interacted with people from a culture significantly different from your own? If so, did they communicate in ways that surprised or puzzled you? How?

?

Have you ever felt like a cultural stranger or minority, even in your home community? If so, how did it affect your communication with others?

?

What could you do to communicate more successfully with people from cultural backgrounds different from your own?

"HOW ARE YOU?" The question seems simple, but it can be tricky to answer, as Bin "Robin" Luo, an exchange student from China, has learned. Where he grew up, people seldom talk to strangers, says Luo. But when people do speak, they tend to focus intently on each other. By contrast, many Americans seem especially outgoing, but they often ask questions without expecting much of a reply.

"I still consider 'How are you?' as more than just a 'Hello,'" Luo says, after five months in the United States. "I want to stop and explain my feelings. But sometimes Americans just ask and then walk past you without waiting for your answer."

Luo, an engineering student from Guangdong Province in southern China, says he was eager to study in the United States because the country and its educational system are highly respected. He thought it would be fascinating to learn about cultural ways so different from his own. Luo's interest in living abroad isn't unique: Around the world, one in seven foreign exchange students is from China, and about half of Chinese exchange students choose the United States.[1] Indeed, the United States hosts more students from China than from any other country.

Adapting to cultural differences isn't always easy. Americans may be surprised to hear that many foreigners find the United States a hard place to make friends. Luo says this is partly because, without knowing the culture well, it's hard to tell when Americans are genuinely interested in friendship and when they are just being polite.

"When I have talked to someone more than 15 minutes, I would think we can go further as friends," Luo says. "But Americans may consider it a casual conversation, and they may say, 'Let's get together' or 'I'll call you' but not follow through." He says he generally admires the optimistic attitude and forthrightness he has encountered, but he is puzzled by some of the communication patterns of his fellow students.

Robin Luo's experiences as an exchange student shed light on how culture influences communication. But you don't have to be a world traveler to encounter people from different places and cultures in today's global village. Without leaving home, we are more connected with people from different backgrounds than at any time in history. That's partly because travel is easier than ever before, and the Internet makes communicating around the world as easy as communicating around town. It's also because society is more diverse than ever—in terms of culture, age, race and ethnicity, physical ability, personal identity, family background, and more.

As you'll see in this chapter, this diversity presents both obvious benefits and unique communication challenges. We will explore both. The main themes in this chapter include:

- When culture does—and when it doesn't—affect communication

- The mostly hidden values and norms that can shape interaction between people from different backgrounds

- How diversity within our own culture shapes communication

- Some ways that culture affects sending and receiving messages

- How to become more competent in communicating with people from different backgrounds

Despite living in Brooklyn, one of the most culturally diverse spots in the United States, the friends on HBO's *Girls* spend much of their time with people who look and speak like them. Creator and star Lena Dunham has been criticized for the show's culturally homogenous characters.

How closely does your circle of acquaintances match the diversity of today's society?

Understanding Cultures and Cocultures

Defining **culture** isn't an easy task. One early survey of scholarly literature revealed 500 definitions, phrasings, and uses of the concept.[2] For our purposes, here is a clear and comprehensive definition of *culture*: "the language, values, beliefs, traditions, and customs people share and learn."[3]

When you think of cultures, you may think of different nationalities. But *within* a society, differences also exist. Social scientists use the term **coculture** to describe the perception of membership in a group that is part of an encompassing culture. The sons and daughters of immigrants, for example, might be immersed in mainstream American culture while still identifying with the customs of their parents' homeland. Members of cocultures often develop unique patterns of communication. The term that describes interaction among members of different cocultures is **intergroup communication**.

Cocultures in today's society include, for example,

- Age (e.g., teen, senior citizen)

- Race/ethnicity (e.g., African American, Latino, white)

culture The language, values, beliefs, traditions, and customs people share and learn.

coculture The perception of membership in a group that is part of an encompassing culture.

intergroup communication The interaction between members of cocultures.

- Sexual orientation (e.g., LGBTQ)
- Physical disability (e.g., wheelchair users, Deaf persons)
- Language (e.g., native and nonnative speakers)
- Religion (e.g., Mormon, Muslim)
- Activity (e.g., biker, gamer)

Membership in cocultures can be a source of identity, enrichment, and pride. But when the group is stigmatized by others, being identified as a member isn't always so fulfilling. Some research suggests that recognizable members of underrepresented groups are disadvantaged in employment interviews and social settings, in which the rules are established by the dominant culture.[4] Studies of Jamaican[5] and Latino[6] children indicate that skin color often influences self-identification and self-esteem. Other research focuses on reactions to people with disabilities. Although almost all of U.S. high school students polled in one survey said they would greet a classmate with a physical disability and would gladly lend the person a pen or pencil, more than a third were reluctant to spend time with the same classmate outside of school, mostly because they felt awkward and they assumed they wouldn't have much in common.[7]

The odds of overturning stereotypes may seem insurmountable. But the news isn't all bad: People who interact daily with people from different cultural backgrounds are less likely to be prejudiced than those who do not.[8] (We'll talk more about this later in the chapter.) Positive media images also have an influence. For example, the popularity of former U.S. First Lady Michelle Obama has helped enhance the attitude that dark skin is beautiful, not only in the United States but around the world.[9]

Membership in a coculture can shape the nature of communication. Sometimes the influence is obvious. For example, it's no surprise that members of the millennial generation rely more on social media to communicate than their grandparents, or even their parents.[10] Children who are native language speakers probably have a very different experience in school than nonnative speakers, and gay men and lesbians can find it awkward to discuss their romantic relationships with some heterosexual people.

In other cases, though, cocultural communication practices aren't so obvious. For example, ethnic background can influence what people consider to be the most important qualities in a friendship. One study found that Latinos were highly likely to value relational support and bonding with friends. By contrast, Asian Americans (as a group, of course) placed a greater value on helping one another achieve personal goals. For African Americans in the study, the most important quality in a friend was respect for and acceptance of the individual. European Americans reported that they valued friends who met their task-related needs, offered advice, shared information, and had common interests.[11]

Intercultural and Intergroup Communication: A Matter of Salience

Intercultural and intergroup communication—at least as we'll use the terms here—don't always occur when people from different backgrounds interact. Those backgrounds must have a significant impact on the exchange before we

? ASK YOURSELF

Which cocultures are most fundamental to your identity?

Jane Villanueva (aka Jane the Virgin) struggles to find the right path in her drama-fraught life. At various times in this telenovela send-up, different parts of Jane's identity (smart, middle class, religious, Latina) are more and less salient.

Does the prominence of cultural and cocultural factors shift in your life, or do some factors almost always seem most salient?

can say that culture has made a difference. Social scientists use the term **salience** to describe how much weight we attach to cultural characteristics. Consider a few examples in which culture has little or no salience:

- A group of preschool children is playing together in a park. These 3-year-olds don't recognize that their parents may come from different countries, or even that they don't speak the same language. At this point we wouldn't say that intercultural or intergroup communication is taking place. Only when cultural factors (e.g., diet, sharing, or parental discipline) become salient do the children begin to think of one another as different.

- Members of a school athletic team—some Asian American, some African American, some Latino, and some white—are intent on winning the league championship. During a game, cultural distinctions aren't salient. There's plenty of communication, but it isn't fundamentally intercultural or intergroup. Away from their games, the players are friendly when they encounter one another, but they rarely socialize. If they did, they might notice some fundamental differences in the ways members of each group communicate.

- A husband and wife were raised in different religious traditions. Most of the time their religious heritage makes little difference and the partners view themselves as a unified couple. Every so often, however—perhaps during the holidays or when meeting members of each other's family—the different backgrounds are more salient. At those times we can imagine the partners feeling quite different from each other—thinking of themselves as members of separate cultures.

These examples show that in order to view ourselves as a member of a culture or coculture, there has to be some distinction between "us" and "them." Social scientists use the label **in-groups** to describe groups with which we identify and are emotionally connected, and **out-groups** to label those we view as different and with whom we have no sense of affiliation.[12]

Cultural Differences Are Generalizations

It's important not to overstate the influence of culture on communication. There are sometimes greater differences *within* cultures than *between* them. Consider the matter of formality as an example: By most measures, U.S. culture, broadly, is far more casual than many others. But Figure 3-1 shows that there may be more common ground between a formal American and a casual member of a formal culture than there is between two Americans with vastly differing levels of formality.

Furthermore, within every culture, members display a wide range of communication styles. For instance, although most Asian cultures tend to be collectivistic, many members of those cultures identify themselves as individualists. It's important to remember that cultural differences are generalizations—broad patterns that do not apply to every member of a group.

Cultural Values and Norms Shape Communication

Some cultural influences on communication are obvious. You don't have to be a researcher to appreciate that different languages and customs can make communication between groups interesting and challenging.

Along with the obvious differences, far less visible values and norms shape how members of cultures think and act. This section will look at several of the subtle

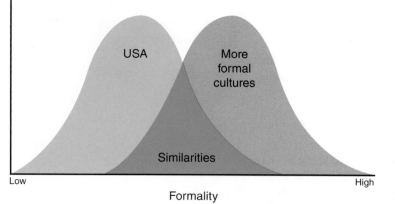

FIGURE 3-1 Formality Differences and Similarities Within and Between Cultures

yet vitally important factors that shape the way members of a culture communicate. Unless communicators are aware of these differences, they may see people from other cultures as unusual—or even offensive—without realizing that their apparently odd behavior comes from following a different set of beliefs and unwritten rules about the "proper" way to communicate.

Individualism and Collectivism

Some cultures value the individual more than the group, whereas others place greater emphasis on the group. Members of **individualistic cultures** view their primary responsibility as helping themselves, whereas communicators in **collectivistic cultures** feel loyalties and obligations to one's extended family, community, or even the organization one works for.[13]

The way names are treated offers an insight into the difference between individualism and collectivism. When asked to identify themselves, individualistic Americans, Canadians, Australians, and Europeans usually respond by giving their first name, surname, street, town, and country. But collectivistic South Asians do it the other way around. If you ask Hindus for their identity, they will give you their caste and village and then their family name, and finally their given name.[14] The Chinese student profiled at the beginning of this chapter writes his name Luo Bin in China, putting his family name first. But in the United States, where individualism is prized, he follows the convention of putting his surname last: Bin (Robin) Luo.

The differences between individualist and collectivist orientations extend beyond naming and are reflected in the very structure of some languages. As mentioned in Chapter 2, the Chinese pronoun for *I* looks very similar to the word for *selfish*.[15] The Japanese language has no equivalent to the English pronoun *I*. Instead, different words are used to refer to one's self depending on the social situation, age, gender, and other social characteristics.[16] Individualistic and collectivistic cultures have very different approaches to communication. For example, individualistic cultures are relatively tolerant of conflicts, using a direct, solution-oriented approach. By contrast, members of collectivistic cultures are less direct, often placing a greater emphasis on harmony.[17] Individualistic cultures are also characterized by self-reliance and competition, whereas collectivistic cultures are more attentive to and concerned with the opinions of significant others.[18] Table 3-1 illustrates some of the main differences between individualistic and collectivistic cultures.

Collectivistic societies predominantly produce team players, whereas individualistic ones are more likely to produce and reward superstars. Some research suggests that collectivist groups have a higher sense of teamwork and are more productive than groups made up of individualistic members.[19]

After growing up in China, Robin Luo says that Americans seem remarkably confident in the way they speak and act. As members of an individualistic culture, they are comfortable talking about their own accomplishments or even saying they are really good or the best at something. Luo estimates that Americans are willing to make decisions, even when they know only "about 70 percent" of the details about a situation. "They are seldom afraid of making mistakes," he observes. By contrast, "in China, when we face a problem, we may be quiet and consider a lot about its consequences before taking actions, even if not much is likely to happen."

individualistic culture A culture in which members focus on the value and welfare of individual members, as opposed to a concern for the group as a whole.

collectivistic culture A culture in which members focus on the welfare of the group as a whole, rather than a concern by individuals for their own success.

TABLE 3-1

The Self in Individualistic Versus Collectivistic Cultures

INDIVIDUALISTIC CULTURES	COLLECTIVISTIC CULTURES
Self is separate, unique, independent; "I" orientation.	Identity is interwoven with extended family and other groups; "we" orientation.
Individual should take care of himself or herself and immediate family.	Person should take care of extended family before self.
There are many flexible group memberships; friends are based on shared interests and activities.	Emphasis is on belonging to a few permanent in-groups, which have a strong influence over the person.
There are rewards for individual achievement and initiative; individual decision is encouraged; individual credit and blame are assigned.	There are rewards for contributing to group goals and well-being; cooperation with in-group members is encouraged; group decisions are valued; credit and blame are shared.
High value is placed on autonomy, change, youth, individual security, and equality.	High value is placed on duty, order, tradition, age, group security, status, and hierarchy.

Source: Adapted by Sandra Sudweeks from H. C. Triandis. (1990). Cross-cultural studies of individualism and collectivism. In J. Berman (Ed.), *Nebraska symposium on motivation* (pp. 41–133). Lincoln: University of Nebraska Press; and E. T. Hall. (1976). *Beyond culture.* New York: Doubleday.

Members of collectivist societies are typically less publicly egotistical because touting personal accomplishments would put the individual ahead of the group. "Chinese tend to describe themselves a little bit lower than what they actually are," Luo explains. "Chinese will say they are so-so even if they are very good." As you might imagine, either of these approaches can be effective, but they may lead to misunderstandings when people have different assumptions about them. Americans may mistakenly assume that people who are humble lack confidence or achievements, whereas people from collectivistic cultures may judge Americans to be overbearing and self-centered.

Cultural differences can also affect the level of comfort or anxiety that people feel when communicating. In societies in which the need to conform is great, there is a higher degree of communication apprehension. For example, South Koreans exhibit more conflict avoidance and more apprehension about speaking in front of people than members of individualistic cultures such as the United States.[20] It's important to realize that different levels of communication apprehension don't mean that shyness is a "problem" in some cultures. In fact, just the just the opposite is true: In these cultures, reticence is valued. When the goal is to avoid being the nail that sticks out, it's logical to feel nervous when you make yourself appear different by calling attention to yourself. A self-concept that includes "assertive" might make a Westerner feel proud, but in much of Asia people may speak forthrightly to members of their in-group but consider it inappropriate or shameful to do so with out-group members.

Culture plays an important role in our ability to understand the perspectives of others. People raised in individualist cultures, which value independence, are often less adept at seeing others' point of view than those from collectivist cultures. In one study, Chinese and American players were paired together in a game that required them to take on the perspective of their partners.[21] By all measures, the collectivist Chinese had greater success in perspective taking than did their American counterparts. Of course, it's important to recall that cultural differences are generalizations. There are elements of collectivism in Western cultures and of individualism in Asian ones.[22]

Singer, songwriter, actress, and model Demi Lovato has used social media to share her experiences with mental illness and addiction.

In what contexts do you consider it appropriate to share personal information? What information would you share only with close friends and loved ones? How do these choices vary by culture?

low-context culture A culture that uses language primarily to express thoughts, feelings, and ideas as directly as possible.

high-context culture A culture that relies heavily on subtle, often nonverbal cues to maintain social harmony.

cultural idiom

beat around the bush: make indirect remarks instead of stating a point clearly and directly.

The difference between individualism and collectivism shows up in everyday interactions. Communication researcher Stella Ting-Toomey has developed a theory that explains cultural differences in important norms, such as honesty and directness.[23] She suggests that in individualistic Western cultures where there is a strong "I" orientation, the low-context norm of speaking directly is honored, whereas in collectivistic cultures where the main desire is to build connections between the self and others, high-context, indirect approaches that maintain harmony are considered more desirable. "I gotta be me" could be the motto of a Westerner, but "If I hurt you, I hurt myself" is closer to the Asian way of thinking.

High and Low Cultural Context

Social scientists have identified two distinct ways that members of various cultures deliver messages.[24] In **low-context cultures,** members use language primarily to express thoughts, feelings, and ideas as directly as possible. By contrast, members of **high-context cultures** rely heavily on subtle, often nonverbal cues to maintain social harmony. Rather than upsetting others by speaking directly, communicators in these societies learn to discover meaning from the context in which a message is delivered: the nonverbal behaviors of the speaker, the history of the relationship, and the general social rules that govern interaction between people.

Americans are likely to state their concerns or complaints up front, whereas people raised in high-context cultures usually hint at them.[25] Luo, the Chinese exchange student profiled on page 68, gives an example:

> Suppose a guy feels bad about his roommate eating his snacks. If he is Chinese, he may try to hide his food secretly or choose a certain time to say, "My snacks run out so fast, I think I need to buy more next time." Before this, he also may think about whether his roommate would hate him if he says something wrong. But Americans may point out directly that someone has been eating their food.

It's easy to see the potential for misunderstanding in situations like this. The roommate from China may feel it is obvious, based on the situation and his indirect statement, that he is upset about his roommate eating his food. But the American—who may expect his friend to say outright if he is upset—may miss the point entirely. Table 3-2 summarizes some key differences in how people from low- and high-context cultures use language.

Mainstream culture in the United States, Canada, northern Europe, and Israel falls toward the low-context end of the scale. Longtime residents generally value straight talk and grow impatient with people who beat around the bush. By contrast, most Asian and Middle Eastern cultures fit the high-context pattern. For them, maintaining harmony is important, so communicators avoid speaking directly if doing so would threaten another person's "face," or dignity.

High- and low-context differences also operate within domestic cocultures. For example, in a classic study, researcher Laura Leets presented European Americans and people of color—Asian American, African American, and Latino—with examples of racist messages.[26] Some of these messages were direct and blatantly offensive, whereas others were indirect and less overtly racist. Leets found that European American participants judged the directly racist messages as more hurtful, whereas Asian American respondents rated the indirectly racist comments as more damaging. She concluded that the traditional Asian tendency to favor high-context messages explains why Asian Americans were more offended by indirectly racist speech.

It's easy to see how the clash between directness and indirectness can present challenges. To members of high-context cultures, communicators with a low-context style can appear overly talkative, lacking in subtlety, and redundant. On

TABLE 3-2

High- and Low-Context Communication Styles

LOW CONTEXT	HIGH CONTEXT
The majority of information is carried in explicit verbal messages, with less focus on the situational context.	Important information is carried in contextual cues such as time, place, relationship, and situation. There is less reliance on explicit verbal messages.
Self-expression is valued. Communicators state opinions and desires directly and strive to persuade others to accept their viewpoints.	Relational harmony is valued and maintained by indirect expression of options. Communicators abstain from saying "no" directly.
Clear, eloquent speech is considered praiseworthy. Verbal fluency admired.	Communicators talk "around" the point, allowing others to fill in the missing pieces. Ambiguity and the use of silence are admired.

the other hand, to people from low-context backgrounds, high-context communicators often seem evasive, or even dishonest.

Uncertainty Avoidance

Uncertainty may be universal, but cultures have different ways of coping with an unpredictable future. The term **uncertainty avoidance** is used to reflect the degree to which members of a culture feel threatened by ambiguous situations and how much they try to avoid them.[27] As a group, residents of some countries (including Singapore, Great Britain, Denmark, Sweden, Hong Kong, and the United States) are relatively unthreatened by change, whereas others (such as natives of Belgium, Greece, Japan, and Portugal) find new or ambiguous situations discomfiting.

A culture's degree of uncertainty avoidance is reflected in the way its members communicate (see Table 3-3). In countries that avoid uncertainty, out-of-the-ordinary people and ideas are considered dangerous, and intolerance is high. People in these cultures are especially concerned with security, so they have a strong need for clearly defined rules and regulations. It's easy to imagine how most relationships in cultures with a high degree of uncertainty avoidance are likely to fit a predictable pattern. By contrast, people in cultures that are less threatened by the new and unexpected are more likely to tolerate—or even welcome—people who don't fit the norm.

> **uncertainty avoidance** The cultural tendency to seek stability and honor tradition instead of welcoming risk, uncertainty, and change.

TABLE 3-3

Differences Between Low- and High-Uncertainty-Avoidance Societies[28]

LOW UNCERTAINTY AVOIDANCE	HIGH UNCERTAINTY AVOIDANCE
Uncertainty inherent in life is accepted.	Uncertainty inherent in life is treated as a continuous threat that must be fought.
Uncertainty is typically not a cause of great anxiety or stress.	Uncertainty often causes high stress, anxiety, emotionality.
Deviant personas and ideas are tolerated.	Personas and ideas that don't fit the norm are treated with intolerance.
Ambiguity and chaos do not cause stress.	Clarity and structure are valued and expected.
Teachers may say "I don't know."	Teachers are supposed to have all the answers.
Rules—written or unwritten—are disliked.	Rules are treated as emotionally necessary, even if they are not obeyed.

Power Distance

power distance The degree to which members of a group are willing to accept a difference in power and status.

Power distance refers to the extent of the gap between social groups who possess resources and influence and those who don't. Cultures with low power distance believe in minimizing the difference between various social classes. Rich and poor, educated and uneducated groups may still exist, but there's a pervasive belief in low-power-difference cultures that one person is as good as another regardless of his or her station in life (see Table 3-4).

Austria, Denmark, Israel, and New Zealand are the most egalitarian countries. At the other end of the spectrum are countries with a high degree of power distance: the Philippines, Mexico, Venezuela, India, and Singapore.[29] Anyone familiar with communication in the United States and Canada knows that those cultures value equality, even if it's not perfectly enacted. For example, they might call their bosses by their first names and challenge the opinions of people in higher status positions.

By contrast, in cultures with high power distance it may seem rude to treat everyone the same way. In the Japanese workplace, for example, new acquaintances exchange business cards immediately, which helps establish everyone's relative status. The oldest or highest ranking person receives the deepest bows from others, the best seat, the most deferential treatment, and so on. This treatment is not regarded as elitist or disrespectful. Indeed, treating a high-status person the same as everyone else would seem rude.

As an exchange student from China, Robin Luo has observed that Americans often interact informally with their bosses and professors, even to the extent of calling them by their first names. That would not happen in China, he says, where it would be considered disrespectful for an employee or student to omit the title or to use a first name. The same rule applies to addressing students of higher rank unless they are very close friends.

Beliefs About Talk and Silence

Beliefs about the very value of talk differ from one culture to another.[32] People in Western cultures tend to view talk as desirable and use it for social purposes as well as to perform tasks. Silence can feel embarrassing and awkward in these cultures. It's likely to be interpreted as lack of interest, unwillingness to communicate, hostility, anxiety, shyness, or a sign of interpersonal incompatibility.

On the other hand, Asian cultures tend to perceive talk quite differently. For thousands of years, people in Asian cultures have favored silence over "excessive" verbal expression of thoughts and feelings. Consider these Taoist sayings: "In much talk there is great weariness" and "One who speaks does not know; one who

TABLE 3-4

Differences Between Low- and High-Power-Distance Societies[30]

LOW POWER DISTANCE	HIGH POWER DISTANCE
Power is usually associated with formal roles; powerful people may be good or evil.	Power is a basic fact of society to be accepted without question.
Parents treat children as equals.	Parents teach children obedience.
Older people are neither respected nor feared.	Older people are both respected and feared.
Education is student centered.	Education is teacher centered.
Hierarchical roles are established for convenience and do not reflect the assumption of inherent equality.	The hierarchy reflects the assumption of inherent inequalities.
Subordinates expect to be consulted.	Subordinates expect to be told what to do.

Power Distance and Culture in the Workplace

On-the-job communication is different in low- and high-power-distance societies.[31] In countries with higher degrees of power distance, employees have much less input into the way they perform their work. In fact, workers from these cultures are likely to feel uncomfortable when given freedom to make their own decisions or when a more egalitarian boss asks for their opinion. They prefer to view their bosses as benevolent decision makers.

The reverse is true when managers from a culture with an egalitarian tradition do business in a country whose workers are used to high power distance. The managers may be surprised that employees do not expect much say in decisions

and do not feel unappreciated when they aren't consulted. They may regard dutiful, submissive, respectful employees as lacking initiative and creativity—traits that helped people gain promotions back home.

Given these differences, it's easy to understand why multinational companies need to consider fundamental differences in communication values and behavior when they set up shop in a new country.

Do you work with people from cultures with beliefs about power distance that differ from yours? How do those differences show up when you work together? Can you think of ways to manage these differences that will improve both personal relationships and workplace effectiveness?

---

knows does not speak." Unlike Westerners, who tend to be uncomfortable with silence, Japanese and Chinese people more often believe that remaining quiet is the proper state when there is nothing to be said. To Asians, a talkative person is often considered a show-off or a fake.

Silence is also valued more in some European cultures. For example, Swedes are more reluctant than Americans to engage in small talk with strangers. With friends, however, their behavior is much more similar to that of Americans.[33]

Members of some Native American communities also honor silence. For example, traditional members of Western Apache tribes maintain silence when others lose their temper. As one member explained, "When someone gets mad at you and starts yelling, then just don't do anything to make him get worse."[34] Apache also consider that silence has a comforting value. The idea is that words are often unnecessary in periods of grief, and it is comforting to have loved ones present without the pressure to maintain conversations with them.

It's easy to see how these different views about speech and silence can lead to communication problems when people from different cultures meet. Both the "talkative" Westerner and the "silent" Asian and Native American are behaving in ways they believe are proper, yet each may view the other with disapproval and mistrust. Only when they recognize the different standards of behavior can they adapt to one another, or at least understand and respect their differences.

Competitive and Cooperative Cultures

Cultures are a bit like people in that they fall somewhere on a spectrum between competitive and cooperative.[35] The cultures of some countries, such as Taiwan, place relatively equal value on competition and cooperation.[36] In other countries, such as Japan, Italy, Nigeria, and Great Britain, qualities such as competitiveness, independence, and assertiveness are highly valued. Gender roles may be clearly differentiated in these more competitive cultures. Women are often expected to take care of home and family life, whereas men have been traditionally expected to shoulder most of the financial responsibilities.

cultural idiom

set up shop: establish a business

The election of Sadiq Khan as the mayor of London was a triumph of multiculturalism. Khan told a reporter that he was more than a Muslim: "I'm a Londoner, I'm a European, I'm British, of Asian origin, of Pakistani heritage.[39]

To what cocultures do you belong? How do these memberships affect your communication?

Gender roles are less differentiated in more cooperative cultures, which emphasize equality, relationships, cooperation, and consensus building.[37] In Iceland, the Netherlands, and Norway, for example, both men and women tend to consider harmony and cooperation to be more important than competition.

Cultural orientations influence how people communicate with one another and what motivates them. Male leaders in competitive cultures tend to focus more on employees' achievements than on their unique qualities and relationships, whereas the opposite is true of leaders in cooperative cultures.[38]

Cocultures and Communication

Much of how we view ourselves and how we relate to others grows from our cultural and cocultural identity—the groups with which we identify. Where do you come from? What's your ethnicity? Your religion? Your sexual orientation? Your age?

In the following pages we will look at some—though by no means all—of the factors that help shape our cultural identity, and hence the way we perceive and communicate with others. As you read on, think about other cocultures that might be added to this list.

Race and Ethnicity

Race is a construct originally created to explain differences between people whose ancestors originated in different regions of the world—Africa, Asia, Europe, and so on. Modern scientists acknowledge that although there are some genetic differences between people with different heritage, they mostly involve superficial qualities such as hair color and texture, skin color, and the size and shape of facial features. As one analyst puts it:

> There is less to race than meets the eye. . . . The genes influencing skin color have nothing to do with the genes influencing hair form, eye shape, blood type, musical talent, athletic ability or forms of intelligence. Knowing someone's skin color doesn't necessarily tell you anything else about him or her.[40]

There are several reasons why race has little use in explaining individual differences. Most obviously, racial features are often misinterpreted. A well-traveled friend of ours from Latin America says that she is often mistaken as Italian, Indian, Spanish, or Native American.

More important, there is more genetic variation within races than between them. For example, some people with Asian ancestry are short, but others are tall. Some have sunny dispositions, while others are more stern. Some are terrific athletes, while others were born clumsy. The same applies to people from every background. Even within a physically recognizable population, personal experience plays a far greater role than superficial characteristics such as skin color.[41] As you read in Chapter 2, stereotyping is usually a mistake.

Ethnicity is another social construct. Ethnicity refers to the degree to which a person identifies with a particular group, usually on the basis of nationality, culture, religion, or some other perspective.[42] This goes beyond physical indicators. For example, a person may have physical characteristics that appear Asian, but they may identify more strongly as a Mormon or a member of the working class.

It is simplistic to think of people as members of a single category. This is true for everyone, but consider Barack Obama as an example. He is generally recognized as the first African American president of the United States, despite the fact that his mother was white. Obama experienced a variety of cultural influences

race A construct originally created to explain differences between people whose ancestors originated in different regions of the world—Africa, Asia, Europe, and so on.

ethnicity A social construct that refers to the degree to which a person identifies with a particular group, usually on the basis of nationality, culture, religion, or some other unifying perspective.

while living in Indonesia, Hawaii, California, New York, Chicago, and Washington, DC. Social scientists use the term **intersectionality** to describe the complex interplay of people's multiple identities. The theory proposes that people are not simply the sum total of the different identities. For example, if you are black, gay, male, rich, and Methodist, your experiences are not a simply an amalgam of these as "separate" identities. Living at the intersection of these factors gives rise to a perspective and collection of experiences all their own.[43,44] Intersectionality theorists argue that it's a mistake to focus on one cultural or cocultural dimension in isolation because people are shaped by all the elements of their identity.[45]

> **intersectionality** The idea that people are influenced in unique ways by the complex overlap and interactions of multiple identities.

Identifying with more than one group can be challenging. Consider someone like Heather Greenwood, who is biracial. Strangers inquire how her children can be fair skinned, because she is dark. Others ask if she is the children's nanny or joke that the kids must have been "switched in the hospital" when they were born. The implication, Greenwood says, is that a legitimate family is either one color or another. She hears comments such as these nearly every day. "Each time is like a little paper cut, and you think, 'Well, that's not a big deal.' But imagine a lifetime of that. It hurts," she says.[46]

Along with the challenges, multiple-group membership can be a bonus. Research indicates such people are more comfortable than usual establishing relationships with a diverse array of people, which increases their options for friendships, romantic partners, and professional colleagues.[47]

Regional Differences

Where you come from can shape feelings of belonging and how others regard you. Accent is a case in point. Speakers of standard ("newscaster") English are typically viewed as more competent and self-confident than others, and people tend to take what they say more seriously.[48] In one experiment, researchers asked human resource professionals to rate the intelligence, initiative, and personality of job applicants after listening to a brief recording of their voices. The speakers with recognizable regional accents—from the southern United States or New Jersey, for example—were tagged for lower level jobs, whereas those with less pronounced speech styles were recommended for higher level jobs that involved more public contact.[49]

The effect of nonnative accents is even stronger. In one study, jurors in the United States found testimony less believable when delivered by witnesses speaking with Mexican, German, or Middle Eastern accents.[50] Not surprisingly, other research shows that speakers with nonnative accents feel stigmatized by the bias against them, often leading to a lower sense of belonging and more communication challenges.[51]

Accents are not the only indicator of regional differences. Sociolinguist Deborah Tannen recorded the conversation at a Thanksgiving Day dinner that included two Christians who grew up in California, three Jewish people from New York, and a British woman.[52] Tannen found that the Jewish New Yorkers

"Anything you say with an accent may be used against you."

Source: Paul Noth The New Yorker Collection/The Cartoon Bank

spoke with what she called a "high-involvement" style that worked well with their companions from the same background but was regarded differently by the Englishwoman and the Californians:

> At times the Californians felt interrupted when their Jewish friends mistook a pause for breath as a turn-relinquishing one. At other times, exclamations like "Wow!" or "That's impossible!" which were intended to encourage the conversation, stopped it instead. The New Yorkers in my study assumed that a speaker who wasn't finished wouldn't stop just because someone else started.[53]

Even facial expressions have a regional basis. In the United States, unwritten rules about smiling vary from one part of the country to another. People from southern and border states smile the most, and Midwesterners smile more than New Englanders.[54] Given these differences, it's easy to imagine how a college freshman from North Carolina might regard a new roommate from Massachusetts as unfriendly, and how the New Englander might view the southerner as overly expressive.

A fascinating series of studies revealed that climate and geographic latitude were remarkably accurate predictors of communication predispositions.[55] Compared to northerners, people living in southern latitudes of the United States were less tolerant of ambiguity, higher in self-esteem, more likely to touch others, and more likely to verbalize their thoughts and feelings. This helps explain why communicators who travel from one part of a country to another often find that their old patterns of communicating do not work as well in their new location. A southerner whose relatively talkative, high-touch style seemed completely normal at home might be viewed as pushy and aggressive in a new northern home.

Sexual Orientation and Gender Identity

"NOBODY KNOWS I'M GAY." That witty saying is emblazoned on thousands of T-shirts sold by former comedian Skyler Thomas, a champion of the LGBTQ movement. *LGBTQ* stands for lesbian, gay, bisexual, transgender, and queer—a collection of adjectives that describes people with diverse sexual orientations and gender identities.

Not all LGBTQ individuals are gay. Transgender individuals don't feel that their biological sex is a good description of who they are. For example, some people who were born boys identify more with a feminine identity, and vice versa. And people typically describe themselves as queer if they don't feel that other gender adjectives describe them well or if they dislike the idea of gender categorizations in general. The Q sometimes stands for "questioning" as well, underlining the idea that gender is not always a fixed or static construct.[56]

Of course, the LGBTQ acronym does not exhaust the spectrum of sexual identities and orientations. Some might add an *I* for intersex individuals, who have both male and female genitalia; an *A* for asexual individuals, who are not interested in sex; another *A* for allies, who fit none of the other descriptions but support the idea of nuanced gender identities; and a *P* for pansexual, people who are attracted to others based on qualities that are not gender specific. Whether or not you think the acronym is getting too long, the point remains: Gender identity is far more diverse than gay or straight, male or female.

The "NOBODY KNOWS I'M GAY" shirts point to a communication dilemma facing many individuals with diverse gender identities. On the one hand, being open about gender identity has advantages—including a sense of being authentic with others and belonging to a supportive coculture. On the other hand, the disclosure can be risky. People may be shocked or judgmental. They may ridicule individuals they consider to be nontraditional, discriminate against them, or even attack them.

Restroom access for transgender people has become a prominent issue.

How have recent developments shaped your views on gender roles and sexual orientation?

UNDERSTANDING DIVERSITY

Gender Pronouns

We usually don't think much about gender-related personal pronouns. But without them, language becomes unwieldy. Consider an example:

> When Mrs. Anne Marie Schreiner came into the room, Mrs. Anne Marie Schreiner thought to Mrs. Anne Marie Schreiner's self, "Is the situation just Mrs. Anne Marie Schreiner, or is the temperature really hot in here?"[61]

This sentence demonstrates the value of pronouns. But as gender identities become more fluid, pronoun choice has become more complex. For example, what pronouns do you use to address a biological female who presents as a man? Most people who fit this description would suggest that words such as *him* and *his* would be most appropriate. Another option is to adopt singular use of the gender-neutral pronoun *they* and its counterparts *them*, *their*, and *themselves* to refer to someone who does not identify as male or female.[62]

It may be hard to guess at pronoun preferences, however, and you may make a mistake. If that happens, simply correct your mistake and move on.[63] Most people who have nontraditional gender identities agree that sincerity and respect are much more important than being a flawless speaker.

On average, one in five hate crimes in the United States targets people on the basis of their sexual orientation.[57]

CNN host Anderson Cooper didn't tell the public he was gay for years because he considered it private information, he felt it might put him and others in danger, and he thought he could do a better job as a journalist if he "blended in."[58] However, Cooper says his silence sometimes felt disingenuous. "I have given some the mistaken impression that I am trying to hide something," he said when he came out. He also felt that he had been missing opportunities to dispel some of the fear and prejudice that surround the issue. "The tide of history only advances when people make themselves fully visible," Cooper said in a 2012 public statement. "The fact is, I'm gay, always have been, and I couldn't be more happy, comfortable with myself, and proud."

For people whose coming out announcements make headlines, going public happens all at once. However, that is not usually the case. "Coming out is a process that never ends," reflects Jennifer Potter, a physician who is lesbian. "Every time I meet someone new I must decide if, how, and when I will reveal my sexual orientation."[59] She says it's often easy to "pass" as heterosexual, but then she experiences the awkwardness of people assuming she has a boyfriend or husband, and it saddens her when she cannot openly refer to her partner or invite her to take part in social gatherings.

Lewis Hancox, a filmmaker and comedian who is transgender, reflects that people who transition to a different gender identity encounter some of the same ambiguities. Having been through a public transition himself, he recommends that trans individuals love and accept themselves, connect with people who are supportive, have patience with those who don't immediately embrace the change, and keep in mind all the qualities that make them unique and special as individuals. "It took me a long time to realize that being transgender didn't make me any less of a guy, or more importantly, any less of a person," Hancox says. "We're all different in our own right and we should embrace those differences."[60]

The social climate has become more receptive to LGBTQ individuals than in the past, at least in most of the developed world. As one report put it,

> Most LGBT people who are now adults can recall feeling that they were "the only one" and fearing complete rejection by their families, friends, and associates

cultural idiom

coming out: declaring one's non-traditional sexual orientation clearly

should their identity become known. Yet today, in this era of GSAs [gay–straight alliances] . . . openly gay politicians and civil unions; debates about same-sex marriage, gay adoption, and gays in the military; and a plethora of websites aimed specifically at LGBT youth, it is hard to imagine many youths who would believe they are alone in their feelings.[64]

Among those websites is ItGetsBetter.org, at which people can post messages to encourage LGBTQ youth that any harassment they may be experiencing is not their fault and that people care about making it better. Since Dan Savage and Terry Miller launched the It Gets Better Project in 2010, people have posted more than 50,000 videos, which have been viewed more than 50 million times.[65] Savage says it shows what can happen when communication technology and good intentions combine.[66]

Religion

In some cultures, religion is the defining factor in shaping in- and out-groups. As fears of terrorism have grown, peace-loving Muslims living in the West have often been singled out and vilified. Yasmin Hussein, who works at the Arab American Institute, reflects on the prejudice and cruelty:

> Many Muslims and individuals of other faiths who were thought to be Muslim have been attacked physically and verbally. Young children have been bullied at schools, others told to go back home and social media has become at times (a lot of the time) an ugly place to be on.[67]

In an effort to dispel unrealistic stereotypes, tens of thousands of Muslims have joined the #NotInMyName social media movement in which they denounce terrorist groups such as ISIS and condemn violence in the name of the religion.

Intergroup hostility abounds in many areas of the world. Whether you belong to the Shia or Sunni sects is enormously important to many Muslims. Tribal divisions have been just as powerful in the Christian world. Within recent memory, Protestants and Catholics in Northern Ireland have fought to the death to defend their beliefs.

In less extreme but still profound ways, religion shapes how and with whom many people communicate. For example, members of the Orthodox Jewish community consider it important to marry within the faith. In one study, some of the young Jewish women interviewed said a man's religious preference is as important or more important than his personality.[68] They also described a cultural gap between dating solely to find a suitable spouse and the "American style" of dating, in which couples may spend time together, and even have sex, although they don't plan to marry. Dating within the Orthodox faith is communication centered and focused, a bit like a job interview. As one woman in the study put it: "When you're not touching the person . . . you're more professional, you take decisions much more seriously."

Other research suggests that, in general, teens who believe that only one religion has merit date less frequently than other teens,

ASK YOURSELF

What are some of the cultural groups you identify with, and how have they shaped your communication style and expectations? In what ways is it helpful or necessary to adapt your communication to fit different cocultural groups?

In the independent film *Arranged*, Rochel and Nasira—an Orthodox Jew and a Muslim, respectively—meet as new teachers at a Brooklyn school. Colleagues and students expect friction, but the women become close friends as they discover they have much in common. Think about others whom you might reflexively regard as different.

What might you discover that you have in common if you learned more about them?

perhaps because their pool of acceptable partners is smaller.[69] However, religious teens who respect the viewpoints of multiple religions typically date more frequently than their nonreligious peers.[70] And the odds are good for interfaith relationships. Studies show that, if interfaith couples communicate openly and respectfully about matters of faith, they are just as likely as other couples to stay together.[71]

Religious beliefs affect family life as well. Members of evangelical churches are likely to view parents as family decision makers and honor children for following their advice without question.[72] Religious activities such as reading scripture at home are most common among Jehovah's Witnesses and members of the Church of Jesus Christ of Latter-Day Saints (Mormons), evangelical churches, and historically black churches.[73] These congregations are also the most likely to believe there is only one true religion.

Physical Ability/Disability

Whereas identities related to ethnicity or nationality may require years of immersion, disability is "a club anyone can join, anytime. It's very easy. Have a stroke and be paralyzed. . . . Be in a car wreck and never walk again."[74]

Although able-bodied people might view disability as an unfortunate condition, many people with disabilities find that belonging to a community of similar people can be rewarding. Deaf culture is a good example: The shared experiences of deafness can create strong bonds. Most notably, distinct languages build a shared worldview and solidarity. There are Deaf schools, Deaf competitions (e.g., Miss Deaf America), Deaf performing arts (including Deaf comedians), and other organizations that bring Deaf people together.

Actor and model Nyle DiMarco won a *Dancing with the Stars* championship, bringing judges to tears with a dance to "The Sound of Silence." Viewers who missed the backstory would never have known he has been Deaf since birth.

In what ways do people stereotype and underestimate you? How can you respond?

UNDERSTANDING DIVERSITY

Communicating with People Who Have Disabilities

Research has revealed some clear guidelines for interacting with people who have disabilities.[76] They include:

1. Speak directly to people with disabilities, rather than looking at and talking to their companions.

2. If you volunteer assistance, wait until your offer is accepted before proceeding. Then listen to or ask for instructions.

3. Treat adults as such. For example, don't address people who have disabilities by their first names unless you're doing so with others.

4. When you're introduced to a person with a disability, offer to shake hands. (People with limited hand use or who wear an artificial limb can usually shake hands.)

5. When meeting a person who is visually impaired, identify yourself and others who may be with you.

6. When speaking with someone who uses a wheelchair, place yourself at eye level in front of the person. Remember that a wheelchair is part of the user's personal body space, so don't touch or lean on it without permission.

7. When you're talking with a person who has difficulty speaking, be patient and wait for him or her to finish. Ask short questions that require a nod, shake of the head, or brief answers. Never pretend to understand if you are having difficulty doing so.

8. With a person who is Deaf, speak clearly, slowly, and expressively to determine whether the person can read your lips. (Not all people who are Deaf can read lips.)

9. Don't be embarrassed if you happen to use common expressions such as "See you later" or "Did you hear about that?" that seem to relate to a person's disability.

10. Relax. Don't be afraid to ask questions when you're unsure what to do.

Notice that these recommendations are mostly similar to the ways we communicate with everyone else. People with disabilities are people first and should be treated as such.

Despite stereotypes, "different than" doesn't have to mean "less than." One former airline pilot who lost his hearing described his trip to China:

> Though we used different signed languages, these Chinese Deaf people and I could make ourselves understood; and though we came from different countries, our mutual Deaf culture held us together. . . . Who's disabled then?[75]

It's important to treat a disability as one feature of a person, not a defining characteristic. Describing someone as "a person who is blind" is both more accurate and less constricting than calling her a "blind person." This difference might seem subtle—until you imagine which label you would prefer if you lost your sight.

Age/Generation

Imagine how odd it would seem to hear an 8-year-old or a senior citizen talking, dressing, or otherwise acting like a 20-something. We tend to think of getting older as a purely biological process. But age-related communication reflects culture at least as much as biology. In many ways, we learn how to "do" being various ages— how to dress, how to talk, and what not to say and do—in the same way we learn how to play other roles in our lives, such as student or employee.

Relationships between older and younger people are shaped by cultural assumptions that change over time. At some points in history, older adults have been regarded as wise, accomplished, and even magical.[77] At others, they have been treated as unwanted surplus and uncomfortable reminders of mortality and decline.[78]

Today, Western cultures mostly honor youth, and attitudes about aging are more negative than positive. On balance, people over age 40 are still twice as likely as younger ones to be depicted in the media as unattractive, bored, and in declining health.[79] And people over age 60 are still underrepresented in the media. Despite negative stereotypes, the data on personal satisfaction present a different story. Studies show that, overall, people in their 60s are just as happy as people in their 20s.[80]

Unfavorable attitudes about aging can show up in interpersonal relationships. Even though gray or thinning hair and wrinkles don't necessarily signify diminished capacity, they may be interpreted that way—with powerful consequences. People who believe older adults have trouble communicating are less likely to interact with them. When they do, they tend to use the mannerisms listed Table 3-5.[81] Even when these speech styles are well intentioned,

---

TABLE 3-5

Patronizing Speech Directed at Older Adults

ELEMENT	DEFINITION AND EXAMPLE
Simplified grammar	Use of short sentences without multiple clauses. "Here's your food. You can eat it. It is good."
Simplified vocabulary	Use of short words rather than longer equivalents. Saying *dog* instead of *Dalmatian*, or *big* instead of *enormous*.
Endearing terms	Calling someone "sweetie" or "love."
Increased volume, reduced rate	Talking LOUDER and s-l-o-w-e-r!
High and variable pitch	Using a slightly squeaky voice style, and exaggerating the pitch variation in speech (a "singsong"-type speech style).
Use of repetition	Saying things over and over again. Repeating. Redundancy. Over and over again. The same thing. Repeated. Again. And again . . .
Use of baby-ish terms	Using words like *doggie* or *choo-choo* instead of *dog* or *train*: "Oh look at the cute little doggie, isn't he a coochie-coochie-coo!"

Source: J. Harwood. (2007). *Understanding communication and aging: Developing knowledge and awareness.* Newbury Park, CA: Sage, p. 76.

they can have harmful effects. Older adults who are treated as less capable than their peers tend to perceive *themselves* as older and less capable.[82] And challenging ageist treatment presents seniors with a dilemma: Speaking up can be taken as a sign of being cranky or bitter, reinforcing the stereotype that older adults are curmudgeons.[83]

Youth doesn't always live up to its glamorous image. Teens and young adults typically experience intense pressure, both internally and from people around them, to establish their identity and prove themselves.[84] At the same time, adolescents typically experience what psychologists call a personal fable (the sense that they are different from everybody else) and an imaginary audience (a heightened self-consciousness that makes it seem as if people are always observing and judging them).[85] These characterize a natural stage of development, but they lead to some classic communication challenges. For one, teens often feel that their parents and other people can't understand them because their situations are different and unique. Couple this with the sense that others are being overly attentive and critical and you have a good recipe for conflict and frustration. Parents may be baffled that their "extensive experience" and "good advice" are summarily rejected. And young people may wonder why people <u>butt into</u> their affairs with "overly critical judgments" and "irrelevant advice."

Communication challenges also can arise when members of different generations work together. For example, millennials (those born between 1980 and 2000) tend to have a much stronger need for affirming feedback than previous generations.[86] Because of their strong desire for achievement, they tend to want clear guidance on how to do a job correctly—but do not want to be micromanaged when they do it. After finishing the task, they have an equally strong desire for praise. To a baby boomer boss, that type of guidance and feedback may feel more like a nuisance. In the boss's experience, "no news is good news," so the absence of negative feedback should be praise enough. Neither perspective is wrong. But when members of these cocultures have different expectations, miscommunication can occur.

The Johnson family in the TV show black-*ish* appears to be living the American dream. They are well off, good looking, and they love one another. But a question nags at the Johnsons: Are they losing touch with their heritage?

How attached are you to your cultural heritage, and how much are you willing to dilute it?

Socioeconomic Status

Social class can have a major impact on how people communicate. People in the United States typically identify themselves as belonging to the working class, middle class, or upper class, and feel a sense of solidarity with people in the same social strata.[87] This is especially true for working-class people, who tend to feel that they are united both by hardship and by their commitment to hard, physical work. One working-class college student put it this way:

> I know that when all is said and done, I'm a stronger and better person than they [members of the upper class] are. That's probably a horrible thing to say and it makes me sound very egotistical, but . . . it makes me more glad that I've been through what I've been through, because at the end of the day, I know I had to bust my a** to be where I want, and that makes me feel really good.[88]

These communication styles can have consequences later in life. College professors often find that working-class and first-generation college students who are raised not to challenge authority can have a difficult time speaking up, thinking

cultural idiom

butt into: meddle in the affairs of others

critically, and arguing persuasively.[89] The effects of social class continue into the workplace, where skills such as assertiveness and persuasiveness are career enhancers. People who come from working-class families and attain middle- or upper-class careers face special challenges. New speech and language, clothing, and nonverbal patterns often are necessary to gain acceptance.[90]

Many of these individuals also must cope with ambivalent emotions related to their career success.[91] **Organizational culture** reflects a relatively stable, shared set of rules about how to behave and a set of values about what is important. In everyday language, culture is the insiders' view of "the way things are around here." See the "@Work" box below for more about workplace cultures.

Even within the same family, educational level can create intercultural challenges. First-generation college (FCG) students feel the intercultural strain of "trying to live simultaneously in two vastly different worlds" of school and home. Communication researchers Mark Orbe and Christopher Groscurth discovered

> **organizational culture** A relatively stable, shared set of rules about how to behave and a set of values about what is important.

@WORK

Organizations Are Cultures, Too

A man stopped by a Nordstrom department store in Oregon to buy an Armani tuxedo for his daughter's wedding. The next day he received a call saying the tux was ready and waiting. He picked it up and left a happy customer.

The kicker is that Nordstrom doesn't sell Armani tuxedos. The salesperson had one overnighted from another supplier and then altered to fit the customer perfectly.[96] It didn't matter that Nordstrom didn't make a penny off the sale. Doing whatever it takes to please customers is part of the company's corporate culture.

As this story illustrates, organizations have cultures that can be just as distinctive as those of larger societies.

Not all the rules and values of an organization are written down. And some that are written down aren't actually followed. Perhaps the workday officially ends at 5 P.M., but you quickly notice that most people stay until at least 6:30. That says something about the culture. Or, even though it doesn't say so in the employee handbook, employees in some companies consider one another as extended family, taking personal interest in their coworkers' lives. That's culture, too.

Because you're likely to spend as much time at work as you do in personal relationships, selecting the right organization is as important as choosing a best friend. Research shows that we are likely to enjoy our jobs and do them well if we believe that the organization's values reflect our own and that its values are consistently and fairly applied.[97] For example, Nordstrom rewards team members for offering great customer service without exception. On the other hand, a boss who talks about customer service but violates those principles cultivates a culture of cynicism and dissatisfaction.

Ask yourself these questions when considering whether a specific organization's culture is a good fit for you. (Notice how important communication is in each case.)

- How do people in the organization present themselves in person and on the telephone? Are they welcoming and inviting?

- Are customers happy with the service and quality provided?

- Do members of the organization have the resources and authority to do a good job?

- Do employees have fun? Are they encouraged to be creative?

- Is there a spirit of cooperation or competition among team members?

- What criteria are used to evaluate employee performance?

- What happens during meetings? Is communication open or highly scripted?

- How often do people leave their jobs to work somewhere else?

- Do leaders make a point of listening, respecting, and collaborating with employees?

- Do people use their time productively or are they bogged down with inefficient procedures or office politics?

Communication is the vehicle through which we both create and embody culture. At a personal and organizational level, effective, consistent, value-based communication is essential to success.

that many FGC students alter their communication patterns dramatically between their two worlds.[92]

Because no one in their family has attended college, FGC students often cope with an unfamiliar environment by trying to assimilate—going out of their way to fit in on campus. Sometimes assimilating requires self-censorship, as FGC students avoid discussions that might reveal their educational or socioeconomic backgrounds. In addition, some FGC students say they overcompensate by studying harder and getting more involved on campus than their non-FGC classmates, just to prove they belong in the college culture.

At home, FGC students may also engage in self-censorship, but for different reasons. They might be cautious when talking about college life for fear of threatening and alienating their families. One exception is that some feel a need to model their new educational status to younger family members so "they can see that it can be done."[93]

At the other end of the socioeconomic spectrum, gangs fit the definition of a coculture.[94] Members have a well-defined identity, both among themselves and according to the outside world. This sense of belonging is often reflected in distinctive language and nonverbal markers such as clothing, tattoos, and hand signals. Gangs provide people who are marginalized by society a sense of identity and security in an often dangerous and hostile world. But the benefits come at considerable cost. When compared to similar youths, gang members have higher rates of delinquency and drug use. They commit more violent offenses and have higher arrest rates.[95]

Developing Intercultural Communication Competence

What distinguishes competent and incompetent intercultural communicators? Before we answer this question, take a moment to complete the self-assessment on page 90 to evaluate your intercultural communication sensitivity.

To a great degree, interacting successfully with strangers calls for the same ingredients of general communicative competence outlined in Chapter 1. It's important to have a wide range of behaviors and to be skillful at choosing and performing the most appropriate ones in a given situation. A genuine concern for others plays an important role. Cognitive complexity and the ability to empathize also help. Finally, self-monitoring is important because it is often necessary to make midcourse corrections in your approach when dealing with strangers.

But beyond these basic qualities, communication researchers have worked hard to identify qualities that are unique, or at least especially important, ingredients of intercultural communicative competence.[98]

Increased Contact

More than a half century of research confirms that, under the right circumstances, spending time with people from different backgrounds leads to a host of positive outcomes: reduced prejudice, greater productivity, and better relationships.[99] The link between exposure and positive attitudes, called the *contact hypothesis*, has been demonstrated in a wide range of contexts.[100]

The benefits of communicating with others from different backgrounds are considerable. People who do so report a greater number of diverse friends than those who are less willing to reach out.[101] They also benefit from more positive relationships, lower stress levels, a more cooperative communication climate, and the chance to disconfirm stereotypes.

Under the right circumstances, increased diversity leads to positive outcomes.

How diverse is your network of relationships? Can you imagine ways to enhance its diversity?

Source: © 2011 Malcolm Evans

It's encouraging to know that increased contact with people from stigmatized groups can transform hostile attitudes. For example, door-to-door canvassers in Los Angeles were able to dramatically change attitudes of area residents by engaging in 10-minute conversations aimed at breaking down stereotypes.[102] By engaging the residents to think more deeply about the rights of transgender people, these conversations resulted in increased empathy.

Along with face-to-face contacts, cyberspace offers a useful way to enhance contact with people from different backgrounds.[103] Online venues make it relatively easy to connect with people you might never meet in person. The asynchronous nature of these contacts reduces the potential for stress and confusion that can easily come in person. It also makes status differences less important: When you're online, gaps in material wealth or physical appearance are much less apparent.

Tolerance for Ambiguity

When we encounter communicators from different cultures, the level of uncertainty is especially high. Consider the basic challenge of communicating in an unfamiliar language. Pico Iyer captures the ambiguity that arises from a lack of fluency when he describes his growing friendship with Sachiko, a Japanese woman he met in Kyoto:

> I was also beginning to realize how treacherous it was to venture into a foreign language if one could not measure the shadows of the words one used. When I had told her, in Asuka, *"Jennifer Beals ga suki-desu. Anata mo"* ("I like Jennifer Beals—and I like you"), I had been pleased to find a way of conveying affection, and yet, I thought, a perfect distance. But later I looked up *suki* and found that I had delivered an almost naked protestation of love. . . .

> Meanwhile, of course, nearly all her shadings were lost to me. . . . Once, when I had to leave her house ten minutes early, she said, "I very sad," and another time, when I simply called her up, she said, "I very happy"—and I began to think her unusually sensitive, or else prone to bold and violent extremes, when really she was reflecting nothing but the paucity of her English vocabulary. . . . Talking in a language not one's own was like walking on one leg; when two people did it together, it was like a three-legged waltz.[104]

Competent intercultural communicators accept—even welcome—this kind of ambiguity. Iyer describes the way the mutual confusion he shared with Sachiko actually helped their relationship develop:

> Yet in the end, the fact that we were both speaking in this pared-down diction made us both, I felt, somewhat gentler, more courteous, and more vulnerable than we would have been otherwise, returning us to a state of innocence.[105]

Without a tolerance for ambiguity, the mass of often confusing and sometimes downright incomprehensible messages that impact intercultural interactions would be impossible to manage. Some people seem to come equipped with this sort of tolerance, whereas others have to cultivate it. One way or the other, that ability to live with uncertainty is an essential ingredient of intercultural communication competence.

ASK YOURSELF

What examples of intercultural or cocultural **in**competence you have witnessed, either in person or on social media? How could the communicator(s) involved have handled the situation more competently?

Open-Mindedness

Being comfortable with ambiguity is important, but without an open-minded attitude a communicator will have trouble interacting competently with people from different backgrounds. To understand open-mindedness, it's helpful to consider three traits that are incompatible with it. **Ethnocentrism** is an attitude that one's own culture is superior to others. An ethnocentric person thinks—either privately or openly—that anyone who does not belong to his or her in-group is somehow strange, wrong, or even inferior. Travel writer Rick Steves describes how an ethnocentric point of view can interfere with respect for other cultural practices:

> We [Americans] consider ourselves very clean and commonly criticize other cultures as dirty. In the bathtub we soak, clean, and rinse, all in the same water. (We would never wash our dishes that way.) A Japanese visitor, who uses clean water for each step, might find our way of bathing strange or even disgusting. Many cultures spit in public and blow their nose right onto the street. They couldn't imagine doing that into a small cloth, called a hanky, and storing that in their pocket to be used again and again.

> Too often we think of the world in terms of a pyramid of "civilized" (us) on the top and "primitive" groups on the bottom. If we measured things differently (maybe according to stress, loneliness, heart attacks, hours spent in traffic jams, or family togetherness) things stack up differently.[106]

Ethnocentrism leads to an attitude of **prejudice**—an unfairly biased and intolerant attitude toward others who belong to an out-group. (Note that the root term in *prejudice* is "pre-judge.") An important element of prejudice is stereotyping (see p. 44)—exaggerated generalizations about a group. Familiarity can change attitudes, however, as American attitudes toward same-sex marriage have shown. As the practice has become legally sanctioned, there has been a dramatic increase in acceptance of gay and lesbian couples tying the knot. By 2016, a majority of Americans (55%) supported same-sex marriage, compared with slightly more than a third who opposed it.[107]

Another barrier to diversity occurs in the form of **hegemony**, the dominance of one culture over another. A common example of this is the impact of Hollywood around the world. People who are consistently exposed to American images and cultural ideas can begin to regard them as desirable. For example, in South Korea, more than 100 women a day undergo surgery on their eyelids, cheekbones, or noses in an effort to look more "Western."[108] The women typically say they will feel "prettier" and enjoy greater success in life if they look like American women they see in the media.

On a more encouraging note, whereas the mass media tend to perpetuate stereotypes, online communication can sometimes help people understand each other better. For example, when students studying German at an American university connected online with students at a German university, most of them were successful in establishing ask-and-share conversations. This result was especially pronounced when they used instant messaging rather than asynchronous posts, which were typically less conversational and interactive.[109]

Knowledge and Skill

Attitude alone isn't enough to guarantee success in intercultural encounters. Communicators need enough knowledge of other cultures to have a clear sense of which approaches are appropriate. The rules and customs that work with one group might be quite different from those that succeed with another. The ability to shift gears and adapt one's style to the norms of another culture or coculture is an essential ingredient of communication competence.[110]

One school of thought holds that uncertainty can motivate relationship development—to a point. For example, you may be interested in a newcomer to

ethnocentrism The attitude that one's own culture is superior to others'.

prejudice An unfairly biased and intolerant attitude toward others who belong to an out-group.

hegemony The dominance of one culture over another.

cultural idioms

tying the knot: getting married
shift gears: change what one is doing

SELF-ASSESSMENT

What Is Your Intercultural Sensitivity?

Below is a series of statements concerning intercultural communication. There are no right or wrong answers. Imagine yourself interacting with people from a wide variety of cultural groups, not just one or two. Record your first impression at each statement by indicating the degree to which you agree or disagree, using the following scale.

5 = strongly agree   4 = agree   3 = uncertain   2 = disagree   1 = strongly disagree

1. _____ I enjoy interacting with people from different cultures.

2. _____ I think people from other cultures are narrow minded.

3. _____ I am pretty sure of myself in interacting with people from different cultures.

4. _____ I find it very hard to talk in front of people from different cultures.

5. _____ I always know what to say when interacting with people from different cultures.

6. _____ I can be as sociable as I want to be when interacting with people from different cultures.

7. _____ I don't like to be with people from different cultures.

8. _____ I respect the values of people from different cultures.

9. _____ I get upset easily when interacting with people from different cultures.

10. _____ I feel confident when interacting with people from different cultures.

11. _____ I tend to wait before forming an impression of culturally distinct counterparts.

12. _____ I often get discouraged when I am with people from different cultures.

13. _____ I am open minded to people from different cultures.

14. _____ I am very observant when interacting with people from different cultures.

15. _____ I often feel useless when interacting with people from different cultures.

16. _____ I respect the ways people from different cultures behave.

17. _____ I try to obtain as much information as I can when interacting with people from different cultures.

18. _____ I would not accept the opinions of people from different cultures.

19. _____ I am sensitive to my culturally distinct counterpart's subtle meanings during our interaction.

20. _____ I think my culture is better than other cultures.

21. _____ I often give positive responses to my culturally different counterpart during our interaction.

22. _____ I avoid those situations in which I will have to deal with culturally distinct persons.

23. _____ I often show my culturally distinct counterpart my understanding through verbal or nonverbal cues.

24. _____ I have a feeling of enjoyment toward differences between my culturally distinct counterpart and me.

To determine your score, begin by reverse-coding (i.e., if you indicated 5, reverse-code to 1; if you indicated 4, reverse-code to 2; and so on) items 2, 4, 7, 9, 12, 15, 18, 20, and 22. Higher scores on the total instrument and each of the five subscales indicate a greater probability of intercultural communication competence.

Sum items 1, 11, 13, 21, 22, 23, and 24 ☐ Interaction Engagement (range is 7–35)

Sum items 2, 7, 8, 16, 18, and 20 ☐ Respect for Cultural Differences (6–30)

Sum items 3, 4, 5, 6, and 10 ☐ Interaction Confidence (5–25)

Sum items 9, 12, and 15 ☐ Interaction Enjoyment (3–15)

Sum items 14, 17, and 19 ☐ Interaction Attentiveness (3–15)

Sum of all the items ☐ (24–120, with a midpoint of 48)

Permission to use courtesy of Guo-Ming Chen. Chen, G. M., & Sarosta, W. J. (2000). The development and validation of the Intercultural Sensitivity Scale. *Human Communication*, 3, 1–14.

your class because he is from another country. However, if attempting a conversation with him heightens your sense of uncertainty and discomfort, you may abandon the idea of making friends. The basic premise of anxiety uncertainty management theory is that, if uncertainty and anxiety are too low or too high, we are likely to avoid communicating.[111]

How can a communicator learn enough about other cultures to feel curious but not overwhelmed? Scholarship suggests three strategies.[112] *Passive observation* involves noticing what behaviors members of a different culture use and applying these insights to communicate in ways that are most effective. *Active strategies* include reading, watching films, and asking experts and members of the other culture how to behave, as well as taking academic courses related to intercultural communication and diversity.[113] The third strategy, *self-disclosure*, involves volunteering personal information to people from the other culture with whom you want to communicate. One type of self-disclosure is to confess your cultural ignorance: "This is very new to me. What's the right thing to do in this situation?" This approach is the riskiest of the three described here, because some cultures may not value candor and self-disclosure as much as others. Nevertheless, most people are pleased when strangers attempt to learn the practices of their culture, and they are usually quite willing to offer information and assistance.

Patience and Perseverance

Becoming comfortable and competent in a new culture or coculture may be ultimately rewarding, but the process isn't easy. After a "honeymoon" phase, it's typical to feel confused, disenchanted, lonesome, and homesick.[114] To top it off, you may feel disappointed in yourself for not adapting as easily as you expected. This stage—which typically feels like a crisis—has acquired the labels *culture shock* or *adjustment shock*.[115]

You wouldn't be the first person to be blindsided by culture shock. Barbara Bruhwiler, who was born in Switzerland and has lived in South Africa for 5 years, says she loves her new home but still experiences moments of confusion and distress.[116] Likewise, when Lynn Chih-Ning Chang came to the United States from Taiwan for graduate school, she cried every day on the way home from class.[117] All her life, she had been taught that it was respectful and ladylike to sit quietly and listen, so she was shocked that American students spoke aloud without raising their hands, interrupted one another, addressed the teacher by first name, and ate food in the classroom. What's more, Chang's classmates answered so quickly that, by the time she was ready to say something, they were already on a new topic. The same behavior that made her "a smart and patient lady in Taiwan," she says, made her seem like a "slow learner" in the United States.

Communication theorist Young Yum Kim has studied cultural adaptation extensively. She says it's natural to feel a sense of push and pull between the familiar and the novel.[118] Kim encourages sojourners to regard stress as a good sign. It means they have the potential to adapt and grow. With patience, the sense of crisis begins to wane, and once again, there's energy and enthusiasm to learn more.

Communication can be a challenge while you're learning how to operate in new cultures, but it can also be a solution.[119] Chang, the Taiwanese student adapting to life in America, learned this firsthand. At first, she says, she was reluctant to approach American students, and they were reluctant to approach her. Gradually, she got up the courage to initiate conversations, and she found that her classmates were friendly and receptive. Eventually, she made friends, began to fit in, and successfully completed her degree.

The transition from culture shock to adaptation and growth is usually successful, but it isn't a smooth, linear process. Instead, people tend to take two steps forward and one step back, and to repeat that pattern many times. Kim[120] calls this a "draw back and leap" pattern. Above all, she says, if people are patient and they keep trying, the rewards are worth it.

ETHICAL CHALLENGE
Civility When Values Clash

Most people acknowledge the importance of treating others from different cultural backgrounds with respect. But what communication obligations do you have when another person's cultural values are different from yours on fundamental matters, such as abortion or gender equity? How should you behave when confronted with views you find shocking or abhorrent?

For his part, our profiled student, Robin Luo, says he is learning a lot more than engineering during his time as an exchange student in the United States. "Americans are friendly, passionate, and inspiring," he says. "They teach me to hold an optimistic attitude against whatever I encounter in life."

MAKING THE GRADE

For more resources to help you understand and apply the information in this chapter, visit the *Understanding Human Communication* website at www.oup.com/us/adleruhc.

OBJECTIVE 3.1 Analyze the influence of cultures and cocultures on intergroup communication, explaining the concepts of salience and generalizations.

- Communicating with people from different backgrounds is more common today than ever before. Some encounters involve people from different cultures, whereas others involve communicating with people from different cocultures within a given society.

- Cultural differences are not a salient factor in every intergroup encounter.

- Although cultural characteristics are real and important, they are generalizations that do not apply equally to every member of a group.

  > Provide two examples of intercultural communication from your own experience—one in which your affiliation with a particular group or culture was salient, and one in which it was not.

  > Think of a cultural group with which you identify. Now describe a conversation that would be interpreted one way by members of that group and another way by out-group members.

  > List three common generalizations about college students. How well does each of those generalizations describe you? What can you learn about generalizations from this exercise?

OBJECTIVE 3.2 Distinguish among the following cultural norms and values, and explain how they influence communication: individualism and collectivism, high and low context, uncertainty avoidance, power distance, talk and silence, competition and cooperation.

- Some cultures value autonomy and individual expression, whereas others are more collectivistic.

- Some pay close attention to subtle, contextual cues (high context), whereas others pay more attention to the words people use (low context).

- Cultures vary in their acceptance of uncertainty.

- Authority figures are treated with more formality and often greater respect in cultures with high power distance than in those with low power distance.

- Depending on the cultural context, people may interpret silence as comforting and meaningful, or conversely, as an indication of awkwardness or disconnection.

- Some cultures value competition, independence, and assertiveness, whereas others emphasize equality, relationships, cooperation, and consensus building.

  > Compare and contrast the communication techniques that job applicants from high- and low-power-distance cultures might use.

  > Analyze the assertion that "young people today have no respect for authority" by applying the notion of power distance to specific examples.

  > Do you identify more with individualism or collectivism? In what ways?

OBJECTIVE 3.3 Evaluate the influence of factors such as race and ethnicity, sexual identity, religion, age, and socioeconomic status on communication.

- People often make assumptions about others based on their apparent race, but appearance is an unreliable indicator of cultural and personal differences.

- People have multiple cultural identities that intersect.

- Gender diversity goes far beyond the simplistic dichotomies of male-or-female and gay-or-straight.

- People from different regions of the same country may uphold quite different beliefs about what is valued and "normal" in terms of communication.

- People may also identify with others of similar religion, age, physical ability, and socioeconomic status in ways that define them as members of a culture.

  > Explain the difference between race and ethnicity and defend why one is a more reliable indicator of cultural identity than the other.

  > List three identities that apply to you. How does living at the intersection of those identities make your life experience different from that of a person who shares one or two of those identities, but not all three?

  > Transcribe a recent conversation as closely as you can, and then replace all the gender pronouns (e.g., he, she, him, hers) with the hypothetical pronouns

xe (which can mean either he or she) and *xyrs* (which can mean either his or hers).[121] Was the meaning still clear, or did it become confusing? How might our attitudes toward people be different if their gender were not evident in the way we talk about them?

OBJECTIVE **3.4** **Use the criteria in this chapter to assess your intercultural and cocultural communication competence, and describe how you could communicate more competently.**

- Cultural competence is typically enhanced by contact with people who have different beliefs.

- It's useful to become comfortable with uncertainty and open to the idea of different, but equally valid, worldviews.

- It often takes patience and skill to connect with people who have diverse perspectives, but the benefits are worth the effort.

> List and explain three communication strategies people might use to learn about different cultures and to share their own ideas with others.

> Think of a time when you met someone who seemed very different from you at first but eventually became a close friend or colleague. Create a timeline that illustrates turning points in your relationship when you learned more about each other and developed rapport.

> How do you rate yourself in terms of intercultural sensitivity and open-mindedness? What steps might you take to keeping growing in this regard?

KEY TERMS

coculture p. 69

collectivistic culture p. 72

culture p. 69

ethnicity p. 78

ethnocentrism p. 89

hegemony p. 89

high-context culture p. 74

in-groups p. 71

individualistic culture p. 72

intergroup communication p. 69

intersectionality p. 79

low-context culture p. 74

organizational culture p. 86

out-groups p. 71

power distance p. 76

prejudice p. 89

race p. 78

salience p. 71

uncertainty avoidance p. 75

ACTIVITY

1. **Recognizing Listening Misconceptions** You can see how listening misconceptions affect your life by identifying important situations when you have fallen for both of the following assumptions. In each case, describe the consequences of believing these erroneous assumptions.

   a. Thinking that because you were hearing a message, you were listening to it.

   b. Believing that listening effectively is natural and effortless.

Language

4

CHAPTER OUTLINE

LEARNING OBJECTIVES

4.1

Describe the symbolic, person-centered, rule-governed nature of language.

4.2

Explain how language both shapes and reflects attitudes.

4.3

Distinguish the main types of troublesome language and their effects, described on pages 107–115.

4.4

Explain how gender and nongender variables, described on pages 116–119, affect communication.

Consider situations in which language has either hindered you or helped you establish a satisfying connection.

?

Think of a statement that people may interpret differently, depending on the situation and people involved. What kinds of misunderstandings arise over differences such as these?

?

How does the way others use language shape your opinion of them? How do you think your language mannerisms shape the way others see you?

?

If you are proficient in more than one language, have you noticed that one language is more expressive than another in certain situations?

WHEN FRIENDS SCOTT H. YOUNG and Vat Jaiswal (pictured here) left their native Canada for a year-long adventure in faraway places, they said *adios* to English. The two vowed to speak only the native languages of the places they would visit—namely Spain (Spanish), Brazil (Portuguese), China (Mandarin), and South Korea (Korean).

So began what the two friends call The Year Without English. Their goal wasn't simply to learn new languages, but to experience the cultures they encountered as deeply and authentically as possible.[1] In this chapter, we explore the idea that language is more than vocabulary words. It is a means of connecting with people and displaying how we feel about and relate to each one another.

Although Young and Jaiswal felt tongue-tied when they first encountered new languages, they found that they could converse reasonably well with the locals after about a month. This allowed them to establish friendships they could not have experienced otherwise.

Learning different ways of speaking is just as important closer to home. "Maybe you hang out with more preppy people and there are some hipsters you see," Young reflects. Do you make an effort to understand them and honor the way they express themselves? If so, Young says, "that is an incredibly powerful skill, perhaps more powerful than learning other languages in other cultures because it works where you live."[2]

By the time you finish this chapter, you should better appreciate the complexity of language, its power to shape our perceptions of people and events, and its potential for incomplete and inaccurate communication. Perhaps more importantly, you will be better equipped to use the tool of language more skillfully to improve your everyday interactions.

The Nature of Language

If a British person has a *chinwag* (chat) with a person from the United States or Canada, he or she might feel *knackered* (exhausted) by how hard it is to share meaning, even within the same language. Sometimes you probably feel the same way when you're speaking with people much closer to home.

To begin understanding what's going on, let's first define some basic terms. A **language** is a collection of symbols governed by rules and used to convey messages between individuals. A **dialect** is a version of the same language that includes substantially different words and meanings.[3] English includes dozens of dialects, as do the other 7,000 or so other languages of the world. A closer look at the nature of language reveals why people often hear something different from what the speaker meant to say. In the end, language is powerful and indispensable, but it is also imprecise and constantly evolving.

Language Is Symbolic

There's nothing natural about calling your loyal, four-footed companion a *dog* or the object you're reading right now a *book*. These words, like virtually all language, are symbols—arbitrary constructions that represent a communicator's thoughts.

Not all linguistic symbols are spoken or written words. Sign language, as "spoken" by most Deaf people, is symbolic in nature and not the pantomime it might seem to nonsigners. There are hundreds of different sign languages spoken around the world. These distinct languages include American Sign Language, British Sign Language, French Sign Language, Danish Sign Language, Chinese Sign Language—even Australian Aboriginal and Mayan sign languages. Each has its own way of representing ideas.[4]

Symbols are more than just labels: They are the way we experience the world. You can prove this by trying a simple experiment.[5] Work up some saliva in your mouth, and then spit it into a glass. Take a good look, and then drink it up. Most people find this process mildly disgusting. But ask yourself why this is so. After all, we swallow our own saliva all the time. The answer arises out of the symbolic labels we use. After the saliva is in the glass, we call it *spit* and think of it in a different way. In other words, our reaction is to the name, not the thing.

The naming process operates in virtually every situation. How you react to a stranger will depend on the symbols you use to categorize him or her: gay (or straight), religious (or not), attractive (or unattractive), and so on.

Meanings Are in People, Not Words

Ask a dozen people what the same symbol means, and you are likely to get 12 different answers. Does an American flag bring up associations of patriots giving their lives for their country? Fourth of July parades? Cultural imperialism? How about a cross: What does it represent? The message of Jesus Christ? Fire-lit gatherings of Ku Klux Klansmen? Your childhood Sunday school? The necklace your sister always wears?

As with physical symbols, the place to look for meaning in language isn't in the words themselves but rather in the way people make sense of them. Linguistic theorists C. K. Ogden and I. A. Richards illustrated that meanings are social constructions in their well-known "triangle of meaning" (Figure 4-1).[6] This model shows that there's only an indirect relationship—indicated by a broken line—between a word and what it claims to represent. Some references are fairly clear, at least to members of the same speech community. In other cases, though, interpretations can be quite different. Consider abstract concepts such as *feminism*, *environmentalism*, and *conservatism*. Part of the person-centered nature of language involves the difference between denotative and connotative meanings. **Denotative meanings** are formally recognized definitions of a term. Most of the time there's little confusion about the denotative meaning of a word, such as *chair*. But consider terms such as *survivor* and *victim*. In reference to violent assaults, these terms are synonymous—they have the same denotative meaning—but each has different connotations. Unlike denotation, the **connotative meaning** involves

Characters in the TV series *Game of Thrones* may appear to be speaking nonsense when they communicate in Dothraki, but it's a complete language that has its own vocabulary and set of rules.

Can you name some rules that govern speech in your native language (for example, subject before verb)? What would it sound like to violate some of those rules?

language A collection of symbols, governed by rules and used to convey messages *between individuals*.

dialect A version of the same language that includes substantially different words and meanings.

denotative meanings Formally recognized definitions for words, as in those found in a dictionary.

connotative meanings Informal, implied interpretations for words and phrases that reflect the people, culture, emotions, and situations involved.

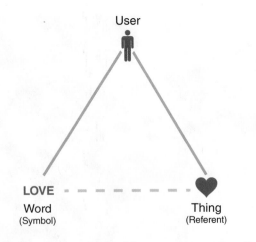

FIGURE 4-1 Ogden and Richards's Triangle of Meaning

phonological rules　Linguistic rules governing how sounds are combined to form words.

syntactic rules　Rules that govern the ways in which symbols can be arranged as opposed to the meanings of those symbols.

the associations, thoughts, and feelings that a statement generates. Ultimately, the meanings people associate with words have far more significance than do their dictionary definitions.

Matters can get even more confusing when you're not fluent in a language. During their Year Without English, Scott Young and Vat Jaiswal learned that calling someone a dog or a "son of a dog" in South Korea is a profane insult. Back home in North America, of course, close friends might call each other *dog* or *dawg*. All in all, understanding how language is used involves much more than vocabulary and grammar. Meaning isn't in words, it's in people—and it can vary substantially depending on the people and situations involved.

Despite the potential for linguistic problems, the situation isn't hopeless. We do, after all, communicate with one another reasonably well most of the time. And with enough effort, we can clear up most of the misunderstandings that occur. The key to more accurate use of language is to avoid assuming that others interpret words the same way we do. In truth, successful communication occurs when we negotiate the meaning of a statement. As one French proverb puts it: *The spoken word belongs half to the one who speaks it and half to the one who hears.*

Language Is Rule Governed

Languages contain several types of rules. **Phonological rules** govern how words sound when pronounced. For instance, the words *champagne*, *double*, and *occasion* are spelled identically in French and English, but all are pronounced differently. Nonnative speakers learning English are plagued by inconsistent phonological rules, as a few examples illustrate:

He could lead if he would get the lead out.

A farm can produce produce.

The dump was so full it had to refuse refuse.

The present is a good time to present the present.

I did not object to the object.

The bandage was wound around the wound.

I shed a tear when I saw the tear in my clothes.

Phonological rules aren't the only ones that govern the way we use language to communicate. **Syntactic rules** govern the structure of language—the way symbols can be arranged. For example, correct English syntax requires that every word contain at least one vowel and prohibits sentences such as "Have you the cookies brought?," which is a perfectly acceptable word order in German. Although most of us aren't able to describe the syntactic rules that govern our language, it's easy to recognize their existence by noting how odd a statement that violates them appears.

Technology has spawned versions of English with their own syntactic rules.[7] For example, people have devised a streamlined version of English for instant messages, texts, and tweets that speeds up typing in real-time communication (although it probably makes teachers of composition grind their teeth in anguish):

A: Hey r u @ home?

B: ys

A: y

B: cuz i need to study for finals c u later tho bye

A: TTYL

Semantic rules deal with the meaning of specific words. They make it possible for us to agree that *bikes* are for riding and *books* are for reading. Without semantic rules, communication would be impossible, because each of us would use symbols in unique ways, unintelligible to one another.

Semantic misunderstandings occur when words can be interpreted in more than one way, as the following humorous church notices demonstrate:

> The peacemaking meeting scheduled for today has been canceled due to a conflict.

> For those of you who have children and don't know it, we have a nursery downstairs.

> The ladies of the Church have cast off clothing of every kind. They may be seen in the basement on Friday afternoon.

> Sunday's sermon topic will be "What Is Hell?" Come early and listen to our choir practice.

Pragmatic rules govern how people use language in everyday interactions, which communication theorists have characterized as a series of *speech acts*.[8] You won't typically find these rules written down, but people familiar with the language and culture rely on them to make sense of what is going on. For example: *Are two people kidding around or being serious? Are they complimenting each other or trading insults?* The challenge is that even the people involved may not answer these questions the same way. Consider the example of a male boss saying, "You look very pretty today" to a female employee. He may interpret the statement very differently than she does based on a number of factors.

EACH PERSON'S SELF-CONCEPT

Boss: Views himself as a nice guy.

Employee: Determined to succeed on merits, not appearance.

THE EPISODE IN WHICH THE COMMENT OCCURS

Boss: Casual remark at the start of the workday.

Employee: A possible come-on?

PERCEIVED RELATIONSHIP

Boss: Views employees like members of the family.

Employee: Depends on boss's goodwill for advancement.

CULTURAL BACKGROUND

Boss: Member of generation in which comments about appearance were common.

Employee: Member of generation sensitive to sexual harassment.

As this example shows, pragmatic rules don't involve semantic issues, because the words themselves are clear. For example, it's common for Americans to ask new acquaintances, "What do you do for a living?" That question might cause offense in France, however, where it may be considered an indirect way of asking, "How much money do you make?" Conversely, Americans typically consider it rude to ask an older person his or her age, but it's more socially acceptable in Japan, where the oldest person in a conversation is typically given the most respect.

Pragmatic rules also govern language that some people find offensive but that others regard positively. For example, consider the word "queer." In earlier

The n-word: When used as a racial epithet, no term can be more inflammatory. But for many black people it has become a term of affection, and even empowerment in certain circumstances. Unless people share the same pragmatic rules, it's best to steer clear.

Is there ever an appropriate time and place to use the n-word?

semantic rules Rules that govern the meaning of language as opposed to its structure.

pragmatic rules Rules that govern how people use language in everyday interaction.

reappropriation The process by which members of a marginalized group reframe the meaning of a term that has historically been used in a derogatory way.

ASK YOURSELF

What pragmatic rules govern the use of swear words in the language communities you belong to?

generations, this term was a slur directed against homosexual men. More recently, however, some gay people have adopted it as a proud expression of their sexual orientation. This is an example of **reappropriation,** a term researchers use to describe how members of marginalized groups sometimes reframe the meaning of a term that has historically been used in a derogatory way.[9] As the photo on the previous page illustrates, the "n-word" is another example of a term that has been reappropriated by some—though by no means all—black people.

The use of profanity is another example of how pragmatic rules govern communication. How do you react when someone unexpectedly swears during normal conversation? Are you offended? Surprised? Intrigued? Like all verbal messages, swear words speak volumes beyond their literal meaning.

A semantic analysis doesn't reveal much about the meaning of most swear words. A closer look at when and how people use them reveals that pragmatic rules govern the use and interpretation of profanity. People who swear are more likely to do so in the company of those they know and trust, because doing so often indicates a level of comfort and acceptance. Swearing can also be a way to enhance solidarity between people. For example, research suggests that in some cases the swearing patterns of bosses and coworkers can help people feel connected on the job.[10]

Of course, swear words may be bad form no matter where they are spoken, and no one appreciates profanity when children can hear it. But looking at the pragmatic rules governing profanity helps explain why some people are offended by language that others consider benign.

The Power of Language

On the most obvious level, language allows us to satisfy basic functions such as describing ideas, making requests, and solving problems. But beyond these functions, the way we use language also influences others and reflects our views in subtle ways, which we will examine now.

Language Shapes Values, Attitudes, and Beliefs

The power of language to shape ideas has been recognized throughout history. In the Bible, for example, Adam demonstrates his dominion over animals by naming them.[11] As we will now see, our speech—sometimes consciously and sometimes not—shapes others' values, attitudes, and beliefs in a variety of ways.

Naming "What's in a name?" Juliet asked rhetorically. A lot, it turns out. Research has demonstrated that names are more than just a simple means of identification: They shape the way others think of us, the way we view ourselves, and the way we act.

At the most fundamental level, some research suggests that even the phonetic sound of a person's name affects the way we regard him or her, at least when we don't have other information available. One study revealed that it's often possible to predict who will win an election based on the candidates' surnames.[12] Voters tend to favor names that are simple, familiar, and easily pronounced. For example, in one series of local elections, candidates Sanders, Reilly, Grady, and Combs attracted more votes than Pekelis, Dellwo, Schumacher, and Bernsdorf. Names don't guarantee victory, but in 78 elections, 48 outcomes supported the value of having a more common name.

Names also play a role in shaping and reinforcing identity. They can create connections across generations, and they can make a statement about cultural identity. Some names may suggest a "black" identity, whereas others sound more "white."[13] The same could be said for Latino, feminine/masculine, Jewish, and other names.

Although names associated with particular groups may support a sense of shared identity, they are sometimes the basis for discrimination. In the United States, job applicants with names such as Mohammed and Lakisha typically receive fewer callbacks from employers than equally qualified candidates whose names sound more European.[14,15] Because of this potential for discrimination, some people advocate for name-blind job applications.[16] (See the "@Work" box for more about names in the workplace.)

Credibility   Speech style—and the credibility associated with it—also influences perception. Scholarly speaking is a good example: Even an impostor who sounds smart and speaks well may impress an audience.

Consider the case of Myron L. Fox, who delivered a talk on "Mathematical Game Theory as Applied to Physical Education." Questionnaires collected after

---

@WORK

What's in a Name?

When it comes to career success, names matter more than you might imagine.

Research suggests that many people pass judgment on prospective workers simply on the basis of their first names. In one study, prospective employers rated applicants with common first names more highly than those with unique or unusual ones.[17] This bias presents challenges for people with unique names and for those from cultures with different naming practices.

Sometimes naming biases reflect stereotypes about gender. People predict career success based on how closely a person's name matches the gender associated with his or her job.[18] When college students were asked how people with various names were likely to do in their careers, they predicted that women with feminine names like Emma or Marta were more likely to be successful in traditionally female occupations such as nursing. By contrast, they estimated that men with masculine names like Hank or Bruno would do better in traditionally male jobs such as plumbing.

Findings like this are worth noting if you hope to succeed in a field in which your identity doesn't match traditional expectations. For example, in the field of law, research suggests that your chances of success are greatest if you have a gender-neutral or traditionally male name. Researchers examined the relationship between the perceived masculinity of a person's name and his or her success in this field.[19] They found that a woman named "Cameron" is roughly three times more likely to become a judge than one named "Sue." A female "Bruce" is five times more likely.

Most people aren't willing to change their name to further career goals. But it's possible to choose variants of a name that have a professional advantage. For example, Christina Jones might use her nickname, Chris, for gender neutrality on job applications. Someone with a hard-to-pronounce name might choose a nickname for work purposes. For example, it's customary in China for businesspeople and students of English to choose a Western name that is similar to theirs: Guanghui may go by Arthur and Junyuan by Joanna. In a competitive job market, little differences can mean a great deal.

> **HELLO**
> my name is

Based on the research described here, do you think your name may affect your career? Have others' names ever shaped your perceptions?

the session indicated that the highly educated audience found the lecture clear and stimulating. [20] Yet Fox was a complete fraud. He was a professional actor whom researchers had coached to deliver a bogus lecture—a patchwork of information from a *Scientific American* article mixed with jokes, illogical conclusions, contradictory statements, and meaningless references to unrelated topics. When wrapped in a linguistic package of high-level professional jargon, however, the presentation succeeded. In other words, the audience's reaction was based more on the credibility that arose from his use of impressive-sounding language than on the ideas he expressed.

Status   In the classic musical *My Fair Lady*, Professor Henry Higgins transforms Eliza Doolittle from a lowly flower girl into a high-society woman by helping her replace her cockney accent with an upper-crust speaking style. Decades of research have demonstrated the power of speech to influence status. Several factors combine to create positive or negative impressions: accent, choice of words, speech rate, and even the apparent age of a speaker. In most cases, speakers of standard dialect are rated higher than nonstandard speakers in a variety of ways: They are viewed as more competent and more self-confident, and the content of their messages is rated more favorably. By contrast, the unwillingness or inability of a communicator to use the standard dialect fluently can have serious consequences. For instance, African American vernacular English is a distinctive dialect with its own accent, grammar, syntax, and semantic rules. Unfortunately, people sometimes assume that people who speak this vernacular are less intelligent, less professional, less capable, less socially acceptable, and less employable that people who speak what scholars call "standard English."[21] Speakers with other nonstandard accents also can be stigmatized.[22]

Worldview   Some scholars and travelers have observed a phenomenon known as **linguistic relativism**, which proposes that the language we speak shapes the way we view the world.[23] For an explanation of this phenomenon, see the "Understanding Diversity" sidebar on page 103.

Sexism and Racism   By now it should be clear that the power of language to shape attitudes goes beyond individual cases and influences how we perceive entire groups of people. Children exposed to words such as *fireman* and *businessman* are typically less likely than other children to think that women can pursue those occupations.[24,25] This assumption is far less prevalent among children exposed to gender-neutral terms such as *firefighter* and *business person*. Based on evidence such as this, many people argue that gender-neutral language is not merely "politically correct," but is a powerful force in shaping opportunities and identities.

Some languages emphasize gender less than others. In Finnish, for example, the same pronoun *hän* refers to both males and females. Finnish speakers sometimes puzzle over whether to call individuals *he* or *she* in English because they aren't accustomed to categorizing people that way.[26] The implications of such language differences are notable. Gender equality is greater in countries such as Finland, where language is nongendered, than in regions where the predominant language (such as Spanish, German, or Russian) attributes a gender to nearly every noun.[27] In Spanish, a spoon (*la cuchara*) is feminine, a fork (*el tenedor*) is masculine, a napkin (*la servilleta*) is feminine, and so on. First names in those languages also tend to be distinctly masculine or feminine. English falls somewhere in the middle, in that people and animals are referred to as *he* and *she*, but groups of people and objects are described in gender-neutral terms such as *they* or *it*. (See the Language and Worldview box for more about the link between language and attitudes.)

(?) ASK YOURSELF

Imagine for a moment that the words **love**, **disappointment**, and **hate** did not exist. Would you be aware of those feelings in the same way? Could you talk about them as effectively?

linguistic relativism The notion that language influences the way we experience the world.

ETHICAL CHALLENGE

One of the most treasured civil liberties is freedom of speech. At the same time, most people would agree that some forms of speech are hateful and demeaning.

• Do you think laws and policies can and should be made that limit certain types of racist and sexist communication?

• If so, how should those limits be drafted to protect open debate?

• If not, how would you justify protecting language that some people find personally degrading?

Language and Worldview

Mudita: Taking delight in the happiness of others (Sanskrit)

Voorpret: The sense of delightful anticipation before a big event (Dutch)

Koi No Yokan: The sense upon first meeting a person that the two of you are going to fall in love (Japanese)

Cavoli Riscaldati: The result of attempting to revive an unworkable relationship (Italian); translates to "reheated cabbage"[28]

Witzelsucht: The tendency to excessively make puns and tell inappropriate jokes and pointless stories (German)[29]

If you feel that life would be a little bit different if you had these phrases in your vocabulary, score one for linguistic relativism, the notion that words influence the way we experience the world.[30]

If not, score one for the opposition.

Whichever team you pick, you're in good company. The best-known declaration of linguistic relativism is the *Sapir-Whorf hypothesis*.[31] Philosophers and scientists have been debating it and theories like it for about 150 years.

In one experiment, researchers asked English speakers and Himba speakers (from a region of southeast Africa) to distinguish between colors. The Himba speakers, whose language includes words for many shades of green, were able to distinguish between green hues that English speakers perceived as being all one color. Conversely, English speakers were able to distinguish

between blue and green, whereas Himba speakers, who use the same word for both, often didn't see a difference.[32]

People who doubt linguistic relativity argue that, when we don't have a specific word for something, we use others in its place. They also point out that perception is not strictly in vocabulary's cage. That is, people often invent things for which they previously had no words. Think Internet, microwave oven, and telephone.[33]

Although the debate about language relativity isn't likely to be resolved any time soon, it reveals a great deal about the power and limitations of words. Keep this in mind the next time words fail you and you think of a witty comeback only after the moment has passed. The Germans have a word for that: *treppenwitz*.

"You'll have to phrase it another way. They have no word for 'fetch.'"

Source: Drew Dernavich The New Yorker Collection/The Cartoon Bank

---

Although English is not entirely gender neutral, it's fairly easy to use nonsexist language. For example, the term *mankind* may be replaced by *humanity, human beings, human race,* or *people; manmade* may be replaced by *artificial, manufactured,* and *synthetic; manpower* may be replaced by *human power, workers,* and *workforce;* and *manhood* may be replaced by *adulthood.*

The use of labels for racist purposes has a long and ugly past. Names have been used throughout history to stigmatize groups that other groups have disapproved of.[34] By using derogatory terms to label some people, the out-group is set apart and pictured in an unfavorable light. Diane Mader provides several examples of this:

> We can see the process of stigmatization in Nazi Germany when Jewish people became vermin, in the United States when African Americans became "niggers" and chattel, in the military when the Vietnam-era enemy became "gooks."[35]

The power of racist language to shape attitudes is difficult to avoid, even when it's obviously offensive. In a classic study, even people who disapproved of a derogatory label used against a member of a minority group tended to think less of the group members after encountering the term.[36] Not only did they rate the minority individual's competence lower when that person performed poorly, but they also found fault with others who associated socially with the minority person—even members of the participant's own ethnic group.

Language Reflects Values, Attitudes, and Beliefs

Besides shaping the way we view ourselves and others, language reflects our values, attitudes, and beliefs. Feelings of control, attraction, commitment, responsibility—all these and more are reflected in the way we use language.

Power   Americans typically consider language powerful when it is clear, assertive, and direct. By contrast, language is often labeled powerless when it suggests that a speaker is uncertain, hesitant, intensely emotional, deferential, or nonassertive.[37]

"Powerful" speech can be an important tool. In employment interviews, for example, people who seem confident and assertive usually fare better than those who stammer and seem unsure of themselves.[38]

It doesn't pay to overdo it, however. Just as an extremely "powerless" approach can feel weak, an overly "powerful" one can come off as presumptuous and bossy. A lot depends on the context. Consider the following statements a student might make to a professor:

> "Excuse me, sir, I hate to say this, but I won't be able to turn in the assignment on time. I had a personal emergency and . . . well . . . it was just impossible to finish it by today. Would that be okay?"

> "I wasn't able to finish the assignment that was due today. I'll have it in your mailbox on Monday."

If you were the professor in this situation, which approach would you prefer? The first sounds tentative and not as fluent, which, on the face of it, seems powerless. In some situations, however, less assertive speakers seem friendlier, more sincere, and less coercive than more assertive ones.[39] You might appreciate the second approach, which is more direct and "powerful," or you may find it presumptuous and disrespectful.

Your reaction to each approach is likely to reflect a number of factors: *Do you know the student well? Is it typical for him or her to miss deadlines? Do you share the same cultural expectations?* People in some cultures admire self-confidence and direct speech. However, in many cultures, helping others <u>save face</u> is a higher priority, so communicators tend to speak in ambiguous terms and use hedge words (e.g., *maybe*) and disclaimers. This is true in many Japanese and Korean cultures. Similarly, in traditional Mexican culture, it's considered polite, rather than powerless, to add *"por favor?"* ("if you please?") to the end of requests, such as when ordering food in a restaurant. By contrast, "powerful" declarative statements, such as "I'll have the fish," are likely to seem bossy, rude, and disrespectful.[40]

As you have probably gathered, the terms *powerful* and *powerless* can be misnomers. In the United States, women's traditional speech patterns have often been described as less powerful than men's. However, there is considerable diversity among men and women. In addition, people who do not seem powerful in traditional or obvious ways may have a great deal of influence, such as through their relationships with others. Since we tend to trust and cooperate most with people who build supportive, friendly relationships with us, sharing power *with* others

cultural idiom
<u>save face</u>: maintain dignity

can be more effective than exercising power *over* them.[41] The best communicators, both male and female, read situations well and combine elements of both "powerful" and "powerless" speech.

Table 4-1 provides examples of speech often categorized as powerless. As you review the list, keep two factors in mind: First, statements may serve several purposes. For example, a phrase may serve as both a hedge and a hesitation. Second, since meaning is personal and situational, these examples are effective in some situations but less so in others.

Affiliation Power isn't the only way language reflects the status of relationships. Language can also be a way of building and demonstrating solidarity with others. During their Year Without English, Scott Young and Vat Jaiswal found that shared language was key to establishing relationships. Young reflects on conversations and relationships he would not have had otherwise: "Discussing mandatory military service over soju and salted fish. Sharing tea and talking politics with a tattooed Buddhist from Tibet."[42]

An impressive body of research has demonstrated that communicators who want to show affiliation with one another adapt their speech in a variety of ways, including their choice of vocabulary, rate of talking, number and placement of pauses, and level of politeness.[43,44] On an individual level, close friends and romantic partners often develop special terms that serve as a way of signifying their relationship. Using the same vocabulary sets these people apart from others, reminding themselves and the rest of the world of their relationship. The same process works among members of larger groups, ranging from street gangs to military personnel. Communication researchers call this linguistic accommodation **convergence**.[45]

Communicators can experience convergence online as well as in face-to-face interactions. Members of online communities often develop a shared language and conversational style, and their affiliation with one another can be seen in

> **convergence** Accommodating one's speaking style to another person, usually a person who is desirable or has higher status.

TABLE 4-1

"Powerless" Language

TYPE OF USAGE	EXAMPLE
Hedges suggest that a speaker is tentative or unsure.	"I'm *kinda* disappointed . . ." "I think *maybe* we should . . ." "I *guess* I'd like to . . ."
Hesitations prolong the length of an utterance.	"*Uh*, can I have a minute of your time?" "*Well, let's see,* we could try this idea . . ." "I wish you would—*er*—try to be on time."
Intensifiers indicate powerful emotion.	"I *really* loved the movie . . ." "You did a fabulous job!!!"
Polite forms display deference or consideration.	"Excuse me, sir . . ." "I hope I'm not bothering you . . ."
Tag questions transform statements into questions.	"It's about time we got started, *isn't it?*" "We should give it another try, *don't you think?*"
Disclaimers display that a statement is atypical for the speaker or is a preliminary conclusion open to revision.	"I don't usually say this, but . . ." "I'm not really sure, but . . ."

Recall a time when you adapted your language to converge with the norms of a particular speech community. What changes did you make? How successful were you in converging?

divergence A linguistic strategy in which speakers emphasize differences between their communicative style and that of others to create distance.

linguistic intergroup bias The tendency to label people and behaviors in terms that reflect their in-group or out-group status.

increased uses of the pronoun *we*.[46] On a larger scale, instant message and email users create and use shortcuts that mark them as Internet-savvy. If you know what ILYSM, OOMF, and HMU mean, you're probably part of that group. (For the uninitiated, those acronyms mean "I like [or love] you so much," "one of my followers," and "hit me up" with a picture or message.)

When two or more people feel equally positive about one another, their linguistic convergence is likely to be mutual. But when communicators want or need the approval of others, they often adapt their speech to suit the others' style, trying to say the "right thing" or speak in a way that will help them fit in. For example, employees who seek advancement tend to speak more like their supervisors. In turn, supervisors tend to adopt the speech style of *their* bosses.[47]

The principle of speech accommodation works in reverse, too. Communicators who want to set themselves apart from others adopt the strategy of **divergence**, speaking in a way that emphasizes their difference from others. For example, members of an ethnic group, even though fluent in the dominant language, might use their own dialect as a way of showing solidarity with one another—a sort of "us against them" strategy. Divergence also operates in other settings. A physician or attorney, for example, who wants to establish credibility with his or her client might speak formally and use professional jargon to create a sense of distance. The implicit message here is "I'm different from (and more knowledgeable than) you."

Convergence and divergence aren't the only ways to express affiliation. **Linguistic intergroup bias** reflects whether we regard others as part of our in-group. A positive bias leads us to describe the personality traits of in-group members in favorable terms and those of out-group members negatively.[48] For example, if an in-group member gives money to someone in need, we are likely to describe her as a generous person. If an out-group member (someone with whom we don't identify) gives to the same person in need, we are likely to describe the behavior as a one-time act of giving away money. The same in-group preferences are revealed when we describe undesirable behaviors. If an in-group member behaves poorly, we are likely to describe the behavior using a concrete action verb, such as "John cheated in the game." In contrast, if the person we are describing is an out-group member, we are more likely to use general disposition adjectives such as "John is a cheater." These selective language choices are so subtle and subconscious that when asked, people being studied reported that there were no differences in their descriptions. We tend to believe we are less biased than we are, but our language reveals the truth about our preferences.

Attraction and Interest   Social customs discourage us from expressing like or dislike in many situations. Only a careless person would respond to the question, "What do you think of the cake I baked for you?" by saying, "It's terrible." Bashful or cautious suitors might not admit their attraction to a potential partner. Even when people are reluctant to speak candidly, the language they use can suggest their degree of interest and attraction toward a person, object, or idea. Morton Weiner and Albert Mehrabian outline a number of linguistic clues that reveal these attitudes.[49]

- **Demonstrative pronoun choice.** "*These* people want our help" (positive) versus "*Those* people want our help" (less positive).

- **Negation.** "It's *good*" (positive) versus "It's *not bad*" (less positive).

- **Sequential placement.** "Dick and Jane" (Dick is more important) versus "Jane and Dick" (Jane is more important). However, sequential placement isn't always significant. You may put "toilet bowl cleaner" at the top of your shopping list simply because of its location in the store.

Responsibility   In addition to suggesting liking and importance, language can reveal the speaker's willingness to accept responsibility for a message.

- **"It" versus "I" statements.** *"It's* not finished" (less responsible) versus *"I* didn't finish it" (more responsible).

- **"You" versus "I" statements.** "Sometimes *you* make me angry" (less responsible) versus "Sometimes *I* get angry when you do that" (more responsible). "I" statements are more likely to generate positive reactions from others as compared to accusatory ones.[50]

- **"But" statements.** "It's a good idea, *but* it won't work." "You're really terrific, *but* I think we ought to spend less time together." (*But* cancels everything before it.)

- **Questions versus statements.** "Do *you* think we ought to do that?" (less responsible) versus "I don't think we ought to do that" (more responsible).

Troublesome Language

Besides being a blessing that enables us to live together, language can be something of a curse. We all have known the frustration of being misunderstood, and most of us have been baffled by another person's overreaction to an innocent comment. In the following pages we will look at several kinds of troublesome language, with the goal of helping you communicate in a way that makes matters better instead of worse. A checklist on page 116 provides a handy reminder of tips for using language well.

The Language of Misunderstandings

The most obvious kind of language problems are semantic: We simply don't understand others completely or accurately. Most misunderstandings arise from some common problems that are easily remedied—after you recognize them.

Equivocal Language   Misunderstandings can occur when words are **equivocal**, meaning that they are open to more than one interpretation. For example, a nurse once told a patient that he "wouldn't be needing" the materials he requested from home. He interpreted the statement to mean he was near death, when the nurse meant he would be going home soon. Some equivocal misunderstandings can be embarrassing, as one woman recalls:

> **equivocal words** Words that have more than one dictionary definition.

> In the fourth grade the teacher asked the class what a period was. I raised my hand and shared everything I had learned about girls getting their period. But he was talking about the dot at the end of a sentence. Oops![51]

Some words are equivocal as a result of cultural or cocultural differences in language usage. While teaching in Ireland, an American friend of ours asked a male colleague if he would give her a ride to the pub. After a few chuckles, he said, "You mean a lift." Our friend was surprised to learn that the word "ride" has sexual connotations in Ireland.

Equivocal misunderstandings can have serious consequences. Equivocation at least partially explains why men may sometimes persist in attempts to become physically intimate when women have expressed unwillingness to do so.[52] Interviews and focus groups with college students revealed

No doubt this sign draws a laugh from many who see it.

Can you think of other words that have multiple meanings?

that rather than saying "no" outright to a man's sexual advances, women often use ambiguous phrases such as, "I'm confused about this," "I'm not sure that we're ready for this yet," and "Are you sure you want to do this?" Whereas women viewed indirect statements as meaning "no," men were more likely to interpret them as meaning "maybe." As the researchers put it, "male/female misunderstandings are not so much a matter of males hearing resistance messages as 'go,' but rather their not hearing them as 'stop.'" Under the law, "no" means precisely that, and anyone who argues otherwise can be in for serious legal problems.

Relative Words   Is the school you attend large or small? This depends on what you compare it to: Alongside a campus like UCLA, with an enrollment of more than 40,000 students, it probably looks small, but compared to a smaller institution, it might seem quite large. **Relative words** gain their meaning by comparison. Other examples include *fast* and *slow*, *smart* and *stupid*, and *short* and *long*.

Some relative words are so common that we mistakenly assume that they have a clear meaning. For instance, if a new acquaintance says, "I'll call you soon," when can you expect to hear from him or her? In the same vein, how much is "a few," and how much is "a lot"? An inquisitive blogger received dozens of replies after posting these questions online. Definitions of "a few" varied from one to a dozen. Most people said "a lot" is at least 20. One respondent suggested that people use "a horde" to describe items that are more numerous than "a lot" but less plentiful than a "swarm" and far less than "zounds."[53]

Using relative words without explaining them can lead to communication problems. Have you been disappointed to learn that classes you've heard were "easy" turned out to be hard, that trips you were told would be "short" were long, that "hilarious" movies were mediocre? The problem in each case came from failing to anchor the relative word to more precise measures for comparison.

Slang and Jargon   Most slang and jargon are related to specialized interests and activities. **Slang** is language used by a group of people whose members belong to a similar coculture or other group. For instance, cyclists who talk about "bonking" are referring to running out of energy.

Other slang consists of *regionalisms*—terms that only people from a relatively small geographic area use and understand. This sort of use illustrates how slang defines insiders and outsiders, creating a sense of identity and solidarity.[54] Residents of the largest U.S. state know that when a fellow Alaskan says, "I'm going outside," he or she is leaving the state. In the East End of London, cockney dialect uses rhyming words as substitutes for everyday expressions: "I haven't a scooby" means "I haven't a clue," derived from the detective canine Scooby Doo, whose name rhymes with clue. Examples such as these illustrate how slang can be used to identify insiders and outsiders. Insiders have no trouble making sense of them, but outsiders are likely to be mystified or to misunderstand.

Slang can also be age related. Your mother might be insulted if you said her dress was "sick," but other people you know may interpret this comment as a compliment.

Almost everyone uses some sort of **jargon**: the specialized vocabulary that functions as a kind of shorthand for people with common backgrounds and experience. Skateboarders and snowboarders have their own language to describe maneuvers: "ollie," "grind," and "shove it." Some jargon consists of *acronyms*—initials of terms that are combined to form a word. Stock traders refer to the NASDAQ (pronounced "naz-dak") securities index, and military people label failure to serve at one's post as being AWOL (absent without leave). The digital age has spawned its own vocabulary of jargon. For instance, "UGC" refers to user-generated content, and "tl;dr" means "too long; didn't read."

relative words   Words that gain their meaning by comparison.

slang   Language used by a group of people whose members belong to a similar coculture or other group.

jargon   Specialized vocabulary used as a kind of shorthand by people with common backgrounds and experience.

Jargon can be a valuable kind of shorthand for people who understand its use. The trauma team in a hospital emergency room can save time, and possibly lives, by speaking in shorthand, referring to "GSWs" (gunshot wounds), "chem 7" lab tests, and so on. But the same specialized vocabulary that works so well among insiders can bewilder and confuse a patient's family members, who don't understand the jargon. The same sort of misunderstandings can arise in less critical settings when insiders use their own language with people who don't share the same vocabulary.

Overly Abstract Language   Most objects, events, and ideas can be described with varying degrees of specificity. Consider the material you are reading. You could call it:

> A book
>
> A textbook
>
> A communication textbook
>
> *Understanding Human Communication*
>
> Chapter 4 of *Understanding Human Communication*
>
> Page 109 of Chapter 4 of *Understanding Human Communication*

In each case your description would be more and more specific. Semanticist S. I. Hayakawa created an **abstraction ladder** to describe this process.[55] This ladder consists of a number of descriptions of the same thing. Lower items focus specifically on the person, object, or event, whereas higher terms are generalizations that include the subject as a member of a larger class. To talk about "college," for example, is more abstract than to talk about a particular school. Likewise, referring to "women" is more abstract than referring to "most of the women I know."

> **abstraction ladder** A range of more-to-less-abstract terms describing an event or object.

UNDERSTANDING COMMUNICATION TECHNOLOGY

Twitter Lingo Incites Controversy

Actor Ralph Fiennes hit on a hot-button issue when he proclaimed that Twitter is dumbing down the English language.[56] He wasn't the first to suggest such a notion. People have weighed in with millions of online posts, some condemning and others applauding the truncated language some call *Twenglish* or *Twitterspeak*.

On one side of the debate are people who argue that English reduced to abbreviations and acronyms loses its luster and intelligence. One professor, for example, bemoans the influx of college admission essays with sentences that lack verbs and are speckled with shorthand symbols such as *4* and *U*.[57]

Proponents of Twitterspeak argue that language is an art form that is meant to be used creatively. People invent and repurpose words because it's fun, because speaking the same language is a sign of group membership, and because they want to talk about phenomena that don't yet have names.[58] *Forbes* writer Alex Knapp says he was a nonbeliever until he became an avid tweeter. Now he admires the artistry and discipline of expressing ideas in under two dozen words or so. "The 140 character restraint not only forces efficiency, but it also lends itself to some really, really fun wordplay," he says.[59] Many people seem to agree. Twitter's stock dropped in 2016 when rumors surfaced that it would soon allow longer tweets.[60]

Because Twenglish changes so rapidly, using it skillfully is both interesting and challenging.[61] New terms arise and others disappear. "Totes jellie" (totally jealous) is passé, and 775 (kiss me) and LPC (lesbian power couple) are in—at least they were 5 minutes ago.

Would Shakespeare be aghast if he knew how drastically language has changed? Maybe not. The Bard was fond of wordplay himself. Words such as *advertising*, *gossip*, and *swagger* (among about 1,700 others) didn't exist until he invented them.[62]

Will formal English morph to be more Twitter-like? As one blogger puts it, we'll just have to W8 N C.

Higher-level abstractions are a useful tool, because without them language would be too cumbersome to be useful. It's faster, easier, and more useful to talk about *Europe* than to list all of the countries on that continent. In the same way, using relatively abstract terms like *friendly* or *smart* can make it easier to describe people than listing their specific actions.

Abstract language—speech that refers to events or objects only vaguely—serves a second, less obvious function. At times it allows us to avoid confrontations by deliberately being unclear. Suppose, for example, your boss is enthusiastic about a new approach to doing business that you think is a terrible idea. Telling the truth might seem too risky, but lying—saying, "I think it's a great idea"—wouldn't feel right either. In situations like this an abstract answer can hint at your true belief without a direct confrontation: "I don't know. . . . It's sure unusual. . . . It *might* work." The same sort of abstract language can help you avoid embarrassing friends who ask for your opinion with questions like, "What do you think of my new haircut?" An abstract response like, "It's really different!" may be easier for you to deliver—and for your friend to receive—than the clear, brutal truth: "It's really ugly!" We will have more to say about this linguistic strategy of equivocation later in this chapter.

Although vagueness does have its uses, highly abstract language can cause several types of problems. The first is *stereotyping*. Consider claims such as, "All whites are bigots," "Men don't care about relationships," "The police are a bunch of pigs," or "Professors around here care more about their research than they do about students." Each of these claims ignores the very important fact that abstract descriptions are almost always too general; they say more than we really mean.

Besides creating stereotypical attitudes, abstract language can lead to the problem of *confusing others*. Imagine the lack of understanding that results from imprecise language in situations like this:

A: We never do anything that's fun anymore.

B: What do you mean?

A: We used to do lots of unusual things, but now it's the same old stuff, over and over.

B: But last week we went on that camping trip, and tomorrow we're going to that party where we'll meet all sorts of new people. Those are new things.

A: That's not what I mean. I'm talking about really unusual stuff.

B: *(becoming confused and a little impatient)* Like what? Taking hard drugs or going over Niagara Falls in a barrel?

A: Don't be stupid. All I'm saying is that we're in a rut. We should be living more exciting lives.

B: Well, I don't know what you want.

The best way to avoid this sort of overly abstract language is to use **behavioral descriptions** instead. (See Table 4-2.) Behavioral descriptions move down the abstraction ladder to identify the specific, observable phenomenon being discussed. A thorough description should answer three questions:

1. **Who is involved?** Are you speaking for just yourself or for others as well? Are you talking about a group of people ("the neighbors," "women") or specific individuals ("the people next door with the barking dog," "Lola and Lizzie")?

---

abstract language Language that lacks specificity or does not refer to observable behavior or other sensory data.

behavioral description An account that refers only to observable phenomena.

cultural idiom

in a rut: having fixed, unsatisfying habits

TABLE 4-2

Abstract Versus Behavioral Descriptions

		BEHAVIORAL DESCRIPTION			
	ABSTRACT DESCRIPTION	WHO IS INVOLVED	IN WHAT CIRCUMSTANCES	SPECIFIC BEHAVIORS	THE DIFFERENCE
PROBLEM	I talk too much . . .	. . . around people I find intimidating	. . . when I want them to like me	I talk (mostly about myself) instead of giving them a chance to speak or asking about their lives.	Behavioral description more clearly identifies behaviors to change.
GOAL	I want to be more constructive . . .	. . . with my roommate	. . . when we talk about household duties	. . . instead of finding fault with her ideas, and suggesting alternatives that might work.	Behavioral description clearly outlines how to act; abstract description doesn't.
APPRECIATION	"You've really been helpful lately."	Coworkers	When I've had to take time off work for personal reasons	"You graciously took on my workload in my absence."	Behavioral descriptions show gratitude more specifically.
REQUEST	"Behave!"	Romantic partner	When we're around my family	"Please don't tell jokes that involve sex."	Behavioral description specifies desired behavioral change.

2. **In what circumstances does the behavior occur?** Where does it occur: everywhere or in specific places (at parties, at work, in public)? When does it occur: when you're tired or when a certain subject comes up? The behavior you are describing probably doesn't occur all the time. In order to be understood, you need to pin down what circumstances set this situation apart from other ones.

3. **What behaviors are involved?** Though terms such as *more cooperative* and *helpful* might sound like concrete descriptions of behavior, they are usually too vague to do a clear job of explaining what's on your mind. Behaviors must be *observable*, ideally both to you and to others. For instance, moving down the abstraction ladder from the relatively vague term *helpful*, you might come to behaviors such as *does the dishes every other day*, *volunteers to help me with my studies*, or *fixes dinner once or twice a week without being asked*. It's easy to see that terms like these are easier for both you and others to understand than are more vague abstractions.

Behavioral descriptions can improve communication in a wide range of situations, as Table 4-2 illustrates. Research also supports the value of specific language. One study found that well-adjusted couples had just as many conflicts as poorly adjusted couples, but the way the well-adjusted couples handled their problems was significantly different. Instead of blaming each other, partners in well-adjusted couples expressed their complaints in behavioral terms.[63] For instance, instead of saying, "You're a slob," an enlightened partner might say, "I wish you wouldn't leave your dishes in the sink."

Disruptive Language

Not all linguistic problems come from misunderstandings. Sometimes people understand one another perfectly and still end up in conflict. Of course, not all disagreements can, or should, be avoided. But eliminating three bad linguistic habits from your communication repertoire can minimize the kind of clashes that don't need to happen, allowing you to save your energy for the unavoidable and important struggles.

Confusing Facts and Opinions   Do you ever state an opinion as if it's a fact? **Factual statements** are claims that can be verified as true or false. By contrast, **opinion statements** are based on the speaker's beliefs. Unlike matters of fact, they can never be proved or disproved. Consider a few examples of the difference between factual statements and opinion statements:

FACT	OPINION
It rains more in Seattle than in Portland.	The climate in Portland is better than in Seattle.
Kareem Abdul-Jabbar is the all-time leading scorer in the National Basketball Association.	Kareem is the greatest basketball player in the history of the game.
Per capita income in the United States is higher than in many other countries.	The United States is the best model of economic success in the world.

When factual statements and opinion statements are set side by side like this, the difference between them is clear. In everyday conversation, we often present our opinions as if they were facts, and in doing so we invite an unnecessary argument. For example:

"That was a dumb thing to say!"

"Spending that much on [ ] is a waste of money!"

"You can't get a <u>fair shake</u> in this country unless you're a white male."

Notice how much less antagonistic each statement would be if it were prefaced by a qualifier like "In my opinion . . ." or "It seems to me. . . ."

Confusing Facts and Inferences   Labeling your opinions can go a long way toward relational harmony, but developing this habit won't solve all linguistic problems. Difficulties also arise when we confuse factual statements with **inferential statements**—conclusions arrived at from an interpretation of evidence. Consider a few examples:

FACT	INFERENCE
He hit a lamppost while driving down the street.	He was daydreaming when he hit the lamppost.
You interrupted me before I finished what I was saying.	You don't care about what I have to say.
You haven't paid your share of the rent on time for the past 3 months.	You're trying to <u>weasel out</u> of your responsibilities.
I haven't gotten a raise in almost a year.	The boss is exploiting me.

---

factual statement A statement that can be verified as being true or false.

opinion statement A statement based on the speaker's beliefs.

inferential statement A conclusion arrived at from an interpretation of evidence.

(?) ASK YOURSELF

Why do you think we so often present our opinions in ways that make them sound like facts?

---

cultural idioms

fair shake: equitable treatment

weasel out: unfairly avoid doing something

Donald Trump's unapologetic use of disruptive language upended notions of civility in the 2016 U.S. presidential campaign. Supporters called it candor, while detractors accused him of being a hateful demagogue.

Do you ever use emotive language? If so, what are the consequences?

There's nothing wrong with making inferences as long as you identify them as such: "She stomped out and slammed the door. It looked to me as if she were furious." The danger comes when we confuse inferences with facts and make them sound like the absolute truth.

One way to avoid fact–inference confusion is to use the perception-checking skill described in Chapter 2 to test the accuracy of your inferences. Recall that a perception check has three parts: a description of the behavior being discussed, your interpretation of that behavior, and a request for verification. For instance, instead of saying, "Why are you laughing at me?" you could say, "When you laugh like that *[description of behavior]*, I get the idea you think something I did was stupid *[interpretation]*. Are you making fun of me *[question]*?"

Emotive Language   Are the words you use neutral or charged? **Emotive language** contains words that sound as if they're describing something when they are really announcing the speaker's attitude toward something. Do you like that old picture frame? If so, you would probably call it "an antique," but if you think it's ugly, you would likely describe it as "a piece of junk." Emotive words may sound like statements of fact but are always opinions.

Barbra Streisand pointed out how some people use emotive language to stigmatize behavior in women that they admire in men:

A man is commanding—a woman is demanding.

A man is forceful—a woman is pushy.

A man is uncompromising—a woman is a ball-breaker.

A man is a perfectionist—a woman's a pain in the ass.

He's assertive—she's aggressive.

He strategizes—she manipulates.

He shows leadership—she's controlling.

> **emotive language** Language that conveys an attitude rather than simply offering an objective description.

He's committed—she's obsessed.

He's persevering—she's relentless.

He sticks to his guns—she's stubborn.

If a man wants to get it right, he's looked up to and respected.

If a woman wants to get it right, she's difficult and impossible.[64]

When feelings run strong, it's easy to use emotive language instead of more objective speech. Recall a time when you used emotive language. What were the consequences?

Problems occur when people use emotive words without labeling them as such. You might, for instance, have a long and bitter argument with a friend about whether a third person was "assertive" or "obnoxious," when a more accurate and peaceable way to handle the issue would be to acknowledge that one of you approves of the behavior and the other doesn't.

Evasive Language

None of the troublesome language habits we have described so far is a deliberate strategy to mislead or antagonize others. Now, however, we'll consider euphemisms and equivocations, two types of language that speakers use by design to avoid communicating clearly. Although both of these have some very legitimate uses, they also can lead to frustration and confusion.

Euphemisms From the Greek meaning "to use words of good omen," a **euphemism** is a mild or indirect term or expression substituted for a more direct but potentially less pleasant one. We are using euphemisms when we say "restroom" instead of "toilet" or "full-figured" instead of "overweight." There certainly are cases in which the euphemistic pulling of linguistic punches can be face saving. It's probably more constructive to question a possible "statistical misrepresentation" than to call someone a liar, for example. Likewise, it may be less disquieting to some to refer to people as "older adults" rather than as "old."

Like many businesses, the airline industry often uses euphemisms.[65] For example, rather than saying "turbulence," pilots and flight attendants use the less frightening term "bumpy air." Likewise, they refer to thunderstorms as "rain showers," and fog as "mist" or "haze." And savvy flight personnel never use the words "your final destination."

Despite their occasional advantages, many euphemisms are best avoided. Some are pretentious and confusing, such as a middle school's labeling of hallways as "behavior transition corridors." Other euphemisms are downright deceptive, such as the U.S. Senate's labeling of a $23,200 pay raise as a "pay equalization concept."

Equivocation It's 8:15 P.M., and you are already a half-hour late for your dinner reservation at the fanciest restaurant in town. Your partner has finally finished dressing and confronts you with the question, "How do I look?" To tell the truth, you hate your partner's outfit. You don't want to lie, but on the other hand you don't want to be hurtful. Just as important, you don't want to lose your table by waiting around for your date to choose something else to wear. You think for a moment and then reply, "You look amazing. I've never seen an outfit like that before. Where did you get it?"

Your response in this situation was an **equivocation**—a deliberately vague statement that can be interpreted in more than one way. Earlier in this chapter we

---

euphemism A mild or indirect term or expression used in place of a more direct but less pleasant one.

equivocation A deliberately vague statement that can be interpreted in more than one way.

cultural idiom
beating around the bush: avoiding a direct expression

Your Use of Language

Circle the number in each continuum that best represents you.

1. I often pause to think carefully about the words   ↔   I like fast-paced conversations in which I don't have to weigh every word.
I use.

1	2		3	4	5

2. I am precise and factual when I tell a story.   ↔   I tend to exaggerate to get a laugh or make a point.

1	2		3	4	5

3. I am careful not to offend anyone when I speak.   ↔   I regularly use curse words and like to tell an off-color joke now and then.

1	2		3	4	5

4. When I am upset, I delay talking until I calm down.   ↔   I am up front about how I feel, even when I'm upset.

1	2		3	4	5

5. I am careful to use gender-neutral language.   ↔   I see no harm in words such as "policeman" and "salesman."

1	2		3	4	5

6. I pride myself on speaking proper English.   ↔   I enjoy using up-to-date slang and jargon.

1	2		3	4	5

7. I'm careful to ask others what they mean rather than   ↔   I'm pretty good at understanding what's on others' minds without
assuming I understand their ideas and motives.   needing to ask for a lot of clarification.

1	2		3	4	5

8. In sensitive situations I try to use diplomatic language.   ↔   I am honest, even when it's not what people want to hear.

1	2		3	4	5

INTERPRETING YOUR RESPONSES

- If you circled mostly 1s, 2s, and 3s, you tend to be a mindful communicator who regards language as a powerful and precise tool.
- Low scores on items 1, 2, and 4 suggest that you typically avoid emotive language and exaggerations.
- Low scores on items 3 and 5 indicate that you are cautious about using language that might be considered offensive.
- A low score on item 7 shows that you recognize the potential for misunderstandings, even in apparently clear statements.
- A low score on item 8 suggests that you are tactful, but your language may seem euphemistic or equivocal at times.
- If you circled mostly 3s, 4s, and 5s, you tend to use language in artistic and spontaneous ways, but you may sometimes go too far.
- High scores on items 1 and 8 suggest that you value open communication and a good debate, but keep in mind that your spontaneity may lead you to risky inferences.
- High scores on items 2 and 6 indicate that you value colorful language and are probably an engaging storyteller, but you may confuse people who aren't familiar with your style or vocabulary.
- If you scored high on item 4, you probably confront issues head-on, but be careful that your language doesn't become overly emotive, which can escalate an argument.
- High scores on items 3 and 5 indicate that you don't take words too seriously, but bear in mind that others may. Your statements may cause offense.

CHECKLIST ✓
Choose Your Words Carefully

It may be tempting sometimes to blurt out statements without thinking, but impetuous word choices can be confusing, hurtful, and downright wrong. Here are some tips to help clarify your language and avoid mix-ups.

☐ Use idioms, slang, jargon, and abbreviations with caution. When you use phrases such as "I'm all in" or "I'm in the weeds," make sure others understand what you mean.

☐ Be specific. "It will take me 30 minutes" is better than "It won't take long."

☐ Clarify whom you represent. If you say "we think . . . ," be clear about who "we" is. Otherwise, use "I" statements.

☐ Explain what you mean when you use abstract words such as *good, bad, helpful,* and *happy.*

☐ Focus on specific behaviors. ("It's important that you arrive by 9 A.M. every day.")

☐ Reference specific situations. ("It meant so much when you texted me before my big exam.")

☐ Don't present opinions as facts. (Instead of "Online classes are best" say "I prefer online classes.")

☐ Don't jump to conclusions. ("I was disappointed when you missed the deadline" is better than "You don't care about the job.")

☐ Avoid emotive language, such as "He runs like a girl," that implies broad or hurtful value judgments.

☐ Euphemisms, such as "He went to a better place," can be gentler than other words, but avoid them if they may also cause confusion or seem dishonest.

☐ Equivocations, such as "Your presentation was unique," can spare people's feelings, but avoid them when you want to convey specific information.

talked about how *unintentional* equivocation can lead to misunderstandings. But our discussion here focuses on *intentionally ambiguous speech* that is used to avoid lying on one hand and telling a painful truth on the other. Equivocations have several advantages.[66] They spare the receiver from the embarrassment that might come from a completely truthful answer, and it can be easier for the sender to equivocate than to suffer the discomfort of being honest.

As with euphemisms, high-level abstractions, and many other types of communication, it's impossible to say that equivocation is always helpful or harmful. As you learned in Chapter 1, competent communication behavior is situational. Your success in relating to others will depend on your ability to analyze yourself, the other person, and the situation when deciding whether to be equivocal or direct.

Gender and Language

"Why do women say they're fine when they're not?"

"Why do men want to talk about sex so much?"

"Why do women talk on the phone for hours?"

"Why won't he tell me how he feels?"

These are common questions on websites with titles such as "I Don't Understand Women"[67] and "8 Things We Don't Understand About Men."[68] Gendered perspectives on language fascinate everyday people as well as researchers. The tricky part can be differentiating between stereotypes and current realities. This section explores 10 true and false statements about how men and women use language.

1. **Metaphorically speaking, men are from Mars, and women are from Venus.** *False.* There's no denying that gender differences *can* be perplexing. But the sexes aren't actually "opposite" or nearly as different as the Mars–Venus metaphor suggests. As you read in Chapter 2, gender is a continuum that concludes a nearly infinite array of identities. Masculine and feminine people are more alike than they are different, and there is immense diversity within every identity group. As you read the following points, keep in mind that generalities don't describe every person, and terms such as *men* and *women,* although common in the literature, don't begin to describe the true diversity among people.

2. **Women talk more than men.** *False.* Men and women speak roughly the same number of words per day, but women generally speak most freely when talking to other women, whereas men usually talk more in professional settings.[69]

3. **Men and women talk about different things.** *True . . . sometimes.* This is most true when women talk to women and men talk to men. Among themselves, women tend to spend more time discussing relational issues such as family, friends, and emotions. Male friends, on the other hand, are more likely to discuss recreational topics such as sports, technology

use, and nightlife.[70] That is not to say that people of different genders always talk about different things. Nearly everyone reports talking frequently about work, movies, and television.[71]

4. **Where romance is concerned, it's complicated.** *Undeniable.* Although social roles have changed dramatically over the decades, powerful vestiges of traditional gender roles persist when it comes to romance. For example, when researchers recently studied an online chat site for teens, they found that young men were more likely than young women to post flirtatious comments and bold sexual invitations, and females were more likely to post friendly comments and to ask about and share their feelings.[72]

"Be honest with me Roger. By 'mid-course correction' you mean divorce, don't you."

Source: Leo Cullum The New Yorker Collection/ The Cartoon Bank

Research about first dates and speed dating shows that men are more inclined than women to bring up the topic of sex, and women are more likely than men to think that a date is successful if conversation flows smoothly and the tone is friendly.[73,74] Take heart: These differences between gender roles tend to moderate as dating partners get to know each other better.

5. **Men and women communicate for different reasons.** *Often true.* People of all genders share a desire to build and maintain social relationships through communication. Their ways of accomplishing this are often different, however. In general, men are more likely than women to emphasize making conversation fun. Their discussions involve a greater amount of joking and good-natured teasing. By contrast, women's conversations focus more frequently on feelings, relationships, and personal problems.[75] Consequently, women may wonder why men "aren't taking them seriously," and men may wonder why women are so "emotionally intense."

6. **Women are emotionally expressive.** *Often true.* Because women frequently use conversation to pursue social needs, they are often said to have an *affective* (emotionally based) style. Female speech typically contains statements showing support for the other person, demonstrations of equality, and efforts to keep the conversation going. Because of these goals, traditionally female speech often contains statements of sympathy and empathy: "I've felt just like that myself," "The same thing happened to me!" Instant messages written by women tend to be more expressive than ones composed by men.[76] They are more likely to contain laughter ("hehe"), emoticons (smiley faces), emphasis (italics, boldface, repeated letters), and adjectives. Women are also inclined to ask lots of questions that invite the other person to share information: "How did you feel about that?" "What did you do next?"

However, women—like men—are typically hesitant to share their feelings when they think others will consider them weak or moody.[77] This is particularly true in professional settings, where women often perceive that they will be judged more harshly than men if they express emotions. This is one rationale behind saying "Nothing's wrong" when something actually is.

7. **Men don't show their feelings.** *Partly true, partly false.* In contrast to women, men have traditionally been socialized to use language *instrumentally*—that is, to accomplish tasks. This is one reason that, when someone shares a problem,

Research shows that male and female speech is different in some respects and similar in others.

How would you characterize your use of language on a stereotypically male–female spectrum?

instead of empathizing, men are prone to offer advice: "That's nothing to worry about . . ." or "Here's what you need to do . . . ."

To compound the issue, men in many cultures are discouraged from crying or showing feelings of sadness. Consequently, they often cope with difficult situations by using humor or distractions to avoid showing grief or sorrow, especially in public.[78]

That's not to say that men never express emotions. They are often demonstrative when they consider it socially acceptable, such as at a sporting event.[79] In their private lives, men typically do express emotion, but often by their *actions* (such as physical affection and favors) rather than in words.[80]

8. **Women's speech is typically powerless, and men's is more powerful.** *False.* As you may remember from the earlier discussion, power is a nebulous term. While it is true that traditionally feminine speech typically includes more hedges and hesitations than traditionally male speech, those less assertive speech patterns can be powerful means of building relationships and collaborating with others.

Moreover, in certain contexts, gender's influence on language is nearly undetectable. For example, researchers asked men and women to describe a health-related episode in their lives.[81] They then created written transcripts of their responses. Detailed analysis revealed that the women used slightly more intense adverbs and personal pronouns than men did. However, everyday people who read the transcripts were largely unable to identify the speaker's gender. The same researchers then asked men, women, and transgender women to describe a painting. Studied closely, the transgender women's word choices were slightly more similar to men's than to women's, but again, most people could not distinguish between them on the basis of word choice.

9. **Men and women are hardwired to communicate differently.** *Partly true, but less than many people assume.* Research shows that men with high testosterone levels are more competitive than those with lower levels of the hormone, and they respond with more emotional language than other men when faced with setbacks.[82,83] Estrogen is associated with heightened emotional experiences and expression of emotion.[84] However, hormones do not correlate perfectly with biological sex. While men typically have more testosterone and women estrogen, these hormones are present in both men and women, and their presence varies by individual. Moreover, hormones' influence is less intense than most people think. For example, only 3% to 8% of women experience hormonal mood swings beyond the range of everyday emotions.[85]

10. **Men and women are socialized to communicate differently.** *True.* One of the main reasons men and women communicate differently is that society expects them to. When researchers asked people to pretend they were either male or female while describing a painting, the differences were pronounced, reflecting the way people *think* men and women communicate, even more than the way they actually do.[86]

Because of preconceived notions, men and women encounter different conversational climates. People in one study interrupted female speakers more than male speakers, even though all the speakers were trained to say much the same thing.[87] Similar effects are notable in the written word. Health news aimed at female audiences typically includes more hedges, hesitations, and tag questions than articles in men's magazines.[88]

By now it's probably clear that neither characteristically male or female styles of speech meet all communication needs. You can improve your linguistic competence by switching and combining styles. If you reflexively take an instrumental approach that focuses on the content of others' remarks, consider paying more attention to the unstated relational messages behind their words. If you generally focus on the feelings part of a message, consider being more task oriented. If your first instinct is to be supportive, consider the value of offering advice; and if advice is your reflexive way of responding, think about whether offering support and understanding might sometimes be more helpful. Research confirms what common sense suggests: Balancing a task-oriented approach with a relationship-oriented approach is usually the most effective strategy. Choosing the approach that is right for the other communicator and the situation can create satisfaction far greater than that which comes from using a single stereotypical style.

This chapter began with Scott Young and Vat Jaiswal's plan to travel the world, adapting to the language of each culture they visited. In the end, they were not able to forsake English entirely. The two friends say they consider the adventure a success, however, in that it made them more appreciative of the diversity that surrounds them every day and more skillful at adapting to it. "A lot of people get souvenirs or they'll get a tattoo or something," Young says, "but having a skill, having this connection to a culture afterwards, I think is so much more rewarding."[89]

MAKING THE GRADE

For more resources to help you understand and apply the information in this chapter, visit the *Understanding Human Communication* website at www.oup.com/us/adleruhc

OBJECTIVE 4.1   Describe the symbolic, person-centered, rule-governed nature of language.

- A language is a collection of symbols governed by a variety of rules and used to convey messages between people.

- Because of its symbolic nature, language is not a precise tool. Meanings rest in people, not in words themselves.

- In order for effective communication to occur, it is necessary to negotiate meanings for ambiguous statements.

  > Test your understanding by explaining the symbolic, rule-governed nature of language to someone unfamiliar with the concepts in this chapter.

  > How has the principle that meanings are in people, not words, created challenges in your life?

> How can you do a better job expressing yourself and understanding the other person in an important relationship?

OBJECTIVE 4.2   Explain how language both shapes and reflects attitudes.

- Language not only describes people, ideas, processes, and events; it also shapes people's perceptions of them in areas such as status, credibility, and attitudes about gender and ethnicity.

- Along with influencing people's attitudes, language reflects them.

- The words we use and our manner of speech reflect power, responsibility, affiliation, attraction, and interest.

  > How might the language a person uses influence his or her social status and credibility? Give examples of statements often categorized as "powerless" and statements considered to be "powerful."

> Record or transcribe from memory a recent conversation in which you were involved. Identify statements that illustrate concepts described on pages 100–107.

> How might you use language more mindfully to impress a prospective employer during a job interview? To show support for a friend?

OBJECTIVE 4.3 Distinguish the main types of troublesome language and their effects, as described on pages 107–115.

- Many types of language have the potential to create misunderstandings.

- Other types of language can result in unnecessary conflicts. In other cases, speech and writing can be evasive, avoiding expression of unwelcome messages.

   > Compare and contrast equivocal language, relative words, and abstract language.

   > When do you consider it acceptable to use euphemisms and equivocal language? When do you think such language is confusing or unfair?

   > Identify examples of troublesome language in a movie or television show. How could you become more mindful of similar patterns in your own communication?

OBJECTIVE 4.4 Explain how gender and nongender variables, described on pages 116–119, affect communication.

- The relationship between gender and language is complex.

- Although there are differences in the ways men and women speak, not all differences in language use can be accounted for by the speaker's gender.

- Occupation, social philosophy, and orientation toward problem solving also influence the use of language, and psychological sex role can be more of an influence than biological sex.

   > In your own words, describe the similarities and differences between characteristically male and female language as outlined on pages 116–119.

   > How do these similarities and differences compare to your use of language? To the linguistic style of people you know well?

   > If the language you use is closest to the "masculine" styles described, when and how might you use "feminine" strategies to expand your repertoire? If your style is more "feminine," when and how might you use "masculine" styles effectively?

KEY TERMS

abstraction ladder p. 109

abstract language p. 110

behavioral description p. 110

connotative meanings p. 97

convergence p. 105

denotative meanings p. 97

dialect p. 97

divergence p. 106

emotive language p. 113

equivocal words p. 107

equivocation p. 114

euphemism p. 114

factual statement p. 112

inferential statement p. 112

jargon p. 108

language p. 97

linguistic intergroup bias p. 106

linguistic relativism p. 102

opinion statement p. 112

phonological rules p. 98

pragmatic rules p. 99

reappropriation p. 100

relative words p. 108

semantic rules p. 99

slang p. 108

syntactic rules p. 98

ACTIVITIES

1. **Powerful Speech and Polite Speech** Increase your ability to achieve an optimal balance between powerful speech and polite speech by rehearsing one of the following scenarios:

   a. Describing your qualifications to a potential employer for a job that interests you

   b. Requesting an extension on a deadline from one of your professors

   c. Explaining to a merchant why you want a cash refund on an unsatisfactory piece of merchandise when the store's policy is to issue credit vouchers

   d. Asking your boss for three days off so you can attend a friend's out-of-town wedding

   e. Approaching your neighbors whose dog barks while they are away from home

Your statement should gain its power by avoiding the types of powerless language listed in Table 4-1 on page 105. You should not become abusive or threatening, and your statement should be completely honest.

2. **Gender and Language** Note differences in the language use of three men and three women you know. Include yourself in the analysis. Your analysis will be most accurate if you record the speech of each person you analyze. Consider the following categories:

> Conversational content

> Conversational style

> Reasons for communicating

> Use of powerful/powerless speech

Based on your observations, answer the following questions:

> How much does gender influence speech?

> What role do other variables play? Consider occupational or social status, cultural background, social philosophy, competitive-cooperative orientation, and other factors in your analysis.

Listening

LEARNING OBJECTIVES

5.1

Describe the benefits that follow from being an effective listener.

5.2

Outline the most common misconceptions about listening, and assess how successfully you avoid them.

5.3

Identify and manage the faulty habits and challenges that make effective listening difficult.

5.4

Know when and how to listen to accomplish a task, enhance relationships, analyze a message, and critically evaluate another's remarks.

5.5

Use the skills introduced on pages 142–149 to offer social support to a friend, partner, or family member.

Erica Nicole

proposes that one key to success is to listen well. As we begin to explore this topic, ask yourself these questions:

?

How do you rate yourself as a listener? How do you think others rate you?

?

What makes it difficult for you to listen effectively?

?

How could you use listening skills to improve your career success?

IF THE MILLENNIALS HAVE A MUSE, she might just be Erica Nicole. The social media aficionado launched her first company at age 25 and quickly became a straight-talking guru for up-and-coming professionals who don't play by the old rules.

Nicole is founder and CEO of a social media and marketing company and the online magazine *YFS*, which stands for "young, fabulous, and self-employed." The magazine provides what one observer calls "fresh, provocative and insanely addictive" advice for young professionals.[1] A common element of that advice is the importance of being a good listener.

In this chapter, we expose some common misconceptions about listening and show what really happens when listening takes place. We'll discuss some poor listening habits, explain why they occur, and suggest better alternatives.

"When you're listening, it shows," declares Nicole.[2] Good listeners have two main advantages, she says: Because they are observant, they can learn how successful people communicate; and because they are tuned in, they are often the first to identify emerging needs and opportunities in the marketplace.

Although Nicole loves social media, she's quick to point out that there is no substitute for face-to-face listening. As Nicole puts it, successful people don't just hear what others say; they "really, really listen."[3] That is to say, they refrain from interrupting, and they pay attention to people's tone of voice, body language, and other cues as well as to their words. In an age dominated by social media, listening is as important as ever.

Really, really listening involves a level of discipline and skill few people stop to consider and even fewer master. Yet the payoffs are enormous, as you will see. Masterful listening can help you make wise decisions, make a positive impression on others, and enrich your relationships.

The Value of Listening

Imagine you are at a career fair where you have access to talent scouts and employers who can help you land the job of your dreams. You might approach these people with a mental list of everything you want to say—a description of your talents, experience, goals, and so on. Or you might take a different approach and listen.

If you chose the listen-more-than-you talk option, give yourself a pat on the back. "Listening is more important than speaking," advises a spokesperson for one of the largest career networks in the United States. In fact she ranks the importance of listening among the "top 5 things recruiters wish you knew."[4] (The other four involve dressing appropriately, handling rejection well, being proactive, and being polite and considerate.)

The need for good listening skills cannot be overemphasized. In his best-selling book, *The 7 Habits of Highly Effective People*, Stephen Covey observes that most people only pretend to listen while they actually rehearse what they want to say themselves.[5] Rare (and highly effective) is the person who listens with the sincere desire to *understand*, observes Covey. An impressive body of evidence backs up this claim, as you will see in the following list of reasons to become a better listener:

Whether in personal or professional settings, listening is a powerful communication tool.

How would others rate you as a listener? What could you do to improve your rating?

1. **People with good listening skills are more likely than others to be hired and promoted.**[6] As you have already read, experts have found that job candidates who listen well have an edge over those who talk most of the time. Listening skills are also important once you get the job. Because good listeners are typically judged to be appealing and trustworthy,[7] they are especially popular with employers and with customers and clients.[8,9]

2. **Listening is a leadership skill.** Leaders who are good listeners typically have more influence and stronger relationships with team members than less attentive leaders do.[10] In fact, leaders' listening skills are even more influential than their talking skills.[11]

3. **Good listeners are not easily fooled.** People who listen carefully and weigh the merits of what they hear are more likely than others to spot what some researchers call "pseudo-profound bullshit"—statements that sound smart but are actually misleading or nonsensical.[12] Mindful listening (a topic we'll discuss more in a moment) is your best defense.

4. **Asking for and listening to advice makes you look good.** "Many people are reluctant to seek advice for fear of appearing incompetent," observes a research team who studied the issue.[13] What they found was the opposite— that people think more *highly* of people who ask them for guidance about challenging issues than those who muddle through on their own. Of course, that's just the first step. Making the most of that advice requires good listening skills and follow-through.

cultural idiom

tuned in: focused, paying attention

5. **Listening makes you a better friend and romantic partner.** While you are getting dressed for an evening out, make sure to clean out your ears, metaphorically speaking. Friends and partners who listen well are considered to be more supportive than those who don't.[14] That probably doesn't surprise you, but this may: Listening well on a date can significantly increase your attractiveness rating.[15,16] The caveat is that you can't *pretend* to listen. Effective listeners are sincerely interested and engaged.

Despite the importance of listening, experience shows that much of the listening we (and others) do is not very effective. We misunderstand others and are misunderstood in return. We become bored and feign attention while our minds wander. We engage in a battle of interruptions in which each person fights to speak without hearing the other's ideas. Some of this poor listening is inevitable, perhaps even justified. But in other cases we can be better receivers by learning a few basic listening skills.

Misconceptions About Listening

In spite of its importance, listening is misunderstood by most people. Because these misunderstandings so greatly affect our communication, let's take a look at three common misconceptions that many communicators hold.

Myth 1: Listening and Hearing Are the Same Thing

hearing The process wherein sound waves strike the eardrum and cause vibrations that are transmitted to the brain.

listening The process wherein the brain reconstructs electrochemical impulses generated by hearing into representations of the original sound and gives them meaning.

Hearing is the process in which sound waves strike the eardrum and cause vibrations that are transmitted to the brain. By contrast, **listening** occurs only when the brain reconstructs these electrochemical impulses into a representation of the original sound and then gives them meaning.

We begin hearing sounds around us even before we're born.[17] Barring illness, injury, or earplugs, hearing can't be stopped. As one neuroscientist put it,

> hearing . . . is easy. You and every other vertebrate that hasn't suffered some genetic, developmental or environmental accident have been doing it for hundreds of millions of years. It's your life line, your alarm system, your way to escape danger and pass on your genes.[18]

Although hearing is automatic, listening is another matter. Many times we hear but do not listen. Sometimes we deliberately tune out unwanted signals—everything from a neighbor's lawn mower or the roar of nearby traffic to a friend's boring remarks or a boss's unwanted criticism.

Source: © 2003 Zits Partnership Distributed by King Features Syndicate, Inc.

A closer look at listening, at least the successful variety, shows that it consists of several stages. After hearing, the next stage is **attending**—the act of paying attention to a signal. An individual's needs, wants, desires, and interests determine what is attended to, or selected, to use the term introduced in Chapter 2.

The next step in listening is **understanding**—the process of making sense of a message. Communication researchers use the term **listening fidelity** to describe the degree of congruence between what a listener understands and what the message sender was attempting to communicate.[19] Chapter 4 discussed many of the ingredients that combine to make understanding possible: a grasp of the syntax of the language being spoken, semantic decoding, and knowledge of the pragmatic rules that help you figure out a speaker's meaning from the context. In addition to these steps, understanding often depends on the ability and effort people put into cognitively organizing the information they hear.[20] An article in Erica Nicole's *YFS* magazine advises managers to "understand you don't have all the answers. The best entrepreneurs know when it is time to shut up and listen to the people around them."[21]

Responding to a message consists of giving observable feedback to the speaker. Offering feedback serves two important functions: It helps you clarify your understanding of a speaker's message, and it shows that you care about what that speaker is saying.

Listeners don't always respond visibly to a speaker—but research suggests that they should. When people are asked to evaluate the listening skills of people around them, the number-one trait they consider is whether the listener offers feedback.[22] Feedback includes eye contact, appropriate facial expressions, asking questions and exchanging relevant ideas, sitting up straight, and facing the speaker. Conversely, it's easy to see how discouraging it is when audience members yawn, slump, or make bored expressions. Adding responsiveness to the listening model demonstrates the fact, discussed in Chapter 1, that communication is transactional in nature. Listening isn't just a passive activity. As listeners, we are active participants in a communication transaction. While we receive messages, we also send them. For example, although some people insist they are listening even when they seem distracted and unresponsive, their demeanor probably puts a damper on the conversation.

The final step in the listening process is **remembering**.[23] It has long fascinated scientists that people remember every detail of some messages but very little of others. For example, you may remember many specifics about gossip you heard, but you may forget what your roommate asked you to buy at the store today. By some accounts, on average, people forget about half of what they hear *immediately after* hearing it, suggesting that they did not truly listen to and store the information.[24]

Given the amount of information we process every day—from instructors, friends, social media, TV, and other sources—it's no wonder that the **residual message** (what we remember) is a small fraction of what we hear. However, with effort, we can increase our ability to remember what is important to us. We'll explore ways of doing that later in the chapter.

Myth 2: Listening Is a Natural Process

Another common myth is that listening is like breathing—a natural activity that people usually do well. The truth is that listening is a skill much like speaking: Everybody does it, though few people do it well.

attending The process of focusing on certain stimuli from the environment.

understanding The act of interpreting a message by following syntactic, semantic, and pragmatic rules.

listening fidelity The degree of congruence between what a listener understands and what the message sender was attempting to communicate.

responding Providing observable feedback to another person's behavior or speech.

remembering The act of recalling previously introduced information. Recall drops off in two phases: short term and long term.

residual message The part of a message a receiver can recall after short- and long-term memory loss.

TING

EAR   MIND
    EYE

HEART

The Chinese word *ting* refers to deep, mindful listening. In its written form, the word combines the symbols for ears, eyes, heart, and mind.

Have you ever felt that someone listened to you with an open mind and heart, attentive to your words and your feelings? How did you respond?

CHECKLIST ✔
Tips for More Mindful Listening

Mindful listening takes effort, but it pays off in terms of self-awareness and stronger connections with others. Here are some tips for becoming a more mindful listener.[30,31,32]

☐ Commit to being fully present.

☐ Minimize distractions, including extraneous thoughts and worries.

☐ Listen for underlying messages as well as surface meanings.

☐ Pay attention to cues about how the speaker feels.

☐ Mentally acknowledge your own feelings. (*"I'm feeling defensive. I'll set that aside for now and try to understand more fully what this person is sharing with me."*)

☐ Acknowledge the other person's feelings. (*"Are you feeling discouraged?"*)

☐ Ask questions and check your understanding. (*"I heard you say that you're confused. Is that a good reflection of how you feel? What are the pros and cons of the situation as you see them?"*)

☐ Be patient. Don't interrupt.

☐ Become comfortable with silence.

☐ Don't rush the speaker. The goal is to understand.

mindful listening Being fully present with people—paying close attention to their gestures, manner, and silences, as well as to what they say.

In a classic study, 144 managers were asked to rate their listening skills. Astonishingly, not one of the managers described himself or herself as a "poor" or "very poor" listener. In fact, 94% rated themselves as "good" or "very good."[25] The favorable self-ratings contrasted sharply with the perceptions of the managers' subordinates, many of whom said their boss's listening skills were weak.

Sometimes it's okay to be mindless about what we hear. Paying attention to every song on the radio or commercial on TV would distract us from more important matters. The problem is being lazy about listening to things that really matter. For example, a college student hurt by his girlfriend's poor listening skills wrote in an online forum, "I have opened up to her about really, really personal things and then two weeks later or within the week . . . she's like, 'oh, you never mentioned it to me.' I just find this really really rude and insulting."[26]

Mindful listening involves being fully present with others—paying close attention to their gestures, manner, and silences, as well as to what they say.[27] It requires a commitment to understanding the other's perspectives without being judgmental or defensive. Consistent with the idea of mindful listening, the Chinese concept of *ting* describes listening with open ears and eyes as well as an open mind and open heart.[28] This type of listening can be difficult, especially when we are busy or when we feel vulnerable ourselves, yet the investment is worthwhile. For more tips on mindful listening, see the checklist at left.

Myth 3: All Listeners Receive the Same Message

When two or more people are listening to a speaker, we tend to assume they all are hearing and understanding the same message. In fact, such uniform comprehension isn't the case. Recall the discussion of perception in Chapter 3, in which we pointed out the many factors that cause each of us to perceive an event differently. Physiological factors, social roles, cultural background, personal interests, and needs all shape and distort the raw data we hear into uniquely different messages.

As we have discussed, some poor listening is inevitable. The good news is that listening can be improved through instruction and training.[29] Despite this fact, the amount of time devoted to teaching listening is far less than that devoted to other types of communication. Table 5-1 reflects this upside-down arrangement.

TABLE 5-1

Comparison of Communication Activities

	LISTENING	SPEAKING	READING	WRITING
LEARNED	First	Second	Third	Fourth
USED	Most	Next to most	Next to least	Least
TAUGHT	Least	Next to least	Next to most	Most

Overcoming Challenges to Effective Listening

Listening well isn't easy. We've all had challenges like these:

- You are binge-watching the long-awaited new season of your favorite TV drama. A neighbor drops by to warn you about some car break-ins nearby. You know that the issue is important, but find yourself becoming irritated at the interruption.

- Over coffee, a friend complains about having a bad day. You want to be supportive, but you are preoccupied with problems of your own, and you need to get back to work soon to meet a deadline.

- Your boss critiques your work. You think the objections are unfair, and you feel the need to defend yourself.

- A family member tells the same story you've heard dozens of times before. You feel obliged to act interested, but your mind is far away.

Situations like these reflect the kinds of challenges we face every day, and your imagined responses probably illustrate some less-than-perfect listening behaviors.

Reasons for Poor Listening

What causes people to listen poorly? There are several reasons, some of which can be avoided or overcome and others that are sad but inescapable facts of life.

Message Overload   The amount of information we hear every day makes careful listening to everything impossible. Along with the deluge of face-to-face messages, we are bombarded by phone calls, emails, tweets, texts, and instant messages. Besides those personal messages, we're awash in programming from the mass media. This deluge of communication has made the challenge of attending tougher than at any time in human history.[33] Background noise typically reduces our ability to listen and to concentrate on cognitive tasks.[34] The "@Work" box on page 130 highlights the dangers of multitasking and argues that managing information overload can lead to better results.

Rapid Thought   Listening carefully is also difficult for a physiological reason. Although we are capable of understanding speech at rates up to 600 words per minute, the average person speaks between 100 and 140 words per minute.[35] Thus, we have a great deal of mental "spare time" to spend while someone is talking. And the temptation is to use this time in ways that don't relate to the speaker's ideas, such as thinking about personal interests, daydreaming, planning a rebuttal, and so on. The trick is to use this spare time to understand the speaker's ideas better rather than to let your attention wander. Try to rephrase the speaker's ideas in your own words. Ask yourself how the ideas might be useful to you. Consider other angles that the speaker might not have mentioned.

Psychological Noise   Another reason why we don't always listen carefully is that we're often wrapped up in personal concerns that are of more immediate importance to us than the messages others are sending. It's hard to pay attention to someone else when you're anticipating an upcoming test or thinking about the wonderful time you had last night with good friends. Yet we still feel we have to "listen" politely to others, and so we continue with our charade. It usually takes a conscious effort to set aside your personal concerns if you expect to give others' messages the attention they deserve.

ASK YOURSELF

Which challenges to effective listening affect you the most?

Multitasking Can Make You Stupid

Multitasking may be a fact of life on the job, but research suggests that dividing your attention has its costs. In one widely reported study, volunteers tried to carry out various problem-solving tasks while being deluged with phone calls and emails.[36] Even though experimenters told the subjects to ignore these distractions, the average performance drop was equivalent to a 10-point decline in IQ. In other words, trying to work on a task while receiving messages about another matter can make you stupid.

You might expect that greater exposure to multiple messages would improve multitasking performance, but just the opposite seems to be the case. Heavy media multitaskers perform worse on task switching than light media multitaskers.[37] Although chronic multitaskers believe they are competent at processing information, in fact they're worse than those who focus more on a single medium.[38]

You may not be able to escape multiple demands at work, but don't hold any illusions about the cost of information overload. When the matter at hand is truly important, the most effective approach may be to turn off the phone, close down the email program or browser, and devote your attention to the single task before you.

Figure 5-1 illustrates four ways in which preoccupied listeners lose focus when distracted by psychological noise. Everyone's mind wanders at one time or another, but excessive preoccupation is both a reason for and a sign of poor listening.

Physical Noise   The world in which we live often presents distractions that make it hard to pay attention to others. The sound of traffic, music, others' speech, and the like interfere with our ability to hear well. Also, fatigue or other forms of discomfort can distract us from paying attention to a speaker's remarks. Consider, for example, how the efficiency of your listening decreases when you are seated in a crowded, hot, stuffy room that is surrounded by traffic and other noises. In such circumstances even the best intentions aren't enough to ensure clear understanding. You can often listen better by insulating yourself from outside distractions. This may involve removing the sources of noise: turning off the television, shutting the book you were reading, closing the window, and so on. In some cases, you and the speaker may need to find a more hospitable place to speak in order to make listening work.

Hearing Problems   Sometimes a person's listening ability suffers from a hearing problem—the most obvious sort of physiological noise, as defined in Chapter 1.

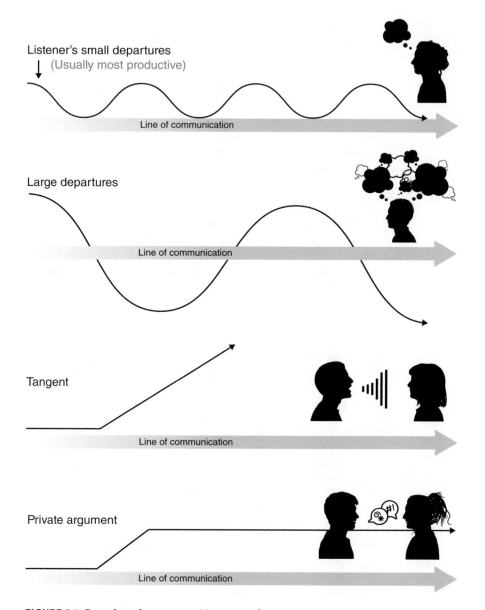

Listener's small departures
(Usually most productive)

Line of communication

Large departures

Line of communication

Tangent

Line of communication

Private argument

Line of communication

FIGURE 5-1 Four thought patterns. It's common for attention to stray when listening. Serious problems arise when we begin thinking about unrelated topics or constructing rebuttals.

One survey explored the feelings of adults who have spouses with hearing loss. Nearly two thirds of the respondents said they feel annoyed when their partner can't hear them clearly. Almost a quarter said that beyond just being annoyed, they felt ignored, hurt, or sad. Many of the respondents believe their spouses are in denial about their condition, which makes the problem even more frustrating.[39]

Older people aren't the only ones affected. The number of young people with hearing loss is on the rise, in part because of earbuds and similar technology that make it possible to blast our eardrums with dangerously loud noise.[40] Medical experts have found that 1 in 8 children and teens and almost 1 in 5 adults have suffered permanent damage to their hearing from excessive exposure to noise.[41]

ASK YOURSELF

In one of your most important relationships, which bad listening habits do you find most annoying? What would the other person say were **your** most annoying listening faults?

CHECKLIST ✓
Techniques for Listening Nondefensively

It's natural to feel uncomfortable when the boss wants to talk about the deadline you missed or when your roommate is upset because you left a mess. However, nondefensive listeners resist the impulse to fight back. Here are some tips for listening nondefensively, even when the heat is on.[45,46]

☐ Take a deep breath and remind yourself that you are a likable person.

☐ Avoid berating yourself with negative self-talk such as *"That was such a dumb mistake"* and *"This person will never forgive me."*

☐ Let go of the idea that you (or anyone) can be perfect.

☐ Listen with an open mind.

☐ Thank the speaker for sharing his or her concerns with you.

☐ Accept that the other person's feelings are real, even if his or her interpretation is different from yours.

☐ Ask questions. (*"Did you feel that I was taking you for granted when you saw my dishes in the sink?"*)

☐ Acknowledge the other person's feelings. (*"I can understand why you were disappointed."*)

☐ Avoid displaying nonverbal cues—such as eye rolls and heavy sighs—that seem defensive or dismissive.

☐ Use "I" language to express your feelings. (*"I wanted to edit the report one more time. I didn't realize the delay would be so costly."*)

☐ If warranted, apologize for your behavior.

☐ Learn from the encounter and move on. And at the very least, congratulate yourself for handling a difficult situation with sincerity and openness.

Cultural Differences   The behaviors that define a good listener vary by culture. Americans are most impressed by listeners who ask questions and make supportive statements.[42] By contrast, Iranians tend to judge people's listening skills based on more subtle indicators such as their posture and eye contact. This is probably because the Iranian culture relies more on context.[43] (As you may remember from Chapter 3, members of high-context cultures are particularly attentive and sensitive to nonverbal cues.) Germans are most likely to think someone a good listener if he or she shows continuous attention to the speaker.[44] One lesson is that, whereas people in some cultures may overlook a quick glance at a cell phone or TV screen, others may interpret that behavior as rudely inattentive.

Media Influences   A final challenge to serious listening is the influence of contemporary mass media, especially television and radio. Programming often consists of short segments: news items, commercials, music videos, and so on. This discourages the kind of focused attention that is necessary for careful listening, especially to complicated ideas and feelings.

Faulty Listening Habits

Compounding these barriers to listening, we may have faulty listening habits. Although we can't listen effectively all the time, most of us possess one or more habits that keep us from understanding truly important messages.

1. **We pretend to listen. Pseudolistening** is an imitation of the real thing. When people pseudolisten, they give the appearance of being attentive: They look you in the eye, nod and smile at the right times, and may even answer occasionally. That appearance of interest, however, is a polite facade to mask thoughts that have nothing to do with what the speaker is saying.

2. **We tune in and out. Selective listeners** respond only to the parts of a speaker's remarks that interest them. All of us are selective listeners from time to time, but it's a habit that can lead to confusion, misunderstandings, and hurt feelings.

3. **We defend ourselves. Defensive listeners** perceive that they are being attacked even when they aren't. As a result, they are more interested in justifying themselves than in understanding how the other person feels. Feelings of guilt or insecurity are often at the root of defensive listening, with the effect that casual remarks may be taken as insults (*"How dare you ask me if I enjoyed lunch? It's my business and no one else's if I cheat on my diet!"*).

   A common defensive response is counterattacking, or even **ambushing** the perceived critic by listening to find faults for later reference. This approach comes in handy for attorneys who are cross-examining witnesses, but in everyday life, it's not likely to make you a lot of friends. Instead of proving your point, you will probably cause the other person to feel defensive and angry.

   At the other end of the spectrum, nondefensive listeners are sincerely interested in understanding the other person's perspective, even when the topic makes them uncomfortable. See the checklist on the previous page for tips on listening nondefensively.

4. **We avoid the issue. Insulated listeners** tend to avoid difficult subjects. If you find yourself tuning out when a subject arises that you'd rather not deal with, you might be guilty of this habit.

5. **We miss the underlying point. Insensitive listeners** tend to take remarks at <u>face value</u> rather than looking below the surface. An insensitive listener might miss the warble in a friend's voice that suggests she is more upset than her words let on. Or when her partner complains, *"I always take out the trash,"* she might miss that what's wanted is a thank-you.

6. **We tend to be self-centered.** The next time you're engaged in conversation, consider who has control. **Conversational narcissists** focus on themselves and their interests instead of listening to and encouraging others.[47] One type of conversational narcissist is the **stage hog**, who actively claims more than his or her fair share of the spotlight. Stage hogs tend to interrupt a lot and switch topics to suit their interests. They may assume that their ideas are better or more important than others' or that they can guess what other people are going to say.[48] Other conversational narcissists are more passive. They may not interrupt, but neither do they encourage others. A passive narcissist is unlikely to make supportive comments such as *"Uh-huh," "That's funny,"* and *"What happened next?"*[49] Whatever their approach, conversational narcissists tend to discourage the equal give-and-take that is the hallmark of mutually satisfying conversation.

7. **We assume that talking is more impressive than listening.** The key to success sometimes seems to be speaking well, but good listening skills are just as important. There is no calculating the esteem and wisdom people earn by listening well rather than talking all the time. As the playwright Wilson Mizner once observed, "a good listener is not only popular everywhere, but after a while he [or she] gets to know something."[50]

You may feel stunned by the egotism behind many of these bad habits but still feel tempted to engage in them sometimes. Focusing on oneself and dismissing others' ideas outright may be justified in rare situations, but it's usually a mistake to <u>rule out</u> what others say. Next, we explore some of the main types of listening, discussing contexts when each is most appropriate.

pseudolistening An imitation of true listening.

selective listening A listening style in which the receiver responds only to messages that interest him or her.

defensive listening A response style in which the receiver perceives a speaker's comments as an attack.

ambushing A style in which the receiver listens carefully to gather information to use in an attack on the speaker.

insulated listening A style in which the receiver ignores undesirable information.

insensitive listening The failure to recognize the thoughts or feelings that are not directly expressed by a speaker, and instead accepting the speaker's words at face value.

conversational narcissists People who focus on themselves and their interests instead of listening to and encouraging others.

stage hogs People who are overly invested in being the center of attention.

Calvin and Hobbes   by Bill Watterson

cultural idioms
at face value: literally
rule out: fail to consider

What Is Your Listening Style?

To discover your listening tendencies, fill in the survey below. Use 1 as "strongly disagree" and 7 as "strongly agree." The section(s) where you marked higher numbers (5, 6, or 7) suggest types of listening that you value. For tips on listening effectively within each of these styles, see pages 134–141.

Task-Oriented Listening							
I am impatient with people who ramble on and on during conversations.	1	2	3	4	5	6	7
I get frustrated when people get off topic during a conversation.	1	2	3	4	5	6	7
I prefer speakers who quickly get to the point.	1	2	3	4	5	6	7
Relational Listening							
I listen to understand the emotions and mood of the speaker.	1	2	3	4	5	6	7
I listen primarily to build and maintain relationships with others.	1	2	3	4	5	6	7
I enjoy listening to others because it allows me to connect with them.	1	2	3	4	5	6	7

Types of Listening

By now you can see that listening well isn't easy. As you'll read in the following pages, listening serves a number of goals. To see which of these are generally most important to you, take the Self-Assessment above before reading further. Each category in the instrument reflects a distinct reason for listening. What did you learn about your own goals for listening from your scores?

Task-Oriented Listening

The goal of **task-oriented listening** is to secure information necessary to get a job done. The situations that call for task-oriented listening are endless and varied: following an instructor's comments in class, hearing a description of a new piece of merchandise or software that you're thinking about buying, getting tips from a coach on how to improve your athletic skill, taking directions from your boss—the list goes on and on.

Task-oriented listening is most concerned with efficiency. Task-oriented listeners view time as a scarce and valuable commodity, and they can grow impatient when they think others are wasting it.

A task orientation can be an asset when deadlines and other pressures demand fast action. It's most appropriate when taking care of business is the primary concern: Such listeners keep a focus on the job at hand and encourage others to be organized and concise.

Despite its advantages, a task orientation can put off others when it seems to disregard their feelings. A no-nonsense task-oriented approach isn't always

Listening is arguably the most underappreciated tool for career success.

How could you enhance your career by listening better?

cultural idiom
put off: displease

Analytical Listening							
I tend to withhold judgment about another's ideas until I have heard everything he or she has to say.	1	2	3	4	5	6	7
When listening to others, I consider all sides of the issue before responding.	1	2	3	4	5	6	7
I fully listen to what a person has to say before forming any opinions.	1	2	3	4	5	6	7
Critical Listening							
I often catch errors in other speakers' logic.	1	2	3	4	5	6	7
I tend to naturally notice errors in what other speakers say.	1	2	3	4	5	6	7
When listening to others, I notice contradictions in what they say.	1	2	3	4	5	6	7

Adapted with the authors' permission from Bodie, G. D., Worthington, D. L., & Gearhart, C. G. (2013). The Revised Listening Styles Profile (LSP–R): Development and validation. *Communication Quarterly, 61,* 72–90.

appreciated by speakers who—by virtue of culture or temperament—lack the skill or inclination to be clear and direct. Also, an excessive focus on getting things done quickly can hamper the kind of thoughtful deliberation that some jobs require. Finally, task-oriented listeners seem to minimize emotional issues and concerns, which may be an important part of business and personal transactions.

You can become more effective as an informational listener by approaching others with a constructive attitude and by using some simple but effective skills. When task-oriented listening is appropriate, the following guidelines will help you be more effective.

Look for Key Ideas   It's easy to lose patience with <u>long-winded</u> speakers who never seem to get to the point—or to have a point, for that matter. Nonetheless, most people do have a central idea in what they say, or what we will call a *thesis* in Chapter 11. By using your ability to think more quickly than the speaker can talk, you may be able to extract the thesis from the surrounding mass of words you're hearing. If you can't figure out what the speaker is driving at, you can always ask in a tactful way by using the skills of questioning and paraphrasing, which we examine now.

Ask Questions   If you are heading to a friend's house for the first time, typical questions might be, *"How is the traffic between here and there?"* or *"Is there anything you'd like me to bring?"* In more emotional situations, questions could include, *"Why do you think that bothers you so much?"* or *"You sound upset—is there something wrong?"* **Questioning** involves asking for additional information to clarify your idea of the sender's message. One key element of these types of questions is that they ask the speaker to elaborate.

Not all questions are equally helpful, however. Whereas **sincere questions** are aimed at understanding others, **counterfeit questions** are not; they are often disguised attempts to send a message.

task-oriented listening A listening style that is primarily concerned with accomplishing the task at hand.

questioning An approach in which the receiver overtly seeks additional information from the sender.

sincere question A question posed with the genuine desire to learn from another person.

counterfeit question A question that is not truly a request for new information.

cultural idiom
long-winded: speaking for a long time

Source: Courtesy of Ted Goff

Counterfeit questions come in several varieties:

- *Questions that make statements.* "Are you finally ready to go now?" "You can't be serious about that, right?" Comments like these are certainly not genuine requests for information. Emphasizing certain words can also turn a question into a statement: "You lent money to Tony?" We also use questions to offer advice. The person who responds with "Are you going to <u>stand up to</u> him and give him what he deserves?" clearly has stated an opinion about what should be done.

- *Questions that carry hidden agendas.* "Are you busy Friday night?" is a dangerous question to answer. If you say, "No," thinking the person has something fun in mind, you won't like hearing, "Good, because I need some help moving my piano."

- *Questions that seek "correct" answers.* Most of us have been victims of question-askers who only want to hear a particular response. "Which shoes do you think I should wear?" can be a sincere question—unless the asker has a predetermined preference. When this happens, the asker isn't interested in listening to contrary opinions, and "incorrect" responses get <u>shot down</u>. Some of these questions may venture into delicate territory. "Honey, do you think I look ugly?" is probably a request for a "correct" answer.

- *Questions that are based on unchecked assumptions.* "Why aren't you listening to me?" assumes the other person isn't paying attention. "What's the matter?" assumes that something is wrong. As Chapter 2 explains, perception checking is a much better strategy: "When you kept looking over at the TV, I thought you weren't listening to me, but maybe I was wrong. *Were* you paying attention?"

Paraphrase Sincere questioning is often a valuable tool for increasing understanding. Sometimes, however, you need to take another step. Now consider another type of feedback—one that would help you confirm your understanding. This sort of feedback, termed **paraphrasing**, involves restating in your own words the message you thought the speaker had just sent. For example:

> (*To a direction giver*) "You're telling me to drive down to the traffic light by the high school and turn toward the mountains, is that it?"

> (*To the boss*) "So you need me both this Saturday *and* next Saturday—right?"

> (*To a professor*) "When you said, 'Don't worry about the low grade on the quiz,' did you mean it won't count against my grade?"

In other cases, a paraphrase will reflect your understanding of the speaker's feelings:

> "You said, 'I've had it with this relationship!' Are you angry or relieved that it's over?"

> "You said you've got a minute to talk, but I'm not sure whether it's a good time for you."

> "You said, 'Forget it,' but it sounds like you're mad. Are you?"

In each case, the key to success is to restate the other person's comments in your own words as a way of cross-checking the information. If you simply repeat the speaker's comments verbatim, you will sound foolish—and you still might be

paraphrasing Feedback in which the receiver rewords the speaker's thoughts and feelings.

cultural idioms

<u>**stand up to:**</u> to confront courageously

<u>**shot down:**</u> rejected or defeated

misunderstanding what has been said and why. Notice the difference between simply parroting (repeating without understanding) a statement and really paraphrasing:

Speaker: I'd like to go, but I can't afford it.

Parroting: You'd like to go, but you can't afford it.

Paraphrasing: So if we could find a way to pay for you, you'd be willing to come. Is that right?

Speaker: What's the matter with you?

Parroting: You think there's something wrong with me?

Paraphrasing: You think I'm mad at you?

As these examples suggest, effective paraphrasing is a skill that takes time to develop. (See the checklist at right for paraphrasing approaches.) It can be worth the effort, however, because it offers two very real advantages:

- First, it boosts the odds that you'll accurately and fully understand what others are saying. We've already seen that using one-way listening or even asking questions may lead you to think that you've understood a speaker when, in fact, you haven't. Paraphrasing, on the other hand, serves as a way of double-checking your interpretation for accuracy.

- Second, paraphrasing guides you toward sincerely trying to understand another person instead of using nonlistening styles such as stage hogging, selective listening, and so on. Listeners who paraphrase to check their understanding are judged to be more socially attractive than listeners who do not.[51]

Take Notes   Understanding others is crucial, of course, but it doesn't guarantee remembering. As you read earlier in this chapter, listeners usually forget about half of what they hear immediately afterward.

Sometimes recall isn't especially important. You don't need to retain many details of the vacation adventures recounted by a neighbor or the childhood stories told by a relative. At other times, though, remembering a message—even minute details—is important. The lectures you hear in class are an obvious example. Likewise, it can be important to remember the details of plans that involve you: the time of a future appointment, the name of a phone caller whose message you took, or the orders given by your boss at work.

At times like these it's smart to take notes instead of relying on your memory. Sometimes these notes may be simple and brief: a name and phone number jotted on a scrap of paper, or a list of things to pick up at the market. In other cases—a lecture, for example—your notes need to be much longer. See the checklist on the next page for note-taking strategies when the details are essential.

Relational Listening

The goal of **relational listening** is to emotionally connect with others. Relationally oriented listeners are typically perceived to be extroverted, attentive, and friendly. They are more focused on understanding people than on trying to control them.[52]

A relational orientation has obvious strengths. But it has some less obvious drawbacks. It's easy to become overly involved with others' feelings. Relational listeners may lose their detachment and ability

---

CHECKLIST ✓
Three Ways to Paraphrase

You can make your paraphrasing sound more natural by taking any of three approaches, depending on the situation:

☐ **Change the speaker's wording.**

Speaker: Bilingual education is just another failed idea.

Paraphrase: You're mad because you think bilingual ed sounds good, but it doesn't work? *(Reflects both the speaker's feeling and the reason for it.)*

☐ **Offer an example of what you think the speaker is talking about.** When the speaker makes an abstract statement, you may suggest a specific example or two to see if your understanding is accurate.

Speaker: Lee is such a jerk. I can't believe the way he acted last night.

Paraphrase: You think those jokes were pretty offensive, huh? *(Reflects the listener's guess about the speaker's reason for objecting to the behavior.)*

☐ **Reflect on the underlying theme of the speaker's remarks.** When you want to summarize the theme that seems to have run through another person's conversation, a complete or partial perception check is appropriate.

Speaker: Remember to lock the door.

Paraphrase: You keep reminding me to be careful. It sounds like you're worried that something bad might happen. Am I right? *(Reflects the speaker's thoughts and feelings and explicitly seeks clarification.)*

---

relational listening A listening style that is driven primarily by the concern to build emotional closeness with the speaker.

CHECKLIST ✓
Taking Detailed Notes

When the details are crucial, a few simple points will help make your note-taking efforts more effective:

☐ **Don't wait too long before beginning to jot down ideas.** If you don't realize that you need to take notes until 5 minutes into a conversation, you're likely to forget much of what has been said and miss out on other information as you <u>scramble to catch up</u>.

☐ **Record only key ideas.** Don't try to capture every word of long messages. If you can <u>pin down</u> the most important points, your notes will be easier to follow and much more useful.

☐ **Develop a note-taking format.** The exact form you choose isn't important. Some people use a formal outlining scheme with headings designated by roman numerals, letters, and numbers, whereas others use simple lists. You might come up with useful symbols: boxes around key names and numbers or asterisks next to especially important information. After you develop a consistent format, your notes will help you not only remember information but structure ideas in a way that's useful to you.

cultural idioms
scramble to catch up: work hard to reach a goal, especially after a period of delay

pin down: identify specifically

to assess the quality of information others are giving in an effort to be congenial and supportive.[53] Less relationally oriented communicators can view them as overly expressive, and even intrusive. Here are some strategies for being an effective relational listener.

Take Time The goal of task-oriented listening is efficiency, but relational listening couldn't be more different. Encouraging others to share their thoughts and feelings can take time. If you're in a hurry, it may be best to reschedule relationally focused conversations for a better time. The gift of attention often speaks for itself, even when you don't know what to say. Medical studies show that, even when doctors cannot cure patients, patients' coping skills are positively linked to the amount of time their doctors spend listening to them.[54]

In some situations, even brief interactions may have relational dimensions. To a harassed customer service rep you might sympathetically say, *"Busy day, huh?"* Or, you might thank an especially patient salesperson by saying, *"I really appreciate you taking time to explain this so patiently."*

Listen for Unexpressed Thoughts and Feelings People often don't say what's on their minds or in their hearts. There are lots of reasons why: Tact, confusion, lack of awareness, fear of being judged negatively . . . the list is a long one.

When relationship building is the goal, it can be valuable to listen for unexpressed messages. Consider a few examples:

STATEMENT	POSSIBLE UNEXPRESSED MESSAGE
"Don't apologize. It's not a big deal."	"I'm angry (or hurt, disappointed) by what you did."
"You're going clubbing tonight? That sounds like fun!"	"I'd like to come along."
"Check out this news story. That's my little sister!"	"I'm proud of what she did."
"That was quite a party you [neighbors] had last night. You were going strong at 2 A.M."	"The noise bothered me."
"You like gaming? I do too!"	"Perhaps we can be friends."

There are several ways to explore unexpressed messages. You can *ask questions* using the guidelines described on pages 135–136: *"How did you feel when he said that?" "Is there something I can do to help?"* You can *paraphrase*, as described on pages 136–137: *"Sounds like that really surprised you,"* or *"So you aren't sure what to do next, right?"* Or, as you'll read in a few pages, you can *prompt* the speaker to volunteer more information: *"Really?" "Is that right?"* You'll read more about listening as a form of social support later in this chapter.

When you consider exploring unexpressed feelings and thoughts, be careful not to pry. Proceed carefully, and phrase your hunches tentatively. For more advice, review the description of perception checking on pages 54 and 55 in Chapter 2.

Encourage Further Comments Even if you don't explore unexpressed messages, you can strengthen relationships simply by encouraging others to say more. Even if you're not a stage hog, it's easy to redirect a conversation back to yourself. Instead, try a simple experiment to prove the value of focusing on the speaker. In your next conversation, focus on drawing out the other person. If you express a sincere desire to learn more, you're likely to be surprised by the positive results. The speaker may be grateful that you have helped him or her work through a problem, when all you have really done is listen and ask questions. Great teachers harness this power regularly. They know that students often learn more when they are asked questions and encouraged to work through problems than when they are given the answers up front.[55]

Analytical Listening

Whereas relational listening may enhance relationships, the goal of **analytical listening** is to understand the message. Analytical listeners explore an issue from a variety of perspectives in order to understand it as fully as possible.

Comforting is one response to another's pain, but it isn't always the best way to help.

What are your common styles of relational listening? How successful are they? What other styles could you use for better results?

Analytical listening is a good approach when your goal is to assess the quality of ideas, and when there is value in looking at issues from a wide range of perspectives. It's especially valuable when the issues at hand are complicated. On the other hand, a thorough analytical approach can be time consuming. So when a deadline is approaching, you may not respond as quickly as the other person would like.

When you want to listen analytically, follow these steps.

Listen for Information Before Evaluating The principle of listening for information before evaluating seems almost too obvious to mention, yet all of us are guilty of judging a speaker's ideas before we completely understand them. The tendency to make premature judgments is especially strong when the idea you are hearing conflicts with your own beliefs. As one writer put it,

> the right to speak is meaningless if no one will listen. . . . It is simply not enough that we reject censorship . . . we have an affirmative responsibility to hear the argument before we disagree with it.[56]

You can avoid the tendency to judge before understanding by following the simple rule of paraphrasing a speaker's ideas before responding to them. The effort required to translate the other person's ideas into your own words will keep you from arguing, and if your interpretation is mistaken, you'll know immediately.

Separate the Message from the Speaker The first recorded cases of blaming the messenger for an unpleasant message occurred in ancient Greece. When messengers reported losses in battles, their generals sometimes responded to the bad news by having the messengers put to death. This sort of irrational reaction is still common (though fortunately less violent) today. Consider a few situations in which there is a tendency to get angry with a communicator bearing unpleasant news: An instructor tries to explain why you did poorly on a major paper; a friend explains what you did to make a fool of yourself at the party last Saturday night; the boss points out how you could do your job better. At times like this, becoming irritated with the bearer of unpleasant information may not only cause you to miss important information but also harm your relationships.

There's a second way that confusing the message and the messenger can prevent you from understanding important ideas. At times you may mistakenly

analytical listening Listening in which the primary goal is to fully understand the message, prior to any evaluation.

ETHICAL CHALLENGE

How Carefully Should You Listen?

How much adaptation is ethical? How far would you go to be effective with an audience?

• What responsibility do communicators have to listen as carefully and thoughtfully as possible?

• Are there ever cases in which we are justified in our faulty listening habits?

• Is it dishonest to fake careful listening?

Evaluating a Speaker's Message

Whatever form of support a speaker uses, you can ask several questions to determine the quality of the evidence and reasoning:[57]

☐ **Is the evidence recent enough?** In many cases, old evidence is worthless. Before you accept even the most credible evidence, be sure it isn't obsolete.

☐ **Is enough evidence presented?** One or two pieces of support may be exceptions and not conclusive evidence. Be careful not to generalize from limited evidence.

☐ **Is the evidence from a reliable source?** Even a large amount of recent evidence may be worthless if the source is weak.

☐ **Can the evidence be interpreted in more than one way?** Evidence that supports one claim might also support others. Alternative explanations don't necessarily mean that the one being argued is wrong, but they do raise questions that need to be answered before you accept an argument.

critical listening Listening in which the goal is to evaluate the quality or accuracy of the speaker's remarks.

cultural idioms

write off: dismiss as worthless or unimportant

tune out: not listen

discount the value of a message because of the person who is presenting it. Even the most boring instructors, the most idiotic relatives, and the most demanding bosses occasionally make good points. If you write off everything a person says before you consider it, you may be cheating yourself out of some valuable information.

Search for Value   Even if you listen with an open mind, sooner or later you will end up hearing information that is either so unimportant or so badly delivered that you're tempted to tune out. Although making a quick escape from such tedious situations is sometimes the best thing to do, there are times when you can profit from paying close attention to apparently worthless communication. This is especially true when you're trapped in a situation in which the only alternatives to attentiveness are pseudolistening or downright rudeness.

Once you try, you probably can find some value in even the worst situations. Consider how you might listen opportunistically when you find yourself locked in a boring conversation with someone whose ideas are worthless. Rather than torture yourself until escape is possible, you could keep yourself amused—and perhaps learn something useful—by listening carefully until you can answer the following (unspoken) questions:

"Is there anything useful in what this person is saying?"

"What led the speaker to come up with ideas like these?"

"What lessons can I learn from this person that will keep me from sounding the same way in other situations?"

Listening with a constructive attitude is important, but even the best intentions won't always help you understand others. The following skills can help you figure out messages that otherwise might be confusing, as well as help you see how those messages can make a difference in your life.

Critical Listening

The goal of **critical listening** is to go beyond trying to understand and analyze the topic at hand, and instead, to assess its quality. At their best, critical listeners apply the tools of analytical listening to see whether an idea holds up under careful scrutiny.

Critical listening can be especially helpful when the goal is to investigate a problem. But people who are critical listeners can also frustrate others, who may think that they nitpick everything people say.

When critical listening is appropriate, follow these guidelines.

Examine the Speaker's Evidence and Reasoning   Speakers usually offer some kind of support to back up their statements. A car dealer who argues that domestic cars are just as reliable as imports might cite frequency-of-repair statistics from *Consumer Reports* or refer you to satisfied customers, for example; and a professor arguing that students don't work as hard as they used to might tell stories about then and now to back up the thesis.

Chapter 12 describes several types of supporting material that can be used to prove a point: definitions, descriptions, analogies, statistics, and so on. See the checklist on page 140 for questions to consider when evaluating a speaker's message.

Besides taking a close look at the evidence a speaker presents, a critical listener will also look at how that evidence is put together to prove a point. Logicians have identified a number of logical fallacies—errors in reasoning that can lead to false conclusions. In fact, logicians have identified more than 200 fallacies.[58] Chapter 14 identifies some of the most common ones.

Evaluate the Speaker's Credibility The acceptability of an idea often depends on its source. If your longtime family friend, the self-made millionaire, invited you to invest your life savings in jojoba fruit futures, you might be grateful for the tip. If your deadbeat brother-in-law made the same offer, you would probably <u>laugh off</u> the suggestion.

Audiences may be enthralled by the rhetoric in political campaigns. But careful listening is essential to sorting out truths and evaluating candidates' logic.

How can you apply critical thinking to political discourse?

Chapter 14 discusses credibility in detail, but two questions provide a quick guideline for deciding whether or not to accept a speaker as an authority:

- *Is the speaker competent?* Does the speaker have the experience or the expertise to qualify as an authority on this subject? Note that someone who is knowledgeable in one area may not be well qualified to comment in another area. For instance, your friend who can answer any question about computer programming might be a terrible advisor when the subject turns to romance.

- *Is the speaker impartial?* Knowledge alone isn't enough to certify a speaker's ideas as acceptable. People who have a personal stake in the outcome of a topic are more likely to be biased. The unqualified praise a commission-earning salesperson gives a product may be more suspect than the mixed review you get from a user. This doesn't mean you should disregard any comments you hear from an involved party—only that you should consider the possibility of intentional or unintentional bias.

Examine Emotional Appeals Sometimes emotion alone may be enough reason to persuade you. You might lend your friend $20 just for old times' sake even though you don't expect to see the money again soon. In other cases, it's a mistake to let yourself be swayed by emotion when the logic of a point isn't sound. The excitement or fun in an ad or the lure of low monthly payments probably isn't good enough reason to buy a product you can't afford. Again, the fallacies described in Chapter 14 will help you recognize flaws in emotional appeals.

As you read about task-oriented, relational, analytical, and critical approaches to listening, you may have noted that you habitually use some more than others. Researchers are still trying to determine how much we rely on different approaches—and how much we should.[59] There's no question that you can control the way you listen to and use the styles that best suit the situation at hand. When your relationship with the speaker needs attention, adopt a relational approach. When clarity is the issue, be an action-oriented listener. If analysis is called for, put on that style. And when efficiency is what matters most, become a model of task-oriented orientation. You can also boost your effectiveness by assessing the listening preferences of your conversational partners and adapting your style to them.

cultural idiom
laugh off: dismiss with a laugh

Listening and Social Support

supportive listening The reception approach to use when others seek help for personal dilemmas.

There's another type of listening and responding that might involve any of the approaches just described. In **supportive listening**, the primary aim is to help the speaker deal with personal dilemmas. Sometimes the problem is a big one: *"I'm not sure this marriage is going to work"* or *"I can't decide whether to drop out of school."* At other times the problem is more modest. A friend might be trying to decide what birthday gift to buy or where to spend a vacation. Supportive listeners are typically judged to be optimistic, honest, understanding, and encouraging.[60]

Blogger Sarah Q remembers an unexpected source of support when she was being bullied in high school. "There were rumors and awful things said about me," she recalls. Her English teacher, Ms. Hodge, noticed Sarah's distress and began spending lunch periods with her. "I told her everything," Sarah says. "She knew what was going on, and she helped me through it all. . . . She sat there and listened. And I needed that." Sarah has since transferred to a different school where she has good friends, but she says she will always be grateful to the teacher who truly listened.[61]

There's no question about the value of receiving support when faced with personal problems. Research shows that supportive communication can reduce loneliness and stress and build self-esteem.[62] There is even evidence that people with good support networks tend to live longer than others.[63] And the benefits go both ways. People who provide social support often feel an enhanced sense of well-being themselves.[64]

Online Social Support

Traditionally, most social support came from personal acquaintances: friends, family, coworkers, neighbors, and so on. However, in online communities, strangers can share interests and concerns and potentially gain support from one another. People who read Sarah Q's online tribute to Ms. Hodge posted comments such as "Thanks for sharing," "I love you Sarah!," and "Awesome article! We're so glad you're at our school now."[65]

Some online groups offer specialized support. Areas of focus include addiction, Asperger's syndrome, codependency, debt problems, domestic violence, eating disorders, gambling, infertility, miscarriage, sexual abuse, and suicide, to name just a few.[66]

In some aspects, online help is similar to the face-to-face variety. The goals are to gain information and emotional support. In other ways, it differs.[67] The most obvious difference is that many members of online communities have not met in person and may not even know each other's real names. This anonymity may be a plus in that it enables people to feel comfortable opening up, but it can also be a drawback, particularly if people are not supportive. The social networking site Reddit has been criticized for allowing users to post racist, sexist, and homophobic jokes and comments anonymously. Because they cannot be identified, some people may be more likely to say things online that they wouldn't say in person.

One difference between in-person and virtual support groups is that online groups often focus specifically on a single issue, whereas in traditional relationships, people are likely to cover a wide range of topics. Another difference involves the rate and amount of self-disclosure: In traditional relationships, people usually reveal personal information slowly and carefully, but with the anonymity of online support groups, they typically open up almost immediately.

Gender and Social Support

Men and women have traditionally defined supportive communication somewhat differently. Linguist Deborah Tannen, who is famous for her research about gender and communication, offers the example of telling one's troubles to another person.

UNDERSTANDING COMMUNICATION TECHNOLOGY

Who Is Listening to You Online?

During a lecture, Australian university student Jonathan Pease tweeted the following message: "Sitting in the back row at syndey uni carving 'Rooney eats it' on the desk . . . feels good to be a rebel:)." He soon discovered that his online audience included Sydney University officials, who soon posted a tweet of their own: "Defacing our desks Jonathan?;) Hope you enjoyed your course."[68] Luckily for Pease, the authorities had a sense of humor. Their response also reinforced an important point: When you go online, your audience may include more viewers than you imagine.

University officials aren't the only ones who track people online. The practice known as "social media listening" has become a sensation among marketing professionals who want to understand consumers' preferences better and develop relationships with them. Robert Caruso realized this when he saw a commercial for Total Bib on TV and tweeted to his friends that it reminded him of a *Saturday Night Live* sketch. "A few hours later, I received a reply tweet from Total Bib thanking me for the mention and engaging me in conversation," Caruso says. "I was pretty amazed since I was not following them previously, they were simply monitoring the stream." Not only that, but Total Bib representatives looked up Caruso online and learned that he was a single father with a toddler, so they sent him a complimentary bib. Caruso said he was touched by the personal attention.[69]

So-called listening software can monitor websites, Facebook, Google searches, blogs, Twitter, and more. That means that marketing professionals monitor people's online search terms and the websites they visit, in addition to the messages and photos they post.

At its best, monitoring technology allows marketers to listen avidly to consumers. At its worst, it can be "unethical and creepy," in the words of Tom Petrocelli, a technology blogger and marketing specialist.[70] Petrocelli is a fan of media listening when it's done well, but he counts the following actions in the creepy column: spying on employees, requiring them to share their "friends" list with the marketing team, and sending out mass emails or tweets uninvited.

One thing is for sure: Whether you consider online monitoring to be an invasion of privacy or marketing genius, the odds are that you are subject to this type of listening.

When women share their troubles with other women, Tannen says, the response is often a matching "me too" disclosure. For example, a woman might say, "I understand. My partner *never* remembers my birthday!" Such a response is usually understood between women as a sign of their connectedness and solidarity. Indeed, women may even dig deep to find a matching experience or emotion to share,[71] which is one reason that happiness (as well as dissatisfaction) often feels contagious.

Men have traditionally been socialized to focus less on emotional connection and more on competition and emotional control. Consequently, if a woman responds to a man's troubles talk with a matching experience, it may feel to him like a <u>one-up</u>, as if she is implying "your problems are not so remarkable" or "mine are even worse." In short, what feels like empathy to her may feel like a <u>put-down</u> to him. And because men are often discouraged from expressing intense emotions, they may consider it supportive to offer a solution or a distraction such as *"Don't worry about it"* or *"Here's what you should do . . ."* As you might predict, women who are accustomed to a different style of social support may feel that men who respond this way are <u>brushing off</u> their concerns or belittling their problems.

The result of these different perspectives, Tannen observes, is often a mutual sense of frustration:

> She blames him for telling her what to do and failing to provide the expected comfort, whereas he thinks he did exactly what she requested and cannot fathom why she would keep talking about a problem if she does not want to do anything about it.[72]

Of course, we must be careful not to overgeneralize. Gender roles continually evolve, and a number of factors interact with gender to shape how people provide social

cultural idioms

one-up: outdoing someone

put-down: an insulting or degrading remark

brushing off: dismissing or ignoring

Factors to Consider Before Offering Advice

Before offering advice, make sure the following four conditions are present:[75]

☐ **Be confident that the advice is correct.** Resist the temptation to act like an authority on matters you know little about. Furthermore, it is both unfair and risky to make suggestions when you aren't positive that they are the best choice. Realize that just because a course of action worked for you doesn't guarantee that it will work for everybody.

☐ **Ask yourself whether the person seeking your advice seems willing to accept it.** In this way you can avoid the frustration of making good suggestions, only to find that the person with the problem had another solution in mind all the time.

☐ **Be certain that the receiver won't blame you if the advice doesn't work out.** You may be offering the suggestions, but the choice and responsibility for accepting them are up to the recipient of your advice.

☐ **Deliver your advice supportively, in a face-saving manner.** Advice that is perceived as being offered constructively, in the context of a solid relationship, is much better than critical comments offered in a way that signals a lack of respect for the receiver.

cultural idioms

face-saving: protecting one's dignity
pin the blame on: claim the fault lies with
wake-up call: a warning or caution to pay attention

support—including cultural background, personal goals, expressive style, and cognitive complexity. All the same, understanding traditional patterns and social mores may help us avoid the assumption that our way is the only way or the right way to offer comfort.

Types of Supportive Responses

Whatever the relationship and topic, you can choose from several styles to respond supportively to another person's remarks.[73] Each of these styles has advantages and disadvantages. As you read them, consider the best style for a particular situation.

Advising　When someone shares a concern with you, you might offer a solution, which scholars call an **advising response**. Although advice is sometimes valuable, often it isn't as helpful as you might think.

There are two main reasons why offering advice doesn't work especially well in general. First, it can be hard to tell when someone actually wants your opinion. Sometimes the request is clear: *"What do you think I should do?"* At other times, though, the speaker's intent isn't as clear. Statements such as *"What do you think of Jeff?,"* *"Would that be an example of sexual harassment?,"* and *"I'm really confused"* may be designed more to solicit information or announce a problem. People often don't want advice, even if they indicate they do. They may not be ready to accept it, needing instead simply to talk out their thoughts and feelings.

Even when someone with a problem asks for advice, offering it may not be helpful. Your suggestion may not offer the best course to follow, in which case it can even be harmful. There's often a temptation to tell others how we would behave in their place, but it's important to realize that what's right for one person may not be right for another. A related consequence of advising is that it often allows others to avoid responsibility for their decisions. A partner who follows a suggestion of yours that doesn't work out can always pin the blame on you.

Advice is most welcome when it has been clearly requested and when the advisor seems concerned with respecting the face needs of the recipient.[74] Before offering advice, consider the checklist on this page.

Judging　A **judging response** evaluates the sender's thoughts or behaviors in some way. The judgment may be favorable (*"That's a good idea"* or *"You're on the right track now"*) or unfavorable (*"An attitude like that won't get you anywhere"*). But in either case it implies that the person doing the judging is in some way qualified to pass judgment on the speaker's thoughts or behaviors.

Sometimes negative judgments are purely critical. How many times have you heard such responses as *"Well, you asked for it!"* or *"I told you so!"* or *"You're just feeling sorry for yourself"*? Although comments like these can sometimes serve as a verbal wake-up call, they usually make matters worse.

At other times negative judgments are less critical. These involve what we usually call *constructive criticism*, which is intended to help a person improve in the future. This is the sort of response given by friends about everything from the choice of clothing to jobs and to friends. Another common setting for constructive criticism occurs in school, where instructors evaluate students' work to help them master concepts and skills. But whether or not it's justified, even constructive criticism runs the risk of arousing defensiveness because it may threaten the self-concept of the person at whom it is directed.

Judgments have the best chance of being received when two conditions exist:

- *The person with the problem has requested an evaluation from you.* Occasionally an unsolicited judgment may bring someone to his or her senses, but more often this sort of uninvited evaluation will trigger a defensive response.

- *Your judgment is genuinely constructive and not designed as a put-down.* If you are tempted to use judgments as a weapon, don't fool yourself into thinking that you are being helpful. Often the statement *"I'm telling you this for your own good"* simply isn't true.

If you can remember to follow these two guidelines, your judgments will probably be less frequent and better received.

Analyzing   In an **analyzing statement**, the listener tries to help by offering an interpretation of a speaker's message. The motive here is different from the kind of analysis described on pages 139–140, in which the goal was to benefit you, the listener. In this case, the analysis is aimed at helping the other person.

Analyses like these are probably familiar to you:

"I think what's really bothering you is . . ."

"She's doing it because . . ."

"I don't think you really meant that."

"Maybe the problem started when she . . ."

Interpretations are often effective ways to help people with problems consider alternative meanings they would not have thought of without your help. Sometimes a clear analysis will make a confusing problem suddenly clear, either suggesting a solution or at least providing an understanding of what is occurring.

At other times, an analysis can create more problems than it solves. There are two problems with analyzing. First, your interpretation may not be correct, in which case the speaker may become even more confused by accepting it. Second, even if your interpretation is correct, saying it aloud might not be useful. There's a chance that it will arouse defensiveness (because analysis can imply superiority and judgment), and even if it doesn't, the person may not be able to understand your view of the problem without working it out personally.

How can you know when it's helpful to offer an analysis? The checklist on this page suggests several guidelines to follow.

Questioning   A few pages ago we talked about questioning as one way to understand others better. A questioning response can also be a way to help others think about their problems and understand them more clearly. For example, questioning can help a conversational partner define vague ideas more precisely. You might respond to a friend with a line of questioning: *"You said Greg has been acting 'differently' toward you lately. What has he been doing?"* Another example of a question that helps clarify is as follows: *"You told your roommates that you wanted them to be more helpful in keeping the place clean. What would you like them to do?"*

Questions can also encourage people to examine situations in more detail by talking either about what happened or about personal feelings—for example, *"How did you feel when they turned you down? What did you do then?"* This type of questioning is particularly helpful when you are dealing with someone who is quiet or is unwilling under the circumstances to talk about the problem very much.

Although questions have the potential to be helpful, they also risk confusing or distracting the person with the problem. The best questioning follows these principles:

- *Don't ask questions just to satisfy your own curiosity.* You might become so interested in the other person's story that you will want to hear more. *"What did he say then?"* you might be tempted to ask. *"What happened next?"*

advising response Helping response in which the receiver offers suggestions about how the speaker should deal with a problem.

judging response A reaction in which the receiver evaluates the sender's message either favorably or unfavorably.

analyzing statement A helping style in which the listener offers an interpretation of a speaker's message.

When and How to Offer an Analysis

☐ *Offer your interpretation in a tentative way rather than as absolute fact.* There's a big difference between saying, *"Maybe the reason is…"* or *"The way it looks to me…"* and insisting, *"This is the truth."*

☐ *Make sure you have a reasonable chance of being correct.* An inaccurate interpretation—especially one that sounds plausible—can leave a person more confused than before.

☐ *Be sure the other person will be receptive to your analysis.* Even if you're completely accurate, your thoughts won't help if the other person isn't ready to consider them.

☐ *Be sure your motive for offering an analysis is truly to help the other person.* It can be tempting to offer an analysis to show how brilliant you are or even to make the other person feel bad for not having thought of the right answer. Needless to say, an analysis offered under such conditions isn't helpful.

comforting A response style in which a listener reassures, supports, or distracts the person seeking help.

Responding to questions like these might confuse the person with the problem, or even leave him or her more agitated than before.

- *Be sure your questions won't confuse or distract the person you're trying to help.* For instance, asking someone, *"When did the problem begin?"* might provide some clue about how to solve it—but it could also lead to a long digression that would only confuse matters. As with advice, it's important to be sure you're on the right track before asking questions.

- *Don't use questions to disguise your suggestions or criticism.* We've all been questioned by parents, teachers, or other figures who seemed to be trying to trap us or indirectly to guide us. In this way, questioning becomes a strategy that can imply that the questioner already has some idea of what direction the discussion should take but isn't willing to tell you directly.

Comforting　A **comforting** response can take several forms:

Agreement	"You're right—the landlord is being unfair."
	"Yeah, that class was tough for me, too."
Offers to help	"I'm here if you need me."
	"Let me try to explain it to him."
Praise	"I don't care what the boss said: I think you did a great job!"
	"You're a terrific person! If she doesn't recognize it, that's her problem."
Reassurance	"The worst part is over. It will probably get easier from here."
	"I know you'll do a great job."
Diversion	"Let's catch a movie and get your mind off this."
	"That reminds me of the time we . . ."
Acknowledgment	"I can see that really hurts."
	"I know how important that was to you."
	"It's no fun to feel unappreciated."

Sometimes comforting words often can be just what the other person needs. In other instances, though, this kind of comment isn't helpful at all; in fact, it can even make things worse. Telling a person who is obviously upset that everything is all right, or joking about a serious matter, can trivialize the problem. People might see your comments as a put-down, leaving them feeling worse than before.

As with the other styles we'll discuss, comforting can be helpful, but only in certain circumstances.[76] For the occasions when comforting is an appropriate response, follow these guidelines:

- *Make sure your comforting remarks are sincere.* Phony agreement or encouragement is probably worse than no support at all, because it adds the insult of your dishonesty to the pain the other person is already feeling.

- *Be sure the other person can accept your support.* Sometimes we become so upset that we aren't ready or able to hear anything positive.

Even if your advice, judgments, and analysis are correct and your questions are sincere, and even if your support comes from the best motives, these responses often fail to help. For example, it's typically hurtful rather than helpful to say, *"There are people worse off than you are"* or *"No one ever said life was fair."*[77] It may also be frustrating to hear *"I understand how you feel"* from people who can't really know what the person is going through.[78] It is usually far more helpful to listen without judgment, acknowledge the person's feelings, and say that you care and will stand by the distressed individual through hard times.[79]

One American Red Cross grief counselor explained to survivors of the September 11, 2001, terrorist attacks on the United States that simply being present can be more helpful than trying to reassure grief-stricken family members who lost loved ones in the tragedy:

> Listen. Don't say anything. Saying *"it'll be okay,"* or *"I know how you feel"* can back-fire. Right now that's not what a victim wants to hear. They want to know people are there and care about them. Be there, be present, listen. The clergy refer to it as a "ministry of presence." You don't need to do anything, just be there or have them know you're available.[80]

Prompting   Advising, judging, analyzing, questioning, and comforting are all active approaches to helping that call for a great deal of input from the respondent. Another approach to problem solving is more passive. **Prompting** involves using silences and brief statements of encouragement to draw others out, and in so doing to help them solve their own problems. Consider this example:

Pablo:	Julie's dad is selling a complete computer system for only $1,200, but if I want it I have to buy it now. He's got another interested buyer. It's a great deal. But buying it would <u>wipe out</u> my savings. At the rate I spend money, it would take me a year to save up this much again.
Tim:	Uh huh.
Pablo:	I wouldn't be able to take that ski trip over winter break . . . but I sure could save time with my schoolwork . . . and do a better job, too.
Tim:	That's for sure.
Pablo:	Do you think I should buy it?
Tim:	I don't know. What do you think?
Pablo:	I just can't decide.
Tim:	*(silence)*
Pablo:	I'm going to do it. I'll never get a deal like this again.

Prompting works especially well when you can't help others make a decision. At times like this your presence can act like a catalyst to help others find their own answers. Prompting will work best when it's done sincerely. Your nonverbal behaviors—eye contact, posture, facial expression, tone of voice—must show that you are concerned with the other person's problem. Mechanical prompting is likely to irritate instead of help.

Reflecting   A few pages ago you read about the value of paraphrasing to understand others. The same approach can be used as a supportive response. We use the term **reflecting** to describe it here, to emphasize that the goal is not as much to clarify your understanding as to help the other person hear and think about the words he or she has just spoken. When you use this approach, be sure to reflect both the *thoughts* and the *feelings* you hear being expressed. This conversation

prompting Using silence and brief statements of encouragement to draw out a speaker.

reflecting Listening that helps the person speaking hear and think about the words just spoken.

cultural idiom
wipe out: deplete

between two friends shows how reflecting can offer support and help a person find the answer to her own problem:

Jill: I've had the strangest feeling about my boss lately.

Mark: What's that? *(A simple question invites Jill to go on.)*

Jill: I'm starting to think maybe he has this thing about women—or maybe it's just about me.

Mark: You mean he's <u>coming on to</u> you? *(Mark paraphrases what he thinks Jill has said.)*

Jill: Oh no, not at all! But it seems like he doesn't take women—or at least me—seriously. *(Jill corrects Mark's misunderstanding and explains herself.)*

Mark: What do you mean? *(Mark asks another simple question to get more information.)*

Jill: Well, whenever we're in a meeting or just talking around the office and he asks for ideas, he always seems to pick men. He gives orders to women—men, too—but he never asks the women to say what they think. But I know he counts on and acknowledges some women in the office.

Mark: Now you sound confused. *(Reflects her apparent feeling.)*

Jill: I am confused. I don't think it's just my imagination. I mean I'm a good producer, but he has never—not once—asked me for my ideas about how to improve sales or anything. And I can't remember a time when he's asked any other women. But maybe I'm overreacting.

Mark: You're not positive whether you're right, but I can tell that this has you concerned. *(Mark paraphrases both Jill's central theme and her feeling.)*

Jill: Yes. But I don't know what to do about it.

Mark: Maybe you should . . . *(Starts to offer advice but catches himself and decides to ask a sincere question instead.)* So what are your choices?

Jill: Well, I could just ask him if he's aware that he never asks women's opinions. But that might sound too aggressive and angry.

Mark: And you're not angry? *(Tries to clarify how Jill is feeling.)*

Jill: Not really. I don't know whether I should be angry because he's not taking women's ideas seriously, or whether he just doesn't take my ideas seriously, or whether it's nothing at all.

Mark: So you're mostly confused. *(Reflects Jill's apparent feeling again.)*

Jill: Yes! I don't know <u>where I stand with</u> my boss, and not being sure is starting to get to me. I wish I knew what he thinks of me. Maybe I could just tell him I'm confused about what is going on here and ask him to clear it up. But what if it's nothing? Then I'll look insecure.

Mark: *(Mark thinks Jill should confront her boss, but he isn't positive that this is the best approach, so he paraphrases what Jill seems to be saying.)* And that would make you look bad.

Jill: I'm afraid maybe it would. I wonder if I could talk it over with anybody else in the office and get their ideas . . .

Mark: . . . see what they think . . .

Jill: Yeah. Maybe I could ask Brenda. She's easy to talk to, and I do respect her judgment. Maybe she could give me some ideas about how to handle this.

Mark: Sounds like you're comfortable with talking to Brenda first.

Jill: *(Warming to the idea.)* Yes! Then if it's nothing, I can calm down. But if I do need to talk to the boss, I'll know I'm doing the right thing.

Mark: Great. Let me know how it goes.

ASK YOURSELF

How can you use supportive listening to help someone work through a problem?

---

cultural idioms

coming on to: making a sexual advance toward

where I stand with: how I am perceived by

Reflecting a speaker's ideas and feelings can be surprisingly helpful. First, reflecting helps the other person sort out the problem. In the dialogue you just read, Mark's paraphrasing helped Jill consider carefully what bothered her about her boss's behavior. The clarity that comes from this sort of perspective can make it possible to find solutions that weren't apparent before. Reflecting also helps the person to unload more of the concerns he or she has been carrying around, often leading to the relief that comes from catharsis. Finally, listeners who reflect the speaker's thoughts and feelings (instead of judging or analyzing, for example) show their involvement and concern. The checklist on this page suggests factors you should consider before you paraphrase while offering support.

When and How to Help

Before committing yourself to helping another person—even someone in obvious distress—make sure your support is welcome. Sometimes, people prefer to handle difficult situations on their own. The presence of a relational partner may make them even more nervous or anxious.[81] And timing is everything. Sometimes the most supportive thing we can do is give someone peace and quiet. Other times, running an errand or offering to help with a task may more helpful than listening.[82]

When help is welcome, there is no single best way to provide it. There is enormous variability in which style will work with a given person in a given situation. This explains why communicators who are able to use a wide variety of helping styles are usually more effective than those who rely on just one or two styles.[83]

You can boost the odds of choosing the best helping style in each situation by considering three factors.

- *The situation:* Sometimes people need your advice. At other times your encouragement and comforting will be most helpful, and at still other times your analysis or judgment may be truly useful. And, as you have seen, there are times when your prompting and reflecting can help others find their own answer.
- *The other person:* Some people are able to consider advice thoughtfully, whereas others use suggestions to avoid making their own decisions. Many communicators are extremely defensive and aren't capable of receiving analysis or judgments without lashing out. Still others aren't equipped to think through problems clearly.
- *Your own strengths and weaknesses:* You may be best at listening quietly, offering a prompt from time to time. Or perhaps you are especially insightful and can offer a truly useful analysis of the problem. Of course, it's also possible to rely on a response style that is unhelpful. You may be overly judgmental or too eager to advise, even when your suggestions aren't invited or productive.

In most cases, the best way to help is to use a combination of responses in a way that meets the needs of the occasion and suits your personal communication style.[84]

In this chapter, we have considered the importance of listening as a means of connecting with others, learning, leading, and developing fulfilling relationships. Erica Nicole and many other successful people operate on a principle stated by the science educator Bill Nye: "Everyone you will ever meet knows something you don't."[85] About life. About love. About career success. About themselves. And maybe about you.

CHECKLIST ✓

Conditions That Make Paraphrasing a Good Option

Reflecting can be helpful, but it is no panacea. A classic study by noted researcher John Gottman revealed that this approach is not an end in itself; rather, it is one way to help others by understanding them better.[86] There are several factors to consider before you decide to paraphrase.

- ☐ **Is the problem complex?** Sometimes people are simply looking for information and not trying to work out their feelings. At times like this, paraphrasing would be out of place.

- ☐ **Do you have the necessary time and concern?** The kind of paraphrasing we've been discussing here takes a good deal of time and can be stressful.[87] Therefore, if you're in a hurry to do something besides listen, it's wise to avoid starting a conversation you won't be able to finish. It's far better to state honestly that you're unable or unwilling to help than to pretend to care when you really don't.

- ☐ **Are you willing to listen rather than share your own experiences?** It isn't always necessary to reciprocate the other person's self-disclosure with information of your own. Sometimes keeping the focus on the other person is the most supportive thing you can do.

- ☐ **Can you withhold judgment?** Use paraphrasing only if you can comfortably paraphrase without injecting your own judgments.

- ☐ **Is your paraphrasing in proportion to other responses?** Although reflecting can be a very helpful way of responding to others' problems, it can become artificial and annoying when it's overused.

MAKING THE GRADE

For more resources to help you understand and apply the information in this chapter, visit the *Understanding Human Communication* website at www.oup.com/us/adleruhc.

OBJECTIVE 5.1 **Describe the benefits that follow from being an effective listener.**

- Listening—the process of giving meaning to an oral message—is a vitally important part of the communication process.

- There are many advantages to being a good listener. People with good listening skills are more likely than others to be hired, promoted, and regarded as leaders. In addition, listening improves relationships.

- Identify people from your own experience whose listening ability illustrates the advantages outlined on pages 125–126.

- Listen, *really* listen, to someone important in your life. Note how that person responds and how you feel about the experience.

- How could better listening benefit your life?

OBJECTIVE 5.2 **Outline the most common misconceptions about listening, and assess how successfully you avoid them.**

- Listening and hearing are not the same thing. Hearing is only the first step in the process of listening. Beyond that, listening involves attending, understanding, responding, and remembering.

- Listening is not a natural process. It takes both time and effort. Mindful listening requires a commitment to understand others' perspectives without being judgmental or defensive.

- It's a mistake to assume that all receivers will hear and understand a message identically. Recognizing the potential for multiple interpretations and misunderstanding can prevent problems.

  > Give examples of how the misconceptions about listening described in this section create problems.

  > Which factors described on pages 129–133 contribute most to your ineffective listening?

  > How can you use the information in this section to improve your own listening?

OBJECTIVE 5.3 **Identify and manage the faulty habits and challenges that make effective listening difficult.**

- Faulty listening habits include pseudolistening, selective listening, defensiveness, avoiding difficult issues, insensitivity, conversational narcissism, and the assumption that talking is more impressive than listening.

- A variety of factors contribute to ineffective listening, including message overload, rapid thought, psychological noise, physical noise, hearing problems, cultural differences, and media influences.

  > From recent experience, identify examples of each type of ineffective listening described in this section.

  > Which of the listening misconceptions listed on pages 126–128 characterize your attitudes about listening?

  > Develop an action plan to overcome your barriers to effective listening and to correct the faulty listening habits you have identified.

OBJECTIVE 5.4 **Know when and how to listen to accomplish a task, enhance relationships, analyze a message, and critically evaluate another's remarks.**

- Task-oriented listening helps people accomplish mutual goals. It involves an active approach in which people often identify key ideas, ask questions, paraphrase, and take notes.

- Relational listening involves the willingness to spend time with people to better understand their feelings and perspectives. It's most effective when the listener is patient, encourages the speaker to continue, and is sensitive to underlying meanings.

- Analytic listening involves the willingness to suspend judgment and consider a variety of perspectives to achieve a clear understanding. This type of listening requires people to discern between what is true and what is not.

- Critical listening is appropriate when the goal is to judge the quality of an idea. A critical analysis is most successful when the listener ensures correct understanding of a message before passing judgment, when the speaker's credibility is taken into account, when the quality of supporting evidence is examined, and when the logic of the speaker's arguments is examined carefully.

  > Identify situations in which each type of listening described in this section is most appropriate.

  > On your own or with feedback from others who know you well, assess your ability to listen for each of the following goals: to accomplish a task, enhance relationships, analyze ideas, and critically evaluate messages. Which types of listening are your strongest and weakest?

  > Develop an action plan to improve your skills in the most critical areas identified above.

OBJECTIVE 5.5 **Use the skills introduced on pages 137–139 to offer social support to a friend, partner, or family member.**

- The aim of supportive listening is to help the speaker, not the receiver.
- Various helping responses include advising, judging, analyzing, questioning, comforting, prompting, and reflecting the speaker's thoughts and feelings.
- Listeners can be most helpful when they use a variety of styles, focus on the emotional dimensions of a message, and avoid being too judgmental.

  > Give examples of each type of social support described on pages 142–149.

  > Which types of social support do you find most helpful? When are certain types of social support not helpful to you?

  > Identify how you can use the information in this section to offer more helpful social support.

KEY TERMS

advising response p. 145

ambushing p. 133

analytical listening p. 139

analyzing statement p. 145

attending p. 127

comforting p. 146

conversational narcissism p. 133

counterfeit question p. 135

critical listening p. 140

defensive listening p. 133

hearing p. 126

insensitive listening p. 133

insulated listening p. 133

judging response p. 145

listening p. 126

listening fidelity p. 127

mindful listening p. 128

paraphrasing p. 136

prompting p. 147

pseudolistening p. 133

questioning p. 135

reflecting p. 147

relational listening p. 137

remembering p. 127

residual message p. 127

responding p. 127

selective listening p. 133

sincere question p. 135

stage hog p. 133

supportive listening p. 142

task-oriented listening p. 135

understanding p. 127

ACTIVITIES

1. **Recognizing Listening Misconceptions** You can see how listening misconceptions affect your life by identifying important situations when you have fallen for each of the following assumptions. In each case, describe the consequences of believing these erroneous assumptions.

   a. Thinking that because you were hearing a message you were listening to it.

   b. Believing that listening effectively is natural and effortless.

   c. Assuming that other listeners were understanding a message in the same way as you.

2. **Supportive Response Styles** This exercise will help improve your ability to listen empathically in the most successful manner. For each of the following statements:

   a. Write separate responses, using each of the following styles:

   > Advising
   > Judging
   > Analyzing
   > Questioning
   > Comforting
   > Prompting
   > Reflecting

   b. Discuss the pros and cons of using each response style.

   c. Identify which response seems most effective and explain your decision.

   > At a party, a guest you have just met for the first time says, "Everybody seems like they've been friends for years. I don't know anybody here. How about you?"

   > Your best friend has been quiet lately. When you ask if anything is wrong, she snaps, "No!" in an irritated tone of voice.

   > A fellow worker says, "The boss keeps making sexual remarks to me. I think it's a come-on, and I don't know what to do."

   > It's registration time at college. One of your friends asks if you think he should enroll in the communication class you've taken.

   > Someone with whom you live remarks, "It seems like this place is always a mess. We get it cleaned up, and then an hour later it's trashed."

Nonverbal Communication

LEARNING OBJECTIVES

6.1

Explain the characteristics of nonverbal communication and the social goals it serves.

6.2

Explain the ways in which nonverbal communication reflects culture and gender differences.

6.3

Describe the functions served by nonverbal communication.

6.4

List the types of nonverbal communication, and explain how each operates in everyday interaction.

6.5

Demonstrate competence in assessing the nonverbal communication of others and managing your own nonverbal messages.

Amy Cuddy's
research highlights the power of nonverbal communication. As you read this chapter, consider these questions:

?

What do your nonverbal cues suggest to other people?

?

How might you alter your nonverbal behavior to feel and appear more confident and powerful?

?

What nonverbal cues catch your attention most? What do you tend to overlook?

WONDER WOMAN AND SUPERMAN might be on to something. Their feet-apart, hands-on-hips stance is a testament to the power of nonverbal communication. Social psychologist Amy Cuddy calls it a power pose. According to her research, it can change the way people think about you and can improve your performance at school and work.

Cuddy became interested in body language as a business professor at Harvard University. She noticed that some students engage in highly assertive nonverbal behaviors. "They get right into the middle of the room before class even starts, like they really want to occupy space. When they sit down, they're sort of spread out," she says.[1] Those are the students, she soon realized, who are most likely to get noticed. They raise their hands high and take part in discussions. By contrast, other students "are virtually collapsing as they come in. They sit in their chairs and they make themselves tiny, and they go like this," Cuddy says, raising her hand no higher than her head, with her arm close to her face.

Cuddy has also noticed that women in her classes are more likely than men to demonstrate low-power behaviors. Slight-built herself, she began to apply the idea to her own behavior and found that people took her more seriously when she demonstrated a confident presence.

Cuddy's research has shown how nonverbal communication affects not just how others perceive us but how we feel about *ourselves*. Try this quick exercise to see for yourself. The next time you anticipate a stressful situation, adopt a power pose for 2 minutes to warm up. Even if no one sees you, "it's going to make you feel more powerful," Cuddy predicts.

Of course, nonverbal communication involves more than body language; some of the things we do vocally are nonverbal. Tone, rate, and volume are nonverbal cues. So are laughter and sighs, since neither of these are words. Likewise, although many verbal messages are vocal, some aren't (e.g., written words). Table 6-1 illustrates these differences.

Types of Communication

	VOCAL COMMUNICATION	NONVOCAL COMMUNICATION
VERBAL COMMUNICATION	Spoken words	Written words
NONVERBAL COMMUNICATION	Tone of voice, sighs, screams, vocal qualities (loudness, pitch, and so on)	Gestures, movement, appearance, facial expression, and so on

Source: Adapted from John Stewart, J., & D'Angelo, G. (1980). *Together: Communicating interpersonally* (2nd ed.). Reading, MA: Addison-Wesley, p. 22. Copyright © 1993 by McGraw-Hill. Reprinted/adapted by permission.

You've probably noticed that there's often a gap between what people say and how they actually feel. An acquaintance says, *"I'd like to get together again"* in a way that leaves you suspecting the opposite. A speaker tries to appear confident but acts in a way that almost screams out, *"I'm nervous!"* You ask a friend what's wrong, and the *"Nothing"* you get in response rings hollow.

Then there are other times when a message comes through even though there are no words at all. A look of irritation, a smile, a sigh—signs like these can say more than a torrent of words.

All situations like these have one element in common: Messages were sent nonverbally. The goal of this chapter is to introduce you to this world of nonverbal communication. Although you have certainly recognized nonverbal messages before, the following pages should introduce you to a richness of information you have never noticed. The experience won't transform you into a mind reader, but it will make you a far more accurate observer of others—and yourself.

> **nonverbal communication** Messages expressed by other than linguistic means.

Characteristics of Nonverbal Communication

Our working definition of **nonverbal communication** is "messages expressed through nonlinguistic means." But this brief definition only hints at the richness of these messages. You can begin to understand their prevalence by trying a simple experiment. Spend time around a group of people who are speaking a language you don't understand. Your goal is to see how much information you can learn about the people you're observing from means other than the verbal messages they transmit. This experiment will reveal several characteristics of nonverbal communication.

What about languages that don't involve spoken words? For example, is American Sign Language considered verbal or nonverbal communication? Most scholars would say sign language is verbal because it largely uses gestures to express particular words.[2] This means that sign language and written words are verbal, but messages transmitted by vocal means that don't involve language—sighs, laughs, and other utterances—are nonverbal. We will discuss this more later in the chapter.

Minions speak a language that's incomprehensible to humans. Nonetheless, their voices, facial expressions, and movements offer plenty of cues about what they are thinking and feeling.

What nonverbal cues do you typically broadcast when your friends are talking to you?

Nonverbal Behavior Has Communicative Value

It's virtually impossible not to communicate nonverbally.[3] Suppose you were instructed to avoid communicating any messages at all. What would you do? Close your eyes? Withdraw into a ball? Leave the room? As the photo of a minion on this page illustrates, the meaning of some nonverbal behavior can be ambiguous. You may not be able to tell exactly what is going on, but the nonverbal cues certainly have communicative value.

cultural idiom

rings hollow: sounds insincere

Of course, we don't always intend to send nonverbal messages. Unintentional nonverbal behaviors differ from intentional ones. For example, we often stammer, blush, frown, and sweat without meaning to do so. Some theorists argue that unintentional behavior may provide information, but it shouldn't count as communication. Others draw the boundaries of nonverbal communication more broadly, suggesting that even unconscious and unintentional behavior conveys messages and thus is worth studying as communication.[4] We take the broad view here because, whether or not nonverbal behavior is intentional, we use it to form impressions about one another.

Even when we intend to communicate nonverbally, we aren't always aware of the cues we are sending. In one study, less than a quarter of experimental subjects who had been instructed to show increased or decreased liking of a partner could describe the nonverbal behaviors they used.[5] Scientists think this is possible because our limbic brain, which encodes and decodes nonverbal cues, also triggers automatic responses to our environment—as when we jump at loud noises or our hearts beat quickly in fear.[6] As a result of this, we can control our nonverbal displays to an extent, but they sometimes occur spontaneously or with minimal conscious thought.

Furthermore, just because communicators are nonverbally expressive doesn't mean that others will tune in to the abundance of unspoken messages that are available. One study comparing the richness of email to in-person communication confirmed the greater amount of information available in face-to-face conversations, but it also showed that some communicators (primarily men) failed to recognize these messages.[7]

The fact that you and everyone around you are constantly sending nonverbal clues is important because it means that you have a constant source of information available about yourself and others. If you can tune in to these signals, you will be more aware of how those around you are feeling and thinking, and you will be better able to respond to their behavior.

Nonverbal Communication Is Primarily Relational

Some nonverbal messages serve utilitarian goals. For example, a police officer uses gestures to direct the flow of traffic, and a conductor leads members of a symphony. But nonverbal communication also serves a far more common (and more interesting) series of social goals.[8]

1. **Nonverbal cues help us manage our identities.** In Chapter 2 we explored the notion that people strive to create images of themselves as they want others to view them. Nonverbal communication plays an important role in this process—in many cases it is more important than verbal communication. Consider, for example, what happens when you attend a party where you are likely to meet strangers you would like to get to know better. Instead of projecting your image verbally (*"Hi! I'm attractive, friendly, and easygoing"*), you behave in ways that will present this identity. You might smile a lot and try to strike a relaxed pose. It's also likely that you will dress carefully—even if the image involves looking as if you hadn't given a lot of attention to your appearance.

2. **Nonverbal cues help define our relationships.** You can appreciate this fact by thinking about the wide range of ways you could behave when greeting another person. Depending on the nature of your relationship (or what you want it to be) you could wave, shake hands, nod, smile, pat the other person on the back, give a hug, or avoid all contact. Even trying to *not* communicate can send a message, as when you avoid talking to someone.

3. **Nonverbal cues convey emotion.** We can convey emotions nonverbally that we may be unwilling or unable to express in words. For example, people who know you well might recognize that you are shocked, happy, stressed, or sad, even when you are trying to hide those feelings or when you haven't fully acknowledged them within yourself.

What emotions do you imagine this couple is feeling?

One reason nonverbal cues are so powerful is that they convey some meanings better than words can. You can prove this for yourself by imagining how you could express each item on the following list nonverbally:

- You're bored.
- You are opposed to capital punishment.
- You are attracted to another person in the group.
- You want to know if you will be tested on this material.
- You are nervous about trying this experiment.

The first, third, and fifth items in this list all involve attitudes; you could probably imagine how each could be expressed nonverbally through what social scientists call **affect displays**—facial expressions, body movements, and vocal traits described in this chapter. By contrast, the second and fourth items involve ideas, and they would be quite difficult to convey without using words. The same principle holds in everyday life: Nonverbal behavior offers many cues about the way people feel—often more than we get from their words alone.

> **affect displays** Facial expressions, body movements, and vocal traits that reveal emotional states.

Nonverbal Communication Is Ambiguous

It is important to realize that nonverbal communication is often difficult to interpret accurately. To appreciate the ambiguous nature of nonverbal communication, study the photo on this page. What emotions do you imagine the couple is feeling: Grief? Anguish? Agony? In fact, none of these is even close. They just learned that they won $1 million in the New Jersey state lottery!

Nonverbal communication can be just as ambiguous in everyday life. For example, relying on nonverbal cues in romantic situations can lead to inaccurate guesses about a partner's interest in a sexual relationship.[9] Workers of the Safeway supermarket chain discovered firsthand the problems with nonverbal ambiguity when they tried to follow the company's new "superior customer service" policy that required them to smile and make eye contact with customers. Twelve employees filed grievances over the policy, reporting that several customers had propositioned them, misinterpreting their actions as come-ons.[10]

Although all nonverbal behavior is ambiguous, some emotions are easier to decode accurately than others. In laboratory experiments, subjects are better at identifying positive facial expressions such as happiness, love, surprise, and interest than negative ones such as fear, sadness, anger, and disgust.[11] In real life, however, spontaneous nonverbal expressions are so ambiguous that observers are able to identify the emotions they convey no more accurately than by blind guessing.[12]

Some people are more skillful than others at accurately decoding nonverbal behavior.[13] For example, those who are better senders of nonverbal messages also are better receivers. Decoding ability also increases with age and training, although there are still differences in ability owing to personality and occupation. For instance, extroverts are more accurate judges of nonverbal behavior than introverts are.[14] Interestingly, in general, women seem to be better than men at

cultural idioms

come-ons: sexual advances

blind guessing: coming to a conclusion without any factual basis for judgment

decoding nonverbal messages.[15] Despite these differences, even the best nonverbal decoders do not approach 100% accuracy.

When you try to make sense out of ambiguous nonverbal behavior, you should consider several factors: the context in which the behaviors occur (e.g., smiling at a joke suggests a different feeling from what is suggested by smiling at another's misfortune); the history of your relationship with the sender (e.g., friendly, hostile); the other's mood at the time; and your feelings (when you're feeling insecure, almost anything can seem like a threat). The important idea is that when you become aware of nonverbal messages, you should think of them not as facts but rather as clues that need to be checked out.

Nonverbal Communication Differs from Verbal Communication

As Table 6-2 shows, nonverbal communication differs in several important ways from spoken and written language. These differences suggest some reasons why it is valuable to focus on nonverbal behavior. For example, whereas verbal messages are almost always intentional, nonverbal cues are often unintended, and sometimes unconscious.

Nonverbal Skills Are Important

It's hard to overemphasize the importance of effective nonverbal expression and the ability to read and respond to others' nonverbal behavior. Nonverbal encoding and decoding skills are strong predictors of popularity, attractiveness, and socioemotional well-being.[16] In general, good nonverbal communicators are more persuasive than people who are less skilled, and they have a greater chance of success in settings ranging from careers to poker to romance. Nonverbal sensitivity is a major part of what some social scientists have called "emotional intelligence," and researchers have come to recognize that it is impossible to study spoken language without paying attention to its nonverbal dimensions.[17]

One way to appreciate the importance of nonverbal skills is to see the challenges faced by people who lack them. Due to a processing deficit in the right hemisphere of the brain, people born with a syndrome called nonverbal learning disorder (NVLD) have trouble making sense of nonverbal cues.[18] People with NVLD often misinterpret humorous or sarcastic messages literally, because those cues are based heavily on nonverbal signals.

TABLE 6-2

Some Differences Between Verbal and Nonverbal Communication

	VERBAL COMMUNICATION	NONVERBAL COMMUNICATION
COMPLEXITY	One dimension (words only)	Multiple dimensions (voice, posture, gestures, distance, etc.)
FLOW	Intermittent (speaking and silence alternate)	Continuous (it's impossible not to communicate nonverbally)
CLARITY	Less subject to misinterpretation	More ambiguous
IMPACT	Has less impact when verbal and nonverbal cues are contradictory	Has stronger impact when verbal and nonverbal cues are contradictory
INTENTIONALITY	Usually deliberate	Often unintentional

People with NVLD also have trouble figuring out how to behave appropriately in new social situations, so they rely on rote formulas that often don't work. For example, a child who has learned the formal way of meeting an adult for the first time by shaking hands and saying, *"Pleased to meet you"* might try this approach with a group of peers. The result, of course, is typically regarded as odd or nerdy. And their disability may lead them to miss nonverbal cues sent by the other children that this isn't the right approach.[19]

Even for those of us who don't live with NVLD, the nuances of nonverbal behavior can be confusing.

Influences on Nonverbal Communication

Much nonverbal communication is universal. For example, researchers have found at least six facial expressions that humans everywhere use and understand: happiness, sadness, fear, anger, disgust, and surprise.[20] Even children who have been blind since birth reveal their feelings using these expressions. Despite these similarities, there are some important differences in the way people use and understand nonverbal behavior. We'll look at some of these differences now.

Culture

Cultures have different nonverbal languages as well as verbal ones. Fiorello La Guardia, the legendary mayor of New York from 1934 to 1945, was fluent in English, Italian, and Yiddish. Researchers who watched films of his campaign speeches with the sound turned off found that they could tell which language he was speaking by the changes in his nonverbal behavior.[21]

The meaning of some gestures varies from one culture to another. The *"okay"* gesture made by joining thumb and forefinger to form a circle is a cheery affirmation to most Americans, but it has less positive meanings in other parts of the world. In France and Belgium it means, *"You're worth zero."* In Greece and Turkey it is a vulgar sexual invitation, usually meant as an insult. Given this sort of cross-cultural ambiguity, an innocent tourist might easily wind up in serious trouble.

Less obvious cross-cultural differences can damage relationships without the parties ever recognizing exactly what has gone wrong. Edward Hall points out that, whereas Americans are comfortable conducting business at a distance of roughly 4 feet, people from the Middle East stand much closer.[22] It is easy to

cultural idiom

wind up in: end up being in

Source: DILBERT © 1991 Scott Adams. Used by permission of UNIVERSAL UCLICK. All rights reserved.

It has become fashionable in some circles to imitate the throaty rumble known as vocal fry, made popular by Katy Perry and other celebrities.

Do you ever change the sound of your voice to make a particular impression on people? If so, how?

visualize the awkward advance-and-retreat pattern that might occur when two diplomats or businesspeople from these cultures meet. The Middle Easterner would probably keep moving forward to close a gap that feels wide to him or her, whereas the American would probably continually back away. Both would feel uncomfortable, although they may not know why.

Like distance, patterns of eye contact vary around the world.[23] A direct gaze is considered appropriate for speakers in Latin America, the Arab world, and southern Europe. On the other hand, Asians, Indians, Pakistanis, and northern Europeans gaze at a listener peripherally or not at all. In either case, deviations from the norm are likely to make a listener uncomfortable.

Culture also affects how nonverbal cues are monitored. In Japan, for instance, people tend to look to the eyes for emotional cues, whereas Americans and Europeans focus on the mouth.[24] These differences can often be seen in the text-based emoticons commonly used in these cultures. (Search for "Western and Eastern emoticons" in your browser for examples.)

Even within a culture, various groups can have different nonverbal rules. For example, many white teachers in the United States use quasi-questions that hint at the information they are seeking. An elementary teacher might encourage the class to speak up by making an incorrect statement that demands refutation: *"So twelve divided by four is six, right?"* Most white students would recognize this behavior as a way of testing their understanding. But this style of questioning is unfamiliar to many students raised in traditional black cultures, who aren't likely to respond until they are directly questioned by the teacher.[25] Given this difference, it is easy to imagine how some teachers might view some children as unresponsive or slow, when in fact they are simply playing by a different set of rules.

Vocal patterns are another nonverbal way to build and demonstrate cocultural solidarity. For instance, younger Americans often use "uptalk" (statements ending with a rise in pitch) and "vocal fry" (words ending with a low guttural rumble). Celebrities such as Kim Kardashian and Zooey Deschanel popularized these vocalic styles. It's therefore not surprising that females use them more than males do,[26] although age of the speaker (i.e., millennial or younger) has more of an impact than gender. There is some debate whether vocal fry diminishes one's credibility[27] or enhances it.[28] Either way, vocal mannerisms are one way that speakers affiliate with their communities.

Communicators become more tolerant of others after they understand that many nonverbal behaviors that seem unusual are the result of cultural differences. In one study, American adults were presented with videotaped scenes of speakers from the United States, France, and Germany.[29] When the sound was cut off, viewers judged foreigners more negatively than their fellow citizens. But when the speakers' voices were added (allowing viewers to recognize that they were from a different country), participants were less critical of the foreign speakers.

Despite differences like these, many nonverbal behaviors have the same meanings around the world. Smiles and laughter are a universal signal of positive emotions, for example, whereas the same sour expressions convey displeasure in every culture.[30] Charles Darwin believed that expressions such as these are the result of evolution, functioning as survival mechanisms that allowed early humans to convey emotional states before the development of language.

Although nonverbal expressions like these may be universal, the way they are used varies widely around the world. In some cultures, display rules discourage the overt demonstration of feelings such as happiness or anger. In other cultures, displaying the same feelings is perfectly appropriate. Thus, a person from Japan might appear much more controlled and placid than an Arab, when in fact their feelings might be identical.

The same principle operates closer to home among cocultures. For example, observational studies have shown that, in general, black women in all-black groups are more nonverbally expressive and interrupt one another more than do white women in all-white groups.[31] This doesn't mean that black women always feel more intensely than their white counterparts. A more likely explanation is that the two groups follow different cultural rules. The researchers found that, in racially mixed groups, both black and white women moved closer to the others' style. This nonverbal convergence shows that skilled communicators can adapt their behavior when interacting with members of other cultures or cocultures to make the exchange smoother and more effective.

Gender

Media depictions often exaggerate stereotypical differences in masculine and feminine styles of nonverbal communication. Think about exaggerated caricatures in animated film classics such as *Beauty and the Beast* and *The Little Mermaid*. Many humorous films and plays have been created around the results that arise when characters try to act like members of another sex.

Although few of us behave like stereotypically masculine or feminine movie characters, there are often recognizable differences in the way men and women look and act. In general, women are more nonverbally expressive than men, and women are typically better at recognizing others' nonverbal behavior. More specifically, research shows that, compared with men, women

- smile more;
- use more facial expressions;
- use more head, hand, and arm gestures (but less expansive gestures);
- touch others more;
- stand closer to others;
- are more vocally expressive; and
- make more eye contact.[32]

Despite these differences, men's and women's nonverbal communication patterns have a good deal in common.[33] Differences such as the ones noted above are noticeable, but they are outweighed by the similar rules we follow in most dimensions of nonverbal behavior. You can prove this by imagining what it would be like to use radically different nonverbal rules: standing only an inch away from others, sniffing strangers, or tapping people's foreheads to get their attention. Those behaviors would mark you as bizarre no matter your gender. Moreover, male–female nonverbal differences are usually less pronounced in conversations involving gay and lesbian participants.[34] Gender and culture certainly have an influence on nonverbal style, but the differences are often a matter of degree rather than kind.

Most communication scholars agree that social factors have more influence than biology does in shaping how men and women behave. For example, the ability to read nonverbal cues may have more to do with women's historically less powerful social status. People in subordinate work positions also have better decoding skills, probably because it's important to be able to read the boss's nonverbal cues.[35]

All in all, although biological sex and cultural norms have an influence on nonverbal style, as we mentioned in Chapter 4, they aren't as dramatic as the "men are from Mars and women are from Venus" thesis suggests.

Functions of Nonverbal Communication

Although verbal and nonverbal messages differ in many ways, the two forms of communication operate together on most occasions. The following discussion explains the many functions of nonverbal communication and shows how nonverbal messages relate to verbal ones.

Repeating

If someone asked you for directions to the nearest drugstore, you could say, *"North of here about two blocks,"* repeating your instructions nonverbally by pointing north. This sort of repetition isn't just decorative: People remember comments accompanied by gestures more than those made with words alone.[36]

Substituting

When a friend asks you what's new, you might shrug your shoulders instead of answering in words. Social scientists use the term **emblems** to describe deliberate nonverbal behaviors that have precise meanings known to everyone within a cultural group. For example, most Americans know that a head nod means *"yes,"* a head shake means *"no,"* a wave means *"hello"* or *"good-bye,"* and a hand to the ear means *"I can't hear you."* (These same gestures mean different things in other cultures, which can cause a great deal of intercultural confusion, as you might imagine.)

Not all substituting consists of emblems, however. Sometimes substituting responses are more ambiguous and less intentional. A sigh, smile, or frown may substitute for a verbal answer to your question, *"How's it going?"* As this example suggests, nonverbal substituting is especially important when people are reluctant to express their feelings in words.

emblems Deliberate nonverbal behaviors with precise meanings, known to virtually all members of a cultural group.

illustrators Nonverbal behaviors that accompany and support verbal messages.

Complementing

Sometimes nonverbal behaviors match the content of a verbal message. Consider, for example, a friend apologizing for forgetting an appointment with you. Your friend's sincerity will be reinforced if the verbal apology is accompanied by the appropriate nonverbal behaviors: the right tone of voice, facial expression, and so on. We often recognize the significance of complementary nonverbal behavior when it is missing. If your friend's apology is delivered with a shrug, a smirk, and a light tone of voice, you will probably doubt its sincerity, no matter how profuse the verbal explanation is.

Much complementing behavior consists of **illustrators**—nonverbal behaviors that accompany and support spoken words. Scratching the head when searching for an idea and snapping your fingers when it occurs are examples of illustrators that complement verbal messages. Research shows that North Americans use illustrators most often when they are emotionally aroused—when they are furious, horrified, very agitated, distressed, or excited, and when trying to explain ideas that are difficult to put into words.[37]

Accenting

Just as we use italics to emphasize an idea in print, we use nonverbal devices to emphasize oral messages. Pointing an accusing finger adds emphasis to criticism (as well as probably creating defensiveness in the receiver). Stressing certain words with the voice ("It was *your* idea!") is another way to add nonverbal accents.

UNDERSTANDING COMMUNICATION TECHNOLOGY

Nonverbal Expressiveness Online

Communication scholars have characterized face-to-face interaction as "rich" in nonverbal cues that convey feelings and attitudes. Even telephone conversations carry a fair amount of emotional information via the speakers' vocal qualities. By comparison, most text-based communication online is relatively lean in relational information. With only words, subtlety is lost. This is why hints and jokes that might work well in person or on the phone often fail when communicated online.

Ever since the early days of email, online correspondents have devised emoticons, using keystrokes to create sad expressions :-(, surprised looks :-0, and more. Even now, when graphic emojis are readily available, it's difficult to keep up with society's craving for more nonverbal cues. Recognizing the need for more than a thumbs-up or thumbs-down image, in 2016 Facebook added a series of new graphics to expand on the familiar but overused "Like" button.

Asterisks   Enclosing a statement in asterisks can add the same sort of light emphasis. Instead of saying

> I really want to hear from you,

you can say

> I *really* want to hear from you.

Notice how changing the placement of asterisks produces a different message:

> I really want to hear from *you*.

Capitalization   Capitalizing a word or phrase can also emphasize the point:

> I hate to be a pest, but I need the $20 you owe me TODAY.

Overuse of capitals can be offensive. Be sure to avoid typing messages in all uppercase letters, which creates the impression of shouting:

> HOW ARE YOU DOING? WE ARE HAVING A GREAT TIME HERE. BE SURE TO COME SEE US SOON.

Multiple Methods of Emphasis   When you want to emphasize a point, you can use multiple methods:

> I can't believe you told the boss that I sleep with a teddy bear! I wanted to *die* of embarrassment. Please don't *EVER* **EVER** do that kind of thing again.

Regulating

Nonverbal behaviors can control the flow of verbal communication. For example, parties in a conversation often unconsciously send and receive turn-taking cues.[38] When you are ready to yield <u>the floor</u>, the unstated rule is as follows: Create a rising vocal intonation pattern, then use a falling intonation pattern, or draw out the final

cultural idiom

the floor: the right or privilege to speak

syllable of the clause at the end of your statement. Finally, stop speaking. If you want to maintain your turn when another speaker seems ready to cut you off, you can suppress the attempt by taking an audible breath, using a sustained intonation pattern (because rising and falling patterns suggest the end of a statement), and avoiding any pauses in your speech. Other nonverbal cues exist for gaining the floor and for signaling that you do not want to speak.

Contradicting

People often simultaneously express different and even contradictory messages in their verbal and nonverbal behaviors. A common example of this sort of mixed message is the experience we've all had of hearing someone with a red face and bulging veins yelling, *"Angry? No, I'm not angry!"*

Even though some of the ways in which people contradict themselves are subtle, mixed messages have a strong impact. Studies suggest that when a receiver perceives an inconsistency between verbal and nonverbal messages, the nonverbal one carries more weight—more than 12.5% more, according to some research.[39]

Many contradictions between verbal and nonverbal messages are unintentional, revealing feelings that the person exhibiting them would rather keep under cover. But there are other times when contradicting can be strategic. One deliberate use of mixed messages is to save face by sending a nonverbal message that might be awkward if expressed verbally. For example, think of a time when you became bored with a conversation while your companion kept rambling on. At such a time the most straightforward statement would be, *"I'm tired of talking to you and want you to stop talking."* Although it might feel good to be so direct, this kind of honesty is impolite for anyone over 5 years of age. Instead of being blunt in situations like this, a face-saving alternative is to express your interest nonverbally. While nodding politely and murmuring *"uh-huh"* and *"no kidding?"* at the appropriate times, you can signal a desire to leave by looking around the room, turning slightly away from the speaker, or even making a point of yawning. In most cases such clues are enough to end the conversation without the awkwardness of expressing outright what's going on.

Deceiving

Some people are better at hiding deceit than others. For example, most people become more successful liars as they grow older. Research shows that this is especially true for women.[40] High self-monitors (those who readily adapt their behavior to suit the circumstances) are usually better at hiding their deception than communicators who are less self-aware, and raters typically judge highly expressive liars as more honest than those who are more subdued.[41] Not surprisingly, people whose jobs require them to act differently than they feel—such as actors, lawyers, diplomats, and salespeople—are more successful at deception than the general population.[42]

Decades of research have revealed that there are no sure-fire nonverbal cues that indicate deception.[43] This fact helps explain why most people have only a coin flip's chance—50%—of accurately identifying a liar.[44] We seem to be worse at catching deceivers when we participate actively in conversations than when we observe from the sidelines.[45] It's easiest to catch liars when they haven't had a chance to rehearse, when they feel strongly about the information being hidden, or when they feel anxious or guilty about their lies.[46] Trust (or lack of it) also plays a role in which deceptive messages are successful: People who are suspicious that a speaker may be lying pay closer attention to the speaker's nonverbal behavior (e.g., talking faster than normal, shifting posture) than do people who are not suspicious.[47] Table 6-3 lists situations in which deceptive messages are most likely to be obvious.

cultural idioms

cut you off: to interrupt you
sure-fire: certain to succeed

Some people are better than others at uncovering deception. For example, women are consistently more accurate than men at detecting lying and what the underlying truth is.[48] The same research showed that, as people become more intimate, their accuracy in detecting lies actually declines. This is a surprising fact. Intuition suggests that we ought to be better at judging honesty as we become more familiar with others. Perhaps an element of wishful thinking interferes with our accurate decoding of these messages. After all, we would hate to think that a lover would lie to us. When intimates *do* become suspicious, however, their ability to recognize deception increases.[49] Despite their overall accuracy at detecting lies, women are more inclined to fall for the deception of intimate partners than men are. No matter how skillful or inept we may be at interpreting nonverbal behavior, training can make us better.

Communication scholars have studied deception detection for years. One review of decades of research on the subject revealed three findings that have been supported time and again in studies:

- We are accurate in detecting deception only slightly more than half the time.

- We overestimate our abilities to detect others' lies.

TABLE 6-3

Leakage of Nonverbal Clues to Deception

DECEPTION CLUES ARE MOST LIKELY WHEN THE DECEIVER . . .	DECEPTION CLUES ARE LEAST LIKELY WHEN THE DECEIVER . . .
Wants to hide emotions being experienced at the moment.	Wants to hide information unrelated to his or her emotions.
Feels strongly about the information being hidden.	Has no strong feelings about the information being hidden.
Feels apprehensive about the deception.	Feels confident about the deception.
Feels guilty about being deceptive.	Experiences little guilt about the deception.
Gets little enjoyment from being deceptive.	Enjoys the deception.
Needs to construct the message carefully while delivering it.	Knows the deceptive message well and has rehearsed it.

Source: Based on material from Ekman, P. (1981). Mistakes when deceiving. In T. A. Sebok & R. Rosenthal (Eds.), *The Clever Hans phenomenon: Communication with horses, whales, apes and people* (pp. 269–278). New York: New York Academy of Sciences. See also Samhita, L., & Gross, H. J. (2013). The "Clever Hans phenomenon" revisited. *Communicative & Integrative Biology, 6*(6), e27122. doi:10.4161/cib.27122

- We have a strong tendency to judge others' messages as truthful—in other words, we want to believe people wouldn't lie to us (which biases our ability to detect deceit).[50]

As one writer put it, "there is no unique telltale signal for a fib. Pinocchio's nose just doesn't exist, and that makes liars difficult to spot." Moreover, some popular prescriptions about liars' nonverbal behaviors simply aren't accurate.[51] For instance, conventional wisdom suggests that liars avert their gaze and fidget more than nonliars. Research, however, shows just the opposite: Liars often sustain *more* eye contact and fidget *less*, in part because they believe that to do otherwise might look deceitful. Popular characterizations of "scientific" lie detection aren't helpful, either. One experiment found that viewers who watched the television show *Lie to Me* (in which the lead character attempts to catch liars) were actually *worse* at detecting lies than nonviewers, in part because the show focused on nonverbal cues and ignored important verbal content.[52]

Before we finish considering how nonverbal behaviors can deceive, it's important to realize that not all deceptive communication is meant to take advantage of the recipient. Sometimes we use nonverbal messages as a polite way to express an idea that would be difficult to handle if expressed in words. In this sense, the ability to deliberately send nonverbal messages that contradict your words can be a kind of communication competence.

Types of Nonverbal Communication

Now that you understand how nonverbal messages operate as a form of communication, we can look at the various forms of nonverbal behavior. The following pages explain how our bodies, artifacts (such as clothing), environments, and the way we use time all send messages.

Body Movements

For many people, the most noticeable elements of nonverbal communication involve visible body movements. In the following pages, we will examine some of the ways both the body and face can convey meanings.

Posture and Gesture   Stop reading for a moment and notice how you are sitting. What does your position say nonverbally about how you feel? Are there other people near you now? What messages do you get from their posture and movements? Tune your television to any program or watch a randomly selected clip on YouTube, and without turning up the sound, see what messages are communicated by the movements and body position of the people on the screen. These simple experiments illustrate the communicative power of **kinesics**, the study of body movement, gesture, and posture.

Posture is a rich channel for conveying nonverbal information. From time to time postural messages are obvious. If you see a person drag him- or herself through the door or slump over while sitting in a chair, it's apparent that something significant is going on. But most postural cues are more subtle. For instance, the act of mirroring the posture of another person can have positive consequences. One experiment showed that career counselors who used "posture echoes" to copy the postures of clients were rated as more empathic than those who did not reflect the clients' postures.[53] Researchers have also found that partners in romantic relationships mirror each other's behaviors.[54]

kinesics The study of body movement, gesture, and posture.

SELF-ASSESSMENT

How Worldly Are Your Nonverbal Communication Skills?

How well can you match the following nonverbal behaviors to their meanings in different cultures? (Note that the same behavior has multiple meanings in different cultures, so you can use it more than once.)

The answers appear at the end of the chapter, on p. 179.

1. Thumb and forefinger form a circle, while the other three fingers are spread out.

2. Two men hold hands in public.

3. Pinkie and pointer finger point straight up, while thumb holds the two middle fingers down.

4. Palm is held flat up toward another person.

5. Hand is in fist with thumb poking out between the forefinger and middle finger.

6. Palm up, thumb holding all fingers except the pointer, which alternately straightens and curls.

7. Hand is in a fist with thumb pointing up.

___ We're a couple. (United States)

___ This is worthless. [vulgar] (France)

___ Hook 'em horns. (United States)

___ Please give me money. (Japan)

___ I defy or reject you. [vulgar] (Australia, Greece, the Middle East)

___ We respect each other. (Arab countries)

___ OK. (United States)

___ Come here. (United States)

___ I've got your nose! (United States)

___ You are as lowly as a dog. [vulgar] (Asia)

___ I'd like to smear excrement on your face. (Greece)

___ A sign of the devil. (Italy)

___ You're an a**hole. [vulgar] (Latin America)

___ Please stop, or tell it to the hand. (United States)

___ I agree. (United States)

___ I cannot do what you have asked. (Turkey)

Posture can communicate vulnerability in situations far more serious than mere social or business settings. One study revealed that rapists sometimes use postural clues to select victims that they believe will be easy to intimidate.[55] Easy targets are more likely to walk slowly and tentatively, stare at the ground, and move their arms and legs in short, jerky motions.

Amy Cuddy, whose work on power poses begins this chapter, has found that people who act confident for 2 minutes tend to *feel* more confident as a result. They even experience physical changes in the way their bodies cope with stress.[56] Consequently, they often perform better during job interviews and other stressful situations.[57] Although you probably wouldn't adopt a Wonder Woman stance during a job interview, it's a good idea to demonstrate in other ways that you are confident and attentive. Over time, by adopting confident body language in public and in private, people tend to become more confident in general. As Cuddy puts it, it's a process of "faking it 'til you become it."[58]

Gestures are a fundamental element of communication—so fundamental, in fact, that people who have been blind from birth use them.[59] One group of ambiguous gestures consists of what we usually call fidgeting—movements in which one part of the body grooms, massages, rubs, holds, pinches, picks, or otherwise manipulates another body part. Social scientists call these behaviors **manipulators**.[60] Social rules may discourage us from performing most manipulators in public, but people still do so without noticing. For example, one study revealed that deceivers bob their heads more often than truth tellers.[61] Research

manipulators Movements in which one part of the body grooms, massages, rubs, holds, pinches, picks, or otherwise manipulates another part.

Despite our best efforts to conceal emotions, the face can offer revealing clues about how others feel.

What can you learn by paying more attention to facial expressions?

affect blend　The combination of two or more expressions, each showing a different emotion.

paralanguage　Nonlinguistic means of vocal expression: rate, pitch, tone, and so on.

confirms what common sense suggests—that increased use of manipulators is often a sign of discomfort.[62] But not all fidgeting signals uneasiness. People also are likely to use manipulators when relaxed. When they let their guard down (either alone or with friends), they will be more likely to fiddle with an earlobe, twirl a strand of hair, or clean their fingernails. Whether or not the fidgeter is hiding something, observers are likely to interpret manipulators as a signal of dishonesty. Because not all fidgeters are liars, it's important not to jump to conclusions about the meaning of manipulators.

Face and Eyes　How important is eye contact? A look at the cereal aisle in the grocery store will suggest an answer. Check out the old favorites—the Quaker Oats man, the Trix rabbit, the Sun-Maid girl, and Aunt Jemima. All of them will be looking back at you. Researchers at Cornell University found that subjects were more likely to choose Trix over competing brands if the rabbit was looking at them rather than away.[63] "Making eye contact even with a character on a cereal box inspires powerful feelings of connection," said Brian Wansink, one of the study's authors.[64]

Our need for eye contact begins at birth. Newborns instinctively lock eyes with their caregivers,[65] and lack of early eye contact can be an early sign of autism.[66] Researchers have also found that children and infants who avoid eye contact are more likely to exhibit antisocial behaviors.[67]

The face and eyes are probably the most noticed parts of the body, and their impact is powerful. For example, looking at a conversational partner enhances evaluations of intelligence.[68] Smiling cocktail waitresses earn larger tips than unsmiling ones, and smiling nuns collect larger donations than ones with glum expressions.[69] The influence of facial expressions and eye contact doesn't mean that their nonverbal messages are always easy to read. The face is a complicated channel of expression for several reasons. One reason is the number of expressions people can produce. Another is the speed with which they can change. For example, slow-motion films have been taken that show expressions fleeting across a subject's face in as short a time as a fifth of a second. Finally, it seems that different emotions show most clearly in different parts of the face: happiness and surprise in the eyes and lower face, anger in the lower face and brows and forehead, fear and sadness in the eyes, and disgust in the lower face.

Expressions reflecting many emotions seem to be recognizable in and between members of all cultures.[70] Of course, **affect blends**—the combination of two or more expressions showing different emotions—are possible. For instance, it's easy to imagine how someone would look who is fearful and surprised or disgusted and angry.

Research indicates that people are quite accurate at judging facial expressions of these emotions.[71] Accuracy increases when judges know the "target" or have knowledge of the context in which the expression occurs, or when they have seen several samples of the target's expressions.

In mainstream Euro-American culture, meeting someone's glance with your eyes is usually a sign of involvement or interest, whereas looking away signals a desire to avoid contact. Prolonged eye contact has been identified by researchers as one of the main ways people indicate attraction.[72]

Solicitors on the street—panhandlers, salespeople, petitioners—try to catch our eye because after they've managed to establish contact with a glance, it becomes harder for the approached person to draw away.

Voice

The voice is another form of nonverbal communication. Social scientists use the term **paralanguage** to describe nonverbal, vocal messages. You can begin to

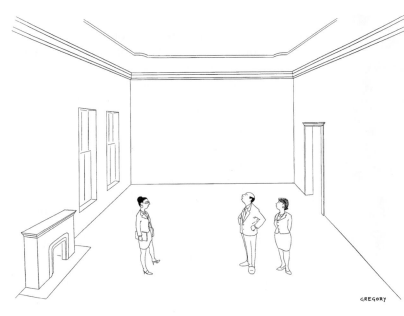

"Wow . . . We could really fill this room with uncomfortable silences."

Source: Alex Gregory The New Yorker Collection/The Cartoon Bank

understand the power of vocal cues by considering how the meaning of a simple sentence can change just by shifting the emphasis from word to word:

- *This* is a fantastic communication book.

    (Not just any book, but this one in particular.)

- This is a *fantastic* communication book.

    (This book is superior, exciting.)

- This is a fantastic *communication* book.

    (The book is good as far as communication goes; it may not be so good as literature or drama.)

- This is a fantastic communication *book*.

    (It's not a play or movie; it's a book.)

@WORK

Vocal Cues and Career Success

Vocal cues "show that we are alive inside—thoughtful, active . . . Text strips that out," said Nicholas Epley, a professor of behavioral science at the University of Chicago Booth School of Business and one of two co authors of the paper "The Sound of Intellect," recently published in *Psychological Science*.

In an experiment presented in the paper, MBA candidates were asked to prepare a 2-minute pitch to a prospective employer. The researchers recorded these pitches and recruited 162 people to evaluate them. Some of the evaluators watched the video; a second group listened to the

audio only; a third group read a transcript of the pitch. The evaluators who heard the pitch—via audio or video—rated the candidates' intellect higher than those who read the transcript. In a second experiment, evaluators read a pitch specifically drafted by candidates to be read, rather than spoken. The result was the same.

One implication is that, whenever possible, it's a good idea to meet with prospective employers in person. Résumés and cover letters, although indispensable, probably don't have the impact of a nonverbally rich interaction.

What counts as attractive varies by geography and coculture. The eclectic look satirized in the TV comedy *Portlandia* (top) is quite different from the New York high fashion styles featured in *Project Runway*.

What qualities are considered attractive in your social circles?

disfluencies Vocal interruptions such as stammering and use of "uh," "um," and "er."

The impact of paralinguistic cues is strong. In fact, research shows that listeners pay more attention to the vocal messages than to the words that are spoken when asked to determine a speaker's attitudes.[73] Furthermore, when vocal factors contradict a verbal message, listeners judge the speaker's intention from the paralanguage, not from the words themselves.[74]

There are many other ways the voice communicates—through tone, speed, pitch, volume, number and length of pauses, and **disfluencies** (such as stammering and use of "uh," "um," "er," and so on). All these factors can do a great deal to reinforce or contradict the message our words convey.

Sarcasm is one instance in which both emphasis and tone of voice help change a statement's meaning to the opposite of its verbal message. Experience this yourself with the following three statements. The first time through, say them literally, and then say them sarcastically.

- Thanks for waking me up.
- I really had a wonderful time on my blind date.
- There's nothing I like better than waking up before sunrise.

Researchers have identified the communicative value of paralanguage through the use of content-free speech—ordinary speech that has been electronically manipulated so that the words are unintelligible, but the paralanguage remains unaffected. (Hearing a foreign language that you do not understand has the same effect.) Subjects who hear content-free speech can consistently recognize the emotion being expressed, as well as identifying its strength.[75]

Paralanguage can affect behavior in many ways, some of which are rather surprising. Researchers have discovered that communicators are most likely to comply with requests delivered by speakers whose speaking rates are similar to their own.[76] Besides complying with same-rate speakers, listeners also feel more positively about people who seem to talk at their own rate.

Some vocal factors influence the way a speaker is perceived by others. For example, communicators who speak loudly and without hesitations are viewed as more confident than those who pause and speak quietly.[77] People who speak more slowly are judged as having greater conversational control than fast talkers.[78] Research has also demonstrated that people with more attractive voices are rated more highly than those whose voices sound less attractive.[79] Along with vocal qualities, accent can shape perceptions. For example, the accents of some nonnative English-speaking job seekers (e.g., those with a pronounced French accent) created favorable impressions with employers, whereas other strong accents (e.g., Japanese) had the opposite effect.[80]

Appearance

How we appear can be just as revealing as how we sound and move. For that reason, let's explore the communicative power of physical attractiveness and clothing.

Physical Attractiveness   Most people claim that looks aren't the best measure of desirability or character, but they typically prefer others whom they find attractive.[81] For example, women who are perceived as attractive have more dates, receive higher grades in college, persuade males with greater ease, and receive shorter court sentences. Both men and women whom others view as attractive are rated as being more sensitive, kind, strong, sociable, and interesting than their less attractive brothers and sisters.

Who is most likely to succeed in business? Place your bet on the attractive job applicant. More than 200 managers in one survey admitted that attractive people get preferential treatment both in hiring decisions and on the job.[82] Height is also a factor. Shorter men have more difficulty finding jobs in the first place, and men over 6 feet 2 inches receive starting salaries that average 12.4% higher than comparable applicants under 6 feet.[83] Based on one study of male earnings, the salary differential is about 2% per inch, which means that a man who is 5 feet 8 inches will earn about $300 a year more (at U.S. minimum wage) than a man who is 5 feet 7 inches, and so on.[84] Consistent with that, tall presidential candidates are historically more likely to win than short ones.

This attractiveness bias has been referred to as "lookism" and can lead to the same kinds of prejudice as racism and sexism.[85] For instance, research shows that women gain an 8% wage bonus for above-average looks and they pay a 4% wage penalty for below-average appearance. For men, the attractiveness wage bonus is only 4%; however, the penalty for below-average looks is a full 13%. Occasionally physical attractiveness has a negative effect: Interviewers may turn down good-looking candidates because they are perceived as threats.[86] While attractiveness generally gets rewarded, glamorous beauty can be intimidating.[87]

Fortunately, attractiveness is something we can control without having to call a plastic surgeon. If you aren't totally gorgeous or handsome, don't despair. Evidence suggests that, as we get to know more about people and like them, we start to regard them as better looking.[88] Moreover, we view others as beautiful or ugly on the basis of not just their "original equipment" but also how they use that equipment. Posture, gestures, facial expressions, and other behaviors can increase the attractiveness of an otherwise unremarkable person. Finally, the way we dress can make a significant difference in the way others perceive us, as you'll now see.

Clothing   Besides protecting us from the elements, clothing is a means of nonverbal communication, providing a relatively straightforward (if sometimes expensive) method of impression management. Clothing can be used to convey, for example, economic status, educational level, social status, moral standards, athletic ability and/or interests, belief system (political, philosophical, religious), and level of sophistication.

Research shows that we make assumptions about people based on their clothing. Communicators who wear special clothing often gain persuasiveness. For example, experimenters dressed in uniforms resembling police officers were more successful than those dressed in civilian clothing in requesting pedestrians to pick up litter and in persuading them to lend a dime for a parking meter to a motorist.[89] Likewise, solicitors wearing a sheriff's or nurse's uniform increased the level of contributions to law enforcement and health care campaigns.[90]

The effects of clothing operate in other contexts as well. Medical patients are significantly more willing to share their social, sexual, and psychological problems with doctors wearing white coats or surgical scrubs than those wearing business dress or casual attire.[91] Along with uniforms, clothing style can shape others' reactions. Pedestrians are more likely to return lost coins to well-dressed people than to those dressed in low-status clothing.[92] Women who are wearing a jacket are rated as being more powerful than those wearing only a dress or skirt and blouse.[93]

As we get to know others better, the importance of clothing shrinks.[94] This fact suggests that clothing is especially important in the early stages of a relationship, when making a positive first impression encourages others to get to know us better. This advice is equally important in personal situations and in employment interviews. In both cases, your style of dress (and personal grooming) can make the difference between the chance to progress further and outright rejection.

ETHICAL CHALLENGE
Clothing and Impression Management

Identify three occasions in which you successfully used clothing to create a favorable but inaccurate impression. What were the consequences of this deception for others? Based on your conclusions, define any situations when clothing may be used as an unethical means of impression management. List both "misdemeanors," in which the consequences are not likely to cause serious harm, and "felonies," in which the deception has the potential to cause serious harm.

cultural idiom
place your bet: predict with confidence

Touch

Physical touch can "speak" volumes. A supportive pat on the back, a high five, or even an inappropriate graze can be more powerful than words, eliciting a strong emotional reaction in the receiver. Social scientists use the term **haptics** when they refer to the study of touch in human behavior.

haptics The study of touch.

Experts argue that one reason actions speak louder than words is because touch is the first language we learn as infants.[95] Besides being the earliest means we have of making contact with others, touching is essential to healthy development.[96] During the 19th and early 20th centuries many babies died from a disease then called *marasmus*, which, translated from Greek, means "wasting away." In some orphanages the mortality rate was quite high, but even children in "progressive" homes, hospitals, and other institutions died regularly from the ailment. When researchers finally tracked down the causes of this disease, they found that many infants suffered from lack of physical contact with parents or nurses rather than poor nutrition, medical care, or other factors. They hadn't been touched enough, and as a result they died. From this knowledge came the practice of "mothering" children in institutions—picking babies up, carrying them around, and handling them several times each day. At one hospital that began this practice, the death rate for infants fell from between 30% and 35% to below 10%.[97]

Touch seems to increase a child's mental functioning as well as physical health. Babies who have been given plenty of physical stimulation by their mothers develop significantly higher IQs than those receiving less contact. By contrast, insufficient physical contact correlates with social problems including communication apprehension and low self-disclosure.[98]

Touch also increases compliance.[99] In one study, people were asked by a male or female confederate to sign a petition or complete a rating scale. People were more likely to cooperate when they were touched lightly on the arm. In one variation of the study, 70% of those who were touched complied, whereas only 40% of the untouched people complied.[100] An additional power of touch is its on-the-job utility. One study showed that fleeting touches on the hand and shoulder resulted in larger tips for restaurant servers.[101]

Touch and Career Success

The old phrase "keeping in touch" takes on new meaning once you understand the relationship between haptics and career effectiveness.

Some of the most pronounced benefits of touching occur in medicine and the health and helping professions. For example, patients are more likely to take their medicines when physicians give a slight touch while prescribing them.[102] In counseling, touch often increases self-disclosure and verbalization of psychiatric patients.[103]

Touch can also enhance success in sales and marketing. Touching customers in a store increases their shopping time, their evaluation of the store, and also the amount of shopping.[104] When an offer to try samples of a product is accompanied by a touch, customers are more likely to try the sample and buy the product.[105]

Touch also has an impact in school. Students are twice as likely to volunteer and speak up in class if they have received a supportive touch on the back or arm from their teacher.[106]

Even athletes benefit from touch. One study of the National Basketball Association revealed that the touchiest teams had the most successful records, whereas in the lowest scoring teams, the team members touched one another the least.[107]

Of course, touch has to be culturally appropriate. Furthermore, touching by itself is no guarantee of success, and too much contact can be bothersome, annoying, or even downright creepy. But research confirms that appropriate contact can enhance your success.

Space

There are two ways that the use of space can create nonverbal messages: the distance we put between ourselves and the territory we consider our own. We'll now look at each of these dimensions.

Distance   The study of the way people and animals use space has been termed **proxemics**. Preferred spaces are largely a matter of cultural norms. For example, people living in hyperdense Hong Kong manage to live in crowded residential quarters that most North Americans would find intolerable.[108] Anthropologist Edward T. Hall has defined four distances used in mainstream North American culture.[109] He says that we choose a particular distance depending on how we feel toward the other person at a given time, the context of the conversation, and our personal goals.

Intimate distance begins with skin contact and ranges to about 18 inches. The most obvious context for intimate distance involves interaction with people to whom we're emotionally close—and then mostly in private situations. Intimate distance between individuals also occurs in less intimate circumstances: visiting the doctor or dentist, at the hairdresser's, and during some athletic contests. Allowing someone to move into the intimate zone usually is a sign of trust.

Personal distance ranges from 18 inches at its closest point to 4 feet at its farthest. The closer range is the distance at which most relational partners stand in public. We are uncomfortable if someone else "moves in" to this area without invitation. The far range of personal distance runs from about 2.5 to 4 feet. This is the zone just beyond the other person's reach—the distance at which we can keep someone "at arm's length." This term suggests the type of communication that goes on at this range: Interaction is still reasonably personal, but less so than communication that occurs a foot or so closer.

Social distance ranges from 4 to about 12 feet. Within it are the kinds of communication that usually occur in business situations. Its closer range, from 4 to 7 feet, is the distance at which conversations usually occur between salespeople and customers and between people who work together. We use the far range of social distance—7 to 12 feet—for more formal and impersonal situations. This is the distance apart we generally sit from the boss.

Public distance is Hall's term for the farthest zone, running outward from 12 feet. The closer range of public distance is the one most teachers use in the classroom. In the farther range of public space—25 feet and beyond—two-way communication becomes difficult. In some cases, it's necessary for speakers to use public distance owing to the size of their audience, but we can assume that anyone who voluntarily chooses to use it when he or she could be closer is not interested in having a dialogue.

Choosing the optimal distance can have a powerful effect on how others respond to us. For example, students are more satisfied with teachers who reduce the distance between themselves and their classes. They also are more

> **proxemics** The study of how people and animals use space.
>
> **intimate distance** One of Hall's four distance zones, ranging from skin contact to 18 inches.
>
> **personal distance** One of Hall's four distance zones, ranging from 18 inches to 4 feet.
>
> **social distance** One of Hall's four distance zones, ranging from 4 to 12 feet.
>
> **public distance** One of Hall's four distance zones, extending outward from 12 feet.

The rules of personal space are revealed most powerfully when they're broken.

Have you ever invaded someone's personal space? Have you ever been unexpectedly standoffish? What were the consequences?

satisfied with the course itself, and they are more likely to follow the teacher's instructions.[110] Likewise, medical patients are more satisfied with physicians who don't "keep their distance."[111]

Territoriality   Whereas personal space is the invisible bubble we carry around as an extension of our physical being, **territory** is fixed space. Any area, such as a room, house, neighborhood, or country, to which we assume some kind of "rights" is our territory. Not all territory is permanent. We often stake out space for ourselves in the library, at the beach, and so on by using markers such as books, clothing, or other personal possessions.

The way people use space can communicate a good deal about power and status relationships. Generally, we grant people with higher status more personal territory and greater privacy.[112] We knock before entering the boss's office, whereas a boss can usually walk into our work area without hesitating. In traditional schools, professors have offices, dining rooms, and even restrooms that are private, whereas the students, who are treated as less important, have no such sanctuaries. In the military, greater space and privacy usually come with rank: Privates sleep 40 to a barracks, sergeants have their own private rooms, and generals have government-provided houses.

Environment

The physical environment people create can both reflect and shape interaction. This principle is illustrated right at home. Researchers showed 99 students slides of the insides or outsides of 12 upper-middle-class homes and then asked them to infer the personality of the owners from their impressions.[113] The students were especially accurate after glancing at interior photos. The decorating schemes communicated accurate information about the homeowners' intellectualism, politeness, maturity, optimism, tenseness, willingness to take adventures, family orientations, and reservedness. The home exteriors also gave viewers accurate perceptions of the owners' artistic interests, graciousness, privacy, and quietness.

Besides communicating information about the designer, an environment can shape the kind of interaction that takes place in it. In a classic experiment, researchers found that the attractiveness of a room influenced the happiness and energy of the people working in it.[114] The experimenters set up three rooms: an "ugly" one, which resembled a janitor's closet in the basement of a campus building; an "average" room, which was a professor's office; and a "beautiful" room, which was furnished with stylish and comfortable furnishings. The subjects in the experiment were asked to rate a series of pictures as a way of measuring their energy and feelings of well-being while at work. Results of the experiment showed that while in the ugly room the subjects became tired and bored more quickly and took longer to complete their task. When they moved to the beautiful room, however, they rated the faces they were judging higher, showed a greater desire to work, and expressed feelings of importance, comfort, and enjoyment. The results teach a lesson that isn't surprising: Workers generally feel better and do a better job when they're in an attractive environment.

The design of an entire building can shape communication among its users. Architects have learned that the way housing projects are designed controls to a great extent the contact neighbors have with one another. People who live in apartments near stairways and mailboxes have many more neighbor contacts than do those living in less heavily traveled parts of the building, and tenants generally have more contacts with immediate neighbors than with people even a few doors away.[115]

territory Fixed space that an individual assumes some right to occupy.

The environment people inhabit makes a statement about the kind of people they are.

What messages does the environment you create say about you?

So far we have talked about how designing an environment can shape communication, but there is another side to consider. Watching how people use an already existing environment can be a way of telling what kind of relationships they want. For example, Robert Sommer watched students in a college library and found that there's a definite pattern for people who want to study alone. While the library was uncrowded, students almost always chose corner seats at one of the empty rectangular tables.[116] At that point, new readers choose seats on the opposite side of occupied tables, thus keeping the maximum distance between themselves and the other readers. One of Sommer's associates tried violating these "rules" by sitting next to, and across from, other female readers when more distant seats were available. She found that the approached women reacted defensively, either by signaling their discomfort through shifts in posture or gesturing or by eventually moving away.

Time

Social scientists use the term **chronemics** for the study of how human beings use and structure time.[117] The way we handle time can express both intentional and unintentional messages.[118] Social psychologist Robert Levine describes several ways that time can communicate.[119] For instance, in a culture that values time highly, like the United States, waiting can be an indicator of status. "Important" people (whose time is supposedly more valuable than that of others) may be seen by appointment only, whereas it is acceptable to intrude without notice on lesser beings. To see how this rule operates, consider how natural it is for a boss to drop in to a subordinate's office unannounced, whereas some employees would never intrude into the boss's office without an appointment. A related rule is that low-status people must never make more important people wait. It would be a serious mistake to show up late for a job interview, although the interviewer might keep you cooling your heels in the lobby. Important people are often whisked to the head of a restaurant or airport line, whereas the presumably less exalted are forced to wait their turn.

Similar principles apply beyond work. For example, when singer Justin Bieber took the stage almost 2 hours late at a sold-out concert in London, his fans were outraged and took to Twitter and other social media to complain. Bieber's management of time sent an unspoken message to many fans: "My life is more important than yours."

The use of time depends greatly on culture.[120] Some cultures (e.g., North American, German, and Swiss) tend to be **monochronic**, emphasizing punctuality, schedules, and completing one task at a time. Other cultures (e.g., South American, Mediterranean, and Arab) are more **polychronic**, with flexible schedules in which multiple tasks are pursued at the same time. One psychologist discovered the difference between North and South American attitudes when teaching at a university in Brazil.[121] He found that some students arrived halfway through a 2-hour class and that most of them stayed put and kept asking questions when the class was scheduled to end. A half hour after the official end of the class, the professor finally closed off discussion, because there was no indication that the students intended to leave. This flexibility of time is quite different from what is common in most North American colleges!

Even within a culture, rules of time vary. Sometimes the differences are geographic. In New York City, the party invitation may say "9 p.m.," but nobody would think of showing up before 9:30. In Salt Lake City, guests are expected to show up on time, or perhaps even a bit early.[122] Even within the same geographic area, different groups establish their own rules about the use of time. Consider your own experience. In school, some instructors begin and end class punctually, whereas others are more casual. With some people you feel comfortable talking for hours in person or on the phone, whereas with others time seems to be precious and not meant to be "wasted."

chronemics The study of how humans use and structure time.
monochronic The use of time that emphasizes punctuality, schedules, and completing one task at a time.

polychronic The use of time that emphasizes flexible schedules in which multiple tasks are pursued at the same time.

cultural idioms
cooling your heels: waiting impatiently
stayed put: remained in place

Building Competence in Nonverbal Communication

By now you should appreciate the wealth of messages expressed nonverbally. You can use this information to develop your communication skills in two respects—by being more attuned to others and by becoming more aware of your own nonverbal messages.

Tune Out Words

It's easy to overlook important nonverbal cues when you're only listening to the words being spoken. As you've already read, words sometimes hide, or even contradict, a speaker's true feelings (e.g., *"I see your point,"* spoken with a frown). Even when spoken words accurately reflect the speaker's thoughts, nonverbal cues can reveal important information about feelings and attitudes.

You can develop skill in recognizing nonverbal cues by tuning out the content of a speaker's language. Because ignoring what your conversational partner is saying can be antagonizing, try focusing on a video or TV program in a language you don't understand. That way you can attend to vocal qualities as well as postures, gestures, facial expressions, and other cues. Once recognizing nonverbal cues has become second nature, you'll find it easier to tune in to them in your everyday conversations

Use Perception Checking

Because nonverbal behaviors are ambiguous, it's important to consider your interpretations as educated guesses, not absolute translations. The yawn that interrupts a story you're telling may signal boredom, but it might also be a sign that the listener is recovering from a sleepless night. Likewise, the impatient tone that greets your suggestion may be aimed at you, or it could mean that your conversational partner is having a bad day.

Perception checking (Chapter 2, pages 54–55) is one way to explore the significance of nonverbal cues. Instead of reading the other person's mind, describe the behavior you've noted, share at least two possible interpretations, and ask for clarification about how to interpret the behavior. With practice, perception checks can sound natural and reflect your genuine desire to understand:

> *To a friend:* At the party last night you said you were tired and left early (behavior). I wasn't sure whether you were bored *(first interpretation)* or whether something else was bothering you *(second interpretation)*. Or maybe you *were* just tired *(third interpretation)*. What was going on?

> *At work:* I need to ask you about something that happened at the end of yesterday's meeting. When I started to ask about the vacation schedule, you interrupted me and said we were running over time and you had to make an important phone call. I'm wondering whether the phone call was the only reason you cut me off, or whether you think I said something wrong. Can you fill me in?

Not every situation is important enough to call for a perception check, and sometimes the meaning of nonverbal cues may seem so clear that you don't need to investigate. But there will certainly be times when exploring alternate interpretations is better than jumping to conclusions.

Melissa McCarthy is known for playing characters who are comically awkward in terms of body movements and facial expressions. The paradox is that playing those roles effectively requires that she have a keen sense for nonverbal cues.

How might you use nonverbal cues to enhance the impact of a joke or a story?

Pay Attention to Your Own Nonverbal Behavior

Along with attending more carefully to the unspoken messages of others, there's value in monitoring your own nonverbal behavior.

You can gain appreciation for this by asking someone to record a video of you when you aren't aware and self-conscious about them doing it. If you're like most people, you're likely to be surprised by at least some of what you see. Most of us have blind spots when it comes to our own communication.[123] For example, we sometimes overestimate how well we are hiding our anxiety, boredom, or eagerness from others. With this in mind, consider the following questions honestly: How does your voice sound? How closely does your appearance match what you've imagined? What messages do your posture, gestures, and face convey?

Once you have a sense of your most notable nonverbal behaviors, you can monitor them without the need for technology.

MAKING THE GRADE

For more resources to help you understand and apply the information in this chapter, visit the *Understanding Human Communication* website at www.oup.com/us/adleruhc.

OBJECTIVE 6.1 **Explain the characteristics of nonverbal communication and the social goals it serves.**

- Nonverbal communication helps us manage our identities, define relationships, and convey emotions.

- It is impossible not to communicate nonverbally. Humans constantly send messages about themselves that are available for others to receive.

- Nonverbal communication is ambiguous. There are many possible interpretations for any behavior. This ambiguity makes it important for the receiver to verify any interpretation before jumping to conclusions about the meaning of a nonverbal message.

- Nonverbal communication is different from verbal communication in complexity, flow, clarity, impact, and intentionality.

  > Describe three messages that qualify as nonverbal communication and one message that is verbal. Explain the difference between the two types.

  > Considering that people cannot *not* communicate, what messages do you think you send nonverbally to strangers who observe you in public?

  > If a friend were preparing for a job interview and asked your advice about appearing and feeling confident, what advice would you give about managing his or her nonverbal communication?

OBJECTIVE 6.2 **Explain the ways in which nonverbal communication reflects culture and gender differences.**

- Gestures may be interpreted entirely differently by members of different cultures.

- In some cultures, eye contact is interpreted as a sign of attentiveness, in others as a challenge or indication of disrespect.

- In general, women are socialized to be more attentive to nonverbal cues than are men.

  > How would you explain the reasons that men and women may use and interpret nonverbal cues differently?

  > Name three rules of appropriateness for making or avoiding eye contact that are familiar to you, but may be unfamiliar to someone from a different culture.

  > What advice would you offer someone who is packing clothing for a trip during which she will encounter people from many different cultures?

OBJECTIVE 6.3 **Describe the functions served by nonverbal communication.**

- Nonverbal communication serves many functions: repeating, substituting, complementing, accenting, regulating, and contradicting verbal behavior, as well as deceiving.

- Emblems describe deliberate nonverbal behaviors that have precise meanings known to everyone within a cultural group.

- Illustrators are nonverbal behaviors that accompany and support spoken words.

- High self-monitors are usually better at hiding their deception than communicators who are less self-aware.
  - > Give an example of each of the following nonverbal behaviors: repeating, substituting for, complementing, accenting, regulating, and contradicting verbal messages.
  - > Observe a conversation for at least 10 minutes, noting as many examples as you can of emblems and illustrators.
  - > Try interacting with people for an hour without using words, then reflect on the experience. What were you able to convey easily through nonverbal means? What was most difficult?

OBJECTIVE 6.4 **List the types of nonverbal communication, and explain how each operates in everyday interaction.**

- We communicate nonverbally in many ways: through posture, gesture, use of the face and eyes, voice, physical attractiveness and clothing, touch, distance and territoriality, environment, and time.
- Kinesics is the study of body movement, gesture, and posture.
- Disfluencies are vocal interruptions such as stammering and using "uh," "um," "er," and so on.
- Affect blends are combinations of two or more expressions showing different emotions.
- Members of some cultures tend to be monochronic, whereas others are more polychronic.
  - > Describe the difference between a monochromic time orientation and a polychromic orientation.
  - > Pause to look at your surroundings. How conducive are they to a positive state of mind? To social interaction or contemplation? How does your environment influence the way you feel right now?
  - > Keep a tally of how many disfluencies you utter in one day. Consider whether you are happy with the results. If not, what would you like to change?

OBJECTIVE 6.5 **Demonstrate competence in assessing the nonverbal communication of others and managing your own nonverbal messages.**

- You can practice by tuning out the content of others' speech and focusing on their behavior, using perception checking, and attending to your own nonverbal cues.
  - > Describe the steps in perception checking, and give an example of each.
  - > Try this experiment: Turn down the sound on a movie or TV show you have never seen. Invite a group of friends to guess what the characters are feeling. See how your interpretations compare.
  - > Notice the nonverbal cues of someone around you right now. Write down two or three interpretations of how that person is feeling. Ask if any of them are accurate.

KEY TERMS

affect blend p. 168

affect displays p. 157

chronemics p. 175

disfluencies p. 170

emblems p. 162

haptics p. 172

illustrators p. 162

intimate distance p. 173

kinesics p. 166

manipulators p. 167

monochronic p. 175

nonverbal communication p. 155

paralanguage p. 168

personal distance p. 173

polychronic p. 175

proxemics p. 173

public distance p. 173

social distance p. 173

territory p. 174

ACTIVITIES

1. **Culture and Nonverbal Communication**

   a. Identify at least three significant differences between nonverbal practices in two cultures or cocultures (e.g., ethnic, age, or socioeconomic groups) within your own society.

   b. Describe the potential difficulties that could arise out of the differing nonverbal practices when members from the cultural groups interact. Are there any ways of avoiding these difficulties?

   c. Now describe the advantages that might come from differing cultural nonverbal practices. How might people from diverse backgrounds profit by encountering one another's customs and norms?

2. **Building Vocal Fluency** You can become more adept at both conveying and interpreting vocal messages by following these directions.

   a. Join with a partner and designate one person A and the other B.

   b. Partner A should choose a passage of 25 to 50 words from a newspaper or magazine, using his or her voice to convey one of the following attitudes:

   > Egotism

   > Friendliness

   > Insecurity

   > Irritation

   > Confidence

c. Partner B should try to detect the emotion being conveyed.

d. Switch roles and repeat the process. Continue alternating roles until each of you has both conveyed and tried to interpret at least four emotions.

e. After completing the preceding steps, discuss the following questions:

> What vocal cues did you use to make your guesses?

> Were some emotions easier to guess than others?

> Given the accuracy of your guesses, how would you assess your ability to interpret vocal cues?

> How can you use your increased sensitivity to vocal cues to improve your everyday communication competence?

SELF-ASSESSMENT

ANSWERS to "How Worldly Are Your Nonverbal Communication Skills?" from p. 167.

1. Thumb and forefinger form a circle, while the other three fingers are spread out.

2. Two men hold hands in public.

3. Pinkie and pointer finger point straight up, while thumb holds the two middle fingers down.

4. Palm is held flat out toward another person.

5. Hand is in fist with thumb poking out between the forefinger and middle finger.

6. Palm up, thumb holding all fingers except the pointer, which alternately straightens and curls

7. Hand is in a fist with thumb pointing up.

The gestures above match up to the meanings listed to the right.

2	We're a couple. (United States)
1	This is worthless. [vulgar] (France)
3	Hook 'em horns. (United States, made popular by University of Texas Longhorns fans)
1	Please give me money. (Japan)
7	I defy or reject you. [vulgar] (Australia, Greece, the Middle East)
2	We respect each other. (Arab countries)
1,7	OK. (United States)
6	Come here. (United States)
5	I've got your nose! (often said playfully to children in the United States)
6	You are as lowly as a dog. [vulgar] (Asia)
4	I'd like to smear excrement over your face. [vulgar] (Greece)
3	A sign of the devil. (Italy)
1	You're an a**hole. [vulgar] (Latin America)
4	Please stop, or tell it to the hand. (United States)
7	I agree. (United States)
5	I cannot do what you have asked. (known as the fig sign in Turkey and some other cultures)

Understanding Interpersonal Communication

7

CHAPTER OUTLINE

LEARNING OBJECTIVES

7.1

Describe characteristics that distinguish interpersonal relationships from impersonal ones and online communication from face-to-face interactions.

7.2

Identify the factors that shape interpersonal attraction.

7.3

Describe the different types of communication dynamics in friendships, family relationships, and romantic relationships.

7.4

Explain the relevance of content and relational meaning, metacommunication, self-disclosure, dialectical tensions, and deception to interpersonal communication.

Reality television may not be real, but it reminds us that interpersonal relationships aren't always easy. As you read this chapter, consider the following questions in light of your own relationships:

?

When has communication felt most personal and most impersonal? What was the difference?

?

When has what *wasn't* said been more important than the topic under discussion?

?

Recall times when you tried to talk about an important relationship with someone. What happened?

"WILL YOU ACCEPT THIS ROSE?" This line from *The Bachelor* and *The Bachelorette* invites a contestant to be part of another round in the reality television game of love. To date, viewers have watched 23 couples say "I will" in marriage proposals at the shows' conclusions. However, only four of those couples have stayed together long enough to actually say "I do."[1] What happens after the show to turn "I love you forever" into "Leave me alone"?

This chapter explores the role of communication in close relationships. As you surely know from experience, these relationships sometimes seem easy and rewarding. At other times, they are challenging and downright confusing.

Sean Lowe, one of the few who found lasting love on *The Bachelor*, describes his wake-up moment: "You leave the show, then you get into the real world and find out like, 'Oh crap! Being in a relationship isn't always easy and it actually takes work."[2] The TV couples who called it quits often say that hurt feelings and unresolved conflicts did them in.[3,4]

Communicating effectively when powerful emotions are involved requires know-how and awareness. This chapter begins by considering what makes some relationships closer than others. It explores how we choose relational partners and the different ways we communicate with friends, family members, and romantic partners. The rest of the chapter is about communication phenomena that affect all of our close relationships. We close with one of the most challenging aspects of interpersonal communication—lies and evasions—which segues into Chapter 8's coverage of effective conflict management.

Characteristics of Interpersonal Communication

If you had to guess, which one of the following five scenarios would you say exemplifies interpersonal communication?

1. Eduardo doesn't know Martina well yet, but he strikes up a conversation so he can learn more about her.

2. Two computer programmers engage in such lively brainstorming that it's difficult to remember later whose idea was whose.

3. Filip smiles at the cashier in the grocery store and says thank you as he leaves with his purchase.

4. After a hard day, Megan looks forward to a conversation with her roommate, whose concern and similar experiences always make her feel better.

5. Aya and Ying established a connection via Aya's blog posts about her experiences in China. Now they text each other nearly every day.

If you picked the scenario with Megan, you're right. If you also selected the online example, score that one as a maybe for now. We'll come back to it as we consider what distinguishes interpersonal communication from other types of interactions.

What Makes Communication Interpersonal?

As these examples show, not all dyadic relationships are interpersonal. Several factors make interpersonal communication unique and precious. Theorist John Stewart defines it as "communication that happens when the people involved talk and listen in ways that maximize the presence of the personal."[5] In other words, **interpersonal communication** involves two-way interactions between people who are part of a close and irreplaceable relationship in which they treat each other as unique individuals. Let's consider the implications of that definition by returning to the examples you just read:

> **interpersonal communication**
> Two-way interactions between people who are part of a close and irreplaceable relationship in which they treat each other as unique individuals.

1. Eduardo and Martina may eventually develop a close relationship, but at this point they can't yet appreciate one another's unique qualities. The emotional intimacy that characterizes interpersonal relationships doesn't occur instantly. Rather, it evolves over time.

2. The computer programmers' brainstorming session is a good example of effective task-related communication. But there's no evidence that the colleagues have a close personal attachment.

3. Although interactions with strangers and casual acquaintances serve an important role in our lives, Filip's exchange with the cashier is not interpersonal. There is no exchange of personal information, and Filip will probably not mind if a different cashier helps him tomorrow. By contrast, Filip's relationship with his best friend is unique and irreplaceable. Part of the sadness when an interpersonal relationship ends is that, even if the relationship had its faults, we will never have another one exactly like it.

4. Megan's conversation with her roommate is sure to be interpersonal in that they are invested in listening to and sharing personal information with each other as unique individuals. By contrast, a therapist may be eager to help a client, but the relationship is based on roles that involve a degree of emotional distance and mostly a one-way flow of personal information. The distinction is so important that therapists typically uphold a code of ethics that prohibits them from developing personal (interpersonal) relationships with their clients.

5. The final example (Aya and Ying) isn't so easy to categorize. In the next section, we explore the nature of online interactions.

cyber relationship An affiliation between people who know each other *only* in the virtual world.

Mediated Interpersonal Communication

Aya and Ying, who connected online, have a **cyber relationship** in that they know each other only in the virtual world. Can two people who have never met in person engage in interpersonal communication? One the one hand, they are strangers in many ways. On the other, their emotional connection may feel even more real and powerful than it does with people they see every day.

Early definitions of interpersonal communication specified that it had to take place in person.[6] That was presumably because theorists considered face-to-face interactions to have a richness lacking in other available channels.

Times have changed. Today, few scholars dispute the idea that interpersonal communication can occur via texts, emails, calls, video chats, and other technical channels.[7] Yet online interpersonal communication is different from face-to-face exchanges. In this section, we consider why people use technology to communicate interpersonally and how that influences their relationships, sometimes for the better and sometimes not.

All in all, it's impossible to say without more information whether Aya and Ying from our example engage in interpersonal communication online. Perhaps their texts are simply about interesting things to do in China, without what Stewart calls "the presence of the personal."[8] On the other hand, they may have a close and personal connection. If that is true, it's a safe assumption that they would like to meet in person one day. The most highly rated friendships are those in which people have both in-person and electronic contact with each other.[9]

Why People Use Communication Technology   Here are four of the most common reasons people communicate via mediated channels with others they care about:

1. **Mediated channels enable communication that would not happen otherwise.** When people are far away from each other or have different routines, technology can help them stay connected. Perhaps for these reasons, adolescents who use online communication in moderation typically have more cohesive friendships than those who do not,[10] and couples who talk frequently via mobile phone feel more loving, committed, and confident about their relationship than couples who don't.[11]

2. **Mediated communication can feel nonthreatening.** For some people, particularly those who are introverted, mediated channels make it easier to build close relationships.[12] Sociolinguist Deborah Tannen remembers when online communication enhanced her relationship with a bashful colleague: "Face to face he mumbled so I could barely tell he was speaking. But when we both got on e-mail, I started receiving long, self-revealing messages; we poured our hearts out to each other."[13] Even if you're not shy, you may find some messages easier to send than to say aloud.

3. **Online communication can be validating.** One appealing quality of online communication is its potential to convey social support. Posting news of the "A" you earned in English may be rewarded almost instantly with "likes" and congratulations. University students who use Facebook typically experience less stress than their peers, especially when they consider their online friends to be supportive, interpersonally attractive, and trustworthy.[14]

4. **Electronic communication often has a pause option.** Many forms of mediated communication are asynchronous, meaning that they allow you to think about messages before sending them. This is a plus when you can catch mistakes or avoid blurting out something you would regret later.[15]

cultural idiom
on the other hand: from the opposite point of view

Given all these advantages, what are the drawbacks of online communication? There are a few, as we discuss next.

Drawbacks of Online Communication   Whether people who communicate online already know one another or not, the virtual world presents challenges in terms of interpersonal communication. For one thing, excessive use of online communication can diminish relationships, as happens when we are more tuned in to texts and tweets than to humans in the same room.[16]

Phubbing refers to episodes in which people snub those around them by paying attention to their phones instead.[17] Researchers in one study found that the mere presence of mobile devices can have a negative effect on closeness, connection, and conversation quality during face-to-face discussions of personal topics.[18]

Another downside is that online communication can encourage quantity over quality. Mobile devices provide a steady stream of information that begs to be read, even though much of it is trivial. And a person can have thousands of online "friends" and followers but few people he or she can count on during hard times. As a result, there is a point of diminishing rewards. In moderation, social media can boost our sense of connection and identity. However, people who take it to extremes tend to be lonelier than their peers.[19] Teen model and social media celebrity Essena O'Neill is one of those people. She minimized her online presence, she says, when she began to feel consumed and isolated by it.[20]

Teen model Essena O'Neill shocked her 612,000 Instagram followers in 2015 when she closed her account, removed thousands of online photos of herself, and vowed to break her social media addiction. "It consumed me," she said. "I wasn't living in a 3D world."[21]

Have you ever felt that online interactions have negatively affected your interpersonal relationships?

How We Choose Relational Partners

Considering the number of people with whom we communicate every day, truly interpersonal interaction is rather scarce. That isn't necessarily unfortunate: Most of us don't have the time or energy to create personal relationships with everyone we encounter—or even to act in a personal way all the time with the people we know and love best. In fact, the scarcity of qualitatively interpersonal communication contributes to its value. Like precious jewels and one-of-a-kind artwork, interpersonal relationships are special because of their scarcity.

Sometimes we don't have a choice about our relationships: Children can't select their parents, and most workers aren't able to choose their bosses or colleagues. In many other cases, though, we seek out some people and actively avoid others. The next section considers the factors behind selecting friends and romantic partners.

phubbing A mixture of the words *phone* and *snubbing*, used to describe episodes in which people pay more attention to their devices than they do to the people around them.

Evaluating Relationship Potential

Following are eight common explanations people offer for why they chose to form a close relationship with someone in particular. As you will see, some factors are more salient to friendships and others to romantic relationships, but they all may figure in both types.

1. **The person is physically attractive.**
    Most people claim that we should judge others on the basis of character, not appearance. The reality, however, is quite the opposite—particularly in the early stages of romantic relationships. People are more likely to show interest in others they consider physically attractive, both in person[22] and online.[23] This may be why physically unattractive people (based on reviewer rankings)

ETHICAL CHALLENGE

Is It Cheating?

If someone in a committed relationship engages in romantic talk online with someone he or she will never meet in person, do you think that counts as cheating? Why or why not?

Similarities may attract us to other people, but there is no evidence that look-alike couples such as Leonardo DiCaprio and Margot Robbie have an advantage in the long run. Long-term relationship success depends far more on compatible personalities and communication styles.

Have you ever been attracted to someone based on looks and then found out you were incompatible in other ways?

are more likely than others to enhance the photos they post on online dating sites, although they usually report truthfully about other details of their lives.[24]

2. **We have a lot in common.**
   In most cases, we like people whose temperament, values, and life goals are similar to our own. For example, the more alike a married couple's personalities are, the more likely they are to report being happy and satisfied in their marriage.[25] Similarity is a factor in the early stages of friendship as well, when all we have go by is appearance. When given a choice of where to sit, we tend to gravitate toward people whose features are similar to our own—those who also wear glasses or have the same color hair we do.[26] Researchers suggest that similarities are comfortable, and they may reduce our fear that the other person will reject us.

3. **We balance each other out.**
   The folk wisdom that "opposites attract" seems to contradict the similarity principle just described. In truth, both are valid. Differences strengthen a relationship when they are *complementary*—when each partner's characteristics satisfy the other's needs. Individuals, for instance, are often likely to be attracted to each other when one partner is dominant and the other passive.[27] Relationships also work well when the partners agree that one will exercise control in certain areas (*"You make the final decisions about money"*), and the other will exercise control in different areas (*"I'll decide how we ought to decorate the place"*). Strains occur when control issues are disputed.

4. **The person likes and appreciates me.**
   It's no mystery why reciprocal liking is appealing. People who approve of us bolster our feelings of self-esteem. Attraction has to be mutual to spark and maintain a relationship, though. And of course, we aren't drawn toward everyone who seems to like us. If we don't find the other person's attributes attractive, their interest can be a turn-off.

5. **I admire the person's abilities.**
   It's natural to admire people who are highly competent in something we care about. Forming relationships with talented and accomplished people can inspire us and provide flattering validation.[28]

6. **The person opens up to me.**
   People who reveal important information about themselves often seem more likable, provided of course that what they share is appropriate to the setting and the stage of the relationship.[29] Self-disclosure is appealing partly because people enjoy a sense of similarity, either in experiences (*"I broke off an engagement, myself"*) or in attitudes (*"I feel nervous with strangers, too"*). And when people share private information, it suggests that they respect and trust us—a kind of liking that we've already seen increases attractiveness. Disclosure plays an even more important role as relationships develop beyond their earliest stages. (We'll talk more about that later in the chapter.)

7. **I see the person frequently.**

   In many cases, proximity leads to liking.[30] For instance, we're more likely to develop friendships with close neighbors than with distant ones, and the chances are good that we'll choose a mate with whom we cross paths often. Proximity allows us to get more information about people and to engage in more relationship-building behaviors together. Also, people in close proximity may be more similar to us than those who live, work, and play in different places. The Internet provides a new means for creating closeness, as users are able to experience "virtual proximity" in cyberspace.[31]

8. **The relationship is rewarding.**

   Some social scientists argue that all relationships—both impersonal and personal—are based on a semi-economic model called **social exchange theory**.[32] This model suggests that we seek out people who can give us rewards that are greater than or equal to the costs we encounter in dealing with them. Rewards may be tangible (a nice place to live, a high-paying job) or intangible (prestige, emotional support, companionship). Costs are undesirable outcomes, such as a sense of obligation, emotional pain, and so on. According to social exchange theorists, we use this formula (usually unconsciously) to decide whether dealing with another person is a "good deal" or "not worth the effort."

   > **social exchange theory** The idea that we seek out people who can give us rewards that are greater than or equal to the costs we encounter in dealing with them.

Relationship Reality Check

Having just pointed out common considerations when choosing relational partners, a few caveats are in order. Read the following before you fall into the trap of thinking you must be supermodel stunning, Mensa smart, or Olympic-level talented for people to find you appealing.

- **First impressions can mislead.** Evidence shows that we befriend people whose interests and attitudes *seem* similar to our own. This is partly an illusion, however. We tend to overestimate how similar we are to our friends and underestimate how similar we are to people we don't know well.[33] In reality, there is strong evidence that superficial similarities such as appearance do not predict long-term happiness with a relationship,[34] and when we are willing to communicate with a range of people, our differences are not usually as great as we thought.[35]

- **Our priorities change.** For example, physical factors that catch our eye at first glance aren't necessarily what we want in the long run. Although women in one study preferred to date muscular men, they considered men with average body shapes to be more appealing candidates for marriage.[36] As one social scientist put it, "attractive features may open doors, but apparently, it takes more than physical beauty to keep them open."[37]

- **Perfection can be a turn-off.** We like people who are attractive and talented, but we are uncomfortable around those who are *too* perfect. Let's face it: No one wants to look bad by comparison. And it's more important to be nice than to be flawless. College students in one study were twice as likely to choose a very nice stranger over a very smart one. The researchers concluded that, if people had to choose, most would rather spend time with a "lovable fool" than a "competent jerk."[38]

- **It's not all about communication, but it's a lot about communication.** The online dating service eHarmony matches couples based on "29 dimensions of compatibility," and other online dating sites make similar promises.

> **?** ASK YOURSELF
>
> Make a list of the qualities that first attracted you to a special person in your life. Did the importance you placed on those characteristics change as the relationship developed? If so, how?

We can imagine such compatibility algorithms for finding friends as well. However, the long-term success of people matched by computer algorithms is no greater than that of people who meet on their own.[39] That's because long-term compatibility relies less on superficial similarities and more on how people interact with each other once they start a relationship and encounter stressful issues.

A few lessons emerge from these observations. One is to break free of your comfort zone and give new people a chance. Another is that when you are the person who seems different from others, you can help reduce the stranger barrier by being friendly and approachable and letting people get to know the real you. Finally, don't be discouraged if you aren't "perfect" by society's standards. Being perfect is overrated. Being nice matters more.

Types of Interpersonal Relationships

A fiendishly cynical Tumblr post defines what happens when one is *friendzoned*: "a person decide[s] that you're just a friend and no longer a dating option. You become this complete non-sexual entity in their eyes, like a sibling or a lamp."[40]

Wait a minute, Tumblr sage! It's natural to be discouraged by flagging romantic interest, but platonic relationships can also be immensely rewarding. In fact the Greek philosopher Plato, whose work (and name) inspired the term *platonic*, would be shocked to hear people use the term "*just* friends." He considered friendship to occupy a higher plane than sexual attraction.[41] Furthermore, the most lasting relationships you are likely to have will be with family members.

In this section, we consider what the Tumblr satirist got right: The communication patterns that define relationships of various types *are* different in many ways. We'll take a closer look at communication with friends, family members, and romantic partners.

Friendship

Good friends keep us healthy, boost our self-esteem, and make us feel loved and supported.[42] They also help us adjust to new challenges and uncertainty.[43] It's not surprising, then, that people with strong and lasting friendships are happier than those without them.[44]

Friendships are special for a number of reasons. First, unlike a parent–child, teacher–student, or doctor–patient relationship, in which one partner has more authority or higher status than the other, friends typically treat each other as equals.[45] Second, unlike family and romantic relationships that may be limited in number, we can have as many friends as we want or have time for. Finally, we are relatively free to design friendships that suit our needs. We may have close friends we talk to every day and others we see only once in a while.

Types of Friendships A quick survey of your social network will confirm that friendships come in many forms. Think of several friends in your life—perhaps a new friend, a longstanding friend, and a colleague at work. Then see how they compare on the dimensions described in this section.

Youthful Versus Mature Some elements of friendship hold true across the life span. For instance, self-disclosure is typical in close relationships from childhood to old age.[46] But in other ways, the nature of friendships varies as the participants mature.[47]

Can Men and Women Be Just Friends?

It's an age-old question. And the answer depends on whom you ask. Women typically say yes. But men give a decidedly iffy answer. In a study of 88 pairs of heterosexual, college-age opposite-sex friends, most women said the friendship is purely platonic, with no romantic interest on either side.[48] The men were more likely to say that they secretly harbor romantic fantasies about their gal pals and they suspect (often wrongly, it seems) that the feeling is mutual.

Researchers speculate that men and women get their wires crossed partly because they communicate differently. Because women usually expect friends to be emotionally supportive and understanding, they engage in self-disclosure and empathy behaviors.[49] From the male perspective, this may feel like the trappings of romance rather than friendship. Men's same-sex friendships typically involve more independence, more friendly competition, and fewer intimate disclosures than women's do.[50] Those behaviors may not strike women as romantic.

So far, it seems there's still a gender gap where friendship is involved. But scientists encourage people to take heart. In the history of human development, male–female friendship is a recent development.[51] It may take a little practice.

Preschoolers rarely have enduring friendships. Instead, they enjoy time with temporary playmates. As they grow older, children usually form more stable friendships, but primarily to meet their own needs and with little sense of empathy. During adolescence, friendships become a central feature of social life—often more important than family. In these teen years, friends begin to be valued for their personal qualities, not just as playmates or activity companions.

As they move away from familiar environments, young adults expand their circle of friends in ways that often prove highly satisfying.[52] By this point in life, the qualities that are important in a friend become stable and mature: helpfulness, support, trust, commitment, and self-disclosure. As the responsibilities of marriage and family grow, the desire to have strong friendships may stay the same, but the time available to support them can decline.[53] In older adulthood, friendships become especially valuable as a means of social support. Having strong relationships contributes to both satisfaction and health.[54]

Short-Term Versus Long-Term   Short-term friends tend to change as our lives do. We say goodbye because we move, graduate, switch jobs, or change lifestyles. Perhaps we party less or spend more time off the ball field than we used to. Our social networks are likely to change, too. On the other hand, long-term friends are with us even when they aren't. These friendships tend to survive changes and distance.[55] Particularly today, with so many different ways to stay in touch, people report that—as long as the trust and a sense of connection are there—they feel as close to their long-term friends who live far away as to those who are nearby.[56]

CHECKLIST ✓
Being a Better Friend

Experts suggest the following communication strategies to keep your friendships strong.

☐ **Be a good listener.** Listen not only to what is said but to what isn't said; pay close attention to your friend's nonverbal cues. Put aside distractions to show how much you care.

☐ **Give advice sparingly.** A better option is to listen attentively and, if appropriate, ask your friend what options he or she imagines and what the pros and cons might be of each.

☐ **Share feelings respectfully.** Although it may be tempting to make a snide remark or say *"it's nothing"* when you feel upset, those strategies are likely to damage friendships.[63]

☐ **Apologize and forgive.** If you slip up, such as by forgetting an important date or saying something that embarrasses the other person, admit the mistake, apologize sincerely, and promise to do better in the future.[64] By the same token, be aware that you are likely to make such mistakes yourself, which may inspire you to offer forgiveness.[65]

☐ **Be validating and appreciative.** Find ways to let your friend know that he or she matters to you.[66]

☐ **Stay true through hard times.** People who believe their friends will be there for them typically experience less everyday stress and more physical and emotional resilience than other people.[67]

☐ **Be trustworthy and loyal.** Two of the most dreaded violations of trust are sharing private information with others and saying unkind things about a friend behind his or her back.[68] Be a good friend by maintaining confidences and standing up for others, even when they aren't around.

☐ **Give and take equally.** We are happiest when there is equal give and take in a friendship. One benefit is the sense that we make a difference in someone's life, not just that someone makes a difference in ours.[69]

Low Disclosure Versus High Disclosure Some of your friends know more about you than others. Self-disclosure is associated with greater levels of intimacy such that only a few confidants are likely to know your deepest secrets. But when it comes to even slightly less personal news, we are experiencing a revolution in terms of self-disclosure. Today, it's quite common for someone to announce personal news to hundreds of friends and acquaintances with a single post or tweet.

Doing-Oriented Versus Being-Oriented Some friends experience closeness "in the doing." That is, they enjoy performing tasks or attending events together and feel closer because of those shared experiences.[57] In these cases, different friends are likely to be tied to particular interests—a golfing buddy or shopping partner, for example. Other friendships are "being-oriented." For these friends, the main focus is on being together, and they might get together just to talk or hang out.[58]

Low Obligation Versus High Obligation There are some friends for whom we would do just about anything. For others, we may feel a lower sense of obligation, both in terms of what we would do for them and how quickly we would do it. Cultural elements may affect this sense. For example, friends raised in a low-context culture such as the United States are more likely than those raised in a high-context culture such as China to express their appreciation for a friend out loud (see Chapter 3). The Chinese are more likely to express themselves indirectly—mostly often by doing favors for friends and by showing gratitude and reciprocity when friends do favors for them.[59] It's easy to imagine the misunderstandings that might occur when one friend puts a high value on words and the other on actions.

Frequent Contact Versus Occasional Contact You probably keep in close touch with some friends. Perhaps you work out, travel, socialize, or Skype daily with them. Other friendships have less frequent contact—maybe an occasional phone call or text message. Of course, infrequent contact doesn't always correlate with levels of disclosure or obligation. Many close friends may see each other only once a year, but they pick right back up in terms of the breadth and depth of their shared information.

Same Sex Versus Other Sex Friendship varies, to some extent, by sex. Same-sex friendships between men typically involve good-natured competition and a focus on tasks and events, whereas female friends tend to treat each other more as equals and to engage in emotional support and self-disclosure.[60] It can work out well when we bring these expectations to our other-sex friendships. Men often say that they find it validating when female friends encourage them to be more emotionally expressive than usual, and women say they appreciate the opportunity to be concrete and direct with their guy friends.[61] Different expectations can lead to misunderstandings, however. See the Understanding Diversity box on this page to explore whether heterosexual men and women can be "just" friends.

Sexual orientation is another factor in friendship. Friendships between people who are gay and those who are straight can lead to feelings of belonging and acceptance. Gay men who have close friends that are straight are less likely than their peers to perceive that society judges them harshly for being gay.[62] And there seems to be some truth to the idea that straight women and gay men make great friends. Gay men are nearly twice as likely as lesbian women to have opposite-sex friendships.[70] This may be because gay men and straight women trust each other's advice about love and romance. Both sides say they enjoy getting an opposite-sex perspective without the complications of a hidden sexual agenda.[71]

In-Person Versus Mediated   The average person has many more online friends than physical ones—double the amount, according to one report.[72] Quantity isn't the only difference between mediated and offline friendships, however. It turns out that online-only friendships may carry a greater risk of deception or hostility.[73]

Research also suggests that face-to-face friends are typically more interdependent than online friends, especially during the early stages of their relationships. In-person friends are more likely to talk about topics in depth, and they typically share a deeper level of understanding and commitment than online friends do. And, not surprisingly, in-person friends tend to have more similar social networks. However, as online friendships develop, the difference in quality when compared with in-person friendships tends to diminish.[74] There is also some evidence that online relationships can become even more personal, as time goes on, than the in-person variety.[75]

To enhance your communication skills, no matter what type of friends you have, see the checklist about being a good friend.

Family Relationships

What makes a family? As your own experiences probably show, a family might encompass bloodline relatives, adopted family members, stepparents and stepsiblings, honorary aunts and uncles, and others. This makes it easy to understand why people can be hurt by questions such as, *"Is he your natural son?"* and *"Is she your real mother?"* Calling some family members *"real"* implies that others are fake or that they don't belong.[76]

In light of the wide range of possibilities, theorist Martha Minnow argues that people who share affection and resources as a family and who think of themselves and present themselves as a family *are* a **family.**[77]

> **family**   A collection of people who share affection and resources and who think of themselves and present themselves as a family

Parents and Children   We learn how to behave largely from our parents. At a more subtle level, we also learn from them how to think about the world around us and how to manage our emotions. If you grew up in a family that emphasized the role of conversation in problem-solving, evidence suggests that you are most likely to engage in that strategy with others as an adult.[78] By contrast, if the emphasis was on conformity ("a rule is a rule"), you are more likely to think people should follow the rules without questioning them.

Siblings   Sibling relationships involve an interwoven, and often paradoxical, collection of emotions. Children are likely to feel both intense loyalty and fierce competition with their brothers and sisters and to be both loving and antagonistic toward them. In the midst of this complexity—what some theorists call the "playing and arguing, joking and bickering, caring and fighting" of sibling life— children learn a great deal about themselves and how to relate to others.[79]

Even in adulthood, sibling relationships vary widely. Supportive brothers and sisters talk regularly and consider themselves to be accessible and emotionally close to one another.[80] On the other end of the spectrum are competitive and even hostile siblings, who never outgrow the rivalry common in childhood and adolescence.

Grandparents and Grandchildren   Less than 100 years ago, adults seldom lived long enough to know their grandchildren. Today, children more often reach adulthood knowing not only their grandparents but their great-grandparents.

Being a Better Family Member

Communicating with family members can be a challenge, but it's vital. Following are some strategies for successful communication based on experts' advice.

☐ **Share family stories.** Family stories contribute to a shared sense of identity. They also convey that adversity is an inevitable part of life, and they can suggest strategies for overcoming it.[92]

☐ **Listen to each other.** People who are involved in reflection and conversation learn how to manage and express their feelings better than people who don't. They tend to have better relationships as a result.[93,94]

☐ **Negotiate privacy rules.** Privacy violations among family members can have serious consequences.[95] At the same time, too much privacy can mean overlooking dangerous behavior and avoiding distressing but important topics. Experts suggest that families talk about and agree on privacy expectations and rules.

☐ **Coach conflict management.** Effective conflict management doesn't just happen spontaneously.[96] It's a sophisticated process that often goes against our fight-or-flight instincts. Families can help by creating safe environments for discussing issues and striving for mutually agreeable solutions.

☐ **Go heavy on confirming messages.** Supportive messages from family members can give us the confidence to believe in ourselves. For example, teens whose parents frequently compliment and encourage them are less likely than others to drop out of high school.[97]

☐ **Have fun.** Happy families make it a point to minimize distractions and spend time together on a regular basis. They establish togetherness rituals that suit their busy lives, such as sharing dessert even when they can't eat dinner together,[98] and they engage in adventures, both large and small.

Grandparents often have the time and inclination to interact with younger members of the family. They can provide loving attention and fun without having to scold or punish, and they can be caring and supportive listeners. It's a positive dynamic both sides can appreciate.[81] Children who interact frequently with their grandparents are more likely than others to remain in close contact with them later in life.[82]

The trick for grandparents, say many theorists, is to manage the balance between "being there" and "interfering."[83] It may be difficult to know when or whether they should serve as disciplinarians or parenting advisors. Bruce Feiler, who researches and writes about family dynamics, and his wife Linda have an understanding with their parents when it comes to the grandkids: *Our house, our rules; your house, your rules.* He says that communicating about expectations openly has strengthened family ties.[84]

See the checklist on this page for tips on communicating well with the people you call family, in all their many roles.

Romantic Partners

Romantic love is the stuff of songs, fairytales, and happy endings. So it might surprise you that the butterflies-in-your-belly sense of romantic bliss isn't a great predictor of happiness. A much better indicator is the effort that couples put into their communication. Factors such as trust, agreeableness, and emotional expressiveness are primarily responsible for long-term relationship success.[85,86] In this section, we explore the role of communication in forming and sustaining romantic relationships.

Male and Female Intimacy Styles By definition, romantic **intimacy** requires that we express ourselves personally through physical contact, shared experiences, intellectual sharing, and emotional disclosures.[87] Being open with another person in this manner involves vulnerability. But intimacy also yields some of life's greatest rewards, including a sense of being understood, accepted, and supported.[88] As you will see, romantic intimacy may mean different things to different people, and communication can be a tool for both enhancing and diminishing it.

Until recently, most social scientists believed that women were better at developing and maintaining intimate relationships than men. This belief grew from the assumption that the most important ingredients of intimacy are sharing personal information and showing emotions. Most research *does* show that women (taken as a group, of course) are more willing than men to share their thoughts and feelings.[89] However, male–female differences aren't as great as they seem,[90] and emotional expression isn't the *only* way to develop close relationships.

Whereas women typically value personal talk, men often demonstrate caring by doing things for their partners and spending time with them. It's easy to imagine the misunderstandings that result from this difference. Indeed, women's most frequent complaint is that men don't stop to focus on "the relationship" enough.[91] Men, however, are more likely to complain about what women do or don't do in an instrumental sense. For example, they may consider it highly significant if a woman doesn't call when she says she will.

Men and women may view sex differently as well. Whereas many women think of sex as a way to express intimacy that has already developed, men are more likely to see it as a way to *create* that intimacy.[99] In this sense, the man who encourages sex early in a relationship or after a fight may view the shared activity as a way to build closeness. By contrast, the woman who views personal talk as the pathway to intimacy may resist the idea of physical closeness before the emotional side of the relationship has been discussed.

What happens when both partners are of the same sex? Research is limited so far, but much of it suggests that, on average, same-sex couples are more satisfied with their relationships than are heterosexual couples. Same-sex couples typically report greater harmony, less emotional distance, and more shared activities than male–female romantic partners.[100] Researchers speculate that this may be because same-sex couples have been socialized to communicate in similar ways and to have similar expectations. And, because they are likely to treat each other as equals, same-sex couples tend to seek common ground when faced with decisions and conflict.[101]

Love Languages   Some intimacy styles have less to do with sex or gender than with personal preferences. Relationship counselor Gary Chapman[102] observes that people typically orient to one of five love languages. You can learn more about love languages in your life by completing the self-assessment on the next page.

Affirming Words   This language includes compliments, thanks, and statements that express love and commitment. Even when you know someone loves and values you, it's often nice to hear it in words. The happiest couples continue to flirt with each other, even after they have been together for many years.[103]

Quality Time   Some people show love by completing tasks together, talking, or engaging in some other mutually enjoyable activity. The good news is that, even when people can't be together physically, talking about quality time can be an important means of expressing love. For example, partners separated by military deployments often say they feel closer to each other just talking about everyday activities and future plans.[104]

Acts of Service   People may show love by performing favors such as caring for each other when they are sick, doing the dishes, making meals, and so on. Committed couples report that sharing daily tasks is the most frequent way they show their love and commitment.[105] Although each person need not contribute in exactly the same ways, an overall sense that they are putting forth equal effort is essential to long-term happiness.[106]

Gifts   It's no coincidence that we buy gifts for loved ones on Valentine's Day and other occasions such as birthdays and anniversaries. For some people, receiving a gift—even an inexpensive or free one such as a flower from the garden or a handmade card—adds to their sense of being loved and valued.[107]

Physical Touch   Loving touch may involve a hug, a kiss, a pat on the back, or having sex. For some people, touch is such a powerful indicator of intimacy that even an incidental touch can spur interest. In one study, a woman asked men in a bar for assistance adding a key to her key ring.[108] She lightly touched some of the men but not others. Afterward, the men who had been touched were more romantically interested in the woman than the other men were. Touch is potent even in long-term relationships. Researchers in one study asked couples to increase the number of times they kissed each other. Six weeks later, the couples' stress levels and relational satisfaction, and even their cholesterol levels, had significantly improved.[109]

> **intimacy** A state of closeness between two (or sometimes more) people. Intimacy can be manifested in several ways: physically, intellectually, emotionally, and via shared activities.

What Is Your Love Language?

Answer these questions to learn more about the love languages you prefer:

1. You have had a stressful time working on a team project. The best thing your romantic partner can do for you is:

    a. Set aside distractions to spend some time with you

    b. Do your chores so you can relax

    c. Give you a big hug

    d. Pamper you with a dessert you love

    e. Tell you the team is lucky to have someone as talented as you

2. What is your favorite way to show that you care?

    a. Go somewhere special together

    b. Do a favor without being asked

    c. Hold hands and sit close together

    d. Surprise your romantic partner with a little treat

    e. Tell your loved one how you feel in writing

3. With which of the following do you most agree?

    a. The most lovable thing someone can do is give you his or her undivided attention.

    b. Actions speak louder than words.

    c. A loving touch says more than words can express.

    d. Your dearest possessions are things your loved one has given you.

    e. People don't say "I love you" nearly enough.

4. Your anniversary is coming up. Which of the following appeals to you most?

    a. An afternoon together, just the two of you

    b. A romantic, home-cooked dinner (you don't have to lift a finger)

    c. A relaxing massage by candlelight

    d. A photo album of good times you have shared

    e. A homemade card that lists the qualities your romantic partner loves about you

For the meaning of your scores, see page 211.

The odds are that you value all of these ways to express love, but you probably give some greater weight than others. Good intentions lead you astray, however, if you assume that your partner feels the same way you do. The golden rule—that we should do unto others as we would have them do unto us—can lead to misunderstandings when our partner's primary love language differs from ours.[110]

So far, we have been speaking about intimacy and love languages in established relationships, but more than most relationships, romantic love tends to develop (and sometimes to decline) in phases.

Stages of Romantic Relationships Some romances ignite quickly, whereas others grow gradually. Either way, couples are likely to progress through a series of stages as they define what they mean to each other and what they should expect in terms of shared activities, exclusivity, commitment, and their public identity. All the while, they are involved in a balancing act as they negotiate between autonomy and togetherness, openness and privacy, and other factors.

One of the best-known explanations of how communication operates in different phases of a relationship was developed by communication scholar Mark Knapp. His **developmental model** depicts five stages of intimacy development (coming together) and five stages in which people distance themselves from each other (coming apart).[111] Other researchers have suggested that the middle phases of the model can also be understood in terms of keeping stable relationships operating smoothly and satisfactorily (relational maintenance).[112] Figure 7-1 shows how Knapp's 10 stages fit into this three-part view of communication in

developmental models (of relational maintenance) Theoretical frameworks based on the idea that communication patterns are different in various stages of interpersonal relationships.

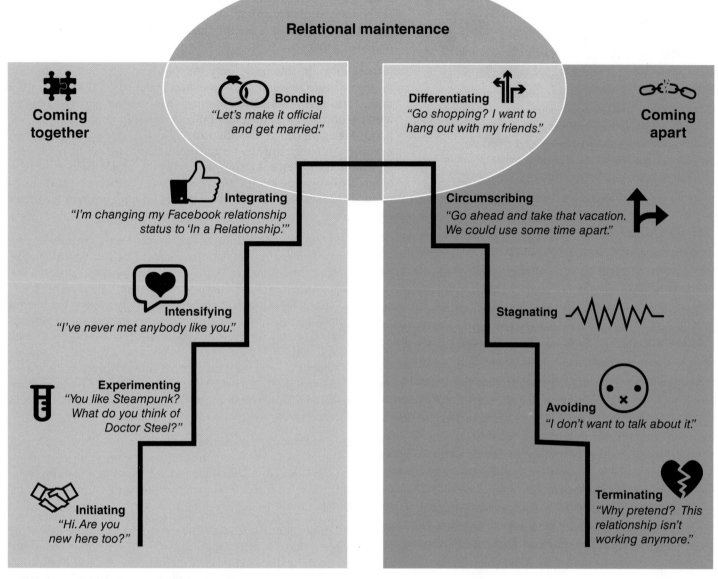

Relational maintenance

Coming together

Bonding
"Let's make it official and get married."

Integrating
"I'm changing my Facebook relationship status to 'In a Relationship.'"

Intensifying
"I've never met anybody like you."

Experimenting
"You like Steampunk? What do you think of Doctor Steel?"

Initiating
"Hi. Are you new here too?"

Differentiating
"Go shopping? I want to hang out with my friends."

Coming apart

Circumscribing
"Go ahead and take that vacation. We could use some time apart."

Stagnating

Avoiding
"I don't want to talk about it."

Terminating
"Why pretend? This relationship isn't working anymore."

FIGURE 7-1 Knapp's Stages of Relational Development

relationships. As you read on, consider how well these stages reflect communication in the close relationships you have experienced.

Initiating The initiating stage occurs when people first encounter each other. Knapp restricts this stage to conversation openers, such as, "It's nice to meet you" and "How's it going?" During this stage, people form first impressions and have the opportunity to present themselves in an appealing manner.

Experimenting People enter the experimental stage when they begin to get acquainted through "small talk." They may ask, *"Where are you from?"* and *"What do you do?"* or *"Do you know Josephine Mandoza? She goes to the same university you do."* Comments during this stage are generally pleasant and uncritical, and commitment is minimal. Though small talk might seem meaningless, Knapp points out that it

cultural idiom
small talk: idle, trivial conversation

CHECKLIST ✓

Meeting an Online Date for the First Time

Researchers call the transition from online-only chatting to a real-time meetup *modality switching*. Daters often use a different term: *awkward*.[113] Expectations may be especially high, and there is a "shopping mentality" about online dating that can turn a first meeting into a sweaty-palms audition.

Following are some tips from communication researchers and dating experts to dial down the awkward and help put you and the other person at ease.

☐ **Be genuine from the beginning.** The computer-enhanced selfie that looks great on screen may cause an online admirer to be disappointed in person. Ultimately, relationships fare best when mediated selves are a close reflection of in-person selves.[114]

☐ **Talk on the phone first.** If your online connection shows promise, see how a phone call goes before meeting in person.

☐ **Be safe.** Arrange to meet in a public setting rather than a private home, and provide your own transportation. (Avoid your favorite hangouts just in case you'd rather not run into each other again in the future.[115])

☐ **Put romantic thoughts aside for now.** It may sound counterintuitive, but as dating advice columnist Jonathan Aslay observes, "most successful long-term relationships are built on a solid friendship."[116] Approaching the encounter as a new friendship can be less anxiety provoking and more realistic than expecting fireworks with someone you barely know.

☐ **Begin with a quick and easy meetup.** "Don't meet for a meal on your first date," recommends Jennifer Flaa, who met her future husband (plus a host of Mr. Wrongs) online.[117] She suggests meeting for coffee or a drink instead. You can schedule a longer date later, if you like.

presents a valuable opportunity: Small talk allows us to interact with a wide range of people to determine who is worth getting to know better. For couples who meet online, experimentation may involve meeting in person for the first time. See the checklist on this page for experts' tips on making that transition.

Intensifying In this stage, truly interpersonal relationships develop as people begin to express how they feel about each other. It's often a time of strong emotions and optimism that may lead either to a higher level of intimacy or to the end of the relationship if, for example, one partner feels pressured and the other rejected. Dating couples often navigate this uncertainty by flirting, hinting around, asking hypothetical questions, giving compliments, and being more affectionate than before. They become bolder and more direct only if their partners seem receptive to these gestures.[118] At this point, couples begin to see themselves as "we" instead of as separate individuals.

Integrating In the integration stage, couples begin to take on an identity as a social unit. Invitations come addressed to the couple. Social circles merge. Couples begin to share possessions and memories—our apartment, our car, our song.[119] The term "Facebook official" applies to this stage. Couples are likely to change their relationship status online, announcing to their friends that they are in a committed, exclusive relationship.[120] As it becomes a given that they will share resources and help each other, partners become comfortable making relatively straightforward requests of each other. Gone are the elaborate explanations, inducements, and apologies. In short, partners in an integrated relationship expect more from each other than they do in less intimate associations.

Bonding The bonding stage is likely to involve a wedding, a commitment ceremony, or some other public means of communicating to the world that this is a relationship meant to last. Bonding generates social support for the relationship and demonstrates a strong sense of commitment and exclusivity.

Differentiating Not all relationships last forever. Even when the bonds between partners are strong and enduring, it is sometimes desirable to create some distance. In the differentiating stage, the emphasis shifts from "how we are alike" to "how we are different." For example, a couple who moves in together may find that they have different expectations about doing chores, sleeping late, what to watch on TV, and so on. This doesn't necessarily mean the relationship is doomed. Differences remind partners that they are distinct individuals. To maintain this balance, couples in this stage may demonstrate verbally and nonverbally that they wish to have space. They may claim different areas of the home for their private use and reduce their use of nicknames, gestures, and words that distinguish the relationship as intimate and unique.[121]

Circumscribing In the circumscribing stage, communication decreases significantly in quantity and quality. Rather than discuss a disagreement, which requires some degree of energy on both parts, partners may withdraw mentally by using silence, daydreaming, or fantasizing. They may also withdraw physically by spending less time together. Circumscribing entails a shrinking of interest and commitment.

Stagnating   If circumscribing continues, the relationship begins to stagnate. Partners behave toward each other in old, familiar ways without much feeling. Like workers who have lost interest in their jobs yet continue to go through the motions, sadly, some couples unenthusiastically repeat the same conversations, see the same people, and follow the same routines without any sense of joy or novelty.

Avoiding   When stagnation becomes too unpleasant, partners distance themselves in more overt ways. They might use excuses, such as *"I've been busy lately,"* or direct requests, such as *"Please don't call. I don't want to see you now."* In either case, the writing about the relationship's future is clearly on the wall.

Terminating   Characteristics of this final stage include summary dialogues about where the relationship has gone and the desire to break up. The relationship may end with a cordial dinner, a note left on the kitchen table, a phone call, or a legal document stating the dissolution. Depending on each person's feelings, this stage can be quite short, or it may be drawn out over time. (See the Understanding Communication Technology box on the next page for a discussion of breaking up online.)

One key difference between couples who get together again after a breakup and those who go their separate ways is how well they communicate about their dissatisfaction and negotiate for a mutually appealing fresh start. Unsuccessful couples deal with their problems by avoidance, indirectness, and reduced involvement with each other. By contrast, couples who repair their relationships more often air their concerns and spend time and effort negotiating solutions to their problems.

A number of practical lessons emerge from the developmental perspective:

1. **Each stage requires different types of communication.** Partners may find that talking about highly personal issues deepens their bond in the intensifying stage but is overwhelming sooner than that. Likewise, the polite behavior of the first two stages may seem cool and distant as intimacy increases.

2. **Relational development involves risk and vulnerability.** At any stage—even those associated with coming together—the relationship may falter. Intimacy only evolves if people are willing to take a chance of becoming gradually more self-disclosive.[124]

3. **Partners can change the direction a relationship is headed.** The direction a relationship takes isn't inevitable. Partners may recognize the early signs of "coming apart" in time to reverse the trend. For example, partners who realize they are differentiating or stagnating can refresh their relationship by focusing energy on the intimacy-enhancing communication of experimenting, intensifying, and integrating. As Knapp puts it, movement is always to a new place.[125]

Having considered the differences between friendship, romance, and family relationships, let's focus on communication patterns common to all types of interpersonal communication.

ASK YOURSELF

In what ways have you established an integrated identity with the people who are close to you? How does that enhance your relationships? Does it ever go too far?

cultural idiom

on the wall: clear evidence of future problems

UNDERSTANDING COMMUNICATION TECHNOLOGY

To End This Romance, Just Press "Send"

It was the middle of a workday two weeks ago, and Larry was deep into a meeting when a text message began scrolling across his cell phone screen. He glanced at it and thought: "You can't be serious."

It was no joke. His girlfriend was breaking up with him . . . again. And she was doing it by email . . . again.

For the sixth time in eight months, she had ended their relationship electronically rather than face-to-face. He had sensed trouble—he had been opening his email with trepidation for weeks—so the previous day he had suggested that they meet in person to talk things over. But she nixed that, instead choosing to send the latest in what Larry had begun to consider part of a virtual genre: "the goodbye email."

Understandably, he'd like to say his own goodbye to that genre. "Email is horrible," says Larry, 36, a U.S. Air Force sergeant from New Hampshire who asked that his last name not be used. "You just get to the point where you hate it. You can't have dialogue. You don't have that person in front of you. You just have that black-and-white text. It's a very cold way of communicating."

Cold, maybe. Popular, though. The use of email and instant messaging to end intimate relationships is gaining popularity because instantaneous communication makes it easy—some say too easy—to just call the whole thing off. Want to avoid one of those squirmy, awkward breakup scenes? Want to control the dialogue while removing facial expressions, vocal inflections, and body language from the equation? A solution is as near as your keyboard or cell phone.

Sometimes there is a legitimate reason for wanting to avoid personal contact. Tara, a 32-year-old woman who lives near Boston, says her ex-husband was intimidating and emotionally abusive during their marriage. So when she wanted to end the marriage several years ago, she felt more comfortable doing so by sending a text message.

Tara says that since then she has ended several other relationships by email. "I'm a softie, and I hate hurting people's feelings," she says. Recently she laid the groundwork for breaking her engagement with a series of emails to her fiancé. After ending the engagement last week, she reached a moment of truth, she says, and has decided that from now on, if she wants to call it quits, "the email option is out."

Do you ever use online communication to say things you would be uncomfortable saying in person? When do you think this strategy is effective? When might it be unfair to the relationship or the other person?

Communication Patterns in Relationships

As you read in Chapter 4, pragmatic rules are often more important in conveying meanings than are formal definitions. Is teasing a form of aggression or a way to express affection? It depends. The same applies to rules of conversation and nonverbal communication. In this section, we consider how a sense of shared meaning unfolds.

Content and Relational Messages

content message A message that communicates information about the subject being discussed.

relational message A message that expresses the social relationship between two or more individuals.

affinity The degree to which people like or appreciate one another. As with all relational messages, affinity is usually expressed nonverbally.

Virtually every verbal statement contains both a **content message**, which focuses on the subject being discussed, and a **relational message**, which makes a statement about how the parties feel toward one another. Following are some of the dimensions communicated on a relational level.

Affinity The degree to which people like or appreciate others is called **affinity**. Sometimes we indicate feelings of affinity explicitly, but more often the clues are nonverbal, such as a pat on the back or a friendly smile.

Respect The degree to which we admire others and hold them in esteem is known as **respect**. Respect and affinity might seem identical, but they are actually different dimensions of a relationship.[126] For example, you might like a 3-year-old child tremendously without respecting her. Likewise, you could respect a boss or teacher's talents without liking him or her. Respect is a tremendously important and often overlooked ingredient in satisfying relationships. It is a better predictor of relational satisfaction than liking, or even loving.[127]

Immediacy Communication scholars use the term **immediacy** to describe the degree of interest and attraction we feel toward and communicate to others. Immediacy is different than affinity. You may like someone, but if you don't communicate or demonstrate that feeling toward the other person, immediacy will be low.

Control In every conversation and every relationship there is some distribution of **control**—the amount of influence communicators seek. Control can be distributed evenly among relational partners, or one person can have more and the other(s) less. An uneven distribution of control won't necessarily cause problems unless people disagree on how control should be distributed.

You can get a feeling for how relational messages operate in everyday life by imagining two ways of saying, *"It's your turn to do the dishes"*—one in a tone that sounds demanding and another that is matter of fact. The demanding tone says, in effect, *"I have a right to tell you what to do around the house,"* whereas the matter-of-fact one suggests, *"I'm just reminding you of something you might have overlooked."*

Most of the time we aren't conscious of the relational messages that bombard us every day, particularly when the messages match our belief about the amount of respect, immediacy, control, and affinity that is appropriate. For example, you probably won't be offended if your boss tells you to do a certain job, because you agree that supervisors have the right to direct employees. However, if your boss delivers the order in a condescending, sarcastic, or abusive tone of voice, you probably will be offended. Your complaint wouldn't be with the order itself but rather would be with the way it was delivered. *"I may work for this company,"* you might think, *"but I'm not a slave or an idiot. I deserve to be treated like a human being."*

As the boss–employee example suggests, relational message are usually expressed nonverbally. To test this fact for yourself, imagine how you could act while saying, *"Can you help me for a minute?"* in a way that communicates each of the following attitudes:

superiority	aloofness	friendliness
helplessness	sexual desire	irritation

Remember, however, that although nonverbal behaviors are a good source of relational messages, they are ambiguous. The sharp tone you take as a personal insult might be due to fatigue, and the interruption you take as an attempt to ignore your ideas might be a sign of pressure that has nothing to do with you.

In addition to the things we say, our actions, expressions, and tone of voice send relational messages about how we feel about relational partners.

Think of a time you said one thing at a content level but sent a different message with the way you acted. Which message do you think the people around you trusted more?

respect The degree to which we hold others in esteem.

immediacy The degree of interest and attraction we feel toward and communicate to others. As with all relational messages, immediacy is usually expressed nonverbally.

control The social need to influence others.

"She's texting me, but I think she's also subtexting me."

Source: Leo Cullum The New Yorker Collection/The Cartoon Bank

Metacommunication

Social scientists use the term **metacommunication** to describe messages that refer to other messages.[128] In other words, metacommunication is communication about communication. Whenever we discuss a relationship with others, we are metacommunicating: *"It sounds like you're angry at me"* or *"I appreciate how honest you've been."* As the cartoon at left shows, even text messages can contain subtexts, otherwise known as metacommunicative dimensions. Given the fact that nonverbal cues are limited in online communication, it's often important to supplement them with metacommunication (such as *"just kidding"*).

Tuning in to metacommunication allows us to look below the surface of a message for underlying meanings where the issue often lies. For example, consider friends bickering because one wants to watch television, while the other wants to talk. Imagine how much better the chances of a positive outcome would be if they used metacommunication: *"Look, it's not the TV watching itself that bothers me. It's that I think you're watching because you're mad at me. Am I right?"*

Metacommunication isn't just a tool for handling problems. It's also a way to reinforce the good aspects of a relationship: *"Thank you for praising my work in front of the boss."* Comments such as this let others know that you value their behavior and boost the odds that the other people will continue the behavior in the future.

Bringing relational issues out in the open does have its risks. Discussing problems can be interpreted as a sign of trouble (*"Our relationship isn't working if we have to keep talking about it"*). Furthermore, metacommunication involves a certain degree of analysis (*"It seems like you're angry at me"*), which can lead to resentment (*"Don't presume to know how I feel"*). This doesn't mean verbal metacommunication is a bad idea, just that it should be used carefully.

Self-Disclosure in Interpersonal Relationships

"We don't have any secrets," some people proudly claim. Opening up certainly is important. Earlier in this chapter you learned that one ingredient in qualitatively interpersonal relationships is disclosure. You've also read that we find others more attractive when they share certain private information with us. Given the obvious importance of self-disclosure, we need to consider the subject carefully. Just what is it? When is it desirable? How can it best be done?

The best place to begin is with a definition. **Self-disclosure** is the process of deliberately revealing information about oneself that is significant and that would not normally be known by others. Let's take a closer look at some parts of this definition. Self-disclosure must be *deliberate*. If you accidentally mentioned to a friend that you were thinking about quitting a job or proposing marriage, that information would not fit into the category we are examining here. Self-disclosure must also be *significant*. Revealing relatively trivial information—the fact that you like fudge, for example—does not qualify as self-disclosure. The third requirement is that the information being revealed would *not be known by others*. There's nothing noteworthy about telling others that you are depressed or elated if they already know how you're feeling.

Under the right conditions, self-disclosure is rewarding.[129] Talking about your feelings and experiences can yield a greater sense of clarity. It's validating

metacommunication Messages (usually relational) that refer to other messages; communication about communication.

self-disclosure The process of deliberately revealing information about oneself that is significant and that would not normally be known by others.

to feel that others know and like the real you, and self-disclosure often inspires a give and take between people that can foster emotional closeness.

Models of Self-Disclosure   Over several decades, social scientists have created various models to represent and understand how self-disclosure operates in relationships. We will look at two of the best-known models here.

Social Penetration Model   Social psychologists Irwin Altman and Dalmas Taylor describe two ways in which communication can be more or less disclosive.[130] Their **social penetration model** (pictured in Figure 7-2) proposes that communication occurs within two dimensions: (a) **breadth**, which represents the range of subjects being discussed; and (b) **depth**, how in-depth and personal the information is.

For example, as you start to reveal to cowork-ers information about your personal life—perhaps what you did over the weekend or stories about your family—the breadth of disclosure in your relation-ship will expand. The depth may also expand if you shift from relatively nonrevealing messages (*"I went out with friends"*) to more personal ones (*"I went on this awful blind date set up by my mom's friend . . ."*).

What makes the disclosure in some messages deeper than others? Some revela-tions are certainly more significant than others. Consider the difference between saying *"I love my family"* and *"I love you."* Other statements qualify as deep dis-closure because they are private. Sharing a secret you've told to only a few close friends is certainly an act of self-disclosure, but it's even more revealing to divulge information that you've never told anyone.

Depending on the breadth and depth of information shared, a relationship can be defined as casual or intimate. The most intimate relationships are those in which disclosure is great in both breadth and depth. Altman and Taylor see the development of a relationship as a progression from the periphery of their model to its center, a process that typically occurs over time.

The Johari Window   Another model that helps represent how self-disclosure operates is the **Johari Window**.[131] Imagine a frame inside which is everything there is to know about you: your likes and dislikes, your goals, your secrets, your needs—everything.

Of course, you aren't aware of everything about yourself. Like most people, you're probably discovering new things about yourself all the time. To represent this, we can divide the frame containing everything about you into two parts: the part you know about (the left quadrants in Figure 7-3) and the part you don't know about (the right quadrants). We can also divide this frame in another way. In this division the first part contains the things about you that others know (the top two quadrants in the diagram), and the second part contains the things about you that you keep to yourself (the bottom quadrants).

When we put this all together we have a Johari Window (Figure 7-3) that pres-ents *everything about you* as a window divided into four parts. One quadrant rep-resents the information of which both you and the other person are aware. This part is your *open area*. Another represents the *blind area*: information of which you are unaware but that the other person knows. You learn about information in the blind area primarily through feedback. A third represents your *hidden area*: information that you know but aren't willing to reveal to others. Items in this hidden area become public primarily through self-disclosure. The fourth

"There's something you need to know about me, Donna. I don't like people knowing things about me."

Source: Leo Cullum The New Yorker Collection/ The Cartoon Bank

social penetration model   A theory that describes how intimacy can be achieved via the breadth and depth of self-disclosure.

breadth (of self-disclosure)   The range of topics about which an individual discloses.

depth (of self-disclosure)   The level of personal information a person reveals on a particular topic.

Johari Window   A model that describes the relationship between self-disclosure and self-awareness.

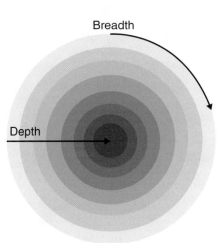

FIGURE 7-2 Social Penetration Model

In her 2016 album *Lemonade*, superstar Beyoncé disclosed feelings of betrayal, jealousy, revenge, and rage surrounding her troubled marriage to rapper Jay Z.

You may not have millions of followers like Beyoncé, but do you ever reveal personal information via social media? If so, what are your motives? What are the results?

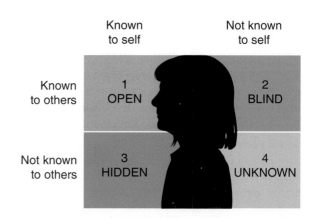

FIGURE 7-3 The Johari Window: Open; Blind; Hidden; Unknown

CHECKLIST ✅
When and How Much to Self-Disclose

Before sharing very important information with someone who does matter to you, you might consider what message is likely to be received. Here are some considerations offered in a study about sharing family secrets. If you can check off many of these, it may be worthwhile to self-disclose a bit and see what happens.

Intimate Exchange

☐ Does the other person have a similar problem?

☐ Would knowing the secret help the other person feel better?

☐ Would knowing the secret help the other person manage his or her problem?

Exposure

☐ Will the other person find out this information, even if I don't tell him or her? Is the other person asking me directly to reveal this information?

represents information that is *unknown* to both you and others. At first, the unknown area seems impossible to verify. After all, if neither you nor others know what it contains, how can you be sure it exists? We can deduce its existence because we are constantly discovering new things about ourselves. It is not unusual to discover, for example, that you have an unrecognized talent, strength, or weakness. Items move from the unknown area into the open area either directly when you disclose your insight or through one of the other areas first.

Interpersonal relationships of any depth are virtually impossible if the individuals involved have little open area. Going a step further, you can see that a relationship is limited by the individual who is less open—that is, who possesses the smaller open area. You have probably found yourself in situations in which you felt the frustration of not being able to get to know someone who was very reserved. Perhaps you have blocked another person's attempts to build a relationship with you in the same way. The fact is that self-disclosure on both sides is necessary for the development of any interpersonal relationship.

Characteristics of Effective Self-Disclosure If you are a *Bachelor* or *Bachelorette* fan, you know the cringeworthy sensation of self-disclosure pushed to its limits. Reality TV manufactures such moments by pressuring strangers to develop intimate relationships very quickly with almost no privacy. But you needn't fall prey to the same traps in real life. As you can imagine, publicly sharing every personal detail of your life with people you barely know usually isn't effective. Here are seven questions you can ask yourself to determine when and how self-disclosing may be beneficial to you and others.

1. **Is the other person important to you?**
   Disclosure may be the path toward developing a more personal relationship with someone. If you value the relationship, sharing more about yourself might bring you closer. However, it can be a mistake to share personal information with people you don't trust or know very well.

2. **Is the disclosure appropriate?**
   This is tricky to answer because appropriateness relies on personal preference and culture. North Americans, with their individualistic orientation, are often comfortable saying things about themselves that people from more collectivistic cultures, such as Japan's, would never dream of blurting out.[132] As a result, North Americans

may come off as exhibitionists who spew personal information to anyone within earshot, and they may assume that people who do not disclose as easily are standoffish or uninterested.

3. **Is the risk of disclosing reasonable?**
   Take a realistic look at the potential risks of self-disclosure. You're asking for trouble when you open up to someone you know is likely to betray your confidences or make fun of you. On the other hand, knowing that your relational partner is trustworthy and supportive makes it more reasonable to speak out.

4. **Is the disclosure relevant to the situation at hand?**
   Even in personal relationships—with close friends or family members—constant disclosure isn't a useful goal. Instead, the level of sharing in successful relationships rises and falls in cycles.

5. **Is the disclosure reciprocated?**
   There's nothing quite like sharing vulnerable information about yourself only to discover that the other person is unwilling to do the same. Unequal self-disclosure creates an unbalanced relationship.

6. **Will the effect be constructive?**
   Self-disclosure can be a vicious tool if it's not used carefully. Psychologist George Bach suggests that every person has a psychological "belt line." Below that belt line are areas about which the person is extremely sensitive. Bach says that jabbing "below the belt" is a surefire way to disable another person, although usually at great cost to the relationship. It's important to consider the effects of your candor before opening up to others. Comments such as "I've always thought you were stupid" may be devastating—to the listener and to the relationship.

7. **Is the self-disclosure clear and understandable?**
   When you express yourself to others, it's important that you do so intelligibly by clearly describing the *sources* of your message. For instance, it's far better to describe another's behavior by saying, "When you don't text me back . . ." than to complain vaguely, "When you avoid me . . . ." It's also vital to express your *thoughts* and *feelings* explicitly. "I feel like you no longer want to spend time with me" is more understandable than "I don't like the way things have been going."

These questions can be useful, but of course every situation and relationship is unique. Effective self-disclosure is a balancing act, defined and managed by people within a relationship. The same concept applies in a larger sense to many aspects of interpersonal communication, a topic we will discuss next.

Dialectical Perspective of Interpersonal Relationships

Consider the following blog posts about friendship:

"I'm an 'all or nothing' type of person on friendships, and I see hanging out as a waste of time if you're not really close friends or if I don't see us turning into close friends."[133]

"[My best friend] is always asking me to Google hangout her to chat but I don't always have the time. I love her and she is is a great friend, I really need to stop giving all my time to her."[134]

CHECKLIST *Continued*

Urgency

☐ Is it important that the other person know this information? Will revealing this information make matters better?

Acceptance

☐ Will the other person still accept me if I reveal this information?

Conversational Appropriateness

☐ Will my disclosure fit into the conversation?

☐ Has the topic of my disclosure come up in this conversation?

Relational Security

☐ Do I trust the other person with this information?

☐ Do I feel close enough to this person to reveal the secret?

Important Reason

☐ Is there a pressing reason to reveal this information?

Permission

☐ Have other people involved in the secret given their permission for me to reveal it? Would I feel okay telling the people involved that I have revealed the secret?

Source: Adapted from Vangelisti, A. L., Caughlin, J. P., & Timmerman, L. (2001). Criteria for revealing family secrets. *Communication Monographs, 68,* 1–27.

---

cultural idioms

within earshot: close enough to be easily overheard
standoffish: aloof, unfriendly
open up: disclose about personal, private subjects

dialectical model The perspective that people in virtually all interpersonal relationships must deal with equally important, simultaneous, and opposing forces such as connection and autonomy, predictability and novelty, and openness versus privacy.

As different as these posts seem to be, the odds are that you have felt both ways before, at least to some degree. The **dialectical model** suggests that relational partners continually must negotiate to satisfy opposing or incompatible forces, both within themselves and with one another.[135] The way we manage these challenges defines the nature of relationships and our communication within them.

Partners face three main types of dialectical tensions throughout the life of their relationship. As you read about each set of opposing needs, consider how they operate in your life.

Connection Versus Autonomy   The conflicting desires for connection and independence are embodied in the *connection–autonomy dialectic*. One of the most common reasons for breaking up is that one partner doesn't satisfy the other's need for connection:[136]

> *"We barely spent any time together."*
> *"She/he wasn't committed to the relationship."*
> *"We had different needs."*

But couples split up for the opposite reason as well:[137]

> *"I felt trapped."*
> *"I needed freedom."*

Even within ourselves, we are faced with the same sort of contradiction. On one hand we desire intimacy, but we often feel the need to maintain some distance as well. Linguist Deborah Tannen explains the tension between connection and autonomy with an analogy from the animal world:

> [Porcupines] huddle together for warmth, but their sharp quills prick each other, so they pull away. But then they get cold. They have to keep adjusting their closeness and distance to keep from freezing and from getting pricked by their fellow porcupines—the source of both comfort and pain.
>
>   We need to get close to each other to have a sense of community, to feel we're not alone in the world. But we need to keep our distance from each other to preserve our independence, so others don't impose on or engulf us. This duality reflects the human condition. We are individual and social creatures. We need other people to survive, but we want to survive as individuals.[138]

Managing dialectical tensions is tricky, because our needs change over time. Author Desmond Morris suggests that each of us repeatedly goes through three stages: "Hold me tight," "Put me down," and "Leave me alone."[139] In marriages and other committed relationships, for example, the "Hold me tight" bonds of the first year are often followed by a desire for independence. This need for autonomy can manifest itself in a number of ways, such as the desire to make friends or engage in activities that don't include the spouse, or making a career move that might disrupt the relationship. Movement toward autonomy may lead to a breakup, but it can also be part of a cycle that redefines the relationship in ways that allow partners to recapture or even surpass the closeness that existed previously. For example, you might find that spending some time apart makes you miss and appreciate your partner more than ever.

Openness Versus Privacy   Disclosure is one characteristic of interpersonal relationships. Yet, along with the need for intimacy, your experience probably shows an equally important need for you to maintain some space between yourself and others. These sometimes-conflicting drives create the *openness–privacy dialectic*.

Source: ©2006 Zits Partnership Distributed by King Features Syndicate Inc.

Even the strongest interpersonal relationships require some distance. Romantic partners may go through periods of sharing and times of relative withdrawal. Likewise, they may experience periods of passion and then periods of little physical contact.

Predictability Versus Novelty   Stability is an important need in relationships, but too much predictability can lead to feelings of staleness and boredom. People differ in their own desire for stability and surprises from one time to another. The classic example is becoming engaged just before graduation or military deployment, when life may seem particularly novel and uncertain. Commitment may balance some of the uncertainty people feel in that situation. However, it may feel too predictable once life settles into a routine. There are a number of strategies people can use to manage contradictory drives such as these.

Strategies for Managing Dialectical Tensions   Dialectical tensions are a fact of life in intimate relationships. But there are a number of ways people can deal with these. Some of these strategies are more productive than others.[140] As you read about them, consider which ones you use and how well they meet your relational needs.

Denial   One of the least functional responses to dialectical tensions is to deny that they exist. People in denial insist that "everything is fine." For example, family members might refuse to deal with conflict, ignoring problems or pretending that they agree about everything.

Disorientation   When communicators feel so overwhelmed and helpless that they are unable to confront their problems they are said to be disoriented. In the face of dialectical tensions they might fight, freeze, or even leave the relationship. A couple who discovers soon after the honeymoon that living a "happily ever after" conflict-free life is impossible might view their marriage as a mistake and seek a divorce.

Selection   When partners employ the strategy of *selection*, they respond to one end of the dialectical spectrum and ignore the other. For example, relational partners caught between the conflicting desires for stability and novelty may decide that predictability is the "right" or "responsible" choice and put aside their longing for excitement.

Alternation   Communicators sometimes alternate between one end of the dialectical spectrum and the other. Friends may spend time apart during the week, but reserve weekends for couple time.

Polarization In some cases, people find a balance of sorts by each staking a claim at opposite ends of a dialectic continuum. One partner might give up nearly all personal interests in the name of togetherness, while the other maintains an equally extreme commitment to being independent. In the classic demand–withdraw pattern,[141] the more one partner insists on closeness, the more the other feels suffocated and craves distance.

Segmentation In *segmentation*, people compartmentalize different areas of the relationship. For example, a couple might manage the openness–privacy dialectic by sharing almost all their feelings about mutual friends with each other but keeping certain parts of their past romantic histories private.

Moderation The moderation strategy is characterized by compromises in which people back off from expressing either end of the dialectical spectrum. A couple might decide that taking separate vacations is too extreme for them, but they will make room for some alone time while they are traveling together.

Reframing Communicators can also respond to dialectical challenges by *reframing* them in terms that redefine the situation so that the apparent contradiction disappears. Consider relational partners who regard the inevitable challenges of managing dialectical tensions as exciting opportunities to grow instead of as relational problems.

Reaffirmation A final strategy for handling dialectical tensions is *reaffirmation*—acknowledging that dialectical tensions will never disappear and accepting or even embracing the challenges they present. Communicators who use reaffirmation view dialectical tensions as part of the ride of life.

People who understand the dialectical perspective can better appreciate several facets of relationship maintenance:

1. **Relationships involve continual change and negotiation.** Relational partners who understand dialectical tensions can give up the unrealistic notion that they will always be in sync or that negotiating relationship options should be effortless.

2. **Partners can be in sync in some ways, but not in others.** Recognizing different dialectical tensions may help people identify the critical issue when things feel out of balance between them.

3. **Some approaches are more conducive to relational satisfaction.** It may be tempting to deny opposing tensions, to polarize, or to exit the relationship altogether. However, other options are usually more effective in satisfying each individual's needs and strengthening the relationship.

Lies and Evasions

People are likely to experience deceit, even in their closest, most intimate relationships. In fact, people lie more than they realize. Research shows that most people lie, on average, once or twice per day[142] and even more when they meet someone new. Upon first meeting, the average is about three lies in the first 10 minutes, especially when romantic attraction is a factor.[143]

Not all lies are equally devastating. As Table 7-1 shows, at least some of the lies we tell are indeed intended to be helpful, or at least relatively benign. The greatest damage occurs when the relationship is most intense, when the importance of the subject is high, and when there have been previous doubts about the deceiver's honesty. Of these three factors, the one most likely to cause a relational crisis is the sense that one's partner lied about something important.[144]

TABLE 7-1

Some Reasons for Lying

REASON	EXAMPLE
Acquire resources	"Oh, please let me add this class. If I don't get in, I'll never graduate on time!"
Protect resources	"I'd like to lend you the money, but I'm short myself."
Initiate and continue interaction	"Excuse me, I'm lost. Do you live around here?"
Avoid conflict	"It's not a big deal. We can do it your way. Really."
Avoid interaction or take leave	"That sounds like fun, but I'm busy Saturday night." "Oh, look what time it is! I've got to run!"
Present a competent image	"Sure, I understand. No problem."
Increase social desirability	"Yeah, I've done a fair amount of skiing."

Source: Adapted from categories originally presented in Camden, C., Motley, M. T., & Wilson, A. White lies in interpersonal communication: A taxonomy and preliminary investigation of social motivations. *Western Journal of Speech Communication, 48,* 315.

Experts suggest that, if you are considering a deception, you consider how others would respond if they knew about it.[145] Would they accept your reasons for being untruthful, or would they be hurt by them? In light of that, we explore three types of lies here: altruistic lies, evasions, and self-serving lies.

altruistic lies Deception intended to be unmalicious, or even helpful, to the person to whom it is told.

Altruistic Lies   You might tell the host of a dinner party that the food was delicious even if it wasn't. Or you might compliment your boyfriend or girlfriend's new haircut to avoid hurting his or her feelings. **Altruistic lies** are defined, at least by the people who tell them, as being harmless, or even helpful, to the person to whom they are told.[146] For the most part, white lies such as these fall in the category of being polite, and effective communicators know how and when to use them without causing offense.

Evasions   Evasions aren't outright mistruths. Rather, they evade full disclosure by being deliberately vague. Often motivated by good intentions, evasions are based on the belief that less clarity can be beneficial for the sender, the receiver, or sometimes both.[147]

One type of evasion is *equivocation*—making deliberately ambiguous statements with two or more equally plausible meanings.[148] As you read in Chapter 4, people sometimes send equivocal messages without meaning to, resulting in confusion. But other times we are deliberately vague. For instance, when your partner asks what you think of an awful outfit, you could say, "It's really unusual—one of a kind!"

Hinting is a second type of evasion. People hint to bring about a desired response without asking for it directly. Some hints are designed to save the receiver from embarrassment. For example, a face-saving guest might hint to her host by saying, "It's getting late," rather than, "I'm bored and want to leave now." Other hints are strategies for saving the sender from embarrassment, as

Altruistic lies are meant to spare people's feelings, but self-serving lies are often hurtful and manipulative.

Have you ever been caught telling a self-serving lie to a loved one or been hurt by a lie told by a someone else? How did it affect your relationship?

when someone says, "I'm pretty sure smoking isn't allowed here" instead of the blunter "Your smoking bothers me." Clearly, hints only work if people pick up on them.

Another type of evasion is *concealment*—failing to reveal information that is pertinent to the conversation at hand. Concealing an important fact or feeling isn't a lie, but it doesn't reveal the full truth either. You might test whether concealment is ethically justified by asking yourself whether the other person would find the information you're not sharing important.

Equivocations, concealments, and hints can be offered in a spirit meant to avoid hurting people's feelings. If your friend hits on you and you are not romantically interested, you might be evasive with an equivocal statement such as, "Your friendship means a lot to me, and I wouldn't want anything to ruin that."

Self-Serving Lies Self-serving lies are attempts to manipulate the listener into believing something that is untrue—not primarily to protect the listener, but to advance the deceiver's agenda. For example, people might lie on their income tax returns or deny that they have been drinking if a cop pulls them over.

Self-serving lies involve an omission or a fabrication—withholding information that another person deserves to know or deliberately misleading another person for one's own benefit. For example, a romantic partner may keep a love affair secret or claim to be somewhere that she or he wasn't.

It's no surprise that such lies can destroy trust. For one thing, they lead the deceived person to wonder if anything the other person says is true and what else might be a lie. However, some couples rebound from serious deceptions, particularly if the lie involves an isolated incident and the wrongdoer's apology seems sincere.[149] The most reliable predictor of what will happen after a deception is whether romantic partners communicate openly about it. Those who avoid the issue typically lose the chance to work through it, even if they want to.[150]

Although few of us will end up on a reality TV show, perhaps we can learn a few lessons from them about relationships: (1) very often, it's not what you say but how you say it; (2) sharing either too much too soon or nothing about yourself can derail a relationship; and (3) the lies you tell today may be publicly revealed tomorrow. Former *Bachelorette* star Jillian Harris adds one more to the list: "It sounds cliché, but be yourself."[151]

cultural idiom

pick up on: recognize

MAKING THE GRADE

For more resources to help you understand and apply the information in this chapter, visit the *Understanding Human Communication* website at www.oup.com/us/adleruhc.

OBJECTIVE 7.1 **Describe characteristics that distinguish interpersonal relationships from impersonal ones and online communication from face-to-face interactions.**

- Interpersonal communication involves two-way communication that "mazimizes the personal" and respects the the uniqueness of the people involved.

- Online communication can facilitate connections and social support that might otherwise be difficult or intimidating. However, people who overuse technology may find that it detracts from their in-person relationships and can lead them to feel lonely and isolated.

  > What does it mean to say that interpersonal relationships are close and irreplaceable?

  > List the people you communicate with in one day. How many of these encounters are relatively impersonal? How many are interpersonal? What functions do both types of relationships serve in your life?

  > How might you maximize the benefits of online communication in relationships that are important to you? How might you minimize the potential drawbacks?

OBJECTIVE 7.2 **Identify the factors that shape interpersonal attraction.**

- We typically gravitate toward people who have a good deal in common with us, have characteristics that complement our own, who like us back, and who offer rewards that are worth the costs required to maintain the relationship.

  > Compare the factors people consider when selecting relational partners with the caveats presented in the "Relationship Reality Check" section (p. 187).

  > Think of your closest friend. Are you mostly similar to each other, or are your characteristics complementary? What is most rewarding about the friendship?

  > How might you use the information in this chapter to help people feel comfortable approaching you and getting to know you?

Objective 7.3 **Describe the different types of communication dynamics in friendships, family relationships, and romantic relationships.**

- Friendships vary in terms of how long they last, how much the friends share with each other, what they do together, how obligated they feel toward one another, and how they communicate.

- Parents have an influence on whether children grow up to value conversation or conformity as a means of solving problems.

- Sibling relationships often involve a complex mixture of camaraderie and competition.

- While it can be challenging to define grandparents' role, they often share a unique and special bond with grandchildren.

- Romantic intimacy can be expressed physically, emotionally, intellectually, and through shared activities.

- Some communication theorists suggest that romantic relationships pass through stages of coming together (initiating, experimenting, intensifying), sustaining the relationship (integrating, bonding, differentiating, and circumscribing), and sometimes, of coming apart (stagnating, avoiding, and terminating).

  > Describe at least three tips for communicating more effectively as a friend and at least three ways to enhance family communication.

  > Which of the love languages (affirming words, quality time, acts of service, gifts, or physical touch) are most meaningful to you? Which are most meaningful to the significant people in your life?

  > Evaluate communication practices used in your own family. In what ways have they made you a better communicator? What do you wish had been different and why?

Objective 7.4 **Explain the relevance of content and relational meaning, metacommunication, self-disclosure, dialectical tensions, and deception to interpersonal communication.**

- Interpersonal communication consists of both content (literal) messages and relational messages that suggest how we feel about the other person in terms of affinity, respect, immediacy, and control.

- Metacommunication involves interpersonal exchanges in which the parties talk about the nature of their interaction.
- Self-disclosure is the process of deliberately revealing information about oneself that is significant and that would not normally be known by others.
  > The social penetration model describes how intimacy can be achieved via the breadth and depth of self-disclosure.
  > The Johari Window describes the relationship between self-disclosure and self-awareness.
- The dialectical perspective calls attention to the way relational partners negotiate a balance between opposing desires such as autonomy and connection, openness and privacy, and predictability and novelty.
- People are likely to experience deceit even in their closest and most intimate relationships.
  > Altruistic lies fall in the category of being polite, and effective communicators know how to use them without causing offense.
  > Evasions are deliberately vague and include equivocation and hinting. They are generally offered in a spirit meant to avoid hurting people's feelings.
  > Self-serving lies are attempts to manipulate the listener into believing something that is untrue. They involve omission or fabrication.
  > Envision and describe a hypothetical encounter that illustrates the role of nonverbal communication in relational-level meaning and metacommunication.
  > Explain the dialectical tensions you have experienced in a close relationship and the strategies you have used to manage them. How successful were those strategies? Would others have been more effective?
  > Create or transcribe a recent conversation, including as much detail as you can about what people said and how they said it. See if you can identify examples of content and relational meaning, metacommunication, dialectical tension, and deception or omission.

KEY TERMS

affinity p. 198

altruistic lies p. 207

breadth (of self-disclosure) p. 201

content message p. 198

control p. 199

cyber relationship p. 184

depth (of self-disclosure) p. 201

developmental models of relational maintenance p. 194

dialectical model p. 204

family p. 191

immediacy p. 199

interpersonal communication p. 183

intimacy p. 193

Johari Window p. 201

metacommunication p. 200

phubbing p. 185

relational message p. 198

respect p. 199

self-disclosure p. 200

social exchange theory p. 187

social penetration model p. 201

ACTIVITIES

1. **Interpersonal Communication: Context and Quality**

   a. Examine your interpersonal relationships in a contextual sense by making two lists. The first should contain all the two-person relationships in which you have participated during the past week. The second should contain all your relationships that have occurred in small group and public contexts. Are there any important differences that distinguish dyadic interaction from communication with a larger number of people?

   b. Now make a second set of two lists. The first one should describe all of your relationships that are interpersonal in a qualitative sense, and the second should describe all the two-person relationships that are more impersonal. Are you satisfied with the number of qualitatively interpersonal relationships you have identified?

   c. Compare the lists you developed in Steps a and b. See what useful information each one contains. What do your conclusions tell you about the difference between contextual and qualitative definitions of interpersonal communication?

2. **Your I.Q. (Intimacy Quotient)** Answer the following questions as you think about your relationship with a person important in your life.

   a. What is the level of physical intimacy in your relationship?

   b. What intellectual intimacy do you share?

   c. How emotionally intimate are you? Is your emotional intimacy deeper in some ways than in others?

   d. Has your intimacy level changed over time? If so, in what ways?

   After answering these questions, ask yourself how satisfied you are with the amount of intimacy in this relationship. Identify any changes you would like to occur, and describe the steps you could take to make them happen.

ANSWERS to "What Is Your Love Language?" from p. 194.

INTERPRETING YOUR RESPONSES

For insight about your primary love languages, see which of the following best describes your answers.

Quality Time

If you answered "a" to one or more questions, you probably feel loved when people set aside life's distractions to spend time with you. Keep in mind that everyone defines quality time a bit differently. It may mean a thoughtful phone call during a busy day, a picnic in the park, or a few minutes every evening to share news about the day. Consider what "quality time" means to you and to the special people in your life.

Acts of Service

Answering "b" means you feel loved when people do thoughtful things for you such as washing your car, helping you with a repair job, bringing you breakfast in bed, or bathing the children so you can put your feet up. Even small gestures say "I love you" to people whose love language involves acts of service.

Physical Touch

Options labeled "c" are associated with the comfort and pleasure we get from physical affection. If your sweetheart texts to say, "Wish we were snuggled up together!" he or she is speaking the love language of touch. Physical touch includes sex but also friendly gestures such as a hug or a pat on the back.

Gifts

If you chose "d," chances are you treasure thoughtful gifts from loved ones. If you get a tear in your eye over things like a necktie, a finger puppet, a crayon drawing bestowed by a child, or a homemade ornament, gifts are probably an important love language to you.

Words of Affirmation

Options labeled "e" refer to words that make us feel loved and valued. These may be conveyed in a note from home, a homemade card or poem, a romantic letter, a song, or an unexpected text that simply says, "I love you." To people who speak this love language, hearing that they are loved (and why) is the sweetest message imaginable.

Managing Conflict in Interpersonal Relationships

8

CHAPTER OUTLINE

LEARNING OBJECTIVES

8.1

Explain the unavoidable but potentially problematic role of conflict in interpersonal relationships.

8.2

Describe the role of communication climate and relational spirals in interpersonal relationships, and practice communication strategies for keeping relationships healthy.

8.3

Identify characteristics of nonassertive, indirect, passive-aggressive, directly aggressive, and assertive communication, and explain how conflict approaches vary.

8.4

Explain the differences among win–lose, lose–lose, compromising, and win–win approaches to conflict resolution, and apply the steps involved in achieving win–win solutions.

Consider the relevance of relational climate and conflict to your own life:

?

Is the emotional tone of your most important relationships warm and welcoming, stagnant, or chilly and unsatisfying? How so?

?

Recall a recent verbal or non-verbal message that made you feel good about yourself. Now think of one that made you feel frustrated or unappreciated. What was different about these episodes?

?

What happened the last time you openly disagreed with someone? Was your relationship with that person better or worse afterward?

IT WAS A MAGIC MOMENT—a couple who met as lifeguards 25 years before, now happily married with children, sharing a nostalgic swim in a beautiful lake. As the couple paused to tread water, "our eyes met," remembers the wife. "I let my sentiments roam freely, tenderly telling Steve, 'I'm so glad we decided to do this together.'" She luxuriated in the moment, expecting "an equally gushing response." Instead, Steve said, "Yeah. Water's good," and starting paddling again.[1]

As quickly and unexpectedly as that, the seeds of conflict can emerge. It's no one's fault, necessarily. Goals and expectations differ. When they do, hurt feelings and frustration can quickly escalate into resentment or arguments.

The woman sharing a nostalgic swim with her husband was Brené Brown, a social work scholar and author of numerous books about embracing one's imperfections and daring to be vulnerable. That doesn't make her impervious to hurt feelings, of course. *"Didn't he hear me?"* she remembers thinking, as her husband swam away. "My emotional reaction was embarrassment, with shame rising."[2]

You've probably found yourself at odds with someone who is important to you. Conflict management is one of the biggest challenges we face in close relationships—whether with romantic partners, friends, coworkers, or family members.

This chapter will help you understand what kinds of communication create a supportive relational environment. It will also give you a toolkit for managing disagreements effectively. You might discover a new appreciation for conflict as a means to transform and strengthen relationships.

Understanding Interpersonal Conflict

Hurt but not defeated, Brené Brown decided to try again when she and her husband reached the opposite shore of the lake. "I flashed a smile in hopes of softening him up and doubled down on my bid for connection," she recalls. She again looked him in the eyes, and this time said, "This is so great. I love that we're doing this. I feel so close to you." Her husband replied, "Yep. Good swim," and swam away toward the original shore. After being twice disappointed, Brené remembers thinking indignantly, "This is total horseshit."[3]

You might like to think that such an experience would never happen in your relationships. But regardless of what we may wish for or dream about, a conflict-free world just doesn't exist. Even the best communicators, the luckiest people, are bound to find themselves in situations in which their needs don't match the needs of others. Money, time, power, sex, humor, and aesthetic taste, as well as a thousand other issues, arise and keep us from living in a state of perpetual agreement.

Many people think that the existence of conflict means that there's little chance for happy relationships with others. Effective communicators know differently, however. They realize that although it's impossible to *eliminate* conflict, there are ways to *manage* it effectively. And those effective communicators know the main point of this chapter—that managing conflict skillfully can lead to healthier, stronger, and more satisfying relationships.

Whatever form it may take, every interpersonal **conflict** involves an expressed struggle between at least two interdependent parties who perceive incompatible goals, scarce resources, and interference from one another in achieving their goals.[4]

A closer look at four parts of this definition helps illustrate the conditions that give rise to interpersonal conflict:

1. **Expressed struggle.** Granted, there are times when we fume to ourselves rather than expressing our frustration. You may be upset for months because a friendly neighbor's loud music keeps you from getting to sleep at night. That's most accurately described as internal conflict. Actual interpersonal conflict requires that both parties know a disagreement exists, such as when you let the neighbor know that you don't appreciate the decibel level. You might say this in words. Or you might use nonverbal cues, as in giving the neighbor a mean look, avoiding him, or slamming your windows shut. One way or another, once both parties know that a problem exists, it's an interpersonal conflict. In Brené Brown's swimming story, the conflict has yet to be expressed, but it will be.

2. **Interdependence.** However antagonistic they might feel toward each other, the parties in a conflict are usually dependent on each other. The welfare and satisfaction of one depend on the actions of another. After all, if they didn't need each other to solve the problem, they could solve it themselves or go their separate ways. Although this seems obvious from a distance, many people don't realize it in the midst of a disagreement. One of the first steps toward resolving a conflict is to take the attitude that "we're in this together."

No matter how satisfying your relationships, some degree of conflict is inevitable.

When do you find yourself most at odds with the people who matter most? How do you handle conflicts when they arise?

conflict An expressed struggle between at least two interdependent parties who perceive incompatible goals, scarce rewards, and interference from the other party in achieving their goals.

? ASK YOURSELF

Think of a time in which you and a relational partner experienced conflict. What goals and resources were involved? Were you able to express your feelings to each other and reach a mutually satisfying conclusion? Why or why not?

3. **Perceived incompatible goals.** Conflicts often look as if one party's gain will be another's loss. If your neighbor turns down his loud music, he loses the enjoyment of hearing it the way he wants, but if he keeps the volume up, then you're still awake and unhappy. It helps to realize that goals often are not as oppositional as they seem. Solutions may exist that allow both parties to get what they want. For instance, you could achieve peace and quiet by closing your windows and getting the neighbor to do the same. You might use earplugs. Or perhaps the neighbor could get a set of headphones and listen to the music at full volume without bothering anyone. If any of these solutions proves workable, then the conflict disappears.

Unfortunately, people often fail to see mutually satisfying answers to their problems. And as long as they *perceive* their goals to be mutually exclusive, they may create a self-fulfilling prophecy in which the conflict is very real.

4. **Perceived scarce resources.** In a conflict, people often believe that there isn't enough of the desired resource to go around. That's one reason conflict so often involves money.

If a person asks for a pay raise and the boss would rather keep the money or use it to expand the business, then the two parties are in conflict.

Time is another scarce commodity. As authors, we constantly struggle about how to use the limited time we have to spend. Should we work on this book? Visit with our partners? Spend time with our kids? Enjoy the luxury of being alone? With only 24 hours in a day, we're bound to end up in conflicts with our families, editors, students, and friends—all of whom want more of our time than we have available to give. You probably know the feeling well.

Having laid out the ingredients for conflict and acknowledged that it's a fact of life, let's turn our attention to ways that we can manage conflict effectively and even use it to strength our relationships. Creating a healthy relational climate is a good first step.

Communication Climates in Interpersonal Relationships

As Brené and Steve swam back across the lake, she envisioned the day unfolding in a pattern they had enacted many times before when they were frustrated with each other. She predicted that Steve would say, "What's for breakfast, babe?" and she would roll her eyes and say: "Gee, Steve. I forgot how vacation works. I forgot that I'm in charge of breakfast. And lunch. And dinner. And laundry. And packing and goggles. And . . ."[5]

communication climate The emotional tone of a relationship as it is expressed in the messages that the partners send and receive.

You get the point. Every relationship has a **communication climate**—an emotional tone. It's a lot like the weather. Some communication climates are fair and warm, whereas others are stormy and cold. Some are polluted and others healthy. Some relationships have stable climates, whereas others change dramatically—calm one moment and turbulent the next. Although the sun was shining, Brené predicted that a metaphorical dark cloud was brewing for her and her husband.

A communication climate doesn't involve specific activities as much as the way people feel about one another as they carry out those activities. Consider two communication classes, for example. Both meet for the same length of time and

How Sunny Is Your Communication Climate?

Think of an important person in your life—perhaps a friend, a roommate, a family member, or a romantic partner. Choose the option in each group in the following list that best describes how you communicate with each other, then see what your answers suggest about your relational climate.

1. When I am upset about something, my relational partner is most likely to:

   a. ignore how I feel

   b. say I should have tried harder to fix or avoid the problem

   c. listen to me and provide emotional support

2. When we are planning a weekend activity and I want to do something my partner doesn't want to do, I tend to:

   a. suggest another option we will both enjoy

   b. beg until I get my way

   c. cancel our plans and engage in the activity with someone else

3. When my partner and I disagree about a controversial subject, we usually:

   a. accuse the other person of using poor judgment or ignoring the facts

   b. ask questions and listen to the other person's viewpoint

   c. avoid the subject

4. If I didn't hear from my partner for a while, I would probably:

   a. call or text to make sure everything was okay

   b. not notice

   c. feel angry about being ignored

5. The statement we are most likely to make during a typical conversation sounds something like this:

   a. "Were you saying something?"

   b. "I appreciate the way you . . ."

   c. "You always forget to . . ."

Evaluating Your Responses

Circle your answers to this self-assessment on the grid below. (Note that they do not appear in alphabetical order.) Then read the forecast on the row where most of your answers appear.

Grouping 1	Grouping 2	Grouping 3	Grouping 4	Grouping 5	Relationship Forecast
c	a	b	a	b	Indications are that your relational climate is warm and sunny, with a high probability of descriptive and supportive communication.
b	b	a	c	c	Your relationship tends to be turbulent, with frequent outbreaks of controlling or defensive behavior. Storm warning: Escalatory conflict spirals can cause serious damage.
a	c	c	b	a	Beware of falling temperatures. It's natural for people to drift apart to some degree, but your relationship shows signs of chilly indifference and neutrality.

follow the same syllabus. It's easy to imagine how one of these classes might be a friendly, comfortable place to learn, whereas the other might be cold and tense—even hostile. The same principle holds for families, coworkers, and other relationships. Communication climates are a function more of the way people feel about one another than of the tasks they perform.

Communication climate influences how people respond when conflict emerges in a relationship. As you will see in the following section, some relationships involve trust and respect, whereas others are steeped in criticism and defensiveness.

Confirming and Disconfirming Messages

What makes some climates positive and others negative? A short but accurate answer is that the communication climate is determined by the degree to which people see themselves as valued. When we believe others view us as important, we are likely to feel good about our relationships with them. By contrast, the relational climate suffers when we think others don't appreciate or care about us.

As you read in Chapter 7, every message has relational dimensions. This means that, whether or not we are aware of the fact, we send and receive confirming and disconfirming messages virtually whenever we communicate. In other words, it isn't *what* we communicate about that shapes a relational climate so much as *how* we speak and act toward one another.

confirming messages Actions and words that express respect and show that we value the other person.

Confirming Messages Messages that show you are valued are called **confirming**.[6] Brené was trying to engage Steve in a confirming exchange when she told him she was glad to be there with him. She remembers how she felt when she didn't receive the validation she had expected in return:

> "*I thought* What's going on? I don't know if I'm supposed to feel humiliated or hostile. *I wanted to cry and I wanted to scream.*"[7]

When we feel hurt, it may be difficult to articulate exactly what we want. And when we're in Steve's shoes, we may be at a loss for how to respond. We can learn some valuable tools from scholars, who have identified three main categories of confirming communication.[8] Here are those categories, in order from the most basic to the most powerful.

1. **Show recognition.** The most fundamental act of confirmation is to recognize the other person. Recognition seems easy and obvious, and yet there are many times when we don't respond to others on this basic level. Brené remembers that when Steve tossed off his "Yep. Good swim" response, "he seemed to be looking through me rather than at me."[9]

   Your friends may feel a similar sense of being invisible or ignored if you don't return phone messages or if you avoid eye contact with them or fail to say hi when you encounter each other at a party or on the street. Of course, this lack of recognition may simply be an oversight. You might not notice your friend, or the pressures of work and school might prevent you from staying in touch. Nonetheless, if the other person *perceives* you as avoiding contact, the message has the effect of being disconfirming.

2. **Acknowledge the person's thoughts and feelings.** Acknowledging the ideas and emotions of others is an even stronger form of confirmation than simply recognizing them. Listening is probably the most common form of acknowledgment. Of course, as we discussed in Chapter 5, pretending to listen when you are actually thinking about something else or gathering ammunition for a rebuttal has the opposite effect of acknowledgment. It's

more confirming to ask questions, paraphrase, and reflect on what people are sharing with you. Not surprisingly, leaders who are supportive of others and their ideas are more successful than leaders who are more concerned with promoting their own image and ideas.[10]

3. **Show that you agree.** Whereas acknowledgment means you are interested in other people's ideas, endorsement means that you agree with them. It's easy to see why endorsement is the strongest type of confirming message: It communicates that we have a lot in common and that we are in sync. Not surprisingly, we tend to be attracted to people who agree with us.[11] Fortunately, it isn't necessary to agree completely with another person in order to endorse her or his message. You can probably find something in the message that you endorse. "I can see why you were so angry," you might say to a friend, even if you don't approve of his or her outburst. Of course, outright praise is a strong form of endorsement and one you can use surprisingly often if you look for opportunities to compliment others.

It's hard to overstate the importance of confirming messages. For example, people who offer confirmation generously are usually considered to be more appealing candidates for marriage than their less appreciative peers.[12] This preference is well founded. One of the most accurate ways to predict whether a marriage will last is to consider how positive a couple's communication is while they are dating.[13] This applies to both spoken words such as "thank you" and "I love you" and to nonverbal cues such as smiles and signs of affection.[14]

Positive, confirming messages are just as important in other relationships. For example, family members are most satisfied when they regularly encourage each other, joke around, and share news about their day.[15] And in the classroom, motivation and learning increase when teachers demonstrate a genuine interest and concern for students.[16]

Of course, confirming messages are only credible if the person delivering them seems sincere.[17] If a parent or teacher says, "You are incredibly smart" with a frustrated look on her face, it may be received as veiled criticism ("So *why* are you acting this way?") rather than a compliment.

Disconfirming Messages   In contrast to confirming communication, **disconfirming messages** deny the value of other people.[18] Disagreement can be disconfirming, especially if it goes beyond disputing the other person's ideas and attacks the speaker personally. However, disagreement is not the most damaging kind of disconfirmation. Personal attacks such as "You're crazy" are even tougher to hear.

disconfirming messages Words and actions that express a lack of caring or respect for another person.

Source: Ted Goff, North America Syndicate, 1994

UNDERSTANDING COMMUNICATION TECHNOLOGY

Can You Hear Me Now?

Thanks to technology, people have never been more connected—or more alienated.

I have traveled 36 hours to a conference on robotic technology in central Japan. The grand ballroom is Wi-Fi enabled, and the speaker is using the Web for his presentation. Laptops are open, fingers are flying. But the audience is not listening. Most seem to be doing their email, downloading files, surfing the Web, or looking for a cartoon to illustrate an upcoming presentation. Every once in a while audience members give the speaker some attention, lowering their laptop screens in a kind of digital curtsy.

In the hallway outside the plenary session, attendees are on their phones or using laptops and PDAs to check their email. Clusters of people chat with one another, making dinner plans, "networking" in that old sense of the term—the sense that implies sharing a meal. But at this conference it is clear that what people mostly want from public space is to be alone with their personal networks. It is good to come together physically, but it is more important to stay tethered to the people who define one's virtual identity, the identity that counts.

We live in techno-enthusiastic times, and we are most likely to celebrate our gadgets. Certainly the advertising that sells us our devices has us working from beautiful, remote locations that signal our status. We are connected, tethered, so important that our physical presence is no longer required. There is much talk of new efficiencies; we can work from anywhere and all the time. But tethered life is complex; it is helpful to measure our thrilling new networks against what they may be doing to us as people.[19]

Sherry Turkle

---

criticism A message that is personal, all-encompassing, and accusatory.

contempt Verbal and nonverbal messages that ridicule or belittle the other person.

defensiveness Protecting oneself by counterattacking the other person.

stonewalling Refusing to engage with the other person.

John Gottman, who has spent more than four decades studying how couples communicate, can predict with a rate of accuracy approaching 90% whether or not a married couple is headed toward divorce, mostly on the basis of their disconfirming behaviors.[20] Gottman calls the most hurtful of these the "Four Horsemen of the Apocalypse" because, when they are present on a regular basis, a relationship is usually is serious trouble and unlikely to survive.[21] Here they are:

1. **Partners criticize each other.** In contrast to complaints, which may focus on specific behaviors, **criticism** is personal, all-encompassing, and accusatory (*"You're lazy."* or *"The only person you think about is yourself."*).

2. **Partners show contempt.** **Contempt** takes criticism to an ever more hurtful level by mocking, belittling, or ridiculing the other person (*"You're pathetic. You disgust me."*). Whereas criticism implies "You are flawed," contempt implies "I hate you."[22] Expressions of contempt may be explicit, but they are more often expressed nonverbally—by sneering, eye rolling, and a condescending tone of voice. Gottman flatly states that the single best single predictor of divorce is contempt.[23]

3. **Partners are defensive.** When faced with criticism and contempt, it's not surprising that partners react with **defensiveness**—protecting their self-worth by counterattacking (*"You're calling me a careless driver? You're the one who got a speeding ticket last month."*). Once an attack-and-defend pattern develops, conflict usually escalates or partners start to avoid each other.

4. **One or both partners engage in stonewalling.** One of the most harmful disconfirming messages is **stonewalling**—a form of avoidance in which one person refuses to engage with the other. Walking away or giving one's partner the silent treatment conveys the message *"You aren't even worth my attention."*

TABLE 8-1

Distancing Tactics

TACTIC	DESCRIPTION
Avoidance	Evading the other person.
Deception	Lying to or misleading the other person.
Degrading	Treating the other person with disrespect.
Detachment	Acting emotionally uninterested in the other person.
Discounting	Disregarding or minimizing the importance of what the other person says.
Humoring	Not taking the other person seriously.
Impersonality	Treating the other person like a stranger; interacting with her or him as a role rather than a unique individual.
Inattention	Not paying attention to the other person.
Nonimmediacy	Displaying verbal or nonverbal clues that minimize interest, closeness, or availability.
Reserve	Being unusually quiet and uncommunicative.
Restraint	Curtailing normal social behaviors.
Restriction of topics	Limiting conversation to less personal topics.
Shortening of interaction	Ending conversations as quickly as possible.

Source: Adapted from Hess, J. A. (2002). Distance regulation in personal relationships: The development of a conceptual model and a test of representational validity. *Journal of Social and Personal Relationships, 19,* 663–683.

These are the big offenders on Gottman's list, but people may engage in a number of other disconfirming messages as well. Table 8-1 lists a variety of tactics people use to create distance between themselves and others. It's easy to see how each of them is inherently disconfirming.

It's important to note that disconfirming messages, like virtually every other kind of communication, are a matter of perception. That's why it can be a good idea to perception check before jumping to conclusions: "Were you laughing at my joke because you think I look stupid, or was it something else?"

How Communication Climates Develop

When we left Brené and Steve on their swim across the lake, she was feeling hurt and was already imagining the bickering that might lay ahead for them. One challenge of conflict management is that we tend to feel defensive and angry when our expectations are thwarted or we don't agree with our relational partners. That's natural. But acting defensively can make a difficult situation even worse. One comment can escalate into hours of snide comments or tense silence.

A **relational spiral** is a reciprocal communication pattern in which each person's message reinforces the other's.[24] This affect is captured in the old saying "what goes around comes around." In positive spirals, one partner's confirming message leads to a positive response from the other person. This positive reaction, in turn, leads the first person to be even more reinforcing, and so on.

relational spiral A reciprocal communication pattern in which each person's message reinforces the other's.

cultural idiom

what goes around comes around: Expect to be treated the way you have treated others.

escalatory spiral A reciprocal pattern of communication in which messages, either confirming or disconfirming, between two or more communicators reinforce one another.

avoidance spiral A communication spiral in which the parties slowly reduce their dependence on one another, withdraw, and become less invested in the relationship.

CHECKLIST ✔
Creating Positive Communication Climates

Here are some ways to initiate positive spirals and avoid negative ones, based on the work of Jack Gibb.[33]

☐ **Avoid judgmental statements.** Don't make "you" statements, such as *"You don't know what you're talking about"* and *"You smoke too much,"* which are likely to cause defensiveness and escalate conflict.

☐ **Use "I" language.** Statements such as *"I get frustrated when you interrupt me"* focus on the speaker's thoughts and feelings instead of judging the listener. The best "I" statements are specific.

☐ **Avoid attempts to control or manipulate the other person.** Be careful not to impose your preferences without regard for the other's needs or interests.

Negative spirals are just as powerful, although they leave the partners feeling worse about themselves and each other. For example, when one partner refuses to talk about a sensitive issue, the other partner is likely to become frustrated and distant as well,[25] and when one person criticizes another, a tit-for-tat pattern of destructive criticism often emerges.[26] Conversely, when one partner shows empathy, the other is more likely to show empathy in return, and conflicts are more likely to be resolved to both people's satisfaction.[27]

Escalatory spirals are the most visible way that disconfirming messages reinforce one another.[28] One attack leads to another until a skirmish escalates into a full-fledged battle. Although they are less obvious, **avoidance spirals** can also be destructive.[29] Rather than fighting, the parties slowly lessen their dependence on one another, withdraw, and become less invested in the relationship.

Spirals rarely go on indefinitely. Even the best relationships can go through periods of conflict and withdrawal. If the spiral is negative, partners may find the exchange growing so unpleasant that they switch from negative to positive messages without discussing the matter. In other cases, they may engage in metacommunication. "Hold on," one might say. "This is getting us nowhere." In still other cases, however, partners may pass the "point of no return," leading to the breakup of a relationship.

It often feels that relational spirals have a life of their own. People may be inclined, even without thinking about it, to mirror and escalate their partners' behaviors, even if they are harmful to the relationship. The best communicators recognize this tendency and make mindful choices instead. To illustrate, let's return to Steve and Brené Brown's experience on the lake.

> When the couple reached the dock where their swim had started, Brené decided to talk about her problem. Rather than blaming Steve for her hurt feelings, which was likely to escalate a conflict spiral, she said to him instead, "I've been trying to connect with you and you keep blowing me off. I don't get it."[30]

Keep reading to see what happened next. In the meantime, take the "How Sunny Is Your Communication Climate?" self-assessment quiz on page 217 and follow along as we take a closer look at different approaches and techniques for managing interpersonal conflict.

Approaches to Conflict

A popular school of thought suggests that conflict involves the dual concerns of (1) pursuing our own goals and agendas and (2) empathizing with other people.[31,32] You might assume that the most effective conflict managers tip the scales in favor of empathy. As you will see, there are situations in which that is a good strategy, just as there are times when pushing for a particular outcome outweighs other concerns. Most of the time, however, effective conflict management involves balancing these dual concerns. In this section, we explore five conflict styles and then consider the influence of gender, culture, and online communication on conflict communication. As you read, ask yourself which of these factors best reflect your conflict approach.

Styles of Expressing Conflict

This section describes five ways people can act when their needs are not met (summarized in Table 8-2). As you will see, some approaches are more productive than others. As you read on, ask yourself which styles you use most often and how these styles affect the quality of your close relationships.

Nonassertion   The inability or unwillingness to express thoughts or feelings in a conflict is known as **nonassertion**. Sometimes nonassertion comes from a lack of confidence. At other times, people lack the awareness or skill to use a more direct means of expression.

Sometimes people know how to communicate in a straightforward way but choose to behave nonassertively. For example, women are less likely to clearly refuse an unwanted request for physical intimacy from a dating partner they would like to see in the future than from one they don't want to see again.[34]

Nonassertion can take a variety of forms. One is *avoidance*—either physically (putting distance between yourself and a friend after an argument) or conversational (changing the topic, joking, or denying that a problem exists). People who avoid conflicts usually believe it's easier to put up with the status quo than to face the problem head-on and try to solve it. *Accommodation* is another type of nonassertive response. Accommodators deal with conflict by giving in, thus putting the other's needs ahead of their own.

Despite the obvious drawbacks of nonassertion, there are situations when accommodating or avoiding is a sensible approach. Avoidance may be the best course if a conflict is minor and short lived. For example, you might let a friend's annoying grumpiness pass without saying anything, knowing that he is having one of his rare bad days. Likewise, you might not complain to a neighbor whose loud music only rarely disturbs you. You may also reasonably choose to say nothing if the conflict occurs in an unimportant relationship, as with an acquaintance whose language you find offensive but whom you don't see often. Finally, you might choose to keep quiet if the risk of speaking up is too great: getting fired from a job you can't afford to lose, being humiliated in public, or even risking physical harm.

Assertiveness is most important when the issue and relationship matter a great deal. In one study, couples volunteered to be videotaped while they talked about sources of conflict in their marriages.[35] Researchers compared the couple's communication techniques to their marital satisfaction scores. They found that when the issue was a relatively minor one, the happiest couples used indirect behaviors such as hinting. However, when the issue was of great concern, relationship satisfaction increased among the couples who addressed the issue assertively rather than indirectly. This was true even when the couples used communication techniques that are typically considered negative, such as blaming each other, making demands, or rejecting their partners' explanations. The researchers concluded that strong words should be used sparingly, but that overreacting is sometimes preferable to downplaying a serious concern.

Indirect Communication   The clearest communication is not necessarily the best approach. **Indirect communication** conveys a message

CHECKLIST *Continued*

☐ **Focus on mutually beneficial problem solving.** You can help build a healthy relational climate by seeking solutions that satisfy both your needs and other people's (see p. 236).

☐ **Be honest.** Think about what you want to say, and plan the wording of your message carefully so that you can express yourself clearly.

☐ **Show empathy.** Empathic messages show that you accept another person's feelings and can put yourself in his or her place.

☐ **Treat others as your equal.** People who convey an attitude of equality communicate that, although they may have greater talent than others in certain areas, other people have just as much worth as they do.

☐ **Be open to others' viewpoints.** Dogmatic and unyielding people stimulate feelings of defensiveness in others. Even when you have strong opinions about a topic, it's important to acknowledge that you don't have a corner on the truth.

---

nonassertion   The inability or unwillingness to express one's thoughts or feelings.

indirect communication   Hinting at a message instead of expressing thoughts and feelings directly.

---

cultural idioms

hold on: wait

face the problem head-on: confront directly

TABLE 8-2					
Individual Styles of Conflict					
	NONASSERTIVE	**INDIRECT**	**PASSIVE-AGGRESSIVE**	**DIRECTLY AGGRESSIVE**	**ASSERTIVE**
APPROACH TO OTHERS	I'm not okay, you're okay.	I'm okay, you're not okay. (But I'll let you think you are.)	I'm okay, you're not okay.	I'm okay, you're not okay.	I'm okay, you're okay.
DECISION MAKING	Lets others choose.	Chooses for others. They don't know it.	Chooses for others. They don't know it.	Chooses for others. They know it.	Chooses for self.
SELF-SUFFICIENCY	Low.	High or low.	Looks high but usually low.	High or low.	Usually high.
BEHAVIOR IN PROBLEM SITUATIONS	Flees, gives in.	Strategic, oblique behavior.	Concealed attack.	Outright attack.	Direct confrontation.
RESPONSE OF OTHERS	Disrespect, guilt, anger, frustration.	Unknowing compliance or resistance.	Confusion, frustration, feelings of manipulation.	Hurt, defensiveness, humiliation.	Mutual respect.
SUCCESS PATTERN	Succeeds by luck or charity of others.	Gains unwitting compliance of others.	Wins by manipulation.	Beats out others.	Attempts "win–win" solutions.

in a roundabout manner in order to save face for the recipient.[36] Although indirect communication lacks the clarity of an aggressive or assertive message, it involves more initiative than nonassertion. It also has none of the hostility of passive-aggressive "crazymaking." The goal is to get what you want without arousing the hostility of the other person. Consider the case of the neighbor's loud music. One indirect approach would be to strike up a friendly conversation with the neighbor and ask if anything you are doing is too noisy for him, hoping he will get the hint.

Because it saves face for the other party, indirect communication is often kinder than blunt honesty. If your guests are staying too long, it's probably kinder to yawn and hint about your big day tomorrow than to bluntly ask them to leave. Likewise, if you're not interested in going out with someone who has asked you for a date, it may be more compassionate to claim that you're busy than to say, "I'm not interested in seeing you."

At other times we communicate indirectly in order to protect ourselves. You might, for example, test the waters by hinting instead of directly asking the boss for a raise, or by letting your partner know indirectly that you could use some affection, instead of asking outright. At times like these, an oblique approach may get the message across while softening the blow of a negative response.

The advantages of protecting oneself and saving face for others help explain why indirect communication is the most common way people make requests, especially if we don't know the person well or if we feel intimidated about asking.[37] The risk of an indirect message, of course, is that the other party will misunderstand you or fail to get the message at all. There are also times when the importance of an idea is so great that hinting lacks the necessary punch. When clarity and directness are your goals, an assertive approach is in order.

cultural idioms

test the waters: try before committing
softening the blow: easing the effect
punch: force or effectiveness

Passive Aggression   **Passive aggression** occurs when a communicator expresses hostility in an obscure way. Psychologist George Bach identified the following varieties of crazymaking[38] behavior:

1. *Pseudoaccommodators* **pretend to agree with you.**
   A passively aggressive person might commit to something (*"I'll be on time from now on"*) but not actually do it.

2. *Guiltmakers* **try to make you feel bad.**
   A guiltmaker will agree to something and then make you feel responsible for the hardship it causes him or her (*"I really should be studying, but I'll give you a ride"*).

3. *Jokers* **use humor as a weapon.**
   An underhanded joker uses humor as an excuse to say unkind things and then claim innocence (*"Where's your sense of humor?"*).

4. *Trivial tyrannizers* **do small things to drive you crazy.**
   Rather than express his or her feelings outright, the trivial tyrannizer does annoying things such as "forgetting" to clean the kitchen or pass along a message as promised.

5. *Withholders* **keep back something valuable.**
   A withholder punishes others by refusing to provide thoughtful gestures such as courtesy, affection, or humor.

It's easy to understand the destructive effects of passive aggression, which can be as hurtful as direct aggression, discussed next.

Direct Aggression   Whereas nonasserters avoid conflicts, communicators who use **direct aggression** embrace them. A directly aggressive message confronts the other person in a way that attacks his or her position—and even the dignity of the receiver. Many directly aggressive responses are easy to spot:

"You don't know what you're talking about."

"That was a stupid thing to do."

"What's the matter with you?"

Other forms of direct aggression come more from nonverbal messages than from words. It's easy to imagine a hostile way of expressing statements such as:

"What is it now?"

"I need some peace and quiet."

Verbal aggressiveness may get you what you want <u>in the short run</u>. Yelling *"Shut up"* might stop the other person from talking, and saying *"Get it yourself"* may save you from some exertion, but the relational damage of this approach probably isn't worth the cost. Direct aggression can be hurtful, and the consequences for the relationship can be long lasting.[39]

Assertion   Assertiveness represents a balance between self-interest and empathy. **Assertive** people handle conflicts by expressing their needs, thoughts, and feelings clearly and directly but without judging others or dictating to them. They have the attitude that most of the time it is possible to resolve problems to everyone's satisfaction.

The introverted scientist Raj Koothrappali (Kunal Nayyar) on *The Big Bang Theory* is comfortable managing disagreements with most of his best friends. When strangers or women (such as Penny, played by Kaley Cuoco) are involved, however, he often resorts to whispering or going silent.

In what situations do you feel confident engaging in conflict management? When do you feel intimidated?

passive aggression An indirect expression of aggression, delivered in a way that allows the sender to maintain a facade of kindness.

direct aggression A message that attacks the position and perhaps the dignity of the receiver.

assertive communication A style of communicating that directly expresses the sender's needs, thoughts, or feelings, delivered in a way that does not attack the receiver.

cultural idiom
in the short run: for a short period of time

Dealing with Sexual Harassment

Sexual harassment takes many forms. It may arise between members of the same sex or between men and women. It can be a blatant sexual overture, or any verbal or nonverbal behavior that creates a hostile work environment. The harasser can be a supervisor, peer, subordinate, or even someone outside the organization.

Thanks to enlightened company policies and government legislation, targets of sexual harassment have legal remedies. Although formal complaints are an assertive and powerful way to protect a target's rights, they can be time consuming, and those lodging the complaints sometimes experience depression, ridicule, isolation, and reprisal.[40]

As Chapter 1 explained, competent communication involves picking the most effective approach. Here are several options to consider if you or someone you care about experience harassment:[41]

1. **Consider dismissing the incident.**

   This nonassertive approach is only appropriate if you truly believe that the remark or behavior is trivial. Dismissing incidents that you believe are important can result in self-blame and diminished self-esteem. Even worse, it can lead to repetition of the offensive behavior.

2. **Tell the harasser to stop.**

   Assertively tell the harasser early that the behavior is unwelcome, and insist that it stop immediately. Your statement should be firm, but it doesn't have to be angry. Remember that

many words or deeds that make you uncomfortable may not be deliberately hostile.

3. **Write a personal letter to the harasser.**

   A written statement may help the harasser to understand what behavior you find offensive. Just as important, it can show that you take the problem seriously. Detail specifics about what happened, what behavior you want stopped, and how you felt. You may want to include a copy of your organization's sexual harassment policy. Keep a record of when you delivered your message.

4. **Ask a trusted third party to intervene.**

   This indirect approach can sometimes persuade the harasser to stop. The person you choose should be someone who you are convinced understands your discomfort and supports your opinion. Also, be sure this intermediary is someone the harasser respects and trusts.

5. **Use company channels.**

   Report the situation to your supervisor, personnel office, or a committee that has been set up to consider harassment complaints.

6. **File a legal complaint.**

   If all else fails or the incident is egregious, you may file a complaint with the federal Equal Employment Opportunity Commission or with your state agency. You have the right to obtain the services of an attorney regarding your legal options.[42]

Possessing this attitude and the skills to bring it about doesn't guarantee that assertive communicators will always get what they want, but it does give them the best chance of doing so. An additional benefit of such an approach is that whether or not it satisfies a particular need, it maintains the self-respect of both the assertors and those with whom they interact. As a result, although people who manage their conflicts assertively may experience feelings of discomfort while they are working through the problem, they usually feel better about themselves and one another afterward—quite a change from the outcomes of nonassertiveness or aggression.

Characteristics of an Assertive Message

Knowing *about* assertive messages isn't the same as being able to express them. The next few pages describe a method for communicating assertively. It works for a variety of messages: your hopes, problems, complaints, and expressions of appreciation. Besides giving you a way to express yourself directly, this format also makes it easier for others to understand you. A complete assertive message has five parts:

1. **Describe the behavior in question.** An assertive description is specific without being evaluative or judgmental.

   > Behavioral description: *"You asked me to tell you what I really thought about your idea, and then when I gave it to you, you told me I was too critical."*

   > Evaluative judgment: *"Don't be so touchy! It's hypocritical to ask for my opinion and then get mad when I give it to you."*

   Judgmental words such as *touchy* and *hypocritical* invite a defensive reaction. The target of your accusation can reply, *"I'm not touchy or hypocritical!"* It's harder to argue with the facts stated in an objective, behavioral description. Furthermore, the neutral language reduces the chances of a defensive reaction.

2. **Share your interpretation of the other person's behavior.** This part is where you can use the perception-checking skill outlined in Chapter 2 (pp. 53–54). Remember that a complete perception check includes two possible interpretations of the behavior:

   > *"Maybe you think I don't care because it took me two days to call you back. Is that it, or is there something else?"*

   The key is to label your hunches as such instead of suggesting that you are positive about what the other person's behavior means.

3. **Describe your feelings.** Expressing your feelings adds a new dimension to a message. For example, consider the difference between these two responses:

   > *"When you kiss me and nibble on my ear while we're watching television* [behavior], *I think you probably want to make love* [interpretation], *and I feel excited."*

   > *"When you kiss me and nibble on my ear while we're watching television* [behavior], *I think you probably want to make love* [interpretation], *and I feel disgusted."*

   Likewise, adding feelings to the situation we have been examining makes the assertive message clearer:

   > *"When you said I was too critical after you asked me for my honest opinion* [behavior], *it seemed to me that you really didn't want to hear a critical remark* [interpretation], *and I felt stupid for being honest* [feeling]."

4. **Describe the consequences.** A consequence statement explains what happens as a result of the behavior you have described, your interpretation, and the ensuing feeling. There are three kinds of consequences: (1) What happens to you, the speaker (*"When you tease me, I avoid you"*), (2) what happens to the target of the message (*"When you drink too much, you start to drive dangerously"*), or (3) what happens to others (*"When you play the radio so loud, it wakes up the baby"*).

5. **State your intentions.** Intention statements are the final element in the assertive format. They can communicate three kinds of messages:

   - Where you stand on an issue: *"I want you to know how much this bothers me"* or *"I want you to know how much I appreciate your support."*

   - Requests of others: *"I'd like to know whether you are angry"* or *"I hope you'll come again."*

   - Descriptions of how you plan to act in the future: *"Don't expect me ever to lend you anything again."*

ETHICAL CHALLENGE

It's Nothing!

Is it ever justifiable to behave as if you are angry with someone but refuse to share your feelings with that person? If so, when? How would you describe this style of conflict management?

cultural idiom
touchy: easily offended

How Assertive Are You?

Circle your answer to each question using the grid below. Note that the answers do not appear in alphabetical order.

1. You feel you deserve the new corner office that has just become available. What would you do?

   a. Hint around that you have outgrown your cubicle.

   b. Tell your coworkers, "You deserve it more than I do," but secretly ask the boss if you can have it.

   c. Meet with your supervisor and lay out the reasons you think you deserve the office.

   d. Threaten to quit if you aren't assigned to the office.

   e. Stay quiet and hope the boss realizes that you deserve the office.

2. Your best friend just called to cancel your weekend trip together at the last minute. This isn't the first time your friend has done this, and you are very disappointed. What do you do?

   a. Announce that the friendship is over. That's no way to treat someone you care about.

   b. Reassure your friend that it's okay and there are no hard feelings.

   c. Resolve to cancel the next trip yourself to teach your friend a lesson.

   d. Declare, "But I've already packed," hoping your friend will take the hint and decide to go after all.

   e. Say, "I feel disappointed, because I enjoy my time with you and because we have made nonrefundable deposits. Can we work this out?"

3. During a classroom discussion, a fellow student makes a comment that you find offensive. What do you do?

   a. Tell the instructor after class that the comment made you uncomfortable.

   b. Ignore it.

   c. Announce that the statement is the stupidest thing you have ever heard.

   d. Say nothing, but tell other people how much you dislike that person.

   e. Join the discussion, mention that you see the issue differently, and invite your classmate to explain why he or she feels that way.

4. You are on a first date when the other person suggests seeing a movie you are sure you will hate. What do you do?

   a. Lie and say you've already seen it.

   b. Say, "Sure!" How bad can it be?

   c. Say, "Okaaay," and raise your eyebrows in a way that suggests your date must be either stupid or kidding.

   d. Suggest that you engage in another activity instead.

   e. Proclaim that you'd rather stay home and watch old reruns than see that movie.

EVALUATING YOUR RESPONSES

	Question 1	Question 2	Question 3	Question 4
ROW 1	e	b	b	b
ROW 2	a	d	a	b
ROW 3	b	c	d	c
ROW 4	c	e	e	d
ROW 5	d	a	c	e

Nonassertive

If the majority of your answers appear on row 1, you rank low on the assertiveness scale. The people around you may be unable to guess when you have a preference or hurt feelings. It may seem that "going with the flow" is the way to go, but research suggests that relationships flounder when people don't share their likes and dislikes with one another. Try voicing your feelings more clearly. People may like you more for it.

Indirect

If the majority of your answers appear on row 2, you tend to know what you want, but you rely on subtlety to convey your preferences. This can be a strength, because you aren't likely to offend people. However, don't be surprised if people sometimes fail to notice when you are upset. Research suggests that indirect communication works well for small concerns, but not for big ones. When the issue is important to you, step up to say so.

Passively Aggressive

If the majority of your answers appear on row 3, you tend to be passive-aggressive. Rather than taking the bull by the horns, you are more likely to seek revenge, complain to people around you, or use snide humor to make your point. These techniques can make the people around you feel belittled and frustrated. Plus, you are more likely to alienate people than to get your way in the long run. Try to break this habit by saying what you feel in a clear, calm way.

Assertive

If the majority of your answers appear on row 4, you have hit the bulls-eye in terms of healthy assertiveness. You tend to say what you feel without infringing on other people's right to do the same. Your combination of respectfulness and self-confidence is likely to serve you well in relationships.

Aggressive

If the majority of your answers appear on row 5, you tend to overshoot assertive and land in the aggressive category instead. Although you may mean well, your comments are likely to offend and intimidate others. Try tuning it back a little by stating your opinions (gently) and encouraging others to do the same. If you refrain from name-calling and accusations, people are likely to take what you say more seriously.

In our ongoing example, adding an intention statement would complete the assertive message:

> *"When you said I was too critical after you asked me for my honest opinion* [behavior], *it seemed to me that you really didn't want to hear a critical remark* [interpretation]. *That made me feel stupid for being honest* [feeling]. *Now I'm not sure whether I should tell you what I'm really thinking the next time you ask* [consequence]. *I'd like to get it clear right now: Do you really want me to tell you what I think or not* [intention]?"

Before you try to deliver messages using the assertive format outlined here, there are a few points to remember. First, it isn't necessary or even wise always to put the elements in the order described here. As you can see from reviewing the examples on the preceding pages, it's sometimes best to begin by stating your feelings. In other cases, you can start by sharing your intentions or interpretations or by describing consequences.

You also ought to word your message in a way that suits your style of speaking. Instead of saying *"I interpret your behavior to mean,"* you might choose to say, *"I think . . ."* or, *"It seems to me . . ."* or perhaps, *"I get the idea. . . ."* In the same way, you can express your intentions by saying, *"I hope you'll understand (or do) . . ."* or perhaps, *"I wish you would. . . ."* It's important that you get your message across, but you should do it in a way that sounds and feels genuine to you.

Realize that there are some cases in which you can combine two elements in a single phrase. For instance, the statement *". . . and ever since then I've been wanting to talk to you"* expresses both a consequence and an intention. In the same way, saying, *". . . and after you said that, I felt confused"* expresses a consequence and a feeling. Whether you combine elements or state them separately, the important point is to be sure that each one is present in your statement.

Finally, realize that it isn't always possible to deliver messages such as the ones here all at one time, wrapped up in neat paragraphs. It will often be necessary to repeat or restate one part many times before your receiver truly understands what

cultural idiom

in the long run: Over an extended period of time

you're saying. As you've already read, there are many types of psychological and physical noise that make it difficult for us to understand one another. Just remember: You haven't communicated successfully until the receiver of your message understands everything you've said. In communication, as in many other activities, patience and persistence are essential.

So far, we have been talking about the dual concerns of conflict management (self and other) as if they are of equal merit. But as you will see in the next sections, gender and cultural expectations sometimes privilege one over the other, and conflict often manifests differently in cyberspace than it does in person.

Gender and Conflict Style

The "Men Are from Mars, Women Are from Venus" theory of conflict has strong intuitive appeal. A body of research seems to support the notion that, in general, men and women typically do approach conflict somewhat differently.

Some differences are evident early on. By the time boys are 6 years old, they tend to gravitate toward large groups in which there is a clear understanding of who outranks whom.[43] In this context, competition is often considered a way to earn respect and status. It's also common for boys to engage in physical tussles with each other, both as a form of play and as a means of settling disputes. Girls, on the other hand, tend to gravitate to one-on-one relationships. The emphasis is less on who outranks whom and more on who is closest to whom. Even in a group, girls typically know who has best friend status. As a result, girls typically engage in more prosocial behaviors (offering compliments, showing empathy, providing emotional support) than boys do and more often shy away from direct confrontations.

Especially because children tend to have mostly same-sex friends, it's easy to imagine the misunderstandings that occur when they form other-sex relationships in their teen years and beyond. Girls may feel that boys are insensitive, boisterous, and emotionally distant, whereas boys may feel that girls are quick to get their feelings hurt but are reluctant to say outright what is bothering them. On the bright side, males typically appreciate the emotional support of their female friends. And women say they enjoy the freedom to be more frank and assertive with their male friends than they usually are with their female friends.[44]

Origins of Gender Differences   Biology explains some of the difference between the way males and females deal with conflict. During disagreements, men tend to experience greater physiological arousal than women, which comes in the form of increased heart rate and blood pressure. This may be why boys exhibit more aggressive behaviors than girls do, even when they are very young. About 1 in 20 male toddlers is frequently aggressive, compared with 1 in 100 female toddlers.[45]

Evolution may play a role as well. Because women are able to bear only a limited number of children, procreation has favored men who can successfully compete for their attention and demonstrate their superiority to other males. Moreover, in their traditional role as hunters and providers, men were challenged to be bold, physical risk takers.[46] It may be that men have evolved to be more physical and competitive than women because—at least in years gone by—that was an advantage. By contrast, women have traditionally nurtured children. In that role, there is an advantage to creating safe environments, working cooperatively with others, and understanding the nuances of nonverbal communication.[47] This may explain why women are often more sensitive to subtle cues and are more aligned to harmony and cooperation than to competition.

Although biology and evolution have some influence, as we grow up we learn to handle our emotions and to mimic our role models. This means that culture plays an important role as well.[48] We all know that there are powerful rewards for "acting like a lady" or "being a real man" and there are serious consequences for defying those expectations. In Western cultures, females are typically expected to be accommodating and males to be competitive. But these expectations can lead to frustrating double binds, as we will discuss next.

Conflict Dilemmas   Women face a double standard: They may be judged more harshly than men if they are assertive, but they may be overlooked if they aren't. Typically, women are more likely than men to use indirect strategies instead of confronting conflict head-on. They are also more likely to compromise and to give in to maintain relational harmony. This style works well in some situations, but not in others. In one study, men and women scored equally well on a set of mathematical challenges, but the men were twice as likely as the women to enter a tournament in which they could compete for cash prizes or raises based on their performance.[49] As a result of their reluctance to compete, women may be overlooked for raises, promotions, and other forms of recognition, even though they are highly qualified.

Cultural norms present a dilemma for men as well. Men are typically rewarded for being competitive and assertive. But those behaviors can seem overly aggressive in close relationships. For example, when conflict arises with coworkers, friends, and loved ones, a competitive stance can make the situation worse.[50] That, coupled with men's high level of physical arousal in conflict conditions, can make interpersonal conflict particularly frustrating for them. They are more likely than women to withdraw if they become uncomfortable or fail to get their way.[51] Women may interpret this as indifference, but men often say they detach to avoid overreacting, physically and verbally, in the heat of the moment.

Gender differences that appear in face-to-face communication also persist online. When researchers compared messages posted by male and female teenagers, they found that the boys typically used assertive language, such as boasts and sexual invitations, whereas the girls used mostly cooperative language, such as compliments and questions.[52] The teens' adherence to traditional gender roles was also evident nonverbally. The girls tended to post photos of themselves in seductive, receptive poses, and the boys in more rugged, dominant poses. These personae are likely to influence how they behave when conflict arises.

Commonalities   General differences aside, it bears emphasizing that social expectations change over time and stereotypes do not always apply. The qualities men and women have in common far outnumber their differences. Put another way, although men and women differ on *average*, most of us live somewhere in the middle, where masculine and feminine styles frequently overlap.[53] For example, men and women are roughly the same in terms of how much closeness they desire in relationships and the value they place on sharing ideas and feelings.[54]

One danger is that we may stereotype others and even ourselves, based on differences that are fairly small or don't actually exist. People who assume that men are aggressive and women are accommodating may notice behavior that fits these stereotypes (*"See how much he bosses her around? A typical man!"*). On the other hand, behavior that doesn't fit these preconceived ideas (accommodating men, pushy women) goes unnoticed or is criticized as "unladylike" or "unmanly."

What, then, can we conclude about the influence of gender on conflict? Research has demonstrated that there are, indeed, some small but measurable differences in the two sexes. But, although men and women may have characteristically different conflict styles, the individual style of each communicator is more important than a person's sex in shaping the way he or she handles conflict.

Cultural Influences on Conflict

The ways in which people communicate during conflicts vary widely from one culture to another. The kind of rational, calm, yet assertive approach that is the ideal for European American disagreements is not the norm in some other cultures. For example, in traditional African American culture, conflict is characterized by a

In the TV series *Modern Family*, Sofia Vergara plays the highly expressive Gloria Pritchett, who has a dramatic flair for conflict. Gloria once warned her husband, "When you're married to me, you're going to get yelled at many times."[57] In real life, cultural expectations influence how people behave in conflict situations.

Are you more inclined to seek out a passionate argument or to avoid one? Why?

UNDERSTANDING DIVERSITY

They Seem to Be Arguing

"The Italian language brings out the passion in me," declares Ewa Niemiec, who is fluent in Italian, English, and Polish.[55] Like many people who are multilingual, she feels that each language evokes a different feeling. Speaking English makes her feel polite, but Italian makes her "mouthy and loud." That's not necessarily bad, she says, unless people mistake the passion for conflict or aggression.

Once, when Niemiec interviewed for a job in Italian, she found herself "bickering" with her would-be boss. She's not unusual in finding that experience enjoyable. Members of many cultures consider verbal disputes to be a form of intimacy and even a game.

Misunderstandings may arise, however, when people from emotionally reserved cultures observe or interact with people from emotionally expressive ones. Americans visiting Greece, for example, often think they are witnessing an argument when they are overhearing a friendly conversation.[56] A comparative study of American and Italian nursery school children showed that one of the Italian children's favorite pastimes was a kind of heated debating that Italians call *discussione*, which Americans would regard as arguing.

Niemiec encourages people to overcome the idea that emotional language is confrontational. "If you're planning to travel or live in Italy," she says, "be prepared for a land of strong feelings, loud voices, and even bigger hand gestures."

greater tolerance for expressions of intense emotions.[58] And ethnicity isn't the only factor that shapes a communicator's preferred conflict style; the degree of assimilation also plays an important role. For example, Latinos with strong cultural identities tend to seek accommodation and compromise more than those with weaker cultural ties.[59]

In individualistic cultures like that of the United States, the goals, rights, and needs of each person are considered important, and most people would agree that it is an individual's right to stand up for himself or herself. People in such cultures typically value direct communication in which you say outright what is bothering you.[60] By contrast, people in collectivist cultures (more common in Latin America and Asia) usually consider the concerns of the group to be more important than those of any individual. Preserving and honoring the face of the other person are prime goals, and communicators go to great lengths to avoid any communication that might embarrass a conversational partner. In these cultures, the kind of assertive behavior that might seem perfectly appropriate to an American or Canadian would seem rude and insensitive.

As you might imagine, low-context cultures like that of the United States place a premium on being direct and literal. By contrast, high-context cultures like that of Japan value self-restraint and avoid confrontation. Communicators in these cultures derive meaning from a variety of unspoken cues, such as context, social conventions, and hints. For this reason, what seems like "beating around the bush" to an American would be polite to an Asian. In Japan, for example, even a simple request like "close the door" may seem too straightforward. A more indirect statement such as "it is somewhat cold today" would be more appropriate. Or a Japanese person may glance at the door or tell a story about someone who got sick in a drafty room.[61] To take a more important example, Japanese are reluctant to simply say *"no"* to a request. A more likely answer would be, *"Let me think about it for a while,"* which anyone familiar with Japanese culture would recognize as a refusal. When indirect communication is a cultural norm, it is unreasonable to expect more straightforward approaches to succeed.

From the examples so far, you might expect the United States to top the charts in terms of directness when it comes to conflict management. However, a mediating factor is at play—emotional expressiveness. In this regard, the United States has a great deal in common with Asian cultures, namely, a preference for calm communication rather than heated displays of emotion.[62] From this perspective, it may seem rude, frightening, or incompetent to show intense emotion during conflict. Indeed, people who become passionate are warned about the danger of saying things they don't mean.

By contrast, in cultures that value emotional expressiveness, people who do *not* show passion are regarded as hiding their true feelings. African Americans, Arabs, Greeks, Italians, Cubans, and Russians are typically considered highly expressive.[63] To them, behaving calmly in a conflict episode may be a sign that a person is unconcerned, insincere, or untrustworthy.

With differences like these, it's easy to imagine how two friends, lovers, or fellow workers from different cultural backgrounds might have trouble finding a conflict style that is comfortable for them both. Sometimes we don't even understand our own reactions to conflict. Many Americans find that, although they usually consider themselves to be direct and individualistic, they are fairly accommodating when it comes to conflict management.[64] This may surprise them as much as it surprises other people.

Conflict in Online Communication

Online communication has changed the nature of interpersonal conflict. Disagreements handled via texting, chatting, email, and blogging can unfold differently from those that play out in person. Some of the characteristics of mediated communication described in Chapter 1 are especially important during conflicts.

Delay The asynchronous nature of most mediated channels means that communicators aren't obliged to respond immediately to one another.

The inherent delays in mediated conflicts present both benefits and risks. On the upside, the chance to cool down and think carefully before replying can prevent aggressive blowups.[65] This is true on a personal and a global level. Researcher Donald Ellis commends inclusive chat rooms and discussion boards, calling them a "new public sphere" and a "deliberative democracy" in which people can share ideas and take their time considering a wide range of viewpoints.[66] On the other hand, participation requires active involvement. People can just as easily ignore online posts and fail to respond to emails, texts, and IMs. When they do reply, there's the temptation to craft insults and jabs that can make matters worse.

Disinhibition The absence of face-to-face contact can make it easy to respond aggressively, without considering the consequences until it's too late. This is especially likely when people don't know one another well and when tensions are high. Researchers describe what they call a "flame war" (based on the term for inflammatory or blunt remarks online) that erupted in an online cancer support group.[67] A flame war erupted when the mother of a cancer patient posted her opinion that hospital beds needed by cancer patients were being taken up by "anorexic girls." Others weighed in to say she should be more compassionate and less judgmental. Eventually, the episode erupted into an online feud comprising nearly 100 posts by 30 of the 42 people in the support group. Capitalized words such as PISSED OFF, ATTACKING, and VULTURES screamed the participants' frustration. The researchers speculate that online communication can facilitate close ties, but at the same time, people may post comments they wouldn't say to one another in person. Although it's easier to overlook the impact of a hostile approach at a virtual distance, remember that communication is irreversible: It's no more possible to retract a message that's been delivered than it is to "unsqueeze" a tube of toothpaste.

ASK YOURSELF

Choose a conflict style that is different from yours, and identify the assumptions on which it is based. Next, suggest how people with different styles can adapt their assumptions and behaviors to communicate more effectively with others.

Musical star Chris Brown has a long record of airing his disagreements in Twitter feuds with other celebrities. This has earned him the reputation of being an out-of-control hothead.

Have you ever used social media to make your conflicts public? If so, what were the consequences?

Permanence Because emails and text messages come in written form, there's a permanent "transcript" that doesn't exist when communicators deal with conflict face to face. This record can help clarify misperceptions and faulty memories. On the other hand, the permanent documents that chronicle a conflict can stir up emotions that make it hard to forgive and forget.

If online communication seems to be making a conflict worse, you may want to consider shifting to a face-to-face approach. Although mediated conflict may feel easier at the time, it may create more lasting damage to the relationship.

Managing Interpersonal Conflicts

It's helpful to understand how conflicts operate, but awareness alone isn't enough. The following pages describe several ways to communicate in the face of disagreement. As you read about them, consider which ones you use now, and whether others might serve you better.

Methods for Conflict Resolution

Regardless of the relational style, gender, or culture of the participants, every conflict is a struggle to have one's goals met. Sometimes that struggle succeeds, and at other times it fails. In the remainder of this chapter we'll look at various approaches to resolving conflicts and see which ones are most promising.

win–lose problem solving An approach to conflict resolution in which one party reaches his or her goal at the expense of the other.

Win–Lose Win–lose conflicts are ones in which one party achieves his or her goal at the expense of the other. People resort to this method of resolving disputes when they perceive a situation as being an "either–or" one: Either I get what I want, or you get your way. The most clear-cut examples of win–lose situations are certain games, such as baseball or poker, in which the rules require a winner and a loser. Some interpersonal issues seem to fit into this win–lose framework: two coworkers seeking a promotion to the same job, for instance, or a couple who disagree on how to spend their limited money.

Power is the distinguishing characteristic in **win–lose problem solving**, because it is necessary to defeat an opponent to get what you want. The most obvious kind of power is physical. Some parents threaten their children with warnings such as *"Stop misbehaving, or I'll send you to your room."* Adults who use physical power to deal with one another usually aren't so blunt, but the legal system is the implied threat: *"Follow the rules, or we'll lock you up."*

Real or implied force isn't the only kind of power used in conflicts. People who rely on authority of many types engage in win–lose methods without ever threatening physical coercion. In most jobs, supervisors have the potential to use authority in the assignment of working hours, job promotions, and desirable or undesirable tasks, and, of course, in the power to fire an unsatisfactory employee. Teachers can use the power of grades to coerce students to act in desired ways.

Even the usually admired democratic principle of majority rule is a win–lose method of resolving conflicts. However fair it may be, this system results in one group getting its way and another group being unsatisfied.

There are some circumstances when win–lose problem solving may be necessary, such as when there are truly scarce resources and when only one party can achieve satisfaction. For instance, if two

"It's not enough that we succeed. Cats must also fail."

Source: Leo Cullum The New Yorker Collection/The Cartoon Bank

suitors want to marry the same person, only one can succeed. And, to return to an earlier example, it's often true that only one applicant can be hired for a job. But don't be too willing to assume that your conflicts are necessarily win–lose: As you'll soon read, many situations that seem to require a loser can be resolved to everyone's satisfaction.

There is a second kind of situation in which win–lose is the best method. Even when cooperation is possible, if the other person insists on trying to defeat you, then the most logical response might be to defend yourself by fighting back. "It takes <u>two to tango</u>," the old cliché goes, and it also often takes two to cooperate.

A final and much less frequent justification for trying to defeat another person occurs when the other person is clearly behaving in a wrong manner and when defeating that person is the only way to stop the wrongful behavior. Few people would deny the importance of restraining a person who is deliberately harming others, even if the aggressor's freedom is sacrificed in the process. Forcing wrong-doers to behave themselves is dangerous because of the wide difference in opinion between people about who is wrong and who is right. Given this difference, it would seem justifiable only in the most extreme circumstances to coerce others into behaving as we think they should.

Lose–Lose   In **lose–lose problem solving**, neither side is satisfied with the outcome. Although the name of this approach is so discouraging that it's hard to imagine how anyone could willingly use it, in truth, lose–lose is a fairly common way to handle conflicts. In many instances both parties strive to be winners, but as a result of the struggle, both end up losers. On the international scene, many wars illustrate this sad point. A nation that gains military victory at the cost of thousands of lives, large amounts of resources, and a damaged national consciousness hasn't truly won much. On an interpersonal level the same principle holds true. Most of us have seen battles of pride in which both parties strike out and both suffer.

Compromise   Unlike lose–lose outcomes, a **compromise** gives both parties at least some of what they wanted, though both sacrifice part of their goals. People usually settle for compromises when they see partial satisfaction as the best they can hope for. Although a compromise may be better than losing everything, this approach hardly seems to deserve the positive image it has with some people. In his valuable book on conflict resolution, management consultant Albert Filley makes an interesting observation about our attitudes toward this approach.[68] Why is it, he asks, that if someone says, *"I will compromise my values,"* we view the action unfavorably, yet we talk admiringly about parties in a conflict who compromise to reach a solution? Although compromises may be the best obtainable result in some conflicts, it's important to realize that both people in a dispute can often work together to find much better solutions. In such cases *compromise* is a negative word.

Most of us are surrounded by the results of bad compromises. Consider the conflict between one person's desire to smoke cigarettes and another's need to breathe clean air. The win–lose outcomes of this conflict are obvious: Either the smoker abstains, or the nonsmoker gets polluted lungs—neither very satisfying. But a compromise in which the smoker gets to enjoy only a rare cigarette or must retreat outdoors and in which the nonsmoker still must inhale some fumes or feel like an ogre is hardly better. Both sides have lost a considerable amount of comfort and goodwill. Of course, the costs involved in other compromises are even greater. For example, if a divorced couple compromises on child care by fighting over custody and then finally, grudgingly agrees to split the time with their children, it's hard to say that anybody has won.

lose–lose problem solving An approach to conflict resolution in which neither party achieves its goals.

compromise An approach to conflict resolution in which both parties attain at least part of what they seek by giving something up.

cultural idiom

two to tango: both parties affect the outcome

win–win problem solving An approach to conflict resolution in which the parties work together to satisfy all their goals.

Win–Win A fourth option, win–win problem solving, is typically the most satisfying and relationship-friendly. In **win–win problem solving**, the goal is to find a solution that satisfies both people's needs. Neither tries to win at the other's expense. Instead, both parties believe that, by working together, it's possible to find a solution that reaches all their goals.

Finding a win–win situation usually involves looking below the surface at what both parties are trying to achieve. Suppose you want a quiet evening at home tonight and your partner wants to go to a party. On the surface, only one of you can win. However, by listening carefully to each other, you realize you can both get your way. You don't feel like getting dressed up and talking to a room full of people. Your partner isn't crazy about that part of it either but would like to connect with two old friends who are going to be at the party. Once you understand the underlying goals, a solution presents itself: Invite those two friends over for a casual dinner at your place before they head off to the party. In this way, neither you nor your partner compromises on what you want to achieve. Indeed, the evening may be more enjoyable than either of you expected.

Although a win–win approach sounds ideal, it is not always possible, or even appropriate. Table 8-3 suggests some factors to consider when deciding which approach to take when facing a conflict. There will certainly be times when compromising is the most sensible approach. You will even encounter instances when pushing for your own solution is reasonable. Even more surprisingly, you will probably discover that there are times when it makes sense to willingly accept the loser's role. All the same, there are so many advantages to win–win problem solving that we devote the remainder of the chapter to the communication strategies involved in achieving it.

Steps in Win–Win Problem Solving

Although win–win problem solving is often the most desirable approach to managing conflicts, it is also one of the hardest to achieve. In spite of the challenge, it's definitely possible to become better at resolving conflicts. The following pages

TABLE 8-3

Choosing the Most Appropriate Method of Conflict Resolution

1. Consider deferring to the other person:
   - When you discover you are wrong
   - When the issue is more important to the other person than it is to you
   - To let others learn by making their own mistakes
   - When the long-term cost of winning may not be worth the short-term gains

2. Consider compromising:
   - When there is not enough time to seek a win–win outcome
   - When the issue is not important enough to negotiate at length
   - When the other person is not willing to seek a win–win outcome

3. Consider competing:
   - When the issue is important and the other person will take advantage of your noncompetitive approach

4. Consider cooperating:
   - When the issue is too important for a compromise
   - When a long-term relationship between you and the other person is important
   - When the other person is willing to cooperate

outline a method to increase your chances of being able to handle your conflicts in a win–win manner, so that both you and others have your needs met. As you learn to use this approach, you should find that more and more of your conflicts end up with win–win solutions. And even when total satisfaction isn't possible, this approach can preserve a positive relational climate.[69]

As it is presented here, win–win problem solving is a highly structured activity. After you have practiced the approach a number of times, this style of managing conflict will become almost <u>second nature</u> to you. You'll then be able to approach your conflicts without the need to follow the step-by-step approach. But for the time being, try to be patient, and trust the value of the following pattern. As you read on, imagine yourself applying it to a problem that's bothering you now.

Step 1: Identify your problem   Before you speak out, it's important to realize that the problem that is causing conflict is yours. Whether you want to return an unsatisfactory piece of merchandise, complain to noisy neighbors because your sleep is being disturbed, or request a change in working conditions from your employer, the problem is yours. Why? Because in each case *you* are the person who is dissatisfied. You are the one who has paid for the defective article; the merchant who sold it to you has the use of your good money. You are the one who is losing sleep as a result of your neighbors' activities; they are probably content to go on as before. You, not your boss, are the one who is unhappy with your working conditions.

Realizing that the problem is yours will make a big difference when the time comes to approach your partner. Instead of feeling and acting in a judgmental way, you'll be more likely to share your problem in a descriptive way, which will not only be more accurate but also will reduce the chance of a defensive reaction.

Step 2: Explore your unmet needs   After you realize that the problem is yours, the next step is to consider what unmet needs have you feeling dissatisfied. Brené Brown calls this process *reckoning with emotion*. At one level, she says, conflict itself can stir up deep-seated fears. "We don't know what to do with the discomfort and vulnerability," Brown says, adding that, "emotion can feel terrible, even physically overwhelming. We can feel exposed, at risk, and uncertain."[70] Considering this, it's no wonder that many people avoid conflict, accommodate other's wishes, or disguise their vulnerability with aggression.

The irony, points out Brown, is that avoiding conflict and handling it badly usually make us feel worse and more disconnected from people, when what we really want is to be understood and accepted. The good news, she says, is that we don't have to be experts at understanding emotions—ours or other people's. We need only to be curious about them in an open and nonjudgmental way. This might involve saying, *"I'm having an emotional reaction to what's happened and I want to understand."*[71]

Sometimes a relational need underlies the content of the issue at hand. Consider these cases:

- A friend hasn't returned some money you lent long ago. Your apparent need in this situation might be to get the cash back. But a little thought will probably show that this isn't the only, or even the main, thing you want. Even if you were <u>rolling in money</u>, you'd probably want the loan repaid because of your most important need: *to avoid feeling victimized by your friend's taking advantage of you.*

- Someone you care about who lives in a distant city has failed to respond to several letters. Your apparent need may be to get answers to the questions you've written about, but it's likely that there's another, more fundamental need: *the reassurance that you're still important enough to deserve a response.*

cultural idioms

second nature: easy and natural

rolling in money: extremely wealthy

As you'll soon see, the ability to identify your real needs plays a key role in solving interpersonal problems. For now, the point to remember is that before you voice your problem to your partner, you ought to be clear about which of your needs aren't being met.

Step 3: Make a date Unconstructive fights often start because the initiator confronts a partner who isn't ready. There are many times when a person isn't in the right frame of mind to face a conflict, perhaps owing to fatigue, being in a hurry, being upset over another problem, or not feeling well. At times like these, it's unfair to insist on an immediate discussion and expect to get full attention for your problem. If you do persist, you'll probably have an ugly fight on your hands.

After you have a clear idea of the problem, approach your partner with a request to try to solve it. For example: *"Something's been bothering me. Can we talk about it?"* If the answer is "yes," then you're ready to go further. If it isn't the right time to confront your partner, find a time that's agreeable to both of you.

Step 4: Describe your problem and needs Your partner can't possibly meet your needs without knowing why you're upset and what you want. Therefore, it's up to you to describe your problem as specifically as possible. When you do so, it's important to use terms that aren't overly vague or abstract. Recall our discussion of behavioral descriptions in Chapter 4 when clarifying your problem and needs. As we have already discussed, it's also essential that you express yourself in ways that don't cause the other person to feel judged and defensive.

When Brown and her husband reached the dock, she told him her feelings had been hurt by his brief responses and she explained her feelings this way:

> *I feel like you're blowing me off, and* the story I'm making up *is either that you looked over at me while I was swimming and thought,* Man, she's getting old. She can't even swim freestyle anymore. *Or you saw me and thought* She sure as hell doesn't rock a Speedo like she did twenty-five years ago.[72]

With this statement, Brown showed the self-awareness and courage to say outright why Steve's half-hearted responses were so painful to her. She recommends the phrase "the story I'm making up" as a way to express oneself without blaming the other person.

Step 5: Check your partner's understanding After you've shared your problem and described what you need, it's important to make sure that your partner has understood what you've said. As you may remember from the discussion of listening in Chapter 5, there's a good chance—especially in a stressful conflict situation—of your words being misinterpreted.

Step 6: Solicit your partner's needs After you've made your position clear, it's time to find out what your partner needs in order to feel satisfied about this issue. There are two reasons why it's important to discover your partner's needs. First, it's fair. After all, the other person has just as much right as you to feel satisfied, and if you expect help in meeting your needs, then it's reasonable that you behave in the same way. Second, just as an unhappy partner will make it hard for you to become satisfied, a happy one will be more likely to cooperate in letting you reach your goals. Thus, it is in your own self-interest to discover and meet your partner's needs.

You can learn about your partner's needs simply by asking about them: *"Now I've told you what I want and why. Tell me what you need to feel okay about this."* After your partner begins to talk, your job is to use the listening skills discussed earlier in this book to make sure you understand.

cultural idiom
frame of mind: mental state

Back at the lake, Brené Brown was surprised to find that her husband was dealing with his own fears. He had suffered a vivid nightmare the previous night in which he had tried desperately to save all five of their children when a boat had come at them suddenly in the water. He told her:

I don't know what you were saying to me today. I have no idea. I was fighting off a total panic attack during that entire swim. I was just trying to stay focused by counting my strokes.[73]

Brené understood. As a lifelong swimmer, she and Steve were aware of the dangers posed by sharing the waterway with motorboats. Such a nightmare would have unnerved her, too.

Step 7: Check your understanding of your partner's needs   Paraphrase or ask questions about your partner's needs until you're certain you understand them. The surest way to accomplish this is to use the paraphrasing skills you learned in Chapter 5.

Step 8: Discuss ways to meet your common goals   Sometimes, sharing your feelings and receiving the comfort you both crave is enough to resolve a conflict. At other times, it's useful also to borrow some tips from skilled negotiators:

- **Identify and define the conflict.** We've discussed this process in the preceding pages. It consists of discovering each person's problem and needs, setting the stage for meeting all of them.

- **Generate a number of possible solutions.** In this step the partners work together to think of as many means as possible to reach their stated ends. The key word here is *quantity*: It's important to generate as many ideas as you can think of without worrying about which ones are good or bad. Write down every thought that comes up, no matter how unworkable; sometimes a far-fetched idea will lead to a more workable one.

- **Evaluate the alternative solutions.** This is the time to talk about which solutions will work and which ones won't. It's important for all concerned to be honest about their willingness to accept an idea. If a solution is going to work, everyone involved has to support it.

- **Decide on the best solution.** Now that you've looked at all the alternatives, pick the one that looks best to everyone. It's important to be sure everybody understands the solution and is willing to try it out. Remember: Your decision doesn't have to be final, but it should look potentially successful.

Step 9: Follow up on the solution   Conflict management is an ongoing process. You can't be sure the solution will work until you try it out. After you've tested it for a while, it's a good idea to set aside some time to talk over how things are going. You may find that you need to make some changes or even rethink the whole problem. The idea is to <u>keep on top of</u> the problem, to keep using creativity to solve it.

Win–win solutions aren't always possible. There will be times when even the best-intentioned people simply won't be able to find a way of meeting all their needs. In cases like this, the process of negotiation has to include some compromise. But even then the preceding steps haven't been wasted. The genuine desire to learn what the other person wants and to try to satisfy those desires will build a climate of goodwill that can help you find the best solution to the present problem and also improve your relationship in the future.

 ASK YOURSELF

Think of a conflict in your life. What are your needs and fears regarding the conflict? What might your partner's be? How might your relational climate be different if you strived to meet both of your needs?

cultural idiom

keep on top of: stay in control

As for Brené Brown, she reflects that the conversation in the lake, which might have ended in bickering and withheld affection, instead resulted in a renewed sense of love and commitment. Expressing their fears and listening to each other brought them closer.

Afterward, as the couple walked back up to the lake house, Steve popped her playfully with his wet towel and said, "Just so you know: You still rock a Speedo."[74]

MAKING THE GRADE

For more resources to help you understand and apply the information in this chapter, visit the *Understanding Human Communication* website at www.oup.com/us/adleruhc.

OBJECTIVE 8.1 Explain the unavoidable but potentially problematic role of conflict in interpersonal relationships.

- Conflict is a fact of life in every relationship, and the way people manage conflict plays a major role in the quality of their relationships.
- Interpersonal conflict is an acknowledged struggle between at least two interdependent people who perceive that they have incompatible goals, scarce resources, and interference from one another in achieving their goals.

  > What distinguishes interpersonal conflict from the frustration you may feel with the behavior of a stranger you will never see again?

  > Think of a significant interpersonal conflict in your own life. What goals and resources were involved?

  > Imagine that someone you care about has begun to say, *"You're right. I'm wrong"* any time conflict between the two of you emerges. Do you think this is a good or a bad sign for your relationship? Why?

OBJECTIVE 8.2 Describe the role of communication climate and relational spirals in interpersonal relationships, and practice communication strategies for keeping relationships healthy.

- Communication climate refers to the emotional tone of a relationship.
- Confirming communication occurs on three increasingly positive levels: recognition, acknowledgment, and endorsement.
- By contrast, disconfirming responses deny the value of others and show a lack of respect. Four particularly damaging forms of disconfirming messages are criticism, contempt, defensiveness, and stonewalling.

- Relational spirals are reciprocal communication patterns that escalate in positive or negative ways.
- Communication strategies that enhance relational climates include using "I" language, striving for mutually satisfying options, being honest, showing empathy, and respecting other people's viewpoints.

  > Contrast confirming and disconfirming messages, and give an example of each.

  > Describe a time when you and a relational partner were involved in an increasingly negative spiral. What did you do (or what might you have done) to help stop the downward spiral?

  > Identify several disconfirming messages from your own experience, and rewrite them as confirming ones using the tips for creating positive communication climates in this chapter (pp. 216–219).

OBJECTIVE 8.3 Identify characteristics of nonassertive, indirect, passive-aggressive, directly aggressive, and assertive communication, and explain how conflict approaches vary.

- Nonassertive behavior reflects a person's inability or unwillingness to express thoughts or feelings in a conflict. It may manifest as avoidance or accommodation.
- Indirect communication involves hinting about a conflict rather than discussing it directly.
- Passive aggressive behavior is somewhat indirect, but it has a hostile tone meant to make the recipient feel bad.
- Directly aggressive communicators attack the problem and sometimes also the individuals involved.
- Assertive communicators share their feelings and goals and encourage others to do the same.
- Although individuals differ greatly, in broad terms, men are typically socialized to approach conflict in a competitive way and women in a cooperative way.

- In different cultures, it may be expected that people will approach conflict indirectly or that they will be expressive and direct about it.

- Online communication offers people an opportunity to consider carefully before responding, but the relatively anonymous and permanent nature of online messages may escalate a conflict's intensity.

  > Describe the pros and cons of each of the following conflict styles: nonassertive, indirect, passive aggressive, directly aggressive, and assertive.

  > Think of a behavior that bothers you. Write out what you might say to the person involved, including the components of assertive messages described on pages 225–230.

  > Scan several Twitter feeds or the comments senctions on a blog or YouTube video. Identify examples in which people said things they might have felt inhibited from saying in person.

OBJECTIVE 8.4   **Explain the differences among win–lose, lose–lose, compromising, and win–win approaches to conflict resolution, and apply the steps involved in achieving win-win solutions.**

- Although win–lose and lose–lose conflicts do not sound appealing, in rare occasions they are the best option.

- Compromise is often heralded in terms of effective conflict management, but it may not always be the best option, considering that it involves less-than-optimal results for everyone involved.

- Win–win outcomes involve goal fullfilment for everyone. This is often possible if the parties involved have the proper attitudes and skills.

  > List and describe the nine steps for win–win conflict management described in the final section of this chapter.

  > Think of a time when you compromised to resolve a conflict. Were you satisfied with the result? Can you think of a way in which both of you might have met your goals completely?

  > Imagine that you and a friend have just signed the lease on an apartment with one regular bedroom and one master suite. Using the principles of win–win conflict management, how might you work together to decide who gets which bedroom?

KEY TERMS

assertive communication p. 225
avoidance spiral p. 222
communication climate p. 216
compromise p. 235

confirming messages p. 218
conflict p. 215
contempt p. 220
criticism p. 220
defensiveness p. 220
direct aggression p. 225
disconfirming message p. 219
escalatory spiral p. 222
indirect communication p. 223
lose–lose problem solving p. 235
nonassertion p. 223
passive aggression p. 225
relational spiral p. 221
stonewalling p. 220
win–lose problem solving p. 234
win–win problem solving p. 236

ACTIVITIES

1. **Your Confirming and Disconfirming Messages** You can gain an understanding of how confirming and disconfirming messages create communication spirals by trying the following exercise.

   a. Think of an interpersonal relationship. Describe several confirming or disconfirming messages that have helped create and maintain the relational climate. Be sure to identify both verbal and nonverbal messages.

   b. Show how the messages you have identified have created either escalatory or de-escalatory conflict spirals. Describe how these spirals reach limits and what events cause them to stabilize or reverse.

2. **Constructing Assertive Messages** Develop your skill at expressing assertive messages by composing responses for each of the following situations:

   a. A neighbor's barking dog is keeping you awake at night.

   b. A friend hasn't repaid the $20 she borrowed 2 weeks ago.

   c. Your boss made what sounded like a sarcastic remark about the way you put school before work.

   d. An out-of-town friend phones at the last minute to cancel the weekend you planned to spend together.

   e. Now develop two assertive messages you could send to a real person in your life. Discuss how you could express these messages in a way that is appropriate for the situation and that fits your personal style.

Communicating in Groups and Teams

9

CHAPTER OUTLINE

LEARNING OBJECTIVES

9.1 Identify the characteristics that distinguish groups and teams from other collections of people.

9.2 Distinguish between group and individual goals.

9.3 Explain how groups are affected by rules, norms, roles, and patterns of interaction.

9.4 Compare and contrast different leadership styles and approaches.

9.5 Evaluate the roles that followers play and the sources of their power.

Nelson Mandela

recognized the importance of encouraging team members by "leading from behind." Your own experience probably demonstrates this principle's importance even in everyday life.

?

Have you worked with a leader who appreciates and supports the contributions of team members? What kinds of communication conveyed the message that followers are important?

?

How has confirmation of your value as a team member shaped your contributions to a group?

THE NEXT TIME you're in a team meeting, imagine what it would be like to work with cows. It sounds far-fetched, but one of the most influential leaders in the modern world made the connection. Nelson Mandela spent his boyhood days tending cattle in the rolling pasture land of South Africa. He quickly learned that it wasn't effective to stand in front of the herd and proclaim, "Follow me!"[1] It was more effective to walk behind and offer encouragement. Mandela explains:

> When you want to get the cattle to move in a certain direction, you stand at the back . . . and then you get a few of the cleverer cattle to go to the front and move in the direction. . . . The rest of the cattle follow the few more-energetic cattle in the front, but you are really guiding them from the back.[2]

As Mandela grew up, he used his boyhood experiences to help him understand people. That's not to say that he considered people to be as slow-witted as cows. He simply understood that that no one is likely to be inspired by leaders who simply issue orders. A more important factor is teamwork. Mandela was fond of saying, "When people are determined, they can overcome anything."[3]

No one could have predicted the degree to which Mandela's early leadership lessons would be put to the test. By the time he was a young man, South Africa was being torn apart by racist policies and poverty.[4] Under apartheid, the country's system of racial segregation and discrimination, white South Africans claimed almost 90% of the land and the vast majority of the country's wealth.

Mandela challenged the apartheid policies of the white South African government and spent 27 years in prison as a result. Yet he persevered, ultimately uniting the country and becoming one of the greatest civil rights leaders in history.[5] Throughout this chapter, we use Mandela's legendary approach to teamwork and leadership as a model to help us succeed in our own endeavors.

In this chapter, we first explore the importance of group and team work, covering communication strategies to maximize the rewards of working closely with others. We then focus on two roles within every team—those of leader and follower. Chapter 10 continues the exploration of teamwork by focusing on shared decision making.

The Nature of Groups and Teams

Although it may seem that leaders are the ones who make things happen, change occurs only when people work together. Mandela saw this dynamic firsthand as a boy, when people would travel many miles on foot or horseback to meet with his guardian, a tribal leader named Jongintaba. The members of the tribe understood the power of coming together to discuss issues, rather than working only as individuals.

Your experiences may be closer to home, but you are probably involved in groups and teams every day. Some groups, such as those made up of friends and family, are informal. Others are part of work and school. Project groups, work teams, and learning groups are common types. You may belong to some groups for fun (a band or athletic team) and others for profit (an investment group). Others center on personal growth (religious study, exercise) or advocacy (Save the Whales or Habitat for Humanity). You can probably think of more examples that illustrate how groups are a central part of life.

Of course, not every group experience is a good one. Group work can be immensely gratifying, but it can also be only vaguely rewarding or even downright miserable. Effective teamwork requires effort, patience, and communication skill.

Mandela remembered that when people assembled at Jongintaba's home, they would sit in a circle and engage in heated debates that often lasted for hours.[6] For some people, one of the most stressful aspects of group work is reconciling different and sometimes highly emotional perspectives. However, Mandela learned from Jongintaba to become comfortable with that process and to respect it. During those debates, the leader usually sat quietly and listened. Only when everyone's voice had been heard and all angles considered did he offer comment, in a calm, measured voice.[7] From this experience, Mandela learned to encourage open discussion and collaboration, even among people who might seem, on the surface, to be enemies.

THE FAR SIDE® By GARY LARSON

"And so you just threw everything together? ... Matthews, a posse is something you have to *organize*."

Source: THE FAR SIDE ©1987 FARWORKS, INC. Used by permission. All rights reserved.

What Is a Group?

Imagine that you're taking a test on group communication. Which of the following would you identify as groups?

- A crowd of onlookers looking at a burning building
- Several passengers at an airline ticket counter discussing their need to find space on a crowded flight
- An army battalion

Because all these situations seem to involve groups, your experience as a canny test taker probably tells you that a commonsense answer will get you in trouble here, and you're right. When social scientists talk about *groups*, they use the word in a special way that excludes each of the preceding examples.

group A small collection of people whose members interact with one another, usually face-to-face, over time in order to reach goals.

For our purposes, a **group** consists of a *small collection of people who interact with one another, usually face-to-face, over time in order to reach goals*. This was true of the people who met regularly at Jongintaba's home to talk about tribal issues. But a closer examination of the definition will show why none of the collections of people described in the previous bulleted list qualifies as a group.

Interaction   Without interaction, a collection of people isn't a group. Consider, for example, the onlookers at a fire. Though they all occupy the same area at a given time, they have virtually nothing to do with one another. Of course, if they should begin interacting—working together to give first aid to or rescue victims, for example—the situation would change. This requirement of interaction highlights the difference between true groups and collections of individuals who merely co-act, simultaneously engaging in a similar activity without communicating with one another. Students who passively listen to a lecture don't technically constitute a group until they begin to exchange messages verbally and nonverbally with one another and their instructor. (This explains why some students feel isolated even though they spend so much time on a crowded campus. Despite being surrounded by others, they really don't belong to any groups.)

Interdependence   In groups, people don't just interact: Group members are *interdependent*.[8] By contrast, when people don't need one another, they are a collection of individuals and not a group. Author John Krakauer captures this distinction in his account of climbers seeking to reach the peak of Mount Everest:

> [An] odd feeling of isolation hung in the air. We were a team in name only, I'd sadly come to realize. Although in a few hours we would leave camp as a group, we would ascend as individuals, linked to one another by neither rope nor any deep sense of loyalty. Each client was in it for himself or herself, pretty much. And I was no different: I sincerely hoped Doug got to the top, for instance, yet I would do everything in my power to keep pushing on if he turned around.[9]

In a true group, the behavior of one person affects all the others in what can be called a "ripple effect."[10] Consider your own experience in family and work groups. When one member behaves poorly, his or her actions shape the way the entire group functions. The ripple effect can be positive as well as negative. Beneficial actions by some members help everyone.

Time   A collection of people who interact for a short while doesn't qualify as a group. As you'll soon read, groups who work together for any length of time begin to take on characteristics that aren't present in temporary aggregations. There are some occasions when a collection of individuals pulls together to tackle a goal quite quickly. A stirring example of this phenomenon occurred on September 11, 2001, when a group of passengers on United Airlines flight 93 banded together in a matter of minutes to thwart the efforts of hijackers who were attempting to crash the plane into a Washington, DC, landmark. Despite examples of ad hoc groups like this, most groups work together long enough to develop a sense of identity and history that shapes their ongoing effectiveness.

Size   Our definition of *groups* includes the word *small*. Most experts in the field set the lower limit of group size at three members.[11] This decision isn't arbitrary, because there are some significant differences between two- and three-person communication. For example, the only ways two people can resolve a conflict are to change each other's minds, give in, or compromise. In a larger group, however, there's a possibility of members forming alliances either to put increased pressure on dissenting members or to outvote them.

There is less agreement about the maximum number of people in a group.[12] Though no expert would call a 500-member army battalion a group in our sense of the word (it would be labeled an organization), most experts are reluctant to set an arbitrary upper limit. Probably the best description of smallness is the ability for each member to be able to know and react to every other member. It's sufficient to say that our focus in these pages will be on collections of people ranging in size from 3 to around 20.

Research suggests that the optimal size for a group is the smallest number of people capable of performing the task at hand effectively.[13] Generally speaking, as a group becomes larger, it is harder to schedule meetings, the members have less access to important information, and they have fewer chances to participate—three ingredients in a recipe for dissatisfaction.

Not every group is an effective team. Identify a team you know that best exhibits the characteristics listed on this page.

What Makes a Group a Team?

Teams share the same qualities as groups, but they take group work to a higher level. You probably know a team when you see it: Members are proud of their identity. They trust and value one another and cooperate. They seek, and often achieve, excellence. Teamwork doesn't come from *what* the group is doing, but *how* they do it.

Communication researchers Carl Larson and Frank LaFasto have spent their careers interviewing members of more than 6,000 teams that were clearly winners.[14] The groups came from a wide range of enterprises, including a successful mountaineering expedition, a top cardiac surgery team, the developers of groundbreaking computer technology, and championship athletic teams. Although the goals of these high-achieving teams were varied, they shared several important characteristics.

- *Clear and inspiring shared goals.* Members of a winning team know why their group exists, and they believe that purpose is important and worthwhile. Ineffective groups have either lost sight of their purpose or do not believe that the goal is truly important.

- *A results-driven structure.* Members of winning teams focus on getting the job done in the most effective manner. They do whatever is necessary to accomplish the task. Less effective groups either are not organized at all or are structured in an inefficient manner, and their members don't care enough about the results to do what is necessary to get the job done.

- *Competent team members.* Members of winning teams have the skills necessary to accomplish their goals. Less effective groups lack people possessing one or more key skills.

- *Unified commitment.* People in successful teams put the group's goals above their personal interests. Although this commitment might seem like a sacrifice to others, for members of winning teams the personal rewards are worth the effort.

- *Collaborative climate.* Another word for collaboration is teamwork. People in successful groups trust and support one another.

ASK YOURSELF

Think of a team you have known. Which of the characteristics listed here did it exhibit? Which did it not, and what were the results?

Building an Effective Team Online

Experts offer the following communication tips to make the most of virtual interactions:[20]

☐ **Encourage socializing.** Building relationships is especially important when members don't have the chance to socialize in person. Make time for members to get acquainted and enjoy one another's company.

☐ **Strive for face time.** The richness of nonverbal cues and face-to-face communication is valuable in building a cohesive team. If possible, create opportunities for members to meet in one place, especially when the group is first formed.

☐ **Allow and encourage side channels.** As in face-to-face groups, members need a chance to work together one on one. Phone calls, emails, and texting can be an efficient way to speed up the group's work.

☐ **Make expectations clear.** *Are deadlines firm or negotiable? How much detail should a job contain? What does excellent work look like?* Setting clear expectations will help members know what is acceptable and what isn't.

☐ **Provide training as necessary.** Not everybody joins a group with the same level of savvy. Make it easy for members to master the communication tools they need to make the team function effectively.

virtual groups People who interact with one another via mediated channels, without meeting face-to-face.

- *Standards of excellence.* In winning teams, doing outstanding work is an important norm. Each member is expected to do his or her personal best. In less successful groups, people try to get by with a minimum amount of effort.

- *External support and recognition.* Successful teams need an appreciative audience that recognizes their effort and provides the resources necessary to get the job done. The audience may be a boss, or it may be the public the group is created to serve.

- *Principled leadership.* Winning teams usually have leaders who can create a vision of the group's purpose and challenge members to finish the job.

- Despite these virtues, not all groups need to function as teams. If the goal is fairly simple, routine, or quickly accomplished, a group may accomplish it quite adequately. For example, you may be effective working alone to solve a math problem or write a press release. But when the job requires a great deal of thought, collaboration, and creativity, nothing beats teamwork. This is because we literally have greater brainpower when we work together and because most people feel more confident tackling complex issues when they share the challenge as a team.[15]

Virtual Groups

Imagine convening a panel of remote peers who help each other reach important goals. That's the goal of Working Out Loud (WOL) circles, an online movement championed by John Stepper, former managing director of Deutsche Bank.[16]

A WOL circle involves four to five peers with similar interests who agree to interact online 1 hour a week for 12 weeks. Some circles are made up entirely of coworkers, while others are broader. Topics for discussion may include effective interviewing, asking for a raise, opening a new company, and many more. Stepper offers free step-by-step guidelines at his Working Out Loud website. As of this writing, there were more than 1,000 WOL circles involving people in 16 countries.[17]

As the WOL movement illustrates, communication technology has led to new networking opportunities in the form of **virtual groups**, whose members interact with one another through mediated channels, without meeting face-to-face.

Virtual communication has clear advantages. Most obviously, a virtual team can meet even if the members are widely separated. This ease of assembly isn't just useful in the business world. For most groups of students working on class projects, finding a convenient time to meet can be a major headache. Virtual groups don't face this challenge to the same degree. A second advantage of virtual teams is that they may partially level status differences, helping promote effective group functioning. This "leveling" effect can also reduce the effects of gender differences, to the benefit of the group.[18]

Yet virtual teams face particular communication challenges. Most of these involve building strong relationships. If part of the team is in one location and others are far away, the more distant members may feel left out and disconnected.[19] It can take a while for virtual groups to work out means of relating to one another. Another danger is that people will feel less committed to the group or less accountable for their actions if they don't know their teammates well. Students who are part of group work in online classes sometimes find that team members are more likely than usual to shirk their responsibilities and to avoid communication attempts.[21] Although virtual teamwork can be challenging, the initial problems can usually be resolved.[22] See the checklist on page 248 for some tips.

Goals of Groups and Their Members

Mandela once said, "If one or two animals stray, you go out and draw them back to the flock. That's an important lesson in politics."[23] Part of his genius was understanding that two forces drive group communication. The first type involves **group goals**, the outcomes you seek to accomplish together. The second involves **individual goals**, the personal motives of each member. He took on one of the most daunting goals imaginable—restoring democracy to a country torn apart by prejudice and brutal oppression. At the same time, he knew that an equally important goal was establishing and nurturing relationships between individuals on all sides of the issue. Even as Mandela vehemently opposed racist policies, he was willing to interact calmly with members of the white establishment, aware that they would all have to work together to make needed changes. In prison, he encouraged teamwork among fellow warders, and he even befriended white prison guards.

Group Goals

Clearly, not all groups have the enormous responsibility that Mandela took on. But the same dynamics of group work apply. Most groups exist to achieve a collective *task*: win a contest, create a product, provide a service, and so on. Along with task goals like these, *social goals* can be equally important reasons for a group's existence: to meet other people and have fun together. Task and social goals aren't mutually exclusive: A working group will probably be more productive when members enjoy one another's company.

Individual Goals

Some individual goals are related to the group's official reason for existing. For example, your primary motive for joining a class study group would probably be to master the course material, and you'd volunteer to help in a local health clinic to make a difference in your community. Other personal goals might be just about you: Your main reason for joining an exercise class or investment group might be to make new friends and have an enjoyable time with them.

Individual goals aren't necessarily harmful. In fact, they can help the larger group. A student seeking a top grade on a team project will probably help the team excel, and an employee aiming for a production bonus is likely to boost the performance of her work team.

Problems arise when individual motives conflict with the group's goal. You've probably experienced this situation. It is especially difficult when the individual goal is a **hidden agenda**. Consider an egocentric group member whose primary goal, which he would never admit, seems to be to hog the discussion. Or visualize an athlete

---

CHECKLIST ✅
Getting Slackers to Do Their Share

Experts offer the following suggestions to make sure everyone on your team does his or her fair share of the work.[24,25,26]

☐ **Focus on the endgame.** True motivation arises from a sense of working together toward an important goal.

☐ **Match the goal to the group size.** Make sure the size and nature of the task match the size and talents of the team.

☐ **Establish clear goals and responsibilities.** Draft a clear action plan to make sure that group members truly know what is expected of them.

☐ **Provide training.** Make sure everyone has the training and tools to deliver.

☐ **Hold people accountable.** Ask team members to regularly share their accomplishments.

☐ **Focus on quality.** Agree on clear guidelines for high-quality work, and offer feedback at every step.

☐ **Ask why.** If team members fall behind, ask them why.

☐ **Don't overlook poor performance.** All team members need to pull their weight, or others may begin to slack off as well.

☐ **Guard against burnout.** Pay attention to members' emotional states and energy levels to make sure that unrealistic demands aren't sapping their strength.

☐ **Celebrate successes.** Make sure that even low-visibility tasks are rewarded with praise and recognition.

---

group goals Goals that a group collectively seeks to accomplish.

individual goals Individual motives for joining a group.

hidden agendas Individual goals that group members are unwilling to reveal.

social loafing The tendency of some people to do less work as group members than they would as individuals.

rule An explicit, officially stated guideline that governs group functions and member behavior.

norms Shared values, beliefs, behaviors, and procedures that govern a group's operation.

social norms Group norms that govern the way members relate to one another.

procedural norms Norms that describe rules for the group's operation.

task norms Group norms that govern the way members handle the job at hand.

whose goal of achieving personal glory might damage the team's overall effectiveness, or the effect of a member who engages in **social loafing**—lazy behavior that some members use to avoid doing their share of the work. (See the checklist on page 249 for suggestions on how to discourage social loafing.)

Characteristics of Groups and Teams

Whatever their goals, all groups have certain characteristics in common. Understanding these characteristics is a first step to behaving more effectively in your own groups.

Rules and Norms

Whether or not members know it, groups and teams have guidelines that govern members' behavior. You can appreciate this fact by comparing the ways you act in class or at work with the way you behave with your friends. The differences show that guidelines about how to communicate do exist.

Rules are official guidelines that govern what the group is supposed to do and how the members should behave. They are usually stated outright. In a classroom, rules include how absences will be treated, the firmness of deadlines, and so on.

Alongside the official rules is an equally powerful set of unspoken standards called **norms**. **Social norms** govern how we interact with one another (e.g., what kinds of humor are/aren't appropriate, how much socializing is acceptable on the job). **Procedural norms** guide operations and decision making (e.g., "We always start on time" or "When there's a disagreement, we try to reach consensus before forcing a vote"). **Task norms** govern how members get the job done (e.g., "Does the job have to be done perfectly, or is an adequate, if imperfect, solution good enough?").

Table 9-1 lists some typical rules and norms that operate in familiar groups. It is important to realize that our norms don't always match ideals. Consider punctuality, for example. A cultural norm in our society is that meetings should begin at the scheduled time, yet the norm in some groups is to delay talking about real business until 10 or so minutes into the meeting. On a more serious level, one cultural norm is that other people should be treated politely and with respect, but in some groups, members' failure to listen, sarcasm, and even outright hostility make the principle of civility a sham.

It's also important to realize that group norms don't emerge immediately or automatically. Groups typically experience a stage of *forming* (coming together) and *storming* (experiencing conflict) before they enter a *norming* phase in which they find a comfortable rhythm together, and ideally, progress to a *performing* stage in which they function cohesively.[27]

It's important to understand a group's norms. On the one hand, following them helps us fit in. On the other hand, we can sometimes help the group operate more effectively by recognizing norms that cause problems. For instance, in some groups a norm is to do things the way they have always been done. Pointing this out to members might be a way to change the unwritten rules and thereby improve the group's work.

Patterns of Interaction

In interpersonal and public speaking settings, two-way information exchange is relatively uncomplicated. But in a group, the possibilities of complications increase exponentially. If there are five members in a group, there are 10 possible combinations for two-person conversations and 75 combinations involving more than two people. Besides the sheer quantity of information exchanged, the more complex structure of groups affects the flow of information in other ways, too.

TABLE 9-1

Typical Rules and Norms in Two Types of Groups

FAMILY RULES (EXPLICIT)	ON-THE-JOB RULES (EXPLICIT)
• If you don't do the chores, you don't get your allowance.	• Attendance is required at meetings held every Monday morning at 9 A.M.
• If you're going to be more than a half hour late, call so the others don't worry about you.	• The job of keeping minutes rotates from person to person.
• If the gas gauge reads "empty," fill up the tank before bringing the car home.	• Meetings last no more than an hour.
• Don't make plans for Sunday nights. That's time for the family to spend together.	• Don't leave the meetings to take phone calls except in emergencies.
• Daniel gets to watch *Sesame Street* from 5 to 6 P.M.	

FAMILY NORMS (UNSTATED)	ON-THE-JOB NORMS (UNSTATED)
• When Dad is in a bad mood, don't bring up problems.	• Use first names.
• Don't talk about Sheila's divorce.	• It's okay to question the boss's ideas, but if she doesn't concede after the first remark, don't continue to object.
• It's okay to tease Lupe about being short, but don't make comments about Shana's complexion.	• It's okay to chat at the beginning of the meeting, but avoid sexual or ethnic topics.
• As long as the kids don't get in trouble, the parents won't ask detailed questions about what they do with their friends.	• It's okay to talk about "gut feelings," but back them up with hard facts.
• At family gatherings, try to change the subject when Uncle Max brings up politics.	• Don't act upset when your ideas aren't accepted, even if you're unhappy.

A look at Figure 9-1 (which features a type of diagram usually called a **sociogram**) will suggest the number and complexity of interactions that can occur in a group. Arrows connecting members indicate remarks shared between them. Two-headed arrows represent two-way conversations, whereas one-headed arrows represent remarks that did not arouse a response. Arrows directed to the center of the circle indicate remarks made to the group as a whole. A network analysis of this sort can reveal both the amount of participation by each member and the recipients of every member's remarks.

sociogram A graphic representation of the interaction patterns in a group.

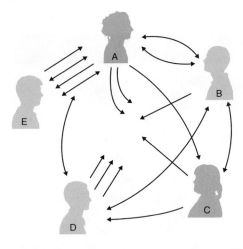

FIGURE 9-1 Patterns of Interaction in a Five-Person Group

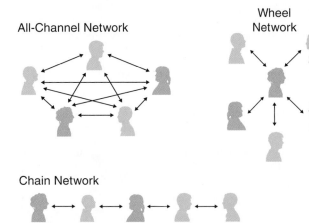

FIGURE 9-2 Small-Group Communication Networks

all-channel network A communication network pattern in which group members are frequently together and share all information with one another.

chain network A communication network in which information passes sequentially from one member to another.

wheel network A communication network in which a gatekeeper regulates the flow of information from all other members.

gatekeeper Person in a small group through whom communication among other members flows.

roles The patterns of behavior expected of group members.

formal role A role assigned to a person by group members or an organization, usually to establish order.

In the group pictured in Figure 9-1, person E appears to be connected to the group only through a relationship with person A; E never addressed any other members, nor did they address E. Also notice that person A is the most active and best connected member. A addressed remarks to the group as a whole and to every other member and was the object of remarks from three individuals as well.

Sociograms don't tell the whole story, because they do not indicate the quality of the messages being exchanged. Nonetheless, they are a useful tool in diagnosing group communication.

Physical arrangement influences communication in groups. It's obviously easier to interact with someone you can see well. Lack of visibility isn't a serious problem in dyadic settings, but it can be troublesome in groups.

Figure 9-2 shows an **all-channel network** in which group members share the same information with everyone on the team. Emails are a handy way to accomplish this. As you probably know from experience, it's nice to be in the loop, but too much sharing can lead to information overload.

Another option is a **chain network** (also in Figure 9-2), in which information moves sequentially from one member to another. Chains are an efficient way to deliver simple verbal messages or to circulate written information when members can't manage to attend a meeting at one time, but they are not very reliable for lengthy or complex verbal messages, because the content of the message can change as it passes from one person to another.

Another communication pattern is the **wheel network**, in which one person acts as a clearinghouse, receiving and relaying messages to all other members. Like chains, wheel networks are sometimes a practical choice, especially if one member is available to communicate with others all or most of the time. This person, called the **gatekeeper**, can become the informational hub who keeps track of messages and people.

Groups sometimes use wheel networks when relationships are strained between two or more members. In such cases, the central member can serve as a mediator or facilitator who manages messages as they flow among others. The success of a wheel network depends heavily on the skill of the gatekeeper. If he or she is a skilled communicator, these mediated messages may help the group function effectively. But if the gatekeeper consciously or unconsciously distorts messages to suit personal goals or plays members off against one another, the group is likely to suffer.

Our communication patterns are shaped partly by the roles we play. You can probably think of people who always have the latest news, people who can be counted on to offer advice and sympathy, people who stay mostly to themselves, and so on. Next, we will look at the roles typically enacted by group and team members.

Roles

Roles define patterns of behavior expected of members. Just like norms, some roles are officially recognized. These **formal roles** are assigned by an organization or group partly to establish order. Formal roles usually come with a label, such as *assistant coach, treasurer,* or *customer service representative.* By contrast, **informal roles** (sometimes called functional roles) are rarely acknowledged by the group in words.[28]

It's important that task-oriented groups include people who suggest ideas, seek and share information, keep the group on task, and encourage others. These are just a few of the task-specific roles that people play. Table 9-2 lists more of them.

TABLE 9-2

Functional Roles of Group Members

TASK ROLES

	TYPICAL BEHAVIORS	EXAMPLES
1. Initiator/contributor	Contributes ideas and suggestions; proposes solutions and decisions; proposes new ideas or states old ones in a novel fashion.	"How about taking a different approach to this chore? Suppose we . . ."
2. Information seeker	Asks for clarification of comments in terms of their factual adequacy; asks for information or facts relevant to the problem; suggests information is needed before making decisions.	"Do you think the others will go for this?" "How much would the plan cost us?" "Does anybody know if those dates are available?"
3. Information giver	Offers facts or generalizations that may relate to the group's task.	"I bet Chris would know the answer to that." *"Newsweek* ran an article on that a couple of months ago. It said . . ."
4. Opinion seeker	Asks for clarification of opinions made by other members of the group and asks how people in the group feel.	"Does anyone else have an idea about this?" "That's an interesting idea, Ruth. How long would it take to get started?"
5. Opinion giver	States beliefs or opinions having to do with suggestions made; indicates what the group's attitude should be.	"I think we ought to go with the second plan. It fits the conditions we face in the Concord plant best. . . ."
6. Elaborator/clarifier	Elaborates ideas and other contributions; offers rationales for suggestions; tries to deduce how an idea or suggestion would work if adopted by the group.	"If we followed Lee's suggestion, each of us would need to make three calls." "Let's see . . . at 35 cents per brochure, the total cost would be $525."
7. Coordinator	Clarifies the relationships among information, opinions, and ideas or suggests an integration of the information, opinions, and ideas of subgroups.	"John, you seem most concerned with potential problems. Mary sounds confident that they can all be solved. Why don't you list the problems one at a time, John, and Mary can respond to each one."
8. Diagnostician	Indicates what the problems are.	"But you're missing the main thing, I think. The problem is that we can't afford . . ."
9. Orienter/summarizer	Summarizes what has taken place; points out departures from agreed-on goals; tries to bring the group back to the central issues; raises questions about the direction in which the group is heading.	"Let's take stock of where we are. Helen and John take the position that we should act now. Bill says, 'Wait.' Rusty isn't sure. Can we set that aside for a moment and come back to it after we . . ."
10. Energizer	Prods the group to action.	"Come on, guys. We've been wasting time. Let's get down to business."
11. Procedure developer	Handles routine tasks such as seating arrangements, obtaining equipment, and handing out pertinent papers.	"I'll volunteer to see that the forms are printed and distributed." "I'd be happy to check on which of those dates are free."
12. Secretary	Keeps notes on the group's progress.	"Just for the record, I'll put these decisions in the memo and get copies to everyone in the group."
13. Evaluator/critic	Constructively analyzes group's accomplishments according to some set of standards; checks to see that consensus has been reached.	"Look, we said we only had two weeks, and this proposal will take at least three. Does that mean that it's out of the running, or do we need to change our original guidelines?"

(Continued)

TABLE 9-2 *(Continued)*

SOCIAL/MAINTENANCE ROLES

	TYPICAL BEHAVIORS	EXAMPLES
1. Supporter/encourager	Praises, agrees with, and accepts the contributions of others; offers warmth, solidarity, and recognition.	"I really like that idea, John." "Priscilla's suggestion sounds good to me. Could we discuss it further?"
2. Harmonizer	Reconciles disagreements; mediates differences; reduces tensions by giving group members a chance to explore their differences.	"I don't think you two are as far apart as you think. Henry, are you saying _____? Benson, you seem to be saying _____. Is that what you mean?"
3. Tension reliever	Jokes or in some other way reduces the formality of the situation; relaxes the group members.	"Let's take a break . . . maybe have a drink." "You're a tough cookie, Bob. I'm glad you're on our side!"
4. Conciliator	Offers new options when his or her own ideas are involved in a conflict; is willing to admit errors so as to maintain group cohesion.	"Looks like our solution is halfway between you and me, John. Can we look at the middle ground?"
5. Gatekeeper	Keeps communication channels open; encourages and facilitates interaction from those members who are usually silent.	"Susan, you haven't said anything about this yet. I know you've been studying the problem. What do you think about _____?"
6. Feeling expresser	Makes explicit the feelings, moods, and relationships in the group; shares his or her own feelings with others.	"I'm really glad we cleared things up today." "I'm just about worn out. Could we call it a day and start fresh tomorrow?"
7. Follower	Goes along with the movement of the group passively, accepting the ideas of others, sometimes serving as an audience.	"I agree. Yes, I see what you mean. If that's what the group wants to do, I'll go along."

informal role A role usually not explicitly recognized by a group that describes functions of group members, rather than their positions. These are sometimes called "functional roles."

task roles Roles group members take on in order to help solve a problem.

social roles Emotional roles concerned with maintaining smooth personal relationships among group members. Also termed "maintenance functions."

dysfunctional roles Individual roles played by group members that inhibit the group's effective operation.

As the list shows, informal roles describe the functions members can fill rather than their formal positions. Many unofficial roles may be filled by more than one member, and some of them may be filled by different people at different times. The important fact is that, at crucial times, the necessary informal roles must be filled by someone.

Notice that the informal roles listed in Table 9-2 fall into two categories: task and maintenance. **Task roles** help the group accomplish its goals, and **social roles** (also called maintenance roles) help the relationships among the members run smoothly. Not all roles are constructive. Table 9-3 lists some of the **dysfunctional roles** that prevent a group from working effectively. As you might expect, research suggests that groups are most effective when people fulfill positive social roles and no one fulfills the dysfunctional ones.[29]

What is the optimal balance between task and social functions? According to Robert Bales, one of the earliest and most influential researchers in the area, the ideal ratio is 2:1, with task-related behavior dominating.[30] This ratio allows the group to get its work done while taking care of the personal needs and concerns of the members.

Groups can suffer from at least three role-related problems. The first occurs when one or more important informal roles (either task or social) go unfilled.

TABLE 9-3

Dysfunctional Roles of Group Members

DYSFUNCTIONAL ROLES

	TYPICAL BEHAVIORS	EXAMPLES
1. Blocker	Interferes with progress by rejecting ideas or taking a negative stand on any and all issues; refuses to cooperate.	"Wait a minute! That's not right! That idea is absurd." "You can talk all day, but my mind is made up."
2. Aggressor	Struggles for status by deflating the status of others; boasts, criticizes.	"Wow, that's really swell! You turkeys have botched things again." "Your constant bickering is responsible for this mess. Let me tell you how you ought to do it."
3. Deserter	Withdraws in some way; remains indifferent, aloof, sometimes formal; daydreams; wanders from the subject, engages in irrelevant side conversations.	"I suppose that's right. . . . I really don't care."
4. Dominator	Interrupts and embarks on long monologues; is authoritative; tries to monopolize the group's time.	"Bill, you're just off base. What we should do is this. First . . ."
5. Recognition seeker	Attempts to gain attention in an exaggerated manner; usually boasts about past accomplishments; relates irrelevant personal experiences, usually in an attempt to gain sympathy.	"That reminds me of a guy I used to know . . ." "Let me tell you how I handled old Marris . . ."
6. Joker	Displays a lack of involvement in the group through inappropriate humor, horseplay, or cynicism.	"Why try to convince these guys? Let's just get the mob to snuff them out." "Hey, Carla, wanna be my roommate at the sales conference?"
7. Cynic	Discounts chances for the group's success.	"Sure, we could try that idea, but it probably won't solve the problem. Nothing we've tried so far has worked."

Source: "Functional Roles of Group Members" and "Dysfunctional Roles of Group Members," adapted from Wilson, G., & Hanna, M. (1986). *Groups in context: Leadership and participation in decision-making groups,* pp. 144–146. Reprinted by permission of McGraw-Hill Companies, Inc.

For instance, there may be no information giver to provide vital knowledge or no harmonizer to smooth things over when members disagree.

Secondly, there are other cases in which the problem isn't an absence of candidates to fill certain roles, but rather an overabundance of them. This situation can lead to unstated competition between members that gets in the way of group effectiveness. You have probably seen groups in which two people both want to be the tension-relieving comedian. Sometimes, members become more concerned with occupying a favorite position than with getting the group's job done.

Finally, even when there is no competition over roles, a group's effectiveness can be threatened when one or more members suffer from "role fixation," acting out a specific role whether or not the situation requires it.[31] As you learned in Chapter 1, a key ingredient of communication competence is flexibility—the ability to choose the right behavior for a given situation. Members who always take the same role (even a constructive one) lack competence, and they hinder the group. As in other areas of life, too much of a good thing can be a

ASK YOURSELF

What unofficial role do you usually play? Does it suit your personality? What would happen in the group if you stopped fulfilling that role?

trait theories of leadership A school of thought based on the belief that some people are born to be leaders and others are not.

problem. You can overcome the potential role-related problems by following these guidelines:

- *Look for unfilled roles.* When a group seems to be experiencing problems, use the list in Table 9-2 to diagnose what roles might be unfilled.
- *Make sure unfilled roles are filled.* After you have identified unfilled roles, you may be able to help the group by filling them yourself. If key facts are missing, take the role of information seeker and try to dig them out. If nobody is keeping track of the group's work, offer to play secretary and take notes. Even if you are not suited by skill or temperament to a job, you can often encourage others to fill it.
- *Avoid role fixation.* Don't fall into familiar roles if they aren't needed. You may be a world-class coordinator or critic, but these talents will only annoy others if you use them when they aren't needed. In most cases your natural inclination to be a supporter might help a group succeed, but if you find yourself in a group in which the members don't need or want this sort of support, your encouragement might become a nuisance.
- *Avoid dysfunctional roles.* Some of these roles can be personally gratifying, especially when you are frustrated with a group, but they do nothing to help the group succeed, and they can damage your reputation as a team player. Nobody needs a blocker, a joker, or any other of the dysfunctional roles listed in Table 9-3. Resist the temptation to indulge yourself by taking on any of them.

Two of the main roles people play are leader and follower. As you will see in the following section, both of them are integral to a team's success.

Leadership and Communication

It may seem that the best leaders are people brimming with charisma and self-importance. Actually, successful leaders are not usually like that. When James Collins[32] and colleagues analyzed the top-performing companies in the United States, they found that none of them were led by charismatic leaders. To the contrary, the leaders they studied were remarkably humble. They were content to let others take the spotlight and quick to say that they were still learning and growing. The same is true of civil rights leaders such as Gandhi and Mandela. In this section, we consider what it takes to be a good leader and explore communication styles and approaches leaders might adopt to help teams succeed.

Understanding Leadership

The question of what gives rise to an effective leader has occupied philosophers, rulers, and, more recently, social scientists, for centuries. The lessons learned by those who came before us can help you become more effective today.

Trait Theories More than 2,000 years ago, Aristotle proclaimed, "From the hour of their birth some are marked out for subjugation, and others for command."[33] This is a radical expression of **trait theories of leadership**, which are sometimes labeled the "great man" or "great woman" approach because they suggest that some people are born to be leaders and others are not. Social scientists began their studies of leader effectiveness by conducting literally hundreds of studies that compared leaders to nonleaders. The results were mixed. As Figure 9-3 shows, a variety of distinguishing characteristics emerged. However, except for intelligence, these mostly reflect skills and awareness that people can build, rather than innate qualities with which they are born. Later research has shown that many other factors are important in determining leader success and that not everyone who possesses these traits becomes a leader. For these reasons, trait theories have limited practical value.

FACTORS APPEARING IN THREE OR MORE STUDIES

- Social nearness, friendliness
- Technical skills
- Task motivation and application
- Group task supportiveness
- Social and interpersonal skills
- Emotional balance and control
- Leadership effectiveness and achievement
- Administrative skills
- General impression
- Ascendance, dominance, decisiveness
- Intellectual skills

0 2 4 6 8 10 12 14 16 18 20
FREQUENCY

FIGURE 9-3 Some Traits Associated with Leaders

Situational Approach   It's more likely that Mandela was right when he said, "The mark of great leaders is the ability to understand the context in which they are operating and act accordingly."[34] In contrast to trait theories of leadership, the principle of **situational leadership** holds that a leader's style should change with the circumstances.

Some circumstances are situation specific. As we will discuss in a moment, inexperienced teams require a different type of leadership than experienced ones do. And in a larger sense, teams and leaders are influenced by factors such as the economy and culture surrounding them. After reviewing current literature about leadership, theorist Montgomery Van Wart proposed that to succeed in today's environment, leaders must be

- good listeners,
- open to innovation,
- able to work well with teams,
- good at facilitating change,
- appreciative of diversity, and
- honest and ethical.[35]

As you might have noticed, the first qualities are accomplished by communicating effectively, and the last two are conveyed through communication. This list reflects overwhelming evidence that change happens too quickly in today's environment for leaders to make most of the decisions themselves. They are more effective when they minimize status differences and encourage open and respectful collaboration among all team members, themselves included.[36]

Leadership Styles   Consistent with the idea of leadership as a situational accomplishment, early scholars in the field identified three basic approaches. The first, an **authoritarian leadership** style, relies on a leader's position and his or her ability to offer rewards or punishment. The second approach, a **democratic leadership** style, encourages members to share in decision making. The third approach, a **laissez-faire leadership** style, reflects a leader's willingness to allow team members to function independently and to make decisions on their own. Some theorists now add a fourth style—**servant leadership**, based on the idea that a leader's job is mostly to recruit outstanding team members and provide the support they need to do a good job.[37] Unlike laissez-faire leaders, who tend to have a hands-off approach, servant leaders are often highly involved with team members and processes.

Each of these styles is effective in some situations. As you might expect, morale tends to be higher in teams with servant leaders than in those with authoritarian leaders.[38] However, an authoritarian approach can sometimes produce faster results. Satisfaction is typically high in teams led by democratic leaders, but inclusive decision making can be time consuming.[39] Highly experienced teams may appreciate the hands-off approach of a laissez-faire leader, but for many teams, the ambiguity involved creates added stress.[40] A Gallup survey of millions of American workers revealed that nearly half of them don't have a clear sense of what their bosses expect of them.[41] Servant leadership has been shown to enhance team members' satisfaction and lead them to feel more self-confident and optimistic.[42]

Although leaders tend to have one or two main styles, they may exhibit a mix, depending on the situation. The next section describes leadership approaches that may involve aspects of all these styles.

Dimensions of Leadership   When *Forbes* writer Travis Bradberry asked people to describe the best and worst bosses they had ever had, he was surprised by what qualities did *not* make the list. "People inevitably ignore innate characteristics (intelligence, extraversion, attractiveness, and so on)," he says. Instead, people

situational leadership A theory that argues that the most effective leadership style varies according to leader–member relations, the nominal leader's power, and the task structure.

authoritarian leadership A style in which the designated leader uses coercive and reward power to dictate the group's actions.

democratic leadership A style in which the leader invites the group's participation in decision making.

laissez-faire leadership A style in which the designated leader gives up his or her formal role, transforming the group into a loose collection of individuals.

servant leadership A style based on the idea that a leader's job is mostly to recruit outstanding team members and provide the support they need to do a good job.

Jack Ma, executive director of the multibillion-dollar international online marketing firm Alibaba, achieved success by recruiting great team members and coaching them to be collaborative, have fun, and let their imaginations guide them.

Have you ever worked with a leader who acted more like a coach than a command-and-control authority figure? If so, how did that influence you as a team member?

focused on "qualities that are completely under the boss's control, such as passion, insight and honesty."[43] In this section, we consider various leadership approaches. As you read, consider which of these best describes you and what qualities you would like to cultivate more as a leader.

Balancing Task and Relational Goals: The Leadership Grid   As you read earlier, teamwork involves the dual goals of accomplishing a task and maintaining relationships. Robert R. Blake and Jane S. Mouton developed a now-famous **Leadership Grid** (Figure 9-4) to explain how leaders manage this balance.[44] The horizontal axis measures a leader's concern for production. This involves a focus on accomplishing the organizational task, with efficiency being the main concern. The vertical axis measures a leader's concern for people's feelings and ideas. Here is an explanation of the five main types of leadership depicted on the grid, roughly in order from least to most effective.

Impoverished Management   Managers who don't display interest in either tasks or relationships exhibit an "impoverished" style. If you have worked with a leader who overlooked poor performance and seemed oblivious to team members' needs, you know how frustrating it can be.

There are a number of reasons leaders may take this approach. Leaders who lack confidence and assertive communication skills may come off as disinterested and aloof. Another cause is burnout. Once-concerned leaders who feel depleted or discouraged may largely give up. Third, some leaders are so concerned with looking good personally that they try do nearly everything themselves.[45] This gives the

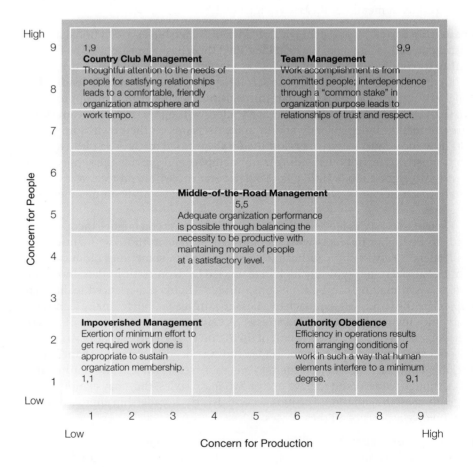

FIGURE 9-4 The Leadership Grid

impression that they don't care much about the team or how it functions. In turn, team members tend to feel resentful and unappreciated, a negative spiral that usually gets worse. Before you write off this approach as something you would never do, consider that inexperienced leaders often behave this way because they haven't learned to trust team members and they are eager to prove themselves to *their* bosses. To avoid the pitfalls of trying to do everything yourself, see the "@ Work" box on delegation on this page.

It's easy to see that that impoverished leadership isn't ideal. Some people equate it with laissez-faire leadership. Indeed, some laissez-faire leaders truly don't care about task fulfillment or relationships. However, other laissez-faire leaders care a great deal about both, but they take a hands-off approach for another reason, such as that they trust the team, or they want them to develop confidence and independence. If that is their motivation, it's important that teams know it. Impoverished leaders take heed: Even when leaders take a hands-off approach, teams perform best when they perceive that leaders are invested in them personally and in their success.[46]

Country Club Management   So-called country club managers exhibit high regard for relationships but give little or no attention to task fulfillment. This style can be effective if team members are highly motivated, experienced, and willing to take responsibility.[47] However, in many cases, country club management fosters an unproductive environment, as you know if you have been part of a project team in which meetings feel like social hours during which little or no actual work gets done.

Country club leaders' primary goal is to keep members happy and maintain harmony. They tend to smile a lot, listen, and offer lots of praise and encouragement. However, they may let misbehavior slide because they dread confrontations, and they typically have a hard time making difficult decisions. In many cases, people led by country club leaders like them personally but feel frustrated because their teams stagnate and underperform.[48]

Authority Obedience   At the other extreme are managers who focus almost entirely on tasks and very little on relationships. This can be useful in small doses, as in emergencies and when inexperienced team members need task-related direction.[49]

> **Leadership Grid** A two-dimensional model that identifies leadership styles as a combination of concern for people and for the task at hand.

@WORK

"I'll Do It Myself"—Or Should I?

It's a rookie mistake some leaders never outgrow—the tendency to tackle a job instead of delegating. The reasons for doing it yourself seem appealing:

1. It's faster to do it myself.
2. Someone else might mess it up.
3. I don't want to ask others to do an unpleasant chore.
4. Other people are busy.
5. No one else knows how to do this.
6. It would take too long to train someone.
7. I like doing this task.
8. I wouldn't ask the team to do something I'm not willing to do myself.
9. If I'm not completing specific tasks, I might seem unnecessary.

The reasons to delegate aren't as numerous, but they are powerful:

1. As a leader, your most important job is to build high-performance, high-capacity teams,[50] and that doesn't happen when you do everything yourself.
2. Trying to do everything yourself slows down the process.
3. Leaders involved in tasks aren't available to lead.
4. Most team members, when asked, say they would gladly perform tasks on the boss's to-do list, if only he or she would listen to them more.
5. Leaders who try to do everything become tired, discouraged, and burned out.

The next time you're tempted to tackle the whole job yourself, ask whether delegating may be better for the team, yourself, and the task at hand.

Melinda and Bill Gates are transformational leaders. They empower followers to achieve an inspiring and impactful goal—improving the lives of people in poverty.

Have you ever worked with a leader whose goal was to help make an important difference in the world? How did their messages motivate others?

transformational leaders   Defined by their devotion to help a team fulfill an important mission.

CHECKLIST ✓
Working with a Difficult Boss

Following are some tips from the pros on keeping a cool head when your boss behaves badly.

☐ **Rise to the challenge.** Meeting your boss's expectations can make your life easier. If your boss is a micromanager, invite her or his input. If your boss is a stickler for detail, invest extra effort into providing more information than you would otherwise do.

☐ **Make up for the boss's shortcomings.** If she's forgetful, diplomatically remind her of important details. If he's disorganized, provide the necessary information before he even asks.

The dark side of this approach is that authoritarian leaders often issue commands and quotas rather than investing in people. Leadership coach Bob Weinstein describes two types of bosses in this category.[51] "The first is the hard-nosed, tough, demanding perfectionist," he says. These bosses can be infuriating, but they are usually open to ideas because they truly care about results. The second type are "unyielding control freaks" who want to micromanage purely for the sense of being in control. Don't leap to conclusions about your boss, or how you should respond, before you read the checklist on this page for tips on managing difficult bosses.

Middle-of-the-Road Management   In the center of Blake and Mouton's grid is an approach characterized by moderate interest in both tasks and relationships. If you have worked with a leader you consider not horrible or great, but simply "okay," you have probably experienced this approach firsthand.

Team Management   Team managers exhibit high regard for both tasks and relationships. Blake and Mouton suggested that this is the most effective leadership approach. It has a great deal in common with the highly revered model of transformational leadership, which we discuss next.

Transformational Leadership   Transformational leaders respect the power of teamwork and positive morale, but they avoid the pitfalls of striving for harmony at all costs. That's because their primary motivation isn't personal glory and it isn't harmony. It's something larger than either of these. **Transformational leaders** are defined by their devotion to help a team fulfill an important *mission*.[52]

Transformational leadership is based on the conviction that people want to accomplish goals with lasting value. Such leaders know that success begins with recruiting great people and supporting them. It also involves a dual focus on relationships and tasks. Transformational leaders listen to team members, authentically care about their feelings, and honor their contributions. They also expect 100% effort from everyone on the team, because otherwise teams cannot live up to their full potential.

Transformational leaders empower teams to make decisions for themselves as much as possible. But when a tough decision by the leader is needed, they aren't afraid to make it.[53] In those circumstances, transformational leaders typically listen to people first, weigh all the factors, and when they announce a decision they explain *why* they made it.[54] The motto of transactional leaders could be, "It's about the team and what we accomplish. It's not about me."

Nelson Mandela exhibited the qualities of a transformational leader. For one, he was deeply committed to a noble vision. He willingly risked his life to advocate for a free and democratic South Africa. He was so committed to that ideal that, even after decades of atrocities against black South Africans and being imprisoned for his beliefs, Mandela did not call for revenge. Instead, his goal was even more transformational—forgiveness and harmony. He reflected in his autobiography that "to be free is not merely to cast off one's chains, but to live in way that respects and enhances the freedom of others."[55] Second, Mandela believed in empowering and supporting others. The style of "leading from behind" that he learned as a boy stayed with him always. Third, despite Mandela's extraordinary achievements, he exhibited humility and gave credit to others. Mandela ultimately served nearly three decades as a political prisoner. Upon his release from prison, he told a crowd of supporters:

> I stand here before you not as a prophet but as a humble servant of you, the people. Your tireless and heroic sacrifices have made it possible for me to be here today. I therefore place the remaining years of my life in your hands.[56]

In 1990, white South African president F. W. de Klerk released Mandela from prison, the two worked together to help the nation heal from more than 50 years of bloodshed, cruelty, and injustice. They were jointly awarded the Nobel Peace Prize in 1993. The following year, Mandela was elected as the first black president of South Africa. He won hundreds of leadership awards and has been lauded as one of the greatest leaders in history.

Becoming a Leader

Even in groups that begin with no official leader, members can take on that role. **Emergent leaders** gain influence without being appointed by higher-ups. Juries elect forepersons, and committees elect chairpersons. Teams choose a captain. Negotiating groups elect spokespeople. The subject of leadership emergence has been studied extensively.

Emergent leaders don't always have official titles. A group of unhappy employees might urge one person to approach the boss and ask for a change. A team of students working on a class project might agree that one person is best suited to take the lead in organizing and presenting their work. Whether or not the role comes with a title, emergent leaders can gain influence in a variety of ways.[57]

Communication researchers have learned that emergent leaders gain influence, especially in newly formed groups, through a process of elimination in which potential candidates are gradually rejected for one reason or another until only one remains.[58] This process occurs in two phases. In the first, members who are clearly unsuitable are rejected. Along with obvious incompetence, lack of involvement and dogmatism are causes for early disqualification. Once clearly unsuitable members have been eliminated, roughly half of the group's members may still be candidates for leadership. During this phase, the following kinds of behavior boost the odds of emerging as the formal or informal leader:

- *Frequent participation.* Talking won't guarantee that you will be recognized as a leader, but failing to speak up will almost certainly knock you out of the running.

- *Demonstrated competence.* Make sure your comments identify you as someone who can help the team succeed. Demonstrate the kinds of power described later in this chapter.

- *Assertion, not aggression.* It's fine to be assertive, but don't try to overpower other members. Even if you are right, your dogmatism is likely to alienate others.

- *Support of other members.* The endorsement of other members (some researchers have called them "lieutenants") increases your credibility and influence.

- *Provide a solution in a time of crisis.* How can the team get the necessary resources? Resolve a disagreement? Meet a deadline? Members who find answers to problems like these are likely to rise to a position of authority.

CHECKLIST *Continued*

☐ **Seek advice.** Gratuitous complaining about your boss is a bad idea. Instead, if other people in your organization have encountered the same problems, you might discover useful information by seeking their advice.

☐ **Have a heart-to-heart with your boss.** If your best efforts don't solve the problem, consider requesting a meeting to discuss the situation. Rather than blaming, use "I" language, such as "I'm feel confused when two managers give me different instructions." Solicit your boss's point of view, and listen nondefensively to what he or she has to say. Paraphrase and use perception checking as necessary to clarify your understanding. As much as possible, seek a win–win outcome.

☐ **Adjust your expectations.** You may not be able to change your boss's behavior, but you can control your attitude about the situation. Sometimes you may need to accept that there are things over which you have little control.

☐ **Maintain a professional demeanor.** Even if your boss has awful interpersonal skills, you will gain nothing by sinking to the same level. It's best to take the high road, practicing the professional communication skills described elsewhere in this chapter.

☐ **If necessary, make a gracious exit.** If you can't fix an intolerable situation, the smartest option may be to look for more rewarding employment. Graciously and diplomatically deliver the news to your boss (e.g., you may be seeking "new opportunities for growth"), and state your appreciation for what you have learned on the job. Help during the transition, and resist the urge to badmouth your boss or the company. Leave on the most positive note you can. [59]

emergent leader A member who assumes leadership roles without being appointed by higher-ups.

Your Leadership Approach

Check the item in each grouping below that *best* characterizes your beliefs as a leader.

1. I believe a leader's most important job is to:

_____ a. Make sure people stay focused on the task at hand.

__✗__ b. Help team members build strong relationships.

_____ c. Make sure the workplace is an enjoyable environment.

2. When it comes to team members, I believe:

_____ a. People have a natural inclination to work hard and do good work.

_____ b. People work best when there are clear expectations and oversight.

__✗__ c. People are most productive when they are happy and enjoying themselves.

3. When a problem arises, I am mostly likely to:

__✗__ a. Solve it myself or smooth things over.

_____ b. Ask team members' input on how to solve it.

_____ c. Implement a new policy or procedure to avoid the same problem in the future.

4. If team members had to describe me in a few words, I would like them to be:

_____ a. Competent and in control.

_____ b. Pleasant and friendly.

__✗__ c. Attentive and trustworthy.

5. When I see team members talking and laughing in the hallway, I am most likely to:

_____ a. Feel frustrated that they are goofing off.

_____ b. Share my latest joke with them.

__✗__ c. Feel encouraged that they get along so well.

EVALUATING YOUR RESPONSES

Circle your answers on the grid below. Note that they do not appear in alphabetical order.

GROUPING 1	GROUPING 2	GROUPING 3	GROUPING 4	GROUPING 5
b	a	b	c	c
a	b	c	a	a
c	c	a	b	b

Relationship Orientation

If the majority of your answers appear in the yellow row, you are a relationship-oriented leader (upper half in the Blake and Mouton Leadership Grid; see Figure 9-4). You are likely to show team members a great deal of respect and attention, which often brings out the best in people. Most people consider this to be the ideal leadership style as long as your focus on relationships does not mean that you neglect task concerns.

Task Orientation

If your answers appear mostly in the green row, you are a task-oriented leader (lower half of the Leadership Grid). You tend to emphasize productivity and may be frustrated by inefficiency. The danger is that you will overlook relationships in your zeal to get the job done, which can be counterproductive in the long run.

Country Club Orientation

If most of your answers are in the blue row, you most closely resemble a country club leader (upper left in the Leadership Grid). Your focus on strong relationships and a pleasant work environment is likely to be appreciated by team members. However, you have a tendency to take that too far. A more moderate focus, in which you emphasize both relationships and tasks, may ultimately be more rewarding for everyone involved.

Mixed Orientation

If your answers are mostly spread all over the grid, you don't show a clear priority for either relationship or task goals. Perhaps you focus on both of them equally (as either a team or middle-of-the-road leader), or you may neglect both of them (an impoverished leader). It's important to consider that both relationships and tasks are highly important. If you pour your energy into both, pat yourself on the back. If you neglect one or both, reconsider your leadership approach.

Followership and Communication

 ASK YOURSELF

Think about the last time you took on a new leadership role. Which of the factors described here played a role in that process?

"What are you, a leader or a follower?" we're asked, and we know which position is generally considered the better one. One reason is a fundamental misunderstanding about what it means to be a follower.

Despite the common belief that leaders are the most important group members, good followers are indispensable. Completing the self-assessment on the next page will help you appreciate why and will also help you gain a sense of the role you can play as a follower.

The self-assessment makes it clear that good followers aren't sheep who blindly follow the herd. According to management consultant Robert Kelley, effective followers "think for themselves, are very active, and have very positive energy."[60] He points out that many leaders have a special term for followers such as these. They call them "my right-hand person" or my "go-to person."

Successful executives agree. In a study of more than 300 senior-level leaders, 94% said that followers help shape leaders, not just the other way around.[61] In their view, effective followers and leaders share many of the same qualities, including honesty, competence, intelligence, and character. The executives also appreciated followers who were loyal, dependable, and cooperative. But they didn't define those qualities in terms of blind obedience. Indeed, almost all of the executives disagreed with the statement that good followers "simply do what they are told." Overall, the lesson seems to be that followership involves a sophisticated array of skills, a good measure of self-confidence, and a strong commitment to teamwork.

As an illustration of this, many people consider that Nelson Mandela's ability to empower followers is his most important quality as a leader. This was demonstrated most dramatically when he was in prison, where he was only allowed one 30-minute visit a year.[62] The civil rights movement might have withered without Mandela, but it didn't. (As a more ordinary test of leadership, consider how well a workplace functions when the boss is away on vacation.) Even in Mandela's absence, people continued to be inspired by his wisdom and vision.

Types of Followers

All followers don't communicate or contribute equally. Barbara Kellerman, a theorist who writes about both leaders and followers, proposes that followers fall into five categories.[63] Which one best describes you?

Isolates   Isolates are indifferent to the overall goals of the organization and communicate very little with people outside their immediate environment.

Bystanders   Bystanders are aware of what's going on around them, but they tend to hang back and watch rather than play an active role. You may find yourself in a bystander role occasionally, especially when you are in a new situation. Because bystanders are usually not as emotionally involved as others, they can sometimes provide an objective, fresh perspective if you encourage them to share their thoughts.

Participants   Participants attempt to have an impact. Some participants support leaders' efforts, whereas others work in opposition. (Opposition isn't necessarily a bad quality in followers. Good followers *should* object when leaders are unethical or ineffective.)

Activists   Activists are more energetically and passionately engaged than participants. They, too, may act either in accordance with, or in opposition to, leaders' efforts. Their commitment is a plus in many ways. At the same time, activists sometimes have difficulty compromising and getting along with others.

How Good a Follower Are You?

Check all of the following that apply to you in your role as a follower.

____ I think for myself.

____ I go above and beyond job requirements.

____ I am supportive of others.

____ I am goal oriented.

____ I focus on the end goal and help others stay focused as well.

____ I take the initiative to make improvements.

____ I realize that my ideas and experiences are essential to the success of the group.

____ I take the initiative to manage my time.

____ I frequently reflect on the job I am doing and how I can improve.

____ I keep learning.

____ I am a champion for new ideas.

If the majority of these statements describe you, pat yourself on the back. These are the qualities of an outstanding follower, according to Robert Kelley, author of *The Power of Followership*.[64]

ASK YOURSELF

Think of a group you have been part of. What type of leader or follower are or were you? What types of power did you use most effectively?

power The ability to influence others' thoughts and/or actions.

legitimate power The ability to influence a group owing to one's position in a group.

nominal leader The person who is identified by title as the leader of a group.

Diehards Diehards will, sometimes literally, sacrifice themselves for the cause. "Being a Diehard is all consuming. It is who you are. It determines what you do," Kellerman says.[65] Soldiers are a classic example, as are people who protest against oppressive rulers or fight for civil rights. Diehards may also work tirelessly in nonprofits or other organizations if they believe the services they provide are essential. Their commitment is unrivaled, but sometimes it's difficult to contain their enthusiasm, even when it runs counter to other peoples' goals.

The Power of Followers

How influential and important are followers? The answer is "More than you might have imagined." To understand why, it's important to understand the nature of power.

Simply put, **power** is the ability to influence others. A few examples show that influence isn't just the domain of leaders:[66]

- In a tense meeting, apartment dwellers are arguing about overcrowded parking and late-night noise. One tenant cracks a joke and lightens up the tense atmosphere.

- A project team at work is trying to come up with a new way to attract customers. The youngest member, fresh from a college advertising class, suggests a winning idea.

- Workers are upset after the boss passes over a popular colleague and hires a newcomer for a management position. Despite their anger, they accept the decision after the colleague persuades them that she is not interested in a career move anyhow.

These examples suggest that power comes in a variety of forms. (See Table 9-4 for a summary.) The most obvious is **legitimate power** (sometimes called "position power")—influence that arises from the title one holds. Jobs like "supervisor," "professor," and "coach" all come with position power. Social scientists use the term **nominal leader** to label the person who is officially designated as being in charge of a group.

TABLE 9-4

Methods for Acquiring Power in Small Groups

Power isn't the only goal to seek in a group. Sometimes being a follower is a comfortable and legitimate role to play. But when you do seek power, the following methods outline specific ways to shape the way others behave and the decisions they make.

LEGITIMATE AUTHORITY

1. Become an authority figure. If possible, get yourself appointed or elected to a position of leadership. Do so by following steps 2–5.

2. Speak up without dominating others. Power comes from visibility, but don't antagonize others by shutting them out.

3. Demonstrate competence on the subject. Enhance legitimate authority by demonstrating information and expertise power.

4. Follow group norms. Show that you respect the group's customs.

5. Gain the support of other members. Don't try to carve out authority on your own. Gain the visible support of other influential members.

INFORMATION POWER

1. Provide useful but scarce or restricted information. Show others that you possess information that isn't available elsewhere.

2. Be certain the information is accurate. One piece of mistaken information can waste the group's time, lead to bad decisions, and destroy your credibility. Check your facts before speaking up.

EXPERT POWER

1. Make sure members are aware of your qualifications. Let others know that you have expertise in the area being discussed.

2. Don't act superior. You will squander your authority if you imply that your expertise makes you superior to others. Use your knowledge for the good of the group, not ego building.

REWARD AND COERCIVE POWER

1. Try to use rewards as a first resort and punishment as a last resort. People respond better to pleasant consequences than unpleasant ones, so take a positive approach first.

2. Make rewards and punishments clear in advance. Let people know your expectations and their consequences. Don't surprise them.

3. Be generous with praise. Let others know that you recognize their desirable behavior.

REFERENT POWER

1. Enhance your attractiveness to group members. Do whatever you can to gain the liking and respect of other members without compromising your principles.

2. Learn effective presentation skills. Present your ideas clearly and effectively in order to boost your credibility.

Source: Adapted from Rothwell, J. D. (1998). *In mixed company: Small group communication* (3rd ed.). Fort Worth, TX: Harcourt Brace, pp. 252–272. Reprinted with permission of Wadsworth, an imprint of the Wadsworth Group, a division of Thomson Learning. Fax 800-730-2215.

Expert Power   **Expert power** comes from what team members know or can do. If you're lost in the woods, it makes sense to follow the advice of a group member who has wilderness experience. If your computer crashes at a critical time, you turn to the team member with IT expertise. In groups it isn't sufficient to be an expert; the other members have to view you as one, too. This means it

expert power The ability to influence others by virtue of one's perceived expertise on the subject in question.

is important to make your qualifications known if you want others to give your opinions weight.

Connection Power   As its name implies, **connection power** comes from a member's ability to develop relationships that help the group reach its goal. For instance, a fundraising group seeking donations from local businesses might profit from the knowledge that one member has about which merchants are hospitable to the group's cause, and a team seeking guest speakers at a seminar might rely on a well-connected member to line up candidates.

Reward Power   **Reward power** exists when others are influenced by the granting or promise of desirable consequences. Rewards come in a variety of forms. Rewards don't come only from the official leader of a group. The goodwill of other members can sometimes be even more valuable. In a class group, for example, having your fellow students think highly of you might be a more powerful reward than the grade you could receive from the instructor. In fact, subordinates sometimes can reward nominal leaders just as much as the other way around. A boss might work hard to accommodate employees in order to keep them happy, for example.

Coercive Power   **Coercive power** comes from the threat or actual imposition of unpleasant consequences. Nominal leaders certainly can coerce members via compensation, assignments, and even termination from the group. But members also possess coercive power. Working with an unhappy, unmotivated teammate can be punishing. For this reason, it's important to keep members feeling satisfied . . . as long as you don't compromise the team's goals.

Referent Power   **Referent power** comes from the respect, liking, and trust others have for a member. If you have high referent power, you may be able to persuade others to follow your lead because they believe in you or because they are willing to do you a favor. Members acquire referent power by behaving in ways others in the group admire and by being genuinely likable. The kinds of confirming communication behaviors described in Chapter 8 can go a long way toward boosting referent power. Listening to others' ideas, honoring their contributions, and taking a win–win approach to meeting their needs lead to liking and respect.

After our look at various ways members can influence one another, three important characteristics of power in groups become clearer.[67]

- **Power is group centered.** Power isn't something an individual possesses. Instead, it is conferred by the group. You may be an expert on the subject being considered, but if the other members don't think you are qualified to talk, you won't have expert power. You might try to reward other people by praising their contributions, but if they don't value your compliments, then all the praise in the world won't influence them.

- **Power is distributed among group members.** Power rarely belongs to just one person. Even when a group has an official leader, other members usually have the power to affect what happens. This influence can be positive, coming from information, expertise, or social reinforcement. It can also be negative, coming from punishing behaviors such as criticizing or withholding the contributions that the group needs to succeed. You can appreciate how power is distributed among members by considering the effect just one member can have by not showing up for meetings or failing to carry out his or her part of the job.

- **Power isn't an either–or concept.** It's incorrect to assume that power is something that members either possess or lack. Rather, it is a matter of degree. Instead of talking about someone as "powerful" or "powerless," it's more accurate to talk about how much influence he or she exerts.

connection power The influence granted by virtue of a member's ability to develop relationships that help the group reach its goal.

reward power The ability to influence others by the granting or promising of desirable consequences.

coercive power The power to influence others by the threat or imposition of unpleasant consequences.

referent power The ability to influence others by virtue of the degree to which one is liked or respected.

It's fitting that we end the chapter where we began, with Nelson Mandela. Like all good followers, Mandela didn't simply do whatever leaders told him to do. He stood his ground when he believed their policies were unjust. Like all good leaders, he understood that true power lies in uniting and empowering people.

Mandela died in 2013 at age 95. As one biographer puts it, "Mandela brought together bitter enemies and unified a nation."[68] Reflecting on his remarkable journey, Mandela told an interviewer:

> Death is something inevitable. When a man has done what he considers to be his duty to his people and his country, he can rest in peace. I believe I have made that effort and that is, therefore, why I will sleep for the eternity.[69]

MAKING THE GRADE

For more resources to help you understand and apply the information in this chapter, visit the *Understanding Human Communication* website at www.oup.com/us/adleruhc.

OBJECTIVE 9.1   Identify the characteristics that distinguish groups and teams from other collections of people.

- Group work involves interaction and interdependence over time among a small number of participants with the purpose of achieving one or more goals.
- Some groups achieve the status of teams, which embody a high level of shared goals and identity, commitment to a common cause, and high ideals.
- The status of "team" may be challenging to achieve, especially when group members are separated geographically, but even virtual teams can excel if they focus on developing strong relationships with one another.
  - > Name at least three similarities and three differences between groups and teams.
  - > Describe the best and worst groups you have ever been part of, and then describe how communication differed in those groups.
  - > Imagine you have to lead a new committee responsible for redesigning your school's grading policy. What will you do to help ensure that the committee functions to its highest potential?

OBJECTIVE 9.2   Distinguish between group and individual goals.

- Groups have their own goals, as do individual members.
- Individual goals that are not known to the group are called *hidden agendas*.

- Social loafing is a common frustration in group work, but there are ways to help ensure that all group members feel accountable.
  - > Describe at least six ways that group members can guard against social loafing.
  - > Think of a group you belong to, then make three lists: (1) your individual goals as a group member, (2) the individual goals of another group member, and (3) the group goals. How do the three lists compare? Are any of the individual goals you listed at odds with the team goals? If so, how?
  - > Think of a time you were tempted to do less than your share as a member of a group. What factors made that choice appealing? What factors motivated (or might have motivated) you do as much as everyone else?

OBJECTIVE 9.3   Explain how groups are affected by rules, norms, roles, and patterns of interaction.

- All groups share the following characteristics: the existence of group norms, individual roles for members, and patterns of interaction that are shaped by the group's structure.
- Members' goals fall into two main categories: task related and social.
- Norms suggest how members should interact with one another, how the group will do business, and who will carry out particular tasks.
- Communication networks reflect who usually communicates with whom and who governs the flow of information among group members.
- Group members play task roles (such as coordinator and diagnostician) and social roles (such as harmonizer and tension reliever).

- Some roles are helpful to the group, whereas others (such as dominator and deserter) can damage performance and member relationships.

  > List the three types of group norms, and give an example of each.

  > Which roles do you most commonly play in groups you belong to? Are these roles mostly conducive or harmful to group performance?

  > Create a sociogram (graphic representation of patterns) to illustrate communication in a group you know well. Does it mostly reflect an all-channel, chain, or network configuration? What are the implications of who communicates with whom most often?

OBJECTIVE 9.4 **Compare and contrast different leadership styles and approaches.**

- Most research suggests that people can learn the skills that contribute to effective leadership.

- No one leadership approach works well in all circumstances. Instead, leaders who understand the relative strengths of various styles are most likely to succeed.

- For the most part, leaders who focus on the overall mission, relationships, and task fulfillment accomplish more than those who are motivated by the desire to achieve personal glory or maintain harmony at all costs.

- Leaders often emerge through a process of elimination, which suggests that, whether or not we know it, we begin "auditioning" for leadership roles as soon as we join a group.

  > Compare the priorities of the five leadership approaches included in Blake and Mouton's Leadership Grid (p. 258).

  > Think of leaders (people you know or public figures) who embody each of the following leadership styles: autocratic, democratic, laissez-faire, and servant leadership. In your opinion, which of these leaders has been most effective, and why?

  > Transformational leaders help people make a significant and valuable contribution in business, science, civil rights, or another arena. Describe a goal that is important to you, and explain how you might embody the qualities of a transformational leader to help people achieve it.

OBJECTIVE 9.5 **Evaluate the roles that followers play and the sources of their power.**

- People often overlook the powerful roles that followers play in shaping innovations, challenging leaders, and pursuing important goals.

- Some followers tend to hang back, either because they are indifferent or because they prefer to watch and learn.

- Other followers take an active role—sometimes to the extent of putting their lives on the line for a cause.

- Followers embody many forms of power. They may be respected as experts, or good relationship builders, or able to reward or coerce others.

  > Choose three of the follower types described by Barbara Kellerman (pp. 263–264). For each one, describe an episode in your life when you embodied that approach. Which did you find most rewarding, and why?

  > Organize the five types of power described in this chapter (pp. 264–266) in order of their importance to you. Give a #1 ranking to the type of power you most want to achieve and a #5 to the type you least want to achieve. Explain why you have rank-ordered them as you have.

  > Why do you think followership often gets less attention than leadership? Present an argument giving three compelling reasons for paying more attention to the roles that followers play.

KEY TERMS

all-channel network p. 252

authoritarian leadership style p. 257

chain network p. 252

coercive power p. 266

connection power p. 266

democratic leadership style p. 257

dysfunctional roles p. 254

emergent leader p. 261

expert power p. 265

formal role p. 252

gatekeepers p. 252

group p. 246

group goals p. 249

hidden agendas p. 249

individual goals p. 249

informal role p. 254

laissez-faire leadership style p. 257

Leadership Grid p. 259

legitimate power p. 264

nominal leader p. 264

norms p. 250

power p. 264

procedural norms p. 250

referent power p. 266

reward power p. 266

roles p. 252

rule p. 250

ACTIVITIES

1. **Group and Individual Goals** Think about two groups to which you belong.

   a. What are your task-related goals in each?

   b. What are your social goals?

   c. Are your personal goals compatible or incompatible with those of other members?

   d. Are they compatible or incompatible with the group goals?

   e. What effect does the compatibility or incompatibility of goals have on the effectiveness of the group?

2. **Norms and Rules in Action** Describe the desirable norms and explicit rules you would like to see established in the following new groups, and describe the steps you could take to see that they are established.

   a. A group of classmates formed to develop and present a class research project

   b. A group of neighbors that is meeting for the first time to persuade the city to install a stop sign at a dangerous intersection

   c. A group of 8-year-olds you will be coaching in a team sport

   d. A group of fellow employees who will be sharing new office space

Solving Problems in Groups and Teams

LEARNING OBJECTIVES

10.1

Weigh the pros and cons of using a group to solve a problem.

10.2

Identify ways that teams can build strong foundations.

10.3

Compare group discussion formats (e.g., breakout groups, focus groups, symposia, reflective thinking, and dialogue), and recommend when to use each.

10.4

Describe elements of Dewey's group problem-solving model, and apply communication strategies to make the most of each stage.

10.5

Identify common obstacles to effective group functioning (e.g., information under- and overload, groupthink), and suggest more effective ways of communicating.

Think about
the importance of teams
in your life.

?

What task-oriented groups
have you belonged to—in
school, in athletics, or at
work?

?

What kinds of
communication distinguished
the effective teams from the
ineffective ones?

?

How would others rate you
as a team player?

IMAGINE IT'S YOUR FIRST DAY on the job at Automattic, the billion-dollar web development company that created WordPress and other online platforms. What should you wear? Where will you sit? Where is the staff meeting? The answers are: Wear anything you like. Sit anywhere you like, as long as you have Wi-Fi. Your living room is fine. So is your deck chair. And the staff meeting? Let's see, is it in Athens or Amsterdam this time?

Automattic is a distributed company. Employees work from different locations—at last count, nearly 400 of them in 60 countries.[1] The company provides all the technology they need and $2,000 each to create their own work environments. Teams work in cyberspace most of the time, but a generous travel budget allows them to schedule face-to-face meetings several times a year anywhere they like (Rome, Hawaii, New Zealand, you name it).

The company's online team approach is the brainchild of Matt Mullenweg, who launched Automattic at age 19. He quickly realized that having a diverse, international workforce was ideal, but that teamwork was essential. Automatticians, as they like to be called, thrive on the stimulation that comes from combining forces to tackle problems.

We revisit Automattic throughout the chapter to consider the communication challenges the company has faced and why its team approach has been so successful. Along the way, we'll consider the advantages and challenges of group problem solving, as well as communication techniques integral to effective teamwork, both in person and in mediated environments.

Problem Solving in Groups: When and Why

Perhaps because most people aren't aware of the problem-solving techniques available to them, groups sometimes get a bad name. You have probably heard the snide remark that "a camel is a horse designed by a committee." This unflattering reputation is at least partly justified. Most of us would wind up with a handsome sum if we had a dollar for every hour wasted in problem-solving groups.

On the other hand, teamwork has brought us computers, cars, space travel, the Internet, and many other innovations. As these examples illustrate, *solving problems*, as we define it here, doesn't refer only to situations in which something is wrong. Perhaps *meeting challenges* and *performing tasks* are better terms. After you recognize this, you can see that problem solving occupies a major part of life.

Employers rank teamwork skills among the 10 most desired traits of people they hire.[2] Employees give teamwork high priority as well. They typically say that effective teamwork makes them feel more powerful and empowered than before, more appreciated, more successful, closer to their colleagues, and more confident that team members will support and encourage them in the future.[3]

Away from work, groups also meet to solve problems. Volunteer groups plan fundraisers, athletic teams work to improve their collective performance, educators and parents work together to improve schools—the list is almost endless.

All in all, groups do have their shortcomings, which we will discuss in a few pages. But extensive research has shown that, when these shortcomings can be avoided, groups are clearly the most effective way to handle many tasks.

Advantages of Group Problem Solving

Years of research show that, in most cases, groups can produce more solutions to a problem than individuals working alone, and the solutions are of higher quality. Groups have proved superior at everything from assembling jigsaw puzzles to solving complex reasoning problems. There are several reasons why groups are effective.[4]

Resources   For many tasks, groups have access to a greater collection of resources than do most individuals. Sometimes the resources involve physical effort. For example, three or four people can put up a tent or dig a ditch better than a lone

Superheroes such as the Avengers come together when civilization is threatened. Each team member has special skills necessary to help solve the problem, and they work together to succeed.

What factors can help you decide when a problem would be more effectively addressed by a group rather than by an individual?

ASK YOURSELF

Think of a difficult task that could not have been accomplished by a person acting alone. Then reflect on the problems you face, and decide which ones could be best tackled by a group.

participative decision making
A process in which people contribute to the decisions that will affect them.

person. Pooled resources can also lead to qualitatively better solutions. Think, for instance, about times when you have studied with other students for a test and you discussed and learned material you might have overlooked if not for the group. Groups not only have more resources than individuals, but, through inter-action among the members, they are better able to mobilize them.

Accuracy Another benefit of group work is the increased likelihood of catching errors. At one time or another, we all make stupid mistakes, like the man who built a boat in his basement and then wasn't able to get it out the door. Working in a group can help prevent foolish errors like this. Sometimes, of course, mistakes aren't so obvious, which makes groups even more valuable as an error-checking mechanism. Another side to the error-detecting story is the risk that group members will support one another in a bad idea. We'll discuss this problem later in this chapter when we focus on conformity.

Commitment Besides coming up with superior solutions, groups also generate a higher commitment to carrying them out. Members are most likely to accept solutions they have helped create and to work harder to carry out those solutions. This fact has led to the principle of **participative decision making**, in which people contribute to the decisions that will affect them. This is an especially important principle for those in authority, such as supervisors, teachers, and parents. As professors, we have seen that students cooperate much more willingly when they help develop a policy for themselves, rather than having it imposed.

Diversity Working with others allows us to consider approaches and solutions we might not think of otherwise. Although we tend to think in terms of "lone geniuses" who make discoveries and solve the world's problems, most break-throughs are actually the result of collective creativity—people working together to create options no one would have thought of alone.[5]

Although diversity is a benefit of teamwork, it requires special effort, especially when members come from different cultural backgrounds. For example, in teams that consist of both Asian-born and American-born members, Americans do most of the talking and are more likely than their Asian teammates to interrupt.[6] And women may be dismayed to find that members of some cultures, including their own, tend to dismiss their comments out of gender bias.[7] To make the most of multiculturalism and avoid some of the common pitfalls, see the "Understanding Diversity" box on page 275.

When to Use Groups for Problem Solving

Despite their advantages, groups aren't always the best way to solve a problem. Many jobs can be tackled more quickly, easily, and even more efficiently by one or more people working independently. The information that follows and the check-list on page 276 will help you decide when to solve a problem using a group and when to tackle it alone.[8]

Is the Job Beyond the Capacity of One Person? Some jobs are simply too big for one person to manage. They may call for more information than a single person possesses or can gather. For example, a group of friends planning a large New Year's party will probably have a better event if they pool their ideas than if one person tries to think of everything. Some jobs also require more time and energy than one person can spare. Putting on the New Year's party could involve a variety of tasks: inviting the guests, hiring a band, finding a place large enough to hold the party, buying food and drinks, and so on. It's both unrealistic and unfair to expect one or two people to do all this work.

UNDERSTANDING DIVERSITY

Maximizing the Effectiveness of Multicultural Teams

In an era of increasing diversity and global commerce, multinational and multicultural teams are more common than ever. Design, engineering, manufacturing, sales, and support teams now span the globe. Work groups include experienced veterans and newcomers with cutting-edge knowledge. In the process, former power structures are being transformed.

Evidence shows that multicultural teams are typically more creative than homogenous groups or individuals.[9] But they also present unique challenges. Communication researchers[10] offer the following tips to maximize the benefits and minimize the pitfalls of multicultural teams:

1. **Allow more time than usual for group development and discussions.**

2. **Agree on clear guidelines for discussions, participation, and decision making.**

3. **Use a variety of communication formats.**

   Based on cultural preferences, people may be more or less comfortable speaking to the entire group, putting their thoughts in writing, speaking one on one, and so on.

4. **If possible, achieve an even distribution of people from various cultures.**

   Research shows that being a "minority member" is especially challenging and not conducive to open communication.[11]

5. **Educate team members about the cultures represented.**

   We are less likely to make unwarranted assumptions (e.g., that a person is lazy, disinterested, or overbearing) if we understand the cultural patterns at play.

6. **Open your mind to new possibilities.**

   Assumptions and too-quick solutions short-circuit the advantage of diverse perspectives.

The results of multicultural teams are usually worth the effort. As one analyst puts it, "diversity makes us smarter."[12]

Are Individuals' Tasks Interdependent?   Remember that a group is more than a collection of individuals working side by side. The best tasks for groups are ones in which members can help one another in some way. Think of a group of disgruntled renters considering how to protest unfair landlords. In order to get anywhere, they realize that they have to assign areas of responsibility to each member, such as researching the law, getting new members, and publicizing their complaints. It's easy to see that these jobs are all interdependent. Getting new members, for example, will require publicity, and publicizing complaints will involve showing how the renters' legal rights are being violated.

Even when everyone is working on the same job, they can be interdependent if different members fulfill the various functional roles described in Chapter 9. Some people might be better at task-related roles such as information giving, diagnosing, and summarizing. Others might contribute by filling social roles such as harmonizing, supporting, or relieving tension. People working independently simply don't have the breadth of resources to perform all these functions.

Is There More Than One Decision or Solution?   Groups are best suited to tackling problems that have no single, straightforward answer. What's the best way to boost

Mixing some social time into group work can help the team stick together and maintain positive relationships.

How cohesive are the groups you belong to? If the level of cohesiveness is not optimal, how can you increase it?

membership in a campus organization? How can funds be raised for a charity? What topic should the group choose for a class project? Gaining the perspectives of every member increases the chances of finding high-quality answers to questions like these.

By contrast, a problem with only one solution won't take full advantage of a group's talents. For example, phoning merchants to get price quotes and looking up a series of books in the library don't require much creative thinking. Jobs like these can be handled by one or two people working alone. Of course, it may take a group meeting to decide how to divide the work to get the job done most efficiently.

Is There Potential for Disagreement?   Tackling a problem as a group is essential if you need the support of everyone involved. Consider a group of friends planning a vacation. Letting one or two people choose the destination, schedule, and budget would be asking for trouble because their decisions would almost certainly disappoint at least some of the people who weren't consulted. It would be far smarter to involve everyone in the most important decisions, even if doing so took more time. After the key decisions were settled, it might be fine to delegate relatively minor issues to one or two people.

Setting the Stage for Problem Solving

Just as a poor blueprint or shaky foundation can lead to a weak house, all the problem-solving tips in the world won't mean much if you don't have a strong team. Every team has its own context and culture. But outstanding teams have one thing in common: the need for talented and committed people to solve problems exceptionally well. Scott Berkun, a former manager at WordPress, proposes that "if you can build trust, provide clarity, and hire well, every other obstacle can be conquered."[13]

Take the "How Effective Is Your Team?" quiz on page 277 to assess the status of your past and present teams. Then read on for guidelines to build relationships and understand the process teams go through as they work on a task.

Maintain Positive Relationships

It's a natural tendency to assume that you understand other members' positions and to interrupt or ignore them. Even if you are right, however, this assumption can lead to ill feelings. On the other hand, careful listening can improve the communication climate, and you may even learn something from your group mates. Following are some strategies for developing strong, mutually respectful relationships among team members.

Build Cohesiveness   The degree to which members feel connected with and committed to their group is known as **cohesiveness**. You might think of cohesiveness as the glue that bonds individuals together, giving them a collective sense of identity.

Highly cohesive groups communicate differently than less cohesive ones. Members spend more time interacting, and there are more expressions of positive feelings for one another. They report more

How Effective Is Your Team?

Think of a team you belong to (or were part of in the past) with the goal of solving a problem or making the most of an opportunity. Select the number on each row that best describes your response.

	Disagree				Agree
1. Team members know and like one another.	1	2	3	4	5
2. We tend to shy away from issues about which we don't agree.	1	2	3	4	5
3. We enjoy tackling challenging situations together.	1	2	3	4	5
4. We enjoy one another's company so much that we often lose focus on what we are trying to achieve.	1	2	3	4	5
5. We trust one another to be responsible and respectful.	1	2	3	4	5
6. Members spend more energy competing with one another than cooperating.	1	2	3	4	5
7. We encourage everyone on the team to have input.	1	2	3	4	5
8. We tend to spend more time complaining about issues than solving them.	1	2	3	4	5
9. We approach challenges in a systematic and creative manner.	1	2	3	4	5
10. Members aren't highly committed to the group or its purpose.	1	2	3	4	5

ANALYZING YOUR RESULTS

Add up the scores you indicated on the odd-numbered questions. Then reverse the scores on the even-numbered questions (5 = 1, 4 = 2, 3 = 3, 2 = 4, 1 = 5) and add them up. Add both totals together and see how you did below.

40–50 points

Congratulations! You have created a team that is cohesive and goal oriented. Although there are likely to be ups and downs, if you maintain your focus on great results and effective teamwork, you are likely to be highly successful together.

30–39 points

You have potential, but this team isn't ready for the big leagues yet. The problem may be that you haven't taken the time to build strong relationships or that not everyone is inspired by the challenge before you. Teams who have high trust and high motivation are typically eager to focus on the issue, and they welcome diverse ideas. You'll find many tips in this chapter for strengthening your team's problem-solving potential.

Less than 30 points

Either your team is very new or you are stuck in an unproductive groove. Over time, the less you accomplish, the less excited and confident members become, which means that commitment and cohesion suffer. Consider how you might turn things around. Perhaps you can host a dialogue session about the group process itself. When you better understand what is holding members back, you may be able to take positive steps to build a more cohesive and productive team. For more about the factors that enhance productivity, see Table 10-1.

satisfaction with the group and its work. In addition, people are more loyal to highly cohesive teams than to other ones. Cohesion keeps people coming back, even when the going is tough.

With characteristics like these, it's no surprise that highly cohesive groups have the potential to be productive. In fact, group cohesion is one of the strongest predictors of innovation, along with effective communication and encouragement.[14]

> **cohesiveness** The totality of forces that causes members to feel themselves part of a group and makes them want to remain in that group.

TABLE 10-1

Communication Factors Associated with Group Productivity

The group contains the smallest number of members necessary to accomplish its goals.

Members care about and agree with the group's goals.

Members are clear about and accept their roles, which match the abilities of each member.

Group norms encourage high performance, quality, success, and innovation.

The group members have sufficient time together to develop a mature working unit and accomplish its goals.

The group is highly cohesive and cooperative.

The group spends time defining and discussing problems it must solve and decisions it must make.

Periods of conflict are frequent but brief, and the group has effective strategies for dealing with conflict.

The group has an open communication structure in which all members may participate.

Source: Adapted from research summarized in S. A. Wheelan, D. Murphy, E. Tsumaura, & S. F. Kline. (1998). Member perceptions of internal group dynamics and productivity. *Small Group Research, 29,* 371–393.

Despite its advantages, cohesiveness is no guarantee of success. If the group is united in supporting unproductive norms, members will feel close but won't get the job done. For example, consider a group of employees who have a boss they think is incompetent and unfair. They might grow quite cohesive in their opposition to the perceived tyranny, spending hours after (or during) work swapping complaints. They might even organize protests, work slowdowns, grievances to their union, or mass resignations. All these responses would boost cohesiveness, but they may not be productive in the long run.

There is a curvilinear relationship between cohesiveness and productivity. Up to a certain point, productivity increases as group members become a unified team. Beyond this point, however, the mutual attraction members feel for one another begins to interfere with the group's efficient functioning. Members may enjoy one another's company, but this enjoyment can keep them from focusing on the job at hand. A study group might become less about studying and more about hanging out as friends.

The goal should be to boost cohesiveness in a way that also helps get the job done. Eight factors can bring about these goals.

1. **Focus on shared or compatible goals.** People draw closer when they share a similar aim or when their goals can be mutually satisfied. For example, members of a conservation group might have little in common until a part of the countryside they all value is threatened by development. Some members might value the land because of its beauty, others because it provides a place to hunt or fish, and still others because the nearby scenery increases the value of their property. As long as their goals are compatible, this collection of individuals will find that a bond exists that draws them together.

2. **Recognize progress toward goals.** While a group is making progress, members feel highly cohesive; when progress stops, cohesiveness decreases. All else being equal, players on an athletic team feel closest when the team is winning. During extended losing streaks, it is likely that players will feel less positive about the team and less willing to identify themselves as members of the group.

3. **Establish shared norms and values.** Although successful groups will tolerate and even thrive on some differences in members' attitudes and behavior, wide variation in the group's definition of what actions or beliefs are proper will reduce cohesiveness. If enough members hold different ideas of what behavior is acceptable, the group is likely to break up. Disagreements over values or norms can fall into many areas, such as humor use, finance, degree of candor, and proportion of time allotted to work and play.

4. **Minimize perceived threats between members.** Cohesive group members see no threat to their status, dignity, and material or emotional well-being. When such interpersonal threats do occur, they can be very destructive. Often competition arises within groups, and as a result members feel threatened. Sometimes there is a struggle over who will be the nominal leader. At other times, members view others as wanting to take over a functional role (problem solver, information giver, and so on), through either competition or criticism. Sometimes the threat is real, and sometimes it's only imagined, but in either case the group must neutralize it or face the consequences of reduced cohesiveness.

5. **Emphasize members' interdependence.** Groups become cohesive when their needs can be satisfied only with the help of other members. When a job can be done just as well by one person alone, the need for membership decreases. This factor explains the reason for food cooperatives, neighborhood yard sales, and community political campaigns. All these activities enable the participants to reach their goal more successfully than if they acted alone.

6. **Recognize threats from outside the group.** Don't create imaginary threats if none exist, but recognize those that do. When members perceive a threat to the group's existence or image (groups have self-concepts, just as individuals do), they grow closer together. Almost everyone knows of a family whose members seem to fight constantly among themselves until an outsider criticizes one of them. At this point, the internal bickering stops, and for the moment the group unites against its common enemy. The same principle often works on a larger scale when nations unite in the face of external aggression.

7. **Develop mutual liking and friendship.** This factor is somewhat circular, because friendship and mutual affinity often are a result of the points just listed, yet groups often do become close simply because the members like one another. Social groups are a good example of a type of group that stays together because its members enjoy one another's company.

8. **Share group experiences.** When members have been through an unusual or trying experience, they draw together. This explains why soldiers who have been in combat together often feel close and stay in touch for years after. Many effective groups devote time to team-building activities, retreats, and social events that involve them in shared experiences before they focus on the main challenge before them.

CHECKLIST ✓
Dealing with Difficult Team Members

Every now and then you are likely to run across a team member who consistently tests your patience. Perhaps he or she is whiny, bossy, aloof, aggressive, overly ingratiating, or a know-it-all. Here are some tips from the experts on coping effectively.[15,16]

☐ **Keep calm.** Some people thrive on goading others and creating drama. Don't play their game.

☐ **Look for underlying reasons.** Consider what factors might have led the person to feel hurt or disrespected.

☐ **Surface the issue.** You might say, *"I noticed that you have interrupted me several times. Do you feel that you didn't get a chance to explain your position?"*

☐ **Lay ground rules.** Agreeing on specific expectations (e.g., no yelling, interrupting, or maintaining side conversations) will make it easier to identify and address issues.

☐ **Write down ideas.** People may be difficult because they don't feel they are being heard or respected. Writing down everyone's ideas on a board or flipchart captures what people are saying and prevents a potential source of team dysfunction.

☐ **Make repercussions clear.** When you are coping with a difficult person whose behavior doesn't improve, it's important to be clear about the consequences. This may involve sharing the problem with a boss or instructor, or if you have the authority, making the repercussions clear yourself.

ASK YOURSELF

What factors influence whether you feel highly committed to a team project? How does communication among team players influence your attitude?

orientation stage When group members become familiar with one another's positions and tentatively volunteer their own.

conflict stage When group members openly defend their positions and question those of others.

emergence stage When a group moves from conflict toward a single solution.

reinforcement stage When group members endorse the decision they have made.

These suggestions work well on most teams. For tips on dealing with especially difficult team members, see page 279.

Recognize Stages of Team Development

When it comes to solving problems in groups, the shortest distance to a solution isn't always a straight line. Communication scholar Aubrey Fisher analyzed tape recordings of problem-solving groups and discovered that many successful groups follow a four-stage process when arriving at decisions.[17] As you read about his findings, visualize how they have applied to problem-solving groups in your experience.

Orientation Stage   In the **orientation stage**, members approach the problem and one another tentatively. Rather than state their own positions clearly and unambiguously, they test possible ideas cautiously and rather politely. This cautiousness doesn't mean that members agree with one another. Rather, they are assessing the situation before asserting themselves. There is little outward disagreement at this stage, but it can be viewed as a calm before the storm.

Conflict Stage   After members understand the problem and become acquainted, a successful group enters the **conflict stage**. Members take strong positions and defend them against those who oppose their viewpoints. Coalitions are likely to form, and the discussion may become polarized. The conflict needn't be personal. It can focus on the issues at hand while preserving the members' respect for one another. Even when the climate does grow contentious, conflict seems to be a necessary stage in group development. The give and take of discussion tests the quality of ideas, and weaker ones may suffer a well-deserved death here.

Emergence Stage   After a period of conflict, effective groups move to an **emergence stage**. One idea might emerge as the best one, or the group might combine the best parts of several plans into a new solution. As they approach consensus, members back off from their dogmatic positions. Statements become more tentative again: "I guess that's a pretty good idea," "I can see why you think that way."

Reinforcement Stage   Finally, an effective group reaches the **reinforcement stage**. At this point not only do members accept the group's decision, they also endorse it. Even if members disagree with the outcome, they do not voice their concerns. There is an unspoken drive toward consensus and harmony.

This isn't a one-time process. Ongoing groups can expect to move through these stages with each new issue, such that their interactions take on a cyclic pattern. They begin discussion of a new issue tentatively, then experience conflict, emergent solutions, and reinforcement. In fact, a group that deals with several issues at once might find itself in a different stage for each problem.

Knowing that these phases are natural and predictable can be reassuring. It can help curb your impatience when the group is feeling its way through an orientation stage. It can also help you feel less threatened when inevitable and necessary conflicts take place. Understanding the nature of emergence and reinforcement can help you know when it is time to stop arguing and seek consensus.

Group Problem-Solving Strategies and Formats

Groups meet to solve problems in a variety of settings and for a wide range of reasons. The formats they use are also varied. Some groups meet before an audience to address a problem. The onlookers may be involved in, and affected by, the topic under discussion, like the citizens who attend a typical city council meeting or voters who attend a candidates' debate. In other cases, the audience members are simply interested spectators, as occurs when people watch C-SPAN or *The People's Court* on TV.

Group Discussion Formats

This list of group discussion formats and approaches is not exhaustive, but it provides a sense of how a group's structure can shape its ability to come up with high-quality solutions.

Breakout Group   When the number of members is too large for effective discussion, **breakout groups** can be used to maximize effective participation. In this approach, subgroups (usually consisting of five to seven members) simultaneously address an issue and then report back to the group at large. The best ideas of each breakout group are then assembled to form a high-quality decision.

Problem Census   When some members are more vocal than others, **problem census** can help equalize participation. Members use a separate card to list each of their ideas. The leader collects all cards and reads them to the group one by one, posting each idea on a board visible to everyone. Because the name of the person who contributed each item isn't listed, issues are separated from personalities. As similar items are read, the leader posts and arranges them in clusters. After all items are read and posted, the leader and members consolidate similar items into a number of ideas that the group needs to address.

Focus Group   Sponsoring organizations often use **focus groups** to learn how potential users or the public at large regards a new product or idea. Unlike some of the other groups discussed here, focus groups don't include decision makers or other members who claim any expertise on a subject. Instead, their comments are used by decision makers to figure out how people in the wider world might react to ideas.

Parliamentary Procedure   Problem-solving meetings can follow a variety of formats. A session that uses **parliamentary procedure** observes specific rules about how topics may be discussed and decisions made. The standard reference book for parliamentary procedure is the revised edition of *Robert's Rules of Order*. Although the parliamentary rules may seem stilted and cumbersome, when well used, they do keep a discussion on track and protect the rights of the minority against domination by the majority.

Panel Discussion   Another common problem-solving format is the **panel discussion**, in which the participants discuss the topic informally, much as they would in an ordinary conversation. A leader (called a "moderator" in public discussions) may help the discussion along by encouraging the comments of some members, cutting off overly talkative ones, and seeking consensus when the time comes for making a decision.

Symposium   In a **symposium** the participants divide the topic in a manner that allows each member to deliver in-depth information without interruption. Although this format lends itself to good explanations of each person's decision, the one-person-at-a-time nature of a symposium won't lead to a group decision. The contributions of the members must be followed by the give-and-take of an open discussion.

Forum   A **forum** allows nonmembers to add their opinions to the group's deliberations before the group makes a decision. This approach is commonly used by public agencies to encourage the participation of citizens in the decisions that affect them.

Some meetings are brief and casual. At other times, formal structure can help a group tackle challenges in a systematic manner.

When would each problem-solving format in this section be most helpful to you?

definitions at left

breakout groups
problem census
focus group
parliamentary procedure
panel discussion
symposium
forum

The Power of Constructive Dialogue

In a dialogue session, about 20 members of a large corporation explored ways to boost the effectiveness of their work team. One participant shared his belief that "people who come to work late are lazy and rude." Even members who thought the complaint was harsh listened respectfully as the speaker explained his ideas about what team members owe one another and the organization.

Another group member acknowledged that she had been arriving late recently and described a series of issues involving child care and transportation. She shared her frustration over the apparently impossible conflict between work obligations and family challenges, and her regret over letting down her colleagues. There was silence as the group pondered the dilemma she described.

Another member spoke out: "I don't have kids. You chose to have them. So your situation is not my problem." The emotional shockwaves in the room were evident on people's faces. The facilitator asked everyone to sit quietly for a few moments before responding.

One man in the group presented a different perspective: "Imagine we are a football team and we have just gotten new uniforms. We look great. We feel great. But the kicker says his feet hurt—his shoes are too small. Would we say, 'That's not our problem'?"

The session adjourned, as dialogues do, without any decisions being made, but leaving members with a good deal to think about. In the days that followed, many people approached the team member with child care problems to express their sympathy for her dilemma. Several said they had the opposite problem: They could come early, but they needed to leave a few minutes before the workday ended. They were able to make a time trade to suit the needs of everyone involved, without compromising the team's mission or inconveniencing the rest of the group.

In a follow-up dialogue, members of the team commented on how easily the situation was resolved once everyone understood one another. "Under different circumstances, we might have created a new policy we didn't need, which wouldn't have helped," said one team member. "Now we say, 'the kicker's feet hurt' any time we are tempted to jump to conclusions without listening first."

---

dialogue A process in which people let go of the notion that their ideas are more correct or superior to others' and instead seek to understand an issue from many different perspectives.

Dialogue Sometimes the best way to tackle a problem is to stop trying to find a solution and listen. **Dialogue** is a process in which people let go of the notion that their ideas are superior to others' and instead try to understand the issue from many perspectives.[18] For example, if the problem is that some children in your community are not being immunized, you might invite a collection of diverse people together to talk about the issue. Perhaps you had assumed that parents were being irresponsible, but you learn by listening that some of them don't have transportation, can't afford the cost, or don't understand medical information very well. You will probably proceed very differently once you realize the complexities of the issue for some people.

In a genuine dialogue, members acknowledge that everything they "know" and believe is an assumption based on their own, unavoidably limited experiences. People engage in curious and open-minded discussion about that assumption. People may ask questions and suspend their own assumptions, but participants guard against either–or thinking. The goal is to understand one another better, not to reach a decision or debate an issue.

Through dialogue, observes theorist David Isaacs, problems are often not so much *solved* as *dissolved*.[19] The issue may cease to be a problem, or it may be so transformed that the solution is obvious or occurs without a formal decision. (For an example of an actual dialogue, see the "@Work" box on this page.)

Solving Problems in Virtual Groups

One advantage of virtual teamwork is autonomy. The company Automattic holds few formal meetings. There are few rules and no set working hours. "I don't care if

UNDERSTANDING COMMUNICATION TECHNOLOGY

Developing Trust Long Distance

Trust is a feeling. We may not be able to say clearly why one person strikes us as trustworthy and another triggers our defenses. But if we could observe trust building in slow motion, we'd find that it's based on a collection of signals, some of them very subtle.

Nonverbal cues are usually the first evidence we consider. Does this person smile, make eye contact, show evidence of being friendly and interested in what we have to say, exhibit appropriate use of space and touch? Over time, we take into account patterns of behavior that suggest the person's underlying character and commitment. Is this person dependable, consistent in his or her actions, committed to the team, honest, and so on?[20]

Some of these factors are difficult to judge at a distance. But emerging research suggests that, with a little effort, trust can flourish among virtual team members. Experts' tips include the following:[21]

1. **Use video technology as often as possible.**
   This provides members with valuable visual and nonverbal cues about one another.

2. **Pay particular attention to your nonverbal communication while you are on camera.**
   Although it is tempting to check your phone or leaf through papers, it may send the signal that you are uninterested.

3. **Show enthusiasm for the group's mission and tasks.**

4. **Encourage members to share information about themselves.**
   As you learned in Chapter 7, self-disclosure is an important way to demonstrate trust in others and to allow them to get to know you.

Trust typically rises in virtual teams who practice these simple principles, and as result, their performance improves as well.[22]

---

you spend the afternoon on the golf course and then work from 2 to 5 A.M.," CEO Matt Mullenweg says. His only question is: "What do you actually produce?"[23] Even the company-wide annual retreat is treated as an occasion for socializing and working on team projects rather than the typical sit-down-and-listen approach.

The lack of meetings doesn't mean communication is limited. Although Automatticians are spread all over the globe, communication among them flows more freely than it might if they shared a roof. They communicate more or less continually in real time through P2, a company-wide blog system similar to Yammer. Anyone can review the posts, comment on them, and use them as an archive of information. "Everyone, from intern to CEO, can weigh in on anything,"[24] observes an industry analyst. The system supports transparency, frequent interaction, open debates, and shared learning—in short, dynamic communication 24 hours a day.

Whether it's a group of students in different locations working on a class project, a sales team sharing ideas and leads from the road, or a military group devising plans from across the globe, technology makes it possible to collaborate from virtually anywhere, either in real time or asynchronously.

Chapter 9 described some benefits of online collaboration. Members who might have kept quiet in face-to-face sessions may be more comfortable "speaking out" online. Also, online meetings often generate a permanent record of the proceedings, which can be convenient.

But where problem solving is concerned, virtual interaction presents some unique challenges as well as benefits. For one thing, if members can't see one another clearly, it may be difficult to convey and understand one another's emotions and attitudes. For another, it may take virtual teams longer to reach decisions than those who meet face-to-face.[25] And whereas high-tech videoconferencing may make it feel that people are in the same room, other forms of interaction (such as typing messages) can be

? ASK YOURSELF

Have your team efforts been enhanced by the ability to communicate online? Have you encountered any difficulties or drawbacks? What lessons about virtual teamwork emerge from your experiences?

laborious. Because typing sometimes takes more time and effort than speaking, messages conveyed via computer can lack the detail of spoken ones. In some cases, members may not bother to type out a message online that they would have shared in person. Finally, the string of separate messages that is generated in a computerized medium can be hard to track, sort out, and synthesize in a meaningful way.

Research comparing the quality of decisions made by face-to-face and online groups is mixed. Evidence suggests that virtual teams can be as effective as others, but only if the members have cultivated trusting relationships with one another.[26] Certain types of mediated communication generally work better than others. For example, asynchronous groups often make better decisions than those functioning in a "chat" mode, perhaps because members have time to digest and synthesize information.[27] Groups who have special decision-support software typically perform better than ones operating without this advantage.[28] Having a moderator also improves the effectiveness of many online groups.[29] (For more on trust building in virtual teams, see the "Understanding Communication Technology" box on page 283.)

Perhaps the most valuable lesson is that online meetings should not entirely replace face-to-face ones, but they can *supplement* in-person sessions. Combining the two forms of interaction can help groups operate both efficiently and effectively.[30]

Approaches and Stages in Problem Solving

Groups may have the potential to solve problems effectively, but they don't always live up to this potential. What makes some groups succeed and others fail? Researchers have spent decades asking this question. To discover their findings, read on.

A Structured Problem-Solving Approach

Although we often pride ourselves on facing problems rationally, intense emotions often hamper our problem-solving ability. This is especially true when frustration or anger leads us to lash out at others[31] or when we don't think things through clearly because we are overly optimistic that we know the right response.[32] Most theorists agree that problem-solving teams benefit from a healthy combination of rationality and creativity. Here, we discuss problem-solving models that incorporate both.

As early as 1910, John Dewey introduced his famous "reflective thinking" method as a systematic approach to solving problems.[33] Since then, other experts have suggested modifications of Dewey's approach. Some emphasize answering key questions, whereas others seek "ideal solutions" that meet the needs of all members. Research comparing various methods has clearly shown that, although no single approach is best for all situations, a structured procedure produces better results than "no pattern" discussions.[34]

The following problem-solving model contains the elements common to most structured approaches developed in the last century:

1. Identify the problem.
   a. Determine the group's goals.
   b. Determine individual members' goals.
2. Analyze the problem.
   a. Word the problem as a broad, open question.
   b. Identify criteria for success.
   c. Gather relevant information.
   d. Identify supporting and restraining forces.

3. Develop creative solutions through brainstorming or the nominal group technique.

  a. Avoid criticism at this stage.

  b. Encourage an unrestricted exchange of ideas.

  c. Develop a large number of ideas.

  d. Combine two or more individual ideas.

4. Evaluate the solutions by asking the following:

  a. Which solution will best produce the desired changes?

  b. Which solution is most achievable?

  c. Which solution contains the fewest serious disadvantages?

5. Implement the plan.

  a. Identify specific tasks.

  b. Determine necessary resources.

  c. Define individual responsibilities.

  d. Provide for emergencies.

6. Follow up on the solution.

  a. Meet to evaluate progress.

  b. Revise the approach as necessary.

Let's consider each step in more detail.

Identify the Problem Sometimes a group's problem is easy to identify. The crew of a sinking ship, for example, doesn't need to conduct a discussion to understand that its goal is to avoid drowning or being eaten by a large fish.

There are many times, however, when the problems facing a group aren't so clear. As an example, think of an athletic team stuck deep in last place well into the season. At first the problem seems obvious: an inability to win any games. But a closer look at the situation might show that there are unmet goals and thus other problems. For instance, individual members may have goals that aren't tied directly to winning: making friends, being recognized as good athletes, or having fun. You can probably see that if the coach or team members took a simplistic view of the situation, looking only at the team's win–loss record, player errors, training methods, and so on, some important problems would probably go overlooked. In this situation, the team's performance could probably be best improved by working on the basic problems of the players' frustration at having their personal needs unmet. The moral is clear: To understand a group's problem, start by identifying the concerns of each member.

What about groups that don't have problems? Several friends planning a surprise birthday party and a family deciding where to go for its vacation don't seem to be in the difficult situation of a losing athletic team—they simply want to have fun. In cases like these, it may be helpful to substitute the word *challenge* for the more gloomy word *problem*. However we express it, members of these teams are also influenced by what each member seeks as a result of belonging to the group.

Analyze the Problem After you have identified the general nature of the problem facing the group, you are ready to look at the problem in more detail. There are several steps you can follow to accomplish this important job.

Word the Problem as a Broad, Open Question If you have ever seen a formal debate, you know that the issue under discussion is worded as a proposition: "The United States should reduce its foreign aid expenditures," for example. Many problem-solving groups define their task in much the same way. "We ought to spend our vacation in the mountains," suggests one family member. The problem with phrasing problems as propositions (positions) is that such wording invites people to take sides. Though this approach is fine for formal debates (which are contests rather like football or card games), taking sides prematurely creates unnecessary conflict in most problem-solving groups.

A far better approach is to state the problem as an open question that encourages exploratory thinking. Asking, "Should we vacation in the mountains or at the beach?" still forces members to choose sides. A far better approach involves asking a question to help define the general goals (interests) that came out during the problem-identification stage: "What do we want our vacation to accomplish?" (perhaps relaxation, adventure, or low cost).

Notice that this question is truly exploratory. It encourages the family members to work cooperatively, not forcing them to make a choice and then defend it. This absence of an either–or situation boosts the odds that members will listen openly to one another rather than listening selectively in defense of their own positions. There is even a chance that the cooperative, exploratory climate that comes from wording the question most broadly will help the family arrive at a consensus about where to vacation, eliminating the need to discuss the matter further.

Identify Criteria for Success   Phrasing the challenge as an open-ended question will help the group identify the criteria for a successful solution. Imagine that a neighborhood task force asks the question, "How can we create a safer environment?" Developing an answer calls for clarifying what would count as the desired outcome. *Fewer incidents reported to police? Less graffiti? More people on the street at night?* Knowing what members want puts a group on the road to achieving its goal. It may even affect the way you engage in problem-solving. Table 10-2 presents some ways to adapt the process to a team's unique circumstances.

Gather Relevant Information   Groups often need to know important facts before they can make decisions or even understand the problem. We remember one group of students who were determined to do well on a class presentation. One of their goals, then, was "to get an A grade." They knew that, to do so, they would have to present a topic that interested both the instructor and the students in the audience. Their first job, then, was to do a bit of background research to find out what subjects would be well received. They interviewed the instructor, asking what topics had been successes and failures in previous semesters. They tested some possible subjects on a few classmates and noted their reactions. From this research they were able to modify their original question, "How can we choose and develop a topic that will earn us

TABLE 10-2	
Adapting Problem-Solving Methods to Special Circumstances	
CIRCUMSTANCES	METHOD
Members have strong feelings about the problem.	Consider allowing a period of emotional ventilation before systematic problem solving.
Task difficulty is high.	Follow the structure of the problem-solving method carefully.
There are many possible solutions.	Emphasize brainstorming.
A high level of member acceptance is required.	Carefully define the needs of all members, and seek solutions that satisfy all needs.
A high level of technical quality is required.	Emphasize evaluation of ideas; consider inviting outside experts.

Source: Adapted from J. Brilhart & G. Galanes. (2001). Adapting problem-solving methods. In *Effective group discussion* (10th ed.), p. 291. Copyright © 2001. Reprinted by permission of McGraw-Hill Companies, Inc.

an A grade?" into a more specific one, "How can we choose and develop a topic that contains humor, action, and lots of information (to demonstrate our research skills to the instructor) and that contains practical information that will improve the audience's social life, academic standing, or financial condition?"

Identify Supporting and Restraining Forces   After members understand what they are seeking, the next step is to see what forces stand between the group and its goals. It might be useful to conduct a **force field analysis**: that is, to list all of the forces that hinder (restrain) the group and all of those that help (impel) it.[35] By returning to our earlier example of the troubled sports team, we can see how the force field operates. Suppose the team defined its problem-question as "How can we (1) have more fun and (2) grow closer as friends?"

One restraining force in Area 1 (having more fun) was clearly the team's losing record. But, more interestingly, discussion revealed that their enjoyment was also diminished by the coach's obsession with winning and his infectiously glum behavior when the team failed. The main restraining force in Area 2 (growing closer as friends) proved to be the lack of socializing among team members in nongame situations. The helping forces in Area 1 included the sense of humor possessed by several members and the confession by most players that winning wasn't nearly as important to them as everyone had suspected. The helping force in Area 2 was the desire of all team members to become better friends. In addition, the fact that members shared many interests was an important plus.

It's important to realize that most problems have many impelling and restraining forces, all of which should be identified during this stage. This may call for another round of research. After the force field is laid out, the group is ready to move on to the next step: deciding how to strengthen the impelling forces and weaken the restraining ones.

Develop Creative Solutions   After the group has established a list of criteria for success, the next job is to develop a number of ways to reach its goal. Considering more than one solution is important, because the first solution may not be the best one. During this development stage, creativity is essential. The biggest danger is the tendency of members to defend their own ideas and criticize others'. This kind of behavior leads to two problems. First, evaluative criticism almost guarantees a defensive reaction from members whose ideas have been attacked. Second, evaluative criticism stifles creativity. People who have just heard an idea rebuked, however politely, will find it hard even to think of more alternatives, let alone share them openly and risk possible criticism. The following strategies can keep groups creative and maintain a positive climate.

Brainstorm   Probably the best known strategy for encouraging creativity and avoiding the dangers just described is **brainstorming**.[36] There are four important rules connected with this strategy:

- **Criticism is forbidden.** As we have said, nothing will stop the flow of ideas more quickly than negative evaluation.

- **Share whatever comes to mind.** Sometimes even the most outlandish ideas prove workable, and even an impractical suggestion might trigger a workable idea.

- **Share a lot of ideas.** The more ideas generated, the better the chance of coming up with a good one.

- **Combine and build upon ideas.** Members are encouraged to "piggyback" by modifying ideas already suggested and to combine previous suggestions.

Use the Nominal Group Technique   Because people in groups often can't resist the tendency to criticize one another's ideas, the **nominal group technique** was developed

force field analysis A method of problem analysis that identifies the forces contributing to resolution of the problem and the forces that inhibit its resolution.

brainstorming A method for creatively generating ideas in groups by minimizing criticism and encouraging a large quantity of ideas without regard to their workability or ownership by individual members.

nominal group technique A method for including the ideas of all group members in a problem-solving session.

to let members brainstorm ideas without being attacked. As the following steps show, the pattern involves alternating cycles of individual work followed by discussion.

- Each member works alone to develop a list of possible solutions.
- In round-robin fashion, each member in turn offers one item from his or her list. The item is listed on a chart visible to everyone. Other members may ask questions to clarify an idea, but no evaluation is allowed during this step.
- Each member privately ranks his or her choice of the ideas in order, from most preferable (5 points) to least preferable (1 point). The rankings are collected, and the top ideas are retained as the most promising solutions.
- A free discussion of the top ideas is held. At this point critical thinking (though not personal criticism) is encouraged. The group continues to discuss until a decision is reached, either by majority vote or by consensus.

Evaluate Possible Solutions   After it has listed possible solutions, the group can evaluate the usefulness of each. One good way of identifying the most workable solutions is to ask three questions.

- *Will this proposal produce the desired changes?* One way to find out is to see whether it successfully overcomes the restraining forces in your force field analysis.
- *Can the proposal be implemented by the group?* Can the members make the most of forces working in their favor and overcome factors that stand in their way? If not, the plan isn't a good one.
- *Does the proposal contain any serious disadvantages?* Sometimes the cost of achieving a goal is too great. For example, one way to raise money for a group is to rob a bank. Although this plan might be workable, it causes more problems than it solves.

Implement the Plan   Everyone who makes New Year's resolutions knows the difference between making a decision and carrying it out. There are several important steps in developing and implementing a plan of action.

- **Identify specific tasks to be accomplished.** What must be done? Even a relatively simple job usually involves several steps. Now is the time to anticipate all the tasks facing the group. Remember everything now, and you will avoid a last-minute rush later.
- **Determine necessary resources.** Identify the equipment, material, and other resources the group will need to get the job done.
- **Define individual responsibilities.** Who will do what? Do all the members know their jobs? The safest plan here is to put everyone's duties in writing, including the due date. This might sound compulsive, but experience shows that it increases the chance of having jobs done on time.
- **Plan ahead for emergencies.** Murphy's Law states that "whatever can go wrong will go wrong." Anyone experienced in group work knows the truth of this. People forget their obligations, get sick, or quit. Machinery breaks down. (One corollary of Murphy's Law is, "The Internet connection will be down whenever it's most needed.") Whenever possible, you ought to develop contingency plans to cover foreseeable problems. Probably the single best suggestion we can give here is to plan on having all work done well ahead of the deadline, knowing that, even with last-minute problems, you can still finish on time.

Follow Up on the Solution   Even the best plans usually require some modifications after they're put into practice. You can improve the group's effectiveness and minimize disappointment by following two steps.

cultural idiom
round-robin fashion: allowing members to express themselves one at a time

- **Meet periodically to evaluate progress.** Follow-up meetings are part of every good plan. The best time to schedule these meetings is as you put the group's plan to work. At that time, a good leader or member will suggest, "Let's get together in a week (or a few days or a month, depending on the nature of the task). We can see how things are going and take care of any problems."

- **Revise the group's approach as necessary.** These follow-up meetings will often go beyond simply congratulating everyone for coming up with a good solution. Problems are bound to arise, and these periodic meetings, in which the key players are present, are the place to solve them.

Although these steps provide a useful outline for solving problems, they are most valuable as a general set of guidelines and not as a precise formula that every group should follow. Certain parts of the model may need emphasis depending on the nature of the specific problem. The general approach will give virtually any group a useful way to consider and solve a problem.

Despite its advantages, the rational, systematic problem-solving approach isn't perfect. The old computer saying, "Garbage in, garbage out" applies here. If the group doesn't possess creative talent, a rational and systematic approach to solving problems won't do much good. Despite this, the rational approach does increase the odds that a group can solve problems successfully. Following the guidelines even imperfectly will help members analyze the problem, come up with solutions, and carry them out better than they could without a plan.

Decision-Making Methods

There are several approaches a group can use to make decisions. We'll look at each of them now, examining their advantages and disadvantages.

Consensus　When all members of a group support a decision we say they have achieved **consensus**. The advantages of consensus are obvious. Full participation can increase the quality of the decision as well as the commitment of the members to support it. Consensus is especially important in decisions on critical or complex matters. In such cases, methods involving less input can diminish the quality of or

> **consensus** Agreement among group members about a decision.

enthusiasm for a decision. Despite its advantages, consensus also has its disadvantages. It can take a great deal of time, which makes it unsuitable for emergencies. In addition, it is often very frustrating. Emotions can run high on important matters, and patience in the face of such pressures is difficult. Because of the need to deal with these emotional pressures, consensus calls for more communication skill than do other decision-making approaches. As with many things in life, consensus has high rewards, which come at a proportionately high cost.

Majority Control　A naive belief of many people (perhaps coming from overzealous high school civics teachers) is that the democratic method of majority rule is always superior. This method does have its advantages in matters in which the support of all members isn't necessary, but in more important matters it is risky. Remember that even if a 51% majority of the members favors a plan, 49% might still oppose it—hardly sweeping support for any decision that needs the support of all members in order to work.

Some contestants on the reality TV show *Survivor* must work together to provide for themselves as they compete with other teams.

In what circumstances might contestants best use each of the decision-making methods described in these pages? When is each method most appropriate in your life?

Besides producing unhappy members, decisions made under majority rule are often inferior to decisions hashed out by a group until the members reach consensus.[37] Under majority rule, members who recognize that they are outvoted often participate less, and the deliberations usually end after a majority opinion has formed, even though minority viewpoints might be worthwhile.

Expert Opinion   Sometimes one group member is defined as an expert and, as such, is given the power to make decisions. This method can work well when that person's judgment is truly superior. For example, if a group of friends is backpacking in the wilderness and one becomes injured, it would probably be foolish to argue with the advice of a doctor in the group. In most cases, however, matters aren't so simple. Who is the expert? There is often disagreement on this question. Sometimes a member thinks he or she is the best qualified to make a decision, but others disagree. In a case like this, the group probably won't support that person's advice, even if it is sound.

Minority Control   Sometimes a few members of a group decide matters. This approach works well with noncritical questions that would waste the whole group's time. In the form of a committee, a minority of members also can study an issue in greater detail than can the entire group. When an issue is so important that it needs the support of everyone, it's best at least to have the committee report its findings for the approval of all members.

Authority Rule   Autocratic leaders most often use authority rule. Though it sounds dictatorial, there are times when such an approach has advantages. This method is quick, so it comes into play when there simply isn't time for a group to decide what to do. The approach is also acceptable with routine matters that don't require discussion in order to gain approval. When overused, however, this approach causes problems. Much of the time, group decisions are of higher quality and gain more support from members than those made by an individual. Thus, failure to consult with members can lead to a decline in effectiveness, even when the leader's decision is a reasonable one.

Which of these decision-making approaches is best? The answer can vary from one culture to another. People in more countries than ever before prefer a democratic style to an authoritarian style, in which leaders make decisions on their own.[38] This is probably because global communication has made human rights a high-priority issue. However, cultural differences still exist. For example, compared to Great Britain and the United States, people in Taiwan,[39] Greece, and Cyprus[40] are more accepting of leaders who make decisions on their own. As you may recall from Chapter 3, people in high-power-distance cultures such as these are more likely than others to revere authority figures and to accept their decisions without question.

When choosing a decision-making approach, weigh the pros and cons of each before you decide which one has the best chance of success in the situation your group is facing. Table 10-3 provides handy reminders.

TABLE 10-3

Choosing an Optimal Decision-Making Method[41]

CIRCUMSTANCES	METHOD
It is important to have everyone's support. The issue is critical or complex. There is adequate time to involve everyone.	Consensus
It isn't necessary to have everyone's buy-in. People are well informed about the issue. Time is of the essence.	Majority control
One person's judgment is truly superior.	Expert opinion
There isn't time or need for everyone to study the issue. It isn't necessary that everyone agree.	Minority control
Time is limited. The issue is a routine matter.	Authority rule

Overcoming Dangers in Group Discussion

Even groups with the best of intentions often find themselves unable to reach satisfying decisions. At other times, they make decisions that later prove to be wrong. Reflecting on Automattic's success, Matt Mullenweg says, "We don't expect perfection—we care more about how quickly they [team members] identify an error, how they communicate about it, and what they learn from it."[42] As CEO, Mullenweg spends about one-third of his time recruiting top-quality team members, which he describes as being self-motivated people with great communication skills and the ability to learn from mistakes.[43] Though there's no foolproof method of guaranteeing high-quality group work, here are several dangers to avoid and some warning signs of danger ahead.

Information Underload and Overload

Information underload occurs when a group lacks information necessary to operate effectively. Sometimes the underload results from overlooking parts of a problem. We know of one group who scheduled a fundraising auction without considering what other events might attract potential donors. They later found that their event was scheduled opposite an important football game, resulting in a loss of sorely needed funds. In other cases, groups suffer from underload because they simply don't conduct enough research. For example, a group of partners starting a new business has to be aware of all the startup costs to avoid going bankrupt in the first months of operation. Overlooking one or two important items can make the difference between success and failure.

Sometimes groups suffer from too much information. **Information overload** occurs when the rate or complexity of material is too great to manage. Having an abundance of information might seem like a blessing, but anyone who has tried to do conscientious library research has become aware of the paralysis that can result from being overwhelmed by an avalanche of books, magazine and newspaper articles, reviews, films, and research studies. When confronted with too much information, it is hard to sort out the essential from the unessential information. See the checklist on this page for useful tips.

Unequal Participation

The value of involving group members in making decisions, especially decisions that affect them, is great.[45] When people participate, their loyalty to the group increases. (Your own experience will probably show that most group dropouts were quiet and withdrawn.) Broad-based participation has a second advantage: It increases the amount of resources focused on the problem. As a result, the quality of the group's decisions goes up. Finally, participation increases members' loyalty to the decisions that they played a part in making.

The key to effective participation is balance. Domination by a few vocal or high-status members can reduce a group's ability to solve a problem effectively. One benefit of group decision making online is that participants tend to be blind to status differences that might sway them in person.[46] The moral to this story? Don't assume that quantity

information underload The decline in efficiency that occurs when there is a shortage of the information necessary to operate effectively.

information overload The decline in efficiency that occurs when the rate or complexity of material is too great to manage.

CHECKLIST ✓
Coping with Information Overload

When you're drowning in data and can't decide what's most important, group expert J. Dan Rothwell offers several tips.[44]

☐ **Specialize whenever possible.** Try to parcel out areas of responsibility instead of expecting each member to explore every angle of the topic.

☐ **Be selective.** Take a quick look at each piece of information to see whether it has real value for your task. If it doesn't, move on to examine more promising material.

☐ **Limit your search.** Information specialists have discovered that there is often a curvilinear relationship between the amount of information a group possesses and the quality of its decision. After a certain point, gathering more material can slow you down without contributing to the quality of your group's decisions.

Source: © Original Artist. Reproduction rights obtainable from www.CartoonStock.com

of speech or the status of the speaker automatically defines the quality of an idea. Instead, seek out and seriously consider the ideas of all members.

Not all participation is helpful, of course. It's better to remain quiet than to act out the dysfunctional roles described in Chapter 9—cynic, aggressor, dominator, and so on. Likewise, the comments of a member who is uninformed can waste time. Finally, downright ignorant or mistaken input can distract a group.

You can encourage the useful contributions of quiet members in a variety of ways:

- Keep the group small. In groups with three or four members, participation is roughly equal, but after the size increases to between five and eight, there is a dramatic gap between the contributions of members.[47]

- Even in a large group, you can increase the contributions of quiet members by soliciting their opinions. This approach may seem obvious, but in their enthusiasm to speak out, more verbal communicators can overlook the people who don't speak up.

- When normally reticent members do offer information, reinforce their contributions. It isn't necessary to excessively praise a quiet person's brilliant remark, but a word of thanks and an acknowledgment of the value of an idea increase the odds that the contributor will speak up again in the future.

- Another strategy is to assign specific tasks to normally quiet members. The need to report on these tasks guarantees that they will speak up.

- Use the nominal group technique described earlier in this chapter to guarantee that the ideas of all members are heard.

- Consider offering communication-friendly gathering areas. Some companies, such as Google, encourage people to move the furniture so that team members can sit work near each other. As you might expect, employees in close proximity engage in more face-to-face communication than they would otherwise. They also email each other more than people who sit farther away.[48] This finding suggests that regular contact is a key factor in stimulating communication at many levels.

Different strategies can help when the problem is one or more members talking too much, especially when their remarks aren't helpful. If the talkative member is at all sensitive, withholding reinforcement can deliver a diplomatic hint that it may be time to listen more and speak less. A lack of response to an idea or suggestion can work as a hint to cut back on speaking. Don't confuse lack of reinforcement with punishment, however. Attacking a member for dominating the group is likely to trigger a defensive reaction and cause more harm than good. If the talkative member doesn't respond to subtle hints, politely expressing a desire to hear from other members can be effective. The next stage in this series of escalating strategies for dealing with dominating members is to question the relevance of remarks that are apparently off topic or unproductive: "I'm confused about what last Saturday's party has to do with the job we have to do today. Am I missing something?"

Pressure to Conform

There's a strong tendency for group members to go along with the crowd, which often results in bad decisions. A classic study by Solomon Asch illustrated this point. College students were shown three lines of different lengths and asked to identify which of them matched a fourth line. Although the correct answer was obvious, the experiment was a setup: Asch had instructed all but one member of the experimental groups to vote for the wrong line. As a result, fully one-third of the uninformed subjects ignored their own good judgment and voted with the majority. If simple tasks like this one generate such conformity, it is easy to see that following the (sometimes mistaken) crowd is even more likely in the much more complex and ambiguous tasks that most groups face.

Even when there's no overt pressure to follow the majority, more subtle influences motivate members, especially in highly cohesive groups, to keep quiet rather than voice any thoughts that deviate from what appears to be the consensus. "Why rock the boat if I'm the only dissenter?" members think. "And if everybody else feels the same way, they're probably right."

With no dissent, the group begins to take on a feeling of invulnerability: an unquestioning belief that its ideas are correct and even morally right. As its position solidifies, outsiders who disagree can be viewed as the enemy, disloyal to what is obviously the only legitimate viewpoint. Social scientists use the term **groupthink** to describe a group's collective striving for unanimity that discourages realistic appraisals of alternatives to its chosen decision.[49] Groupthink has led to a number of disasters, including the United States's botched Bay of Pigs invasion of Cuba in the 1960s, the *Challenger* Space Shuttle disaster in 1986, and the corporate culture that led to the downfall of energy giant Enron in 2001. A more recent example is the Pennsylvania State University sex abuse scandal. For more than 14 years, numerous people who knew about the abuse didn't report it because they feared they would lose their jobs or damage the institution's reputation if they went public.[50] To avoid groupthink disasters yourself, see the checklist on this page.

We opened this chapter by describing Matt Mullenweg's phenomenal success building high-tech teams. In the end, he says, it doesn't matter whether teams meet in person or cyberspace. Either way, excellent results come from having the right people on board and great communication. As Mullenweg puts it, ". . . if you give people autonomy to execute on something meaningful, and bias the environment to moving quickly, amazing things can happen."[51]

CHECKLIST ✓
Avoiding Groupthink

Several group practices can discourage the troublesome force of groupthink.[52]

- [ ] **Recognize early warning signs of groupthink.** It may seem like a good sign if agreement comes quickly and easily, but under the surface, the group may be avoiding the tough but necessary search for alternatives. Considering all the options now may save time and hardship later.

- [ ] **Minimize status differences.** Group members sometimes fall into the trap of agreeing with anything the leader says. To minimize this tendency, leaders should make it clear that they encourage open debate rather than blind obedience. They might also encourage members to conduct initial brainstorming sessions on their own, among peers.

- [ ] **Make respectful disagreement the norm.** After members recognize that questioning one another's positions doesn't signal personal animosity or disloyalty, a constructive exchange of ideas can lead to top-quality solutions. Sometimes it can be helpful to designate a person or subgroup as a "devil's advocate" who reminds the others about the dangers of groupthink and challenges the trend toward consensus.

---

groupthink A group's collective striving for unanimity that discourages realistic appraisals of alternatives to its chosen decision.

---

cultural idioms

rock the boat: disturb a stable condition

devil's advocate: one who argues against a widely held view in order to clarify issues

MAKING THE GRADE

For more resources to help you understand and apply the information in this chapter, visit the *Understanding Human Communication* website at www.oup.com/us/adleruhc.

OBJECTIVE 10.1 Weigh the pros and cons of using a group to solve a problem.

- Despite the bad reputation of groups, research shows that they are often the most effective setting for problem solving.
- Groups possess greater resources, both quantitatively and qualitatively, than do either individuals or collections of people working in isolation.
- Teamwork can result in greater accuracy than individual efforts, and people may feel more committed to solutions they have helped to produce.
- Groups are probably the best option if (1) the problem is beyond the capacity of one person to solve, (2) tasks are interdependent, (3) there is more than one desired solution or decision, and (4) the agreement of all members is essential.

  > Name six conditions in which it is usually more effective to address a problem as a group rather than as an individual.

  > Describe a group experience in which you felt it would have been more productive to work alone. What factors caused you to feel that way?

  > Think of a challenge you have faced while belonging to a group (e.g., at work, with roommates or family, on a team). Evaluate whether it would be better to tackle the problem collectively or to let members deal with it individually.

OBJECTIVE 10.2 Identify ways that teams can build strong foundations.

- Effective groups strive to build a sense of cohesiveness among members.
- Strong teams move through the following stages as they solve a problem: orientation, conflict, emergence, and reinforcement. Describe at least five methods of building team cohesiveness.

  > Summarize the development of a team that is familiar to you, from orientation through reinforcement, giving an example of communication in each stage.

  > Assess the level of cohesiveness on a 1 (low) to 10 (high) scale in a group to which you currently belong or have belonged in the past. Using the factors in this chapter, describe the reasons for the cohesiveness level you identified, and develop recommendations for moving the group's cohesiveness toward the optimal level.

OBJECTIVE 10.3 Compare group discussion formats (e.g., breakout groups, focus groups, symposia, reflective thinking, and dialogue), and recommend when to use each.

- Groups use a variety of discussion formats when solving problems, and each format has advantages and disadvantages depending on the size of the group, the behavior of members, and the nature of the problem. Breakout groups, focus groups, symposia, forums, and panel discussions are well suited to large gatherings, whereas dialogue and problem consensus are well suited to medium-sized and small groups.
- Because face-to-face meetings can be time consuming and difficult to arrange, virtual teamwork is a good alternative for some group tasks. Mediated meetings provide a record of discussion, and they can make it easier for normally quiet members to participate. They can take more time, however, and they lack the nonverbal richness of face-to-face conversation.

  > Describe at least five problem-solving formats, and give an example of each.

  > Pretend that you are advising a newly formed team whose members will interact with each other online. What advice from this chapter would you share with them for developing trust at a distance?

  > Imagine that you and five other students have been asked to engage in a team project to benefit a nonprofit organization of the team's choice. What problem-solving formats would you be most likely to use while deciding what your project might involve? If we expanded the project to include the entire class (let's say 30 people), would your choice of formats change? Why or why not? How might your communication strategies differ if members met virtually rather than in person?

OBJECTIVE 10.4 Describe elements of Dewey's group problem-solving model, and apply communication strategies to make the most of each stage.

- Problem-solving groups should begin by identifying the problem and recognizing the unexpressed needs of individual members.
- The next step is to analyze the problem, including identification of forces both favoring and blocking progress. Only at this point should the group begin to develop possible solutions, taking care not to stifle creativity by evaluating any of them prematurely.

- During the implementation phase of the solution, the group should monitor the situation carefully and make any necessary changes to its plan.

  > List and explain the six main components of the structured problem-solving format described in this chapter.

  > Describe a tough challenge faced by a team to which you have belonged. Construct a force field diagram (see p. 287) that lists factors that acted in the team's favor and factors that made it more difficult to succeed. Explain what your group did (or could have done) to maximize the beneficial forces and minimize the blocking ones.

  > What decision-making methods would you be most likely to use when working on a team project to benefit a nonprofit organization of the team's choice? Explain how you would take into account factors such as the size of the team, the expertise of members, and the time available to make the decision.

OBJECTIVE **10.5** **Identify common obstacles to effective group functioning (e.g., information under- and overload, groupthink), and suggest more effective ways of communicating.**

- Groups function best when they get the information they need without feeling overloaded.

- Members of effective teams make sure that they participate equally by encouraging the contributions of quiet members and by keeping more talkative people on topic.

- Effective team members also guard against groupthink by minimizing pressure on members to conform for the sake of harmony or approval.

  > Describe three strategies for avoiding groupthink.

  > Recall a group in which one or more members talked too much and others said very little. Describe an approach you might take to ensure more equal participation in the future.

  > Teamwork involves some inherent risks. Explain what strategies you would adopt in a new team to make sure that (1) members get the right amount of information, (2) everyone participates equally, and (3) the team avoids groupthink.

KEY TERMS

brainstorming p. 287

breakout groups p. 281

cohesiveness p. 277

conflict stage p. 280

consensus p. 289

dialogue p. 282

emergence stage p. 280

focus group p. 281

force field analysis p. 287

forum p. 281

groupthink p. 293

information overload p. 291

information underload p. 291

nominal group technique p. 287

orientation stage p. 280

panel discussion p. 281

parliamentary procedure p. 281

participative decision making p. 274

problem census p. 281

reinforcement stage p. 280

symposium p. 281

ACTIVITIES

1. **When to Use Group Problem Solving** Explain which of the following tasks would best be managed by a group:

   a. Collecting and editing a list of films illustrating communication principles.

   b. Deciding what the group will eat for lunch at a one-day meeting.

   c. Choosing the topic for a class project.

   d. Finding which of six companies had the lowest auto insurance rates.

   e. Designing a survey to measure community attitudes toward a subsidy for local artists.

2. **The Pros and Cons of Cohesiveness** Based on the information on pages 276–279 and your own experiences, give examples of groups who meet each of the following descriptions:

   a. A level of cohesiveness so low that it interferes with productivity

   b. An optimal level of cohesiveness

   c. A level of cohesiveness so high that it interferes with productivity

   d. For your answers to a and c, offer advice on how the level of cohesiveness could be adjusted to improve productivity.

   e. Are there ever situations in which maximizing cohesiveness is more important than maximizing productivity? Explain your answer, supporting it with examples.

Preparing and Presenting Your Speech

11

LEARNING OBJECTIVES

11.1
Describe the importance of topic, purpose, and thesis in effective speech preparation.

11.2
Analyze both the audience and occasion in any speaking situation.

11.3
Gather information on your chosen topic from a variety of sources.

11.4
Assess and manage debilitative speaking apprehension.

11.5
Make effective choices in the delivery of your speech.

MALALA YOUSAFZAI has been called "the most famous teenager in the world."[1] She doesn't have a hit song or a gig on reality television. Rather, she has become famous for speaking out on behalf of human rights, even when her opponents have tried to kill her.

Malala grew up in Pakistan, where the Taliban, an extremist militant group, opposes education for girls. The group has gone so far as to bomb schools and to kill and kidnap children. To date, at least 838 school bombings in Pakistan are attributed to Taliban members and sympathizers,[2] which points to the danger of speaking out against them.

Even as a young girl, Malala was courageous in declaring publicly that girls should be educated. At age 11, she presented a speech called "How Dare the Taliban Take Away My Basic Right to Education?" before media professionals in Pakistan. "I speak not for myself, but so those without a voice can be heard," she has said.[3]

Malala's commitment to public speaking has not wavered, even through very difficult times. When she was 15, Taliban representatives intending to silence her boarded her school bus and shot her in the head. She was hospitalized for 4 months and underwent numerous surgeries, but she lived, and as soon as she could, she resumed advocating for public education. At age 17, she became the youngest person ever to receive a Nobel Peace Prize. Her acceptance speech for that honor serves as the extended example at the end of this chapter.

In this chapter, we explore the process of creating and delivering an effective speech. The consequences of your speech may not be dire, as they were for Malala, but speaking in front of an audience can still be nerve racking. By using some of the techniques she uses, you can develop the habits and skills to become a confident speaker. You might be advocating for social change, making a job-related presentation, or speaking on something more personal, such as at a wedding toast or a eulogy. You might find yourself speaking in favor of a civic-improvement project in your hometown or trying to persuade members of your club to work toward solving global problems such as war, religious strife, or environmental threats.

Despite the potential benefits of effective speeches, the prospect of standing before an audience terrifies many people. In fact, giving a speech seems to be one of the most anxiety-producing things we can do: When asked to list their common fears, research subjects mention public speaking more often than they do insects, heights, accidents, and even death.[4]

There's no guarantee that the following chapters will make you love the idea of giving speeches, but we can promise that the information these chapters contain will give you the tools to design and deliver remarks that will be clear, interesting, and effective. And it's very likely that, as your skill grows, your confidence will too. This chapter will deal with your first steps in that process, through careful speech planning.

When accepting his 2016 Oscar for Best Actor, Leonardo DiCaprio tried to persuade his audience that climate change is an urgent matter.

What goal would you have pursued in this situation?

Getting Started

Your first tasks are generally choosing a topic, determining your purpose, and finding information.

Choosing Your Topic

The first question many student speakers face is, "What should I talk about?" When you need to choose a topic, you should try to pick one that is right for you, your audience, and the situation. You should try to choose a topic that interests you and that your audience will care about. Decide on your topic as early as possible. Those who wait until the last possible moment usually find that they don't have enough time to research, outline, and practice their speech.

Defining Your Purpose

No one gives a speech—or expresses *any* kind of message—without having a reason to do so. Your first step in focusing your speech is to formulate a clear and precise statement of that purpose.

Writing a Purpose Statement

Your **purpose statement** should be expressed in the form of a complete sentence that describes your **specific purpose**—exactly what you want your speech to accomplish. It should stem from your **general purpose**, which might be to inform, persuade, or entertain. Beyond that, though, there are three criteria for an effective purpose statement:

1. **A purpose statement should be result oriented**. Having a *result orientation* means that your purpose is focused on the outcome you want to accomplish with your audience members. For example, if you were giving an informative talk on the high cost of college, this would be an inadequate purpose statement:

    My purpose is to tell my audience about high college costs.

    As that statement is worded, your purpose is "to tell" an audience something, which suggests that the speech could be successful even if no one

purpose statement A complete sentence that describes precisely what a speaker wants to accomplish.

specific purpose The precise effect that the speaker wants to have on an audience. It is expressed in the form of a purpose statement.

general purpose One of three basic ways a speaker seeks to affect an audience: to entertain, inform, or persuade.

ASK YOURSELF

Think of someone who is an effective public speaker. What makes this person so effective?

listened! A result-oriented purpose statement should refer to the response you want from your audience: It should tell what the audience members will know or be able to do after listening to your speech.

2. **A purpose statement should be specific.** To be effective, a purpose statement should be worded specifically, with enough details so that you would be able to measure or test your audience, after your speech, to see if you had achieved your purpose. In the example given earlier, simply "knowing about high college costs" is too vague; you need something more specific, such as:

> *After listening to my speech, my audience will be able to reduce college costs.*

This is an improvement, but it can be made still better by applying a third criterion:

3. **A purpose statement should be realistic.** It's fine to be ambitious, but you need to design a purpose that has a reasonable chance of success. You can appreciate the importance of having a realistic goal by looking at some unrealistic ones, such as "My purpose is to convince my audience to make federal budget deficits illegal." Unless your audience happens to be a joint session of Congress, it won't have the power to change U.S. fiscal policy. But any audience can write its congressional representatives or sign a petition. In your speech on college costs, it would be impossible for your audience members to change the entire structure of college financing. So a better purpose statement for this speech might sound something like this:

> *After listening to my speech, my audience will be able to list four simple steps to lower their college expenses.*

Consider the following sets of purpose statements:

LESS EFFECTIVE	MORE EFFECTIVE
To talk about professional wrestling (not receiver oriented)	After listening to my speech, my audience will understand that kids who imitate professional wrestlers can be seriously hurt.
To tell my audience about gun control (not specific)	After my speech, the audience will sign my petition calling for universal background checks.

You probably won't include your purpose statement word-for-word in your actual speech. Rather than being aimed at your listeners, a specific purpose statement usually is a tool to keep you focused on your goal as you plan your speech.

Stating Your Thesis

After you have defined your purpose, you are ready to start planning what is arguably the most important sentence in your entire speech. The **thesis statement** tells your listeners the central idea of your speech. It is the one idea that you want your audience to remember after it has forgotten everything else you had to say. The thesis statement for a speech about winning in small claims court might be worded like this:

> *Arguing a case on your own in small claims court is a simple, five-step process that can give you the same results you would achieve with a lawyer.*

Unlike your purpose statement, your thesis statement is almost always delivered directly to your audience. The thesis statement is usually formulated later in the speech-making process, after you have done some research on your topic. The progression from topic to purpose to thesis is, therefore, another focusing process, as you can see in the following example:

thesis statement A complete sentence describing the central idea of a speech.

Topic: Organ donation

Specific Purpose: After listening to my speech, audience members will recognize the importance of organ donation and will sign an organ donor's card for themselves.

Thesis: Because not enough of us choose to become organ donors, thousands of us needlessly die every year. You can help prevent this needless dying.

Analyzing the Speaking Situation

There are two components to analyze in any speaking situation: the audience and the occasion. To be successful, every choice you make in putting together your speech—your purpose, topic, and all the material you use to develop your speech—must be appropriate to both of these components.

The Listeners

Audience analysis involves identifying and adapting your remarks to the most pertinent characteristics of your listeners.

Audience Purpose　Just as you have a purpose for speaking, audience members have a reason for gathering. Sometimes virtually all the members of your audience will have the same, obvious goal. Expectant parents at a natural childbirth class are all seeking a healthy delivery, and people attending an investment seminar are looking for ways to increase their net worth.

There are other times, however, when audience purpose can't be so easily defined. In some instances, different listeners will have different goals, some of which might not be apparent to the speaker. Consider a church congregation, for example. Whereas most members might listen to a sermon with the hope of applying religious principles to their lives, a few might be interested in being entertained or in merely appearing pious. In the same way, the listeners in your speech class probably have a variety of motives for attending. Becoming aware of as many of these motives as possible will help you predict what will interest them. Observing audience demographics helps you make that prediction.

Demographics　Your audience has a number of characteristics that you may be able to learn about in advance. These factors, known as **demographics**, include cultural differences, age, gender, group membership, number of people, and so on. Demographic characteristics might affect your speech planning in a number of ways.[5] For example:

- **Cultural diversity.** Do audience members differ in terms of race, religion, or national origin? The guideline here might be, *Do not exclude or offend any portion of your audience on the basis of cultural differences.* If there is a dominant cultural group represented, you might decide to speak to it, but remember that the point is to analyze, not stereotype, your audience. If you talk down to any segment of your listeners, you have probably stereotyped them.

- **Gender.** Although masculine and feminine stereotypes are declining, it is still important to think about how gender can affect the way you choose and approach a topic. Every speech teacher has a horror story about a student getting up in front of a class composed primarily, but not entirely, of men and speaking on a subject such as "Picking Up Babes."

- **Age.** Our interests vary and change with our age. These differences may run relatively deep; our approach to literature, films, finance, health, and

It's essential to tailor your remarks to the audience you will be addressing.

What demographic characteristics do you recognize in this audience? How does this group differ from the ones you will be addressing?

audience analysis　A consideration of characteristics, including the type, goals, demographics, beliefs, attitudes, and values of listeners.

demographics　Audience characteristics that can be analyzed statistically, such as age, gender, education, and group membership.

cultural idioms

talk down to: speak to in a condescending way

picking up babes: offensive term for making the acquaintance of women or girls with sexual purposes in mind

long-term success may change dramatically over just a few years, perhaps from graphic novels to serious literature, from punk to classical music, or from hip-hop to epic poetry.

- **Group membership.** Groups generally form around shared interests. By examining the groups to which your audience members belong, you can potentially surmise their political leanings, religious beliefs, or occupation. Group membership is often an important consideration in college classes. Consider the difference between a daytime class and one that meets in the evening. At many colleges the evening students are generally older and tend to belong to civic groups, church clubs, and the local chamber of commerce. Daytime students are more likely to belong to sororities and fraternities, sports clubs, and social action groups.[6]

- **Number of people.** Topic appropriateness varies with the size of an audience. With a small audience you can be less formal and more intimate; you can, for example, talk more about your feelings and personal experiences. If you gave a speech before 5 people as impersonally as if they were a standing-room-only crowd in a lecture hall, they would probably find you stuffy. On the other hand, if you talked to 300 people about your unhappy childhood, you'd probably make them uncomfortable.

You have to decide which demographics of your audience are important for a particular speech. For example, when Sneha Polisetti, a student at James Madison University in Virginia, gave a speech on the loss of Native American culture, she knew she had to broaden the appeal of her topic beyond the small demographic referred to in her speech. She adapted to her broader audience this way:

> When the Native American cultures that are tied closely to America's story are lost, we all lose a part of our identity and history, whether we're Native Americans or not.[7]

These five demographic characteristics are important examples, but the list goes on. Other demographic characteristics that might be important in a college classroom include the following:

- Educational level
- Economic status
- Hometown
- Year in school
- Major subject
- Ethnic background

A final factor to consider in audience analysis concerns members' attitudes, beliefs, and values.

Attitudes, Beliefs, and Values Audience members' feelings about you, your subject, and your intentions are central issues in audience analysis. One way to approach these issues is through a consideration of attitudes, beliefs, and values.[8] Attitudes, beliefs, and values reside in human consciousness like layers of an onion (see Figure 11-1). **Attitudes**, which are closest to the surface, reflect a predisposition to view you or your topic in a favorable or unfavorable way. **Beliefs** lie a little deeper and deal with the truth of something. **Values** are deeply rooted feelings about a concept's inherent worth or worthiness. You can begin to appreciate the usefulness of these concepts by considering an example. Suppose you were a dentist trying to persuade a group of patients to floss their teeth more often. Consider how audience analysis would help you design the most promising approach:

attitude The predisposition to respond to an idea, person, or thing favorably or unfavorably.

belief An underlying conviction about the truth of an idea, often based on cultural training.

value A deeply rooted belief about a concept's inherent worth.

Attitudes. How do your listeners feel about the importance of dental hygiene? If they recognize its importance, you can proceed confidently, knowing they'll probably want to hear what you have to say. On the other hand, if they are vaguely disgusted by even thinking about the topic, you will need to begin by making them want to listen.

Beliefs. Does your audience accept the relationship between regular flossing and dental health? Or do you need to inform them about the consequences of neglecting this daily ritual?

Values. Which underlying values matter most to your listeners: Health? Attractiveness? Career success? The approach you'll use will depend on the answer to these questions.

ASK YOURSELF

How might your own attitudes, beliefs, and values affect how you react to a speech?

Experts in audience analysis, such as professional speechwriters, often try to concentrate on values. As one team of researchers pointed out, "values have the advantage of being comparatively small in number, and owing to their abstract nature, are more likely to be shared by large numbers of people."[9] Stable American values include the ideas of good citizenship, a strong work ethic, tolerance of differing political views, individualism, and justice for all. Brianna Mahoney, a student at the University of Florida, appealed to her audience's values when she wanted to make the point that anti-homelessness laws were inhumane:

> Extreme poverty in the United States is a shockingly overlooked issue, making the homeless one of our most vulnerable populations. Recent legislation has worsened this by denying the homeless the ability to help themselves. People experiencing homelessness exist in every community; it's time to stop accusing them of becoming a burden when they are already struggling to carry their own.[10]

Mahoney pointed out that discriminating against homeless people was unfair and impractical; her analysis had suggested that the value of fairness would be important to this audience. She also surmised that they would be offended by unfairness combined with impracticality, as in this piece of evidence she used:

> Ninety-year-old Arnold Abbott made international headlines when he was arrested in Fort Lauderdale, Florida and faced 60 days in prison for feeding hot meals to people who were homeless. When police arrived they grabbed trays of food and shoved them into the trash, while lines of hungry people were forced to just look on.[11]

You can often make an inference about audience members' attitudes by recognizing the beliefs and values they are likely to hold. In this example, Brianna knew that her audience, made up mostly of idealistic college students and professors, would dislike the idea of unfair and impractical discrimination.

The analysis of hidden psychological states can be extremely helpful in audience analysis. For example, a religious group might hold the value of "obeying God's word." For some fundamentalists this might lead to the belief, based on their religious training, that women are not meant to perform the same functions in society as men. This, in turn, might lead to the attitude that women ought not to pursue careers as firefighters, police officers, or construction workers.

You can also make a judgment about one attitude your audience members hold based on your knowledge of other attitudes they hold. If your audience is made up of undergraduates who have a positive attitude toward liberation movements, it is a good bet they also have a positive attitude toward civil rights and ecology. If they have a negative attitude

Values

Beliefs

Attitudes

FIGURE 11-1 Structure of Attitudes, Beliefs, and Values

It can be fun to go for a quick laugh when making a speech. But keep in mind the impression that will linger.

Have you ever made remarks inappropriate for the occasion? What can you do to avoid this kind of mistake in the future?

toward collegiate sports, they may also have a negative attitude toward fraternities and sororities. This should suggest not only some appropriate topics for each audience but also ways that those topics could be developed.

The Occasion

The second phase in analyzing a speaking situation focuses on the occasion. The occasion of a speech is determined by the circumstances surrounding it. Three of these circumstances are time, place, and audience expectations.

Time Your speech occupies an interval of time that is surrounded by other events. For example, other speeches might be presented before or after yours, or comments might be made that set a certain tone or mood. In the sample speech at the end of this chapter, Malala talks about a speech that occurred before hers. External events such as elections, the start of a new semester, or even the weather can color the occasion in one way or another. The date on which you give your speech might have some historical significance. If that historical significance relates in some way to your topic, you can use it to help build audience interest.

The time available for your speech is also an essential consideration. You should choose a topic that is broad enough to say something worthwhile but brief enough to fit your limits. "Wealth," for example, might be an inherently interesting topic to some college students, but it would be difficult to cover such a broad topic in a 10-minute speech and still say anything significant. However, a topic like "The problem of income inequality in America today" could conceivably be covered in 10 minutes in enough depth to make it worthwhile. All speeches have limits, whether or not they are explicitly stated. If you are invited to say a few words, and you present a few volumes, you might not be invited back.

Place Your speech also occupies a physical space. The beauty or squalor of your surroundings and the noise or stuffiness of the room should all be taken into consideration. These physical surroundings can be referred to in your speech if appropriate. If you were talking about world poverty, for example, you could compare your surroundings to those that might be found in a poorer country.

Audience Expectations Finally, your speech is surrounded by audience expectations. A speech presented in a college class, or a TED Talk such as the one that appears at the end of Chapter 13, is usually expected to reflect a high level of thought and intelligence. This doesn't necessarily mean that it has to be boring or humorless; wit and humor are, after all, indicative of intelligence. But it does mean that you have to put a little more effort into your presentation than if you were discussing the same subject with friends over coffee.

When you are considering the occasion of your speech, it pays to remember that every occasion is unique. Although there are obvious differences among the occasions of a college class, a church sermon, and a <u>bachelor party</u> "roast," there are also many subtle differences that will apply only to the circumstances of each unique event. In the sample speech at the end of this chapter, Malala recognized several unique aspects of the occasion of her speech.

cultural idiom
bachelor party: a men-only gathering of the groom and his friends just prior to his wedding

Gathering Information

This discussion about planning a speech purpose and analyzing the speech situation makes it apparent that it takes time, interest, and knowledge to develop a topic

Sample Analysis of a Speaking Situation

Audience: Employees in the Production Department of my company

Situation: Management has realized that our company is at risk of being involved in a sexual harassment lawsuit.

Purpose: I want audience members to view sexual harassment as a legitimate concern and to avoid communicating in a way that might be perceived as harassing.

Analysis: This is a tricky situation for me: First of all, this is a captive audience, forced to listen to talk on an uncomfortable subject. In addition, I am younger than anyone in my audience, and I'm a woman. In fact, I've been the target of some of the behavior I'm being asked to discourage. There's a strong risk that the men who are the primary target of my remarks won't take me seriously, so I have to change their attitude about the subject and me.

I know that scolding and threatening would be a big mistake. Even if the men involved didn't object out loud, they would probably dismiss my stance as "politically correct" and out of touch with the way the real world operates.

To avoid this sort of negative reaction, I need to separate myself from the law that they dislike, instead taking the position of sharing with them: "Here's what I've learned about how it works." I might even give them a few examples of

harassment suits that I think were frivolous, so we can agree that some people are much too sensitive. That common ground will help put us on the same side. Then I can emphasize that they don't have to agree with the law to follow it. I'll tell stories of people like them who suffered as targets of harassment suits, pointing out that even an unfair accusation of harassment could make all of our lives miserable. My basic argument will be that potentially harassing behavior "isn't worth it."

I also hope to use my age and gender as advantages. A couple of the men have told me that they have daughters my age. I could ask the group to imagine how they would feel if their daughters were the targets of suggestive comments and sexual jokes. I'll tell them that I know how angry and protective my dad would feel, and I'll tell them that I know that, as good fathers, they'd feel the same way. I could also ask them to think about how they would feel if their wives, sisters, or mothers were the targets of jokes that made those women feel uncomfortable.

I don't think any speech will totally reverse attitudes that were built over these men's lifetimes, but I do think that getting on their side will be much more effective than labeling them as insensitive sexist pigs and threatening them with lawsuits.

---

well. Setting aside a block of time to reflect on your own ideas is essential. However, you will also need to gather information from outside sources.

By this time you are probably familiar with both web searches and library research as forms of gathering information. Sometimes, however, speakers overlook interviewing, personal observation, and survey research as equally effective methods of gathering information. Let's review all these methods here and perhaps provide a new perspective on one or more of them.

Online Research

The ease of using search engines like Google has made them the popular favorite for speech research. But students are sometimes so grateful to have found a website dealing with their topic that they forget to evaluate it. Like any other written sources you would use, websites should be accurate and rational. Beyond that, there are four specific criteria that you can use to evaluate the quality of a website. They are listed in the checklist on page 306.

In the case of some special search engines, like Google Scholar, the criteria of credibility, objectivity, and currency will be practically guaranteed. However, these guidelines are especially important when accessing information from Wikipedia, the popular online encyclopedia. Because anyone can edit a

ASK YOURSELF

Why is effective research so important to a college-level audience?

Consider the following four criteria when choosing a website for online research:

☐ **Credibility.** Anyone can establish a website, so it is important to evaluate where your information is coming from. Who wrote the page? Are their names and contact information listed? Anonymous sources should not be used. If the sources *are* listed, are their credentials listed? What institution publishes the document? Remember that although an attractive site design doesn't guarantee high-quality information, obvious mistakes such as misspellings are clear signs of low quality.

☐ **Objectivity.** What opinions (if any) does the site express? Are these opinions backed up with facts, or are they purely based on inferences? Is the site trying to sell something, including a candidate or a political idea?

☐ **Currency.** When was the page produced? When was it updated? Are the links working? If any of the links are dead, the information might not be current.

☐ **Functionality.** Is the site easy to use, so you can locate information you are looking for? Are options to return to the home page and tops of pages provided? Is the site searchable?

database A computerized collection of information that can be searched in a variety of ways to locate information that the user is seeking.

survey research Information gathering in which the responses of a population sample are collected to disclose information about the larger group.

Wikipedia article at any time, many professors forbid the use of it as a primary resource. Others allow Wikipedia to be used for general information and inspiration. Most will allow its use when articles have references to external sources (whether online or not) and the student reads the references and checks whether they really do support what the article says.

Library experts help you make sense of and determine the validity of the information you find, whether online or in print. And a library can be a great environment for concentration, a rare quiet place with minimal distractions.

Library Research

Libraries, like people, tend to be unique. Although many of your library's resources will be available online through your school's website, it can be extremely rewarding to get to know your library in person, to see what kind of special collections and services it offers, and just to find out where everything is. A few resources are common to most libraries, including the library catalog, reference works, periodicals, nonprint materials, and databases.

Databases, which can be particularly useful, are computerized collections of highly credible information from a wide variety of sources. One popular collection of databases is LexisNexis, which contains millions of articles from news services, magazines, scholarly journals, conference papers, books, law journals, and other sources. Other popular databases include ProQuest, Factiva, and Academic Search Premier, and there are dozens of specialized databases, such as Communication and Mass Media Complete. Database searches are slightly different from web searches; they generally don't respond well to long strings of terms or searches worded as questions. With databases it is best to use one or two key terms with a connector such as AND, OR, or NOT.[12] Once you learn this technique and a few other rules (perhaps with a librarian's help), you will be able to locate dozens of articles on your topic in just a few minutes.

Interviewing

An information-gathering interview allows you to view your topic from an expert's perspective, to take advantage of that expert's experience, research, and thought. You can also use an interview to stimulate your own thinking. Often the interview will save you hours of Internet or library research and allow you to present ideas that you could not have uncovered any other way. And because an interview is an interaction with an expert, many ideas that otherwise might be unclear can become more understandable through questions and answers. Interviews can be conducted face-to-face, by telephone, or by email. If you do use an interview for research, you might want to read the section on that type of interview in the appendix.

Survey Research

One advantage of **survey research**—the distribution of questionnaires for people to respond to—is that it can give you up-to-date answers concerning "the way things are" for a specific audience. For example, if you handed out questionnaires a week or so before presenting a speech on the possible dangers of body piercing, you could present information like this in your speech:

> According to a survey I conducted last week, 90 percent of the students in this class believe that body piercing is basically safe. Only 10 percent are familiar with the scarring and injury that can result from this practice. Two of you, in fact, have experienced serious infections from body piercing: one from a pierced tongue and one from a simple pierced ear.

That statement would be of immediate interest to your audience members because *they* were the ones who were surveyed. Another advantage of conducting your own survey is that it is one of the best ways to find out about your audience: It is, in fact, *the* best way to collect the demographic data mentioned earlier. The one disadvantage of conducting your own survey is that, if it is used as evidence, it might not have as much credibility as published evidence found in the library. But the advantages seem to outweigh the disadvantages of survey research in public speaking.

No matter how you gather your information, remember that it is the *quality* rather than the quantity of the research that is most important. The key is to determine carefully what type of research will answer the questions you need to have answered. Sometimes only one type of research will be necessary; at other times every type mentioned here will have to be used. Generally, you will collect far more information than you'll use in your speech, but the winnowing process will ensure that the research you do use is of high quality.

Along with improving the quality of what you say, effective research will also minimize the anxiety of actually giving a speech. Let's take a close look at that form of anxiety.

Online and library research are essential to speech preparation. In addition, talking to credible sources can provide information that makes a speech more interesting and compelling.

Who can you interview to enrich the content of your speech?

Managing Communication Apprehension

The terror that strikes the hearts of so many beginning speakers is commonly known as *stage fright* or *speech anxiety* and is called *communication apprehension* by communication scholars.[13] Whatever term you choose, the important point to realize is that fear about speaking can be managed in a way that works for you rather than against you.

Facilitative and Debilitative Communication Apprehension

Although communication apprehension is a very real problem for many speakers, it can be overcome. The first step in feeling less apprehensive about speaking is to realize that a certain amount of nervousness is not only natural but also facilitative. That is, **facilitative communication apprehension** is a factor that can help improve your performance. Just as totally relaxed actors or musicians aren't likely to perform at the top of their potential, speakers think more rapidly and express themselves more energetically when their level of tension is moderate.

By contrast, **debilitative communication apprehension** inhibits effective self-expression. Intense fear causes trouble in two ways. First, the strong emotion keeps you from thinking clearly.[14] This has been shown to be a problem even in the preparation process: Students who are highly anxious about giving a speech will find the preliminary steps, including research and organization, to be more difficult.[15] Second, intense fear leads to an urge to do something, anything, to make the problem go away. This urge to escape often causes a speaker to speed up

facilitative communication apprehension A moderate level of anxiety about speaking before an audience that helps improve the speaker's performance.

debilitative communication apprehension An intense level of anxiety about speaking before an audience, resulting in poor performance.

Even a flamboyant star like Rihanna can experience communication apprehension. She has confessed that she gets nervous and nauseous before live performances. The lesson is that anxiety does not necessarily keep you from performing well.

How can you apply the information in this chapter to manage—and benefit from—your natural apprehension?

delivery, which results in a rapid, almost machine-gun style. As you can imagine, this boost in speaking rate leads to even more mistakes, which only add to the speaker's anxiety. Thus, a relatively small amount of nervousness can begin to feed on itself until it grows into a serious problem.

Sources of Debilitative Communication Apprehension

Before we describe how to manage debilitative communication apprehension, let's consider why people are afflicted with the problem in the first place.[16]

Previous Negative Experience   People often feel apprehensive about speech giving because of unpleasant past experiences. Most of us are uncomfortable doing *anything* in public, especially if it is a form of performance in which our talents and abilities are being evaluated. An unpleasant experience in one type of performance can cause you to expect that a future similar situation will also be unpleasant.[17] These expectations can be realized through the self-fulfilling prophecies discussed in Chapter 3. A traumatic failure at an earlier speech and low self-esteem from critical parents during childhood are common examples of experiences that can cause later communication apprehension.

You might object to the idea that past experiences cause communication apprehension. After all, not everyone who has bungled a speech or had critical parents is debilitated in the future. To understand why some people are affected more strongly than others by past experiences, we need to consider another cause of communication apprehension.

Irrational Thinking   Cognitive psychologists argue that it is not events that cause people to feel nervous but rather the beliefs they have about those events. Certain irrational beliefs leave people feeling unnecessarily apprehensive. Psychologist Albert Ellis lists several such beliefs, or examples of **irrational thinking**, which we will call "fallacies" because of their illogical nature.[18]

irrational thinking Beliefs that have no basis in reality or logic; one source of debilitative communication apprehension.

fallacy of catastrophic failure The irrational belief that the worst possible outcome will probably occur.

fallacy of perfection The irrational belief that a worthwhile communicator should be able to handle every situation with complete confidence and skill.

- **Catastrophic failure.** People who succumb to the **fallacy of catastrophic failure** operate on the assumption that if something bad can happen, it probably will. Their thoughts before a speech resemble these:

  "As soon as I stand up to speak, I'll forget everything I wanted to say."

  "Everyone will think my ideas are stupid."

  "Somebody will probably laugh at me."

  Although it is naive to imagine that all your speeches will be totally successful, it is equally naive to assume they will all fail miserably. One way to escape the fallacy of catastrophic failure is to take a more realistic look at the situation. Would your audience members really hoot you off the stage? Will they really think your ideas are stupid? Even if you did forget your remarks for a moment, would the results be a genuine disaster? It helps to remember that nervousness is more apparent to the speaker than to the audience.[19] Beginning public speakers, when congratulated for their poise during a speech, are apt to say, "Are you kidding? I was dying up there."

- **Perfection.** Speakers who succumb to the **fallacy of perfection** expect themselves to behave flawlessly. Whereas such a standard of perfection might serve as a target and a source of inspiration (like the desire to make a hole in one while golfing), it is totally unrealistic to expect that you will write

cultural idiom

hole in one: hitting the golf ball in the hole with one swing of the club; a perfect shot

and deliver a perfect speech, especially as a beginner. It helps to remember that audiences don't expect you to be perfect.

- **Approval.** The mistaken belief called the **fallacy of approval** is based on the idea that it is vital—not just desirable—to gain the approval of everyone in the audience. It is rare that even the best speakers please everyone, especially on topics that are at all controversial. To paraphrase Abraham Lincoln, you can't please all the people all the time, and it is irrational to expect you will.

- **Overgeneralization.** The **fallacy of overgeneralization** might also be labeled the fallacy of exaggeration, because it occurs when a person <u>blows one experience out of proportion</u>. Consider these examples:

> "I'm so stupid! I mispronounced that word."
>
> "I completely blew it—I forgot one of my supporting points."
>
> "My hands were shaking. The audience must have thought I was crazy."

A second type of exaggeration occurs when a speaker treats occasional lapses as if they were the rule rather than the exception. This sort of mistake usually involves extreme labels, such as "always" or "never."

> "I always forget what I want to say."
>
> "I can never come up with a good topic."
>
> "I can't do anything right."

fallacy of approval The irrational belief that it is vital to win the approval of virtually every person a communicator deals with.

fallacy of overgeneralization Irrational beliefs in which (1) conclusions (usually negative) are based on limited evidence or (2) communicators exaggerate their shortcomings.

Overcoming Debilitative Communication Apprehension

There are five strategies that can help you manage debilitative communication apprehension:

- **Use nervousness to your advantage.** Paralyzing fear is obviously a problem, but a little nervousness can actually help you deliver a successful speech. Being completely calm can take away the passion that is one element of a good speech. Use the strategies below to control your anxiety, but don't try to completely eliminate it.

- **Understand the difference between rational and irrational fears.** Some fears about speaking are rational. For example, you ought to be worried if you haven't properly prepared for your speech. But fears based on the fallacies you just read about aren't constructive. It's not realistic to expect that you'll deliver a perfect speech, and it's not rational to indulge in catastrophic fantasies about what might go wrong.

- **Maintain a receiver orientation.** Paying too much attention to your own feelings—even when you're feeling good about yourself—will take energy away from communicating with your listeners. Concentrate on your audience members rather than on yourself. Focus your energy on keeping them interested, and on making sure they understand you.

- **Keep a positive attitude.** Build and maintain a positive attitude toward your audience, your speech, and yourself as a speaker. Some communication consultants suggest that public speakers should concentrate on three statements immediately before speaking. The three statements are as follows:

> I'm glad I have the chance to talk about this topic.
>
> I know what I'm talking about.
>
> I care about my audience.

cultural idiom

blows . . . out of proportion: exaggerates

SELF-ASSESSMENT

Speech Anxiety Symptoms

To what degree do you experience the following anxiety symptoms while speaking?

1. Sweating
   a. Nonexistent
   b. Moderate
   c. Severe

2. Rapid breathing
   a. Nonexistent
   b. Moderate
   c. Severe

3. Difficulty catching your breath
   a. Nonexistent
   b. Moderate
   c. Severe

4. Rapid heartbeat
   a. Nonexistent
   b. Moderate
   c. Severe

5. Restless energy
   a. Nonexistent
   b. Moderate
   c. Severe

6. Forgetting what you wanted to say
   a. Nonexistent
   b. Moderate
   c. Severe

Evaluating Your Responses

Give yourself one point for every "a," two points for every "b," and three points for every "c." If your score is:

6 to 9 You have nerves of steel. You're probably a natural public speaker, but you can always improve.

10 to 13 You are the typical public speaker. Practice the strategies discussed in this chapter to improve your skills.

14 to 18 You tend to have significant apprehension about public speaking. You need to consider each strategy in this chapter carefully. Although you will benefit from the tips provided, you should keep in mind that some of the greatest speakers of all time have considered themselves highly anxious.

---

visualization A technique for rehearsal that involves the successful completion of a speech.

Repeating these statements (until you believe them) can help you maintain a positive attitude.

Another technique for building a positive attitude is known as **visualization**.[20] This technique has been used successfully with athletes. It requires you to use your imagination to visualize the successful completion of your speech. Visualization can help make the self-fulfilling prophecy discussed in Chapter 3 work in your favor.

- **Be prepared!** Preparation is the most important key to controlling communication apprehension. You can feel confident if you know from practice that your remarks are well organized and supported and your delivery is smooth. Researchers have determined that the highest level of communication apprehension occurs just before speaking, the second highest level at the time the assignment is announced and explained, and the lowest level during the time you spend preparing your speech.[21] You should take advantage of this relatively low-stress time to work through the problems that would tend to make you nervous during the actual speech. For example, if your anxiety is based on a fear of forgetting what you are going to say, make sure that your note cards are complete and effective, and that you have practiced your speech thoroughly (we'll go into speech practice in more detail in a moment). If, on the other hand, your great fear is "sounding stupid," then getting started early with lots of research and advance thinking is the key to relieving your communication apprehension.

Presenting Your Speech

Once you have done all the planning and analysis that precedes speechmaking, you can prepare for your actual presentation. Your tasks at this point include choosing an effective type of delivery, formulating a plan for practicing your speech, and thinking carefully about the visual and auditory choices you will make.

Choosing an Effective Type of Delivery

There are four basic types of delivery: extemporaneous, impromptu, manuscript, and memorized. Each type creates a different impression and is appropriate under different conditions. Any speech may incorporate more than one of these types of delivery. For purposes of discussion, however, it is best to consider them separately.

The best type of speech delivery depends on the situation.

What styles are most appropriate in the situations where you might deliver remarks?

1. An **extemporaneous speech** is planned in advance but presented in a direct, spontaneous manner. Extemporaneous speeches are conversational in tone, which means that they give the audience members the impression that you are talking to them, directly and honestly. Extemporaneous speaking is the most common type of delivery in both the classroom and the "outside" world.

2. An **impromptu speech** is given off the top of one's head, without preparation. This type of speech is spontaneous by definition, but it is a delivery style that is necessary for informal talks, group discussions, and comments on others' speeches. It is also a highly effective training aid that teaches you to think on your feet and to organize your thoughts quickly.

3. **Manuscript speeches** are read word for word from a prepared text. They are necessary when you are speaking for the record, such as at legal proceedings or when presenting scientific findings. The greatest disadvantage of a manuscript speech is, of course, the lack of spontaneity.

4. **Memorized speeches**—those learned by heart—are the most difficult and often the least effective. They often seem excessively formal. However, like manuscript speeches, they may be necessary on special occasions. They are used in oratory contests and as training devices for memory.

One guideline holds for each type of speech: Practice.

extemporaneous speech A speech that is planned in advance but presented in a direct, conversational manner.

impromptu speech A speech given "off the top of one's head," without preparation.

manuscript speech A speech that is read word for word from a prepared text.

memorized speech A speech learned and delivered by rote without a written text.

Practicing Your Speech

A smooth and natural delivery is the result of extensive practice. Get to know your material until you feel comfortable with your presentation. One way to do that is to go through some or all of the steps listed in the checklist on page 312. In each of these steps, critique your speech according to the guidelines that follow.

Guidelines for Delivery

Let's examine some nonverbal aspects of presenting a speech. As you read in Chapter 6, nonverbal behavior can change, or even contradict, the meaning of the words a speaker utters. If audience members want to interpret how you feel about something, they are likely to trust your nonverbal communication more than the words you speak. If you tell them, "It's great to be here today," but you stand before them slouched over with your hands in your pockets and an expression on your face like you're about to be shot, they are likely to discount what you say. This might cause your audience members to react negatively to your speech, and their negative

cultural idioms

off the top of one's head: with little time to plan or think about

for the record: word-for-word documentation

Practicing Your Presentation

Be sure to give yourself plenty of time to practice. Here are a few suggestions.

☐ First, present the speech to yourself. Talk through the entire speech, including your examples and forms of support. Don't skip parts by using placeholders. Make sure you have a clear plan for presenting your statistics and explanations.

☐ Record the speech on your phone, and listen to it. Because we are sometimes surprised at what we sound like and how we appear, video recording has been shown to be an especially effective tool for rehearsals.[22]

☐ Present the speech in front of a small group of friends or relatives.[23]

☐ Present the speech to at least one listener in the room where you will present the final speech (or, if that room is not available, a similar room).

reaction might make you even more nervous. This cycle of speaker and audience reinforcing each other's feelings can work for you, though, if you approach a subject with genuine enthusiasm. Enthusiasm is shown through both the visual and auditory aspects of your delivery.

Visual Aspects of Delivery

Visual aspects of delivery include appearance, movement, posture, facial expression, and eye contact.

Appearance This is not a presentation variable as much as a preparation variable. Some communication consultants suggest new clothes, new glasses, and new hairstyles for their clients. In case you consider any of these, be forewarned that you should be attractive to your audience but not flashy. Research suggests that audiences like speakers who are similar to them, but they prefer the similarity to be shown conservatively.[24] Speakers, it seems, are perceived to be more credible when they look businesslike. Part of looking businesslike, of course, is looking like you took care in the preparation of your wardrobe and appearance.

Movement The way you walk to the front of your audience will express your confidence and enthusiasm. And after you begin speaking, nervous energy can cause your body to shake and twitch, and that can be distressing both to you and to your audience. One way to control involuntary movement is to move voluntarily when you feel the need to move. Don't feel that you have to stand in one spot or that all your gestures need to be carefully planned. Simply get involved in your message, and let your involvement create the motivation for your movement. That way, when you move, you will emphasize what you are saying in the same way you would emphasize it if you were talking to a group of friends.

Movement can also help you maintain contact with all members of your audience. Those closest to you will feel the greatest contact. This creates what is known as the "action zone" in the typical classroom, within the area of the front and center of the room. Movement enables you to extend this action zone, to include in it people who would otherwise remain uninvolved. Without overdoing it, you should feel free to move toward, away from, or from side to side in front of your audience.

Remember: Move with the understanding that it will add to the meaning of the words you use. It is difficult to bang your fist on a podium or take a step without conveying emphasis. Make the emphasis natural by allowing your message to create your motivation to move.

Posture Generally speaking, good posture means standing with your spine relatively straight, your shoulders relatively squared off, and your feet angled out to keep your body from falling over sideways. In other words, rather than standing at military attention, you should be comfortably erect.

Good posture can help you control nervousness by allowing your breathing apparatus to work properly; when your brain receives enough oxygen, it's easier for you to think clearly. Good posture also increases your audience contact because the audience members will feel that you are interested enough in them to stand formally, yet relaxed enough to be at ease with them.

Facial Expression The expression on your face can be more meaningful to an audience than the words you say. Try it yourself with a mirror. Say, "You're a terrific audience," for example, with a smirk, with a warm smile, with a deadpan expression, and then with a scowl. It just doesn't mean the same thing. But don't try to fake it. Like your movement, your facial expressions will reflect your genuine involvement with your message.

Eye Contact Eye contact is perhaps the most important nonverbal facet of delivery. Eye contact not only increases your direct contact with your audience but also can be used to help you control your nervousness. Direct eye contact is a form of reality testing. The most frightening aspect of speaking is the unknown. How will the audience react? What will it think? Direct eye contact allows you to test your perception of your audience as you speak. Usually, especially in a college class, you will find that your audience is more "with" you than you think. By deliberately establishing contact with any apparently bored audience members, you might find that they are interested, they just aren't showing that interest because they don't think anyone is looking.

To maintain eye contact, you could try to meet the eyes of each member of your audience squarely at least once during any given presentation. After you have made definite eye contact, move on to another audience member. You can learn to do this quickly, so you can visually latch on to every member of a good-sized class in a relatively short time.

The characteristics of appearance, movement, posture, facial expression, and eye contact are visual, nonverbal facets of delivery. Now consider the auditory nonverbal messages that you might send during a presentation.

Auditory Aspects of Delivery

As you read in Chapter 6, your paralanguage—the way you use your voice—says a good deal about you, especially about your sincerity and enthusiasm. In addition, using your voice well can help you control your nervousness. It's another cycle: Controlling your vocal characteristics will decrease your nervousness, which will enable you to control your voice even more. But this cycle can also work in the opposite direction. If your voice is out of control, your nerves will probably be in the same state. Controlling your voice is mostly a matter of recognizing and using appropriate volume, rate, pitch, and articulation.

Volume The loudness of your voice is determined by the amount of air you push past the vocal folds in your throat. The key to controlling volume, then, is controlling the amount of air you use. The key to determining the right volume is audience contact. Your delivery should be loud enough so that your audience members can hear everything you say but not so loud that they feel you are talking to someone in the next room. Too much volume is seldom the problem for beginning speakers. Usually they either are not loud enough or have a tendency to fade off at the end of a thought. Sometimes, when they lose faith in an idea in midsentence, they compromise by mumbling the end of the sentence so that it isn't quite coherent.

Rate There is a range of personal differences in speaking speed, or **rate**. Daniel Webster, for example, is said to have spoken at around 90 words per minute, whereas one actor who is known for his fast-talking commercials speaks at about 250. Normal speaking speed, however, is between 120 and 150 words per minute. If you talk much more slowly than that, you may tend to lull your audience to sleep. Faster speaking rates are stereotypically associated with speaker competence,[25] but if you speak too rapidly, you will tend to be unintelligible. Once again, your involvement in your message is the key to achieving an effective rate.

> **rate** The speed at which a speaker utters words.

pitch The highness or lowness of one's voice.

articulation The process of pronouncing all the necessary parts of a word.

deletion An articulation error that involves leaving off parts of words.

substitution The articulation error that involves replacing part of a word with an incorrect sound.

addition The articulation error that involves adding extra parts to words.

slurring The articulation error that involves overlapping the end of one word with the beginning of the next.

Pitch The highness or lowness of your voice—**pitch**—is controlled by the frequency at which your vocal folds vibrate as you push air through them. Because taut vocal folds vibrate at a greater frequency, pitch is influenced by muscular tension. This explains why nervous speakers have a tendency occasionally to "squeak," whereas relaxed speakers seem to be more in control. Pitch will tend to follow rate and volume. As you speed up or become louder, your pitch will have a tendency to rise. If your range in pitch is too narrow, your voice will have a singsong quality. If it is too wide, you may sound overly dramatic. You should control your pitch so that your listeners believe you are talking with them rather than performing in front of them. Once again, your involvement in your message should take care of this naturally for you.

When considering volume, rate, and pitch, keep emphasis in mind. Remember that a change in volume, pitch, or rate will result in emphasis. If you pause or speed up, your rate will suggest emphasis. Words you whisper or scream will be emphasized by their volume.

Articulation The final auditory nonverbal behavior, articulation, is perhaps the most important. For our purposes here, **articulation** means pronouncing all the parts of all the necessary words and nothing else.

It is not our purpose to condemn regional or ethnic dialects within this discussion. It is true that a considerable amount of research suggests that regional dialects can cause negative impressions,[26] but our purpose here is to suggest careful, not standardized, articulation. Incorrect articulation is usually nothing more than careless articulation. It is caused by (1) leaving off parts of words (deletion), (2) replacing parts of words (substitution), (3) adding parts to words (addition), or (4) overlapping two or more words (slurring).

Deletion The most common mistake in articulation is **deletion**, or leaving off part of a word. As you are thinking the complete word, it is often difficult to recognize that you are saying only part of it. The most common deletions occur at the ends of words, especially *-ing* words. *Going, doing,* and *stopping* become *goin', doin',* and *stoppin'.* Parts of words can be left off in the middle, too, as in *terr'iss* for *terrorist, Innernet* for *Internet,* and *asst* for *asked.*

Substitution **Substitution** takes place when you replace part of a word with an incorrect sound. The ending *-th* is often replaced at the end of a word with a single *t,* as when *with* becomes *wit.* The *th-* sound is also a problem at the beginning of words, as *this, that,* and *those* have a tendency to become *dis, dat,* and *dose.* (This tendency is especially prevalent in many parts of the northeastern United States.)

Addition The articulation problem of **addition** is caused by adding extra parts to words, such as *incentative* instead of *incentive, athalete* instead of *athlete,* and *orientated* instead of *oriented.* Sometimes this type of addition is caused by incorrect word choice, as when *irregardless* is used for *regardless.*

Another type of addition is the use of "tag questions," such as *you know?* or *you see?* or *right?* at the end of sentences. To have every other sentence punctuated with one of these barely audible superfluous phrases can be annoying.

Probably the worst type of addition, or at least the most common, is the use of *uh* and *anda* between words. *Anda* is often stuck between two words when *and* isn't even needed. If you find yourself doing that, you might want just to pause or swallow instead.[27]

Slurring This is caused by trying to say two or more words at once—or at least overlapping the end of one word with the beginning of the next. Word pairs ending with *of* are the worst offenders in this category. *Sort of* becomes *sorta, kind of* becomes *kinda,* and *because of* becomes *becausa.* Word combinations ending with

UNDERSTANDING DIVERSITY

A Compendium of American Dialects

The following is a short glossary of examples of regionalized pronunciation (with apologies to all residents who find them exaggerated).

Appalachian Hill Country

Bile To bring water to 212 degrees

Cowcumber A vittle you make pickles out of

Hern Not his'n

Tard Exhausted

Bawlamerese (Spoken around Baltimore)

Arn What you do with an arnin board

Blow The opposite of above

Pleece Two or more po-leece

Torst Tourist

Boston

Back The outer covering of a tree trunk

Had licka Hard liquor

Moa The opposite of less

Pahk To leave your car somewhere, as in, "Pahk the cah in Haavaad Yahd"

NooYorkese

Huh The opposite of him

Mel pew? May I help you?

Reg you la caw fee Coffee with milk and sugar

Pock A place with trees and muggers

Philadelphia

Fluffya The name of the city

Mayan The opposite of yours

Pork A wooded recreational area

Tail What you use to dry off with after a shower

Southern

Abode A plank of wood

Bidness Such as, "Mistah Cottah's paynut bidness"

Shurf A local law enforcement officer

Watt The color of the Watt House in Wushinton

Texas

Ah stay Iced tea

Bayer A beverage made from hops

Pars A town in Texas. Also, the capital of France

Awful Tar The famous tall structure in Pars, France

Other interesting regionalisms can be found at the Slanguistics website, www.slanguage.com.

---

to are often slurred, as when *want to* becomes *wanna*. Sometimes even more than two words are blended together, as when *that is the way* becomes *thatsaway*. Careful articulation means using your lips, teeth, tongue, and jaw to bite off your words, cleanly and separately, one at a time.

Sample Speech

Malala Yousafzai, whose profile began this chapter, was born in the Swat Valley of Pakistan in 1997. She began blogging about her life under the Taliban at the age of 11. That blog made her famous and publicized her cause, leading to a documentary film and many other opportunities to speak out. Unfortunately, that publicity also led to the nearly successful attempt on her life when she was 15.

In 2014, Malala was named as the co-recipient of that year's Nobel Peace Prize, for her struggle for the right of all children to an education. Aged 17 at the time, Yousafzai became the youngest-ever Nobel Prize laureate. Part of receiving the Nobel Prize is the presentation of a lecture.

As Malala analyzed her audience, she had to contend with the idea that she was speaking not just for the people who were sitting in front of her, but also for a much wider global audience who would view her on television and online, and hear her on podcasts.

Therefore, she had at least four purposes for speaking, which might be worded as follows:

After listening to my speech, my audience members will:

1. Understand that I am grateful for the honor that has been bestowed.
2. Be able to visualize the plight of girls who are denied an education.
3. Consider joining and/or supporting my cause.
4. Understand that the Taliban do not represent the majority of Muslims.

Her audience analysis, in fact, would have to be profound. In-person attendees included the king and queen of Norway, members of the Norwegian parliament, members of the Nobel jury, and a number of international artists and celebrities, such as Steven Tyler from the rock group Aerosmith.

It was a distinguished audience, and not a homogeneous one. For example, there were those before her who might be considered dissenters. One Nobel Prize jurist, in fact, had publicly declared that he disagreed with the decision to give Malala the award. "This is not for fine people who have done nice things and are glad to receive it," he said. "All of that is irrelevant. What Nobel wanted was a prize that promoted global disarmament."[28]

Malala's award also drew some skeptical responses from members of her global audience. One conservative Pakistani politician had stated, "There are lots of girls in Pakistan who have been martyred in terrorist attacks, women who have been widowed, but no one gives them an award. So these out of the box activities are suspicious."[29] There was even a private school organization that declared that Malala's activism was "anti-Pakistan and anti-Islam."[30]

Malala also had to consider her occasion carefully. For example, the day that she was to give her speech was also the anniversary of Alfred Nobel's death, as well as International Human Rights Day. In the final analysis, she chose not to refer to those aspects of the occasion in her speech, but she would refer to speeches that came before her and to relevant members of her audience.

One can only imagine the communication apprehension Malala faced. Yet she practiced until she only needed to glance down at her manuscript occasionally. She knew the speech well enough that she could speak partially extemporaneously by repeating key phrases.

Malala also had to consider the visual and auditory aspects of delivery. She decided to speak slowly and carefully, given the formal occasion, and to use only simple, natural gestures, such as pointing out guests she referred to in her speech, or adjusting her traditional headscarf.

The following is her speech.

SAMPLE SPEECH | **Malala Yousafzai**

Nobel Lecture[31]

1 *Bismillah hir rahman ir rahim. In the name of God, the most merciful, the most beneficent.*

Malala received a standing ovation as she got up to give her speech, which she begins with a traditional blessing from the Quran. It is important that her worldwide audience realize that she is religious, because her enemies will claim otherwise.

2 Your Majesties, Your Royal Highnesses, distinguished members of the Norwegian Nobel Committee, dear sisters and brothers, today is a day of great happiness for me. I am humbled that the Nobel Committee has selected me for this precious award. Thank you to everyone for your continued support and love. Thank you for the letters and cards that I still receive from all around the world. Your kind and encouraging words strengthen and inspire me. I would like to thank my parents for their unconditional love. Thank you to my father for not clipping my wings and for letting me fly. Thank you to my mother for inspiring me to be patient and to always speak the truth—which we strongly believe is the true message of Islam. And also thank you to all my wonderful teachers, who inspired me to believe in myself and be brave.

> She then proceeds to give the thanks that are expected for this special occasion.

3 I am proud to be the first Pashtun, the first Pakistani, and the youngest person to receive this award. Along with that, I am pretty certain that I am also the first recipient of the Nobel Peace Prize who still fights with her younger brothers. I want there to be peace everywhere, but my brothers and I are still working on that.

> She establishes her credentials but also uses humor to relax her audience and establish a personal connection with them.

4 I am also honoured to receive this award together with Kailash Satyarthi, who has been a champion for children's rights for a long time. Twice as long, in fact, than I have been alive. I am proud that we can work together and show the world that an Indian and a Pakistani can work together and achieve their goals of children's rights.

> She acknowledges her co-recipient, Kailash Satyarthi, 60, the Indian activist who is credited with saving 80,000 children from slave labor. Malala includes a message of peace between their two nations, which are often in conflict.

5 Dear brothers and sisters, I was named after the inspirational Malalai of Maiwand who is the Pashtun Joan of Arc. The word Malala means "grief stricken, sad" but in order to lend some happiness to it, my grandfather would always call me Malala—"The happiest girl in the world"—and today I am very happy that we are together fighting for an important cause.

> She again uses a personal touch to introduce her thesis statement, one that deals with the importance of her cause. She includes everyone in her audience as an active participant in that cause.

6 This award is not just for me. It is for those forgotten children who want education. It is for those frightened children who want peace. It is for those voiceless children who want change.

I am here to stand up for their rights, to raise their voice. It is not time to pity them. It is time to take action so it becomes the last time that we see a child deprived of education.

> She repeats the phrases "It is not time to pity them" and "the last time" for emphasis.

7 I have found that people describe me in many different ways. Some people call me the girl who was shot by the Taliban. And some, the girl who fought for her rights. Some people, call me a "Nobel Laureate" now. However, my brothers still call me that annoying bossy sister. As far as I know, I am just a committed and even stubborn person who wants to see every child getting quality education, who wants to see women having equal rights and who wants peace in every corner of the world.

8 Education is one of the blessings of life—and one of its necessities. That has been my experience during the 17 years of my life. In my paradise home, Swat, I always loved learning and discovering new things. I remember when my friends and I would decorate our hands with henna on special occasions. And instead of drawing flowers and patterns we would paint our hands with mathematical formulas and equations.

> She repeats the phrase "We had a thirst for education."

9 We had a thirst for education because our future was right there in that classroom. We would sit and learn and read together. We loved to wear neat and tidy school uniforms and we would sit there with big dreams in our eyes. We wanted to make our parents proud and prove that we could also excel in our studies and achieve those goals, which some people think only boys can.

> She does not use statistics or expert testimony to prove her points, relying instead on her personal credibility.

10 But things did not remain the same. When I was in Swat, which was a place of tourism and beauty, it suddenly changed into a place of terrorism. I was just ten when more than 400 schools were destroyed. Women were flogged. People were killed. And our beautiful dreams turned into nightmares.

11 Education went from being a right to being a crime. Girls were stopped from going to school. When my world suddenly changed, my priorities changed too. I had two options. One was to remain silent and wait to be killed. And the second was to speak up and then be killed. I chose the second one. I decided to speak up.

12 We could not just stand by and see those injustices of the terrorists denying our rights, ruthlessly killing people and misusing the name of Islam. We decided to raise our voice and tell them: Have you not learnt that in the Holy Quran Allah says: if you kill one person it is as if you kill the whole of humanity? Do you not know that Mohammad, peace be upon him, the prophet of mercy, says, "do not harm yourself or others." And do you not know that the very first word of the Holy Quran is the word "Iqra," which means "read"?

One of her purposes is to show that the Taliban who terrorize her people are not the true interpreters of Islam. She makes that point directly here. For emphasis, she repeats the phrase "have you not learnt."

13 The terrorists tried to stop us and attacked me and my friends who are here today, on our school bus in 2012, but neither their ideas nor their bullets could win. We survived. And since that day, our voices have grown louder and louder. I tell my story, not because it is unique, but because it is not. It is the story of many girls. Today, I tell their stories too. I have brought with me some of my sisters from Pakistan, from Nigeria and from Syria, who share this story. My brave sisters Shazia and Kainat who were also shot that day on our school bus. But they have not stopped learning. And my brave sister Kainat Soomro who went through severe abuse and extreme violence, even her brother was killed, but she did not succumb.

She points to her friends who are in the audience.

14 Also my sisters here, whom I have met during my Malala Fund campaign. My 16-year-old courageous sister, Mezon from Syria, who now lives in Jordan as a refugee and goes from tent to tent encouraging girls and boys to learn. And my sister Amina, from the North of Nigeria, where Boko Haram threatens, and stops girls and even kidnaps girls, just for wanting to go to school.

She points out her guests, who represent the worldwide struggle against terrorism and for human rights, with specific examples of their deeds.

15 Though I appear as one girl, one person, who is 5 foot 2 inches tall, if you include my high heels. (This means I am 5 foot only.) I am not a lone voice. I am many. I am Malala. But I am also Shazia. I am Kainat. I am Kainat Soomro. I am Mezon. I am Amina. I am those 66 million girls who are deprived of education. And today I am not raising my voice, it is the voice of those 66 million girls.

She repeats the phrases, "Though I appear as one girl," and "I am not a lone voice." She jokes about her height, and then follows up with a dramatic statement of purpose, raising her voice for emphasis. Enthusiastic applause follows.

16 Sometimes people like to ask me, Why should girls go to school, why is it important for them? But I think the more important question is, Why shouldn't they, why shouldn't they have this right to go to school?

17 Dear sisters and brothers, today, in half of the world, we see rapid progress and development. However, there are many countries where millions still suffer from the very old problems of war, poverty, and injustice. We still see conflicts in which innocent people lose their lives and children become orphans. We see many people becoming refugees in Syria, Gaza and Iraq. In Afghanistan, we see families being killed in suicide attacks and bomb blasts.

18 Many children in Africa do not have access to education because of poverty. And as I said, we still see girls who have no freedom to go to school in the north of Nigeria. Many children in countries like Pakistan and India, as Kailash Satyarthi mentioned, many children, especially in India and Pakistan, are deprived of their right to education because of social taboos, or they have been forced into child marriage or into child labor.

She refers to the earlier speech by her co-recipient. This reference shows how she is adapting to the circumstances of her occasion.

19 One of my very good school friends, the same age as me, who had always been a bold and confident girl, dreamed of becoming a doctor. But her dream remained a dream. At the age of 12, she was forced to get married. And then soon she had a son, she had a child when she herself was still a child—only 14. I know that she could have been a very good doctor. But she couldn't . . . because she was a girl. Her story is why I dedicate the Nobel Peace Prize money to the Malala Fund, to

She reveals here that she will donate her half of the Nobel Prize money (around $500,000) to the foundation that supports her cause. Applause follows, and she calmly waits for it to die down so her next remarks can be heard.

help give girls quality education, everywhere, anywhere in the world and to raise their voices. The first place this funding will go to is where my heart is, to build schools in Pakistan—especially in my home of Swat and Shangla.

20 In my own village, there is still no secondary school for girls. And it is my wish and my commitment, and now my challenge, to build one so that my friends and my sisters can go there to school and get quality education and to get this opportunity to fulfill their dreams. This is where I will begin, but it is not where I will stop. I will continue this fight until I see every child in school.

She repeats "every child."

21 Dear brothers and sisters, great people, who brought change, like Martin Luther King and Nelson Mandela, Mother Teresa and Aung San Suu Kyi, once stood here on this stage. I hope the steps that Kailash Satyarthi and I have taken so far and will take on this journey will also bring change—lasting change. [Applause]

She recognizes former Nobelists, including Aung San Suu Kyi, the Burmese activist who had spent more than 15 of the previous 21 years under house arrest in Myanmar. Applause follows.

22 My great hope is that this will be the last time we must fight for education. Let's solve this once and for all. We have already taken many steps. Now it is time to take a leap. It is not time to tell world leaders to realize how important education is—they already know it—their own children are in good schools. Now it is time to call them to take action for the rest of the world's children. We ask the world leaders to unite and make education their top priority.

23 Fifteen years ago, world leaders decided on a set of global goals, the Millennium Development Goals. In the years that have followed, we have seen some progress. The number of children out of school has been halved, as Kailash Satyarthi said. However, the world focused only on primary education, and progress did not reach everyone. In year 2015, representatives from all around the world will meet in the United Nations to set the next set of goals, the Sustainable Development Goals. This will set the world's ambition for the next generations.

She refers again to the earlier speech.

24 The world can no longer accept that basic education is enough. Why do leaders accept that for children in developing countries, only basic literacy is sufficient, when their own children do homework in Algebra, Mathematics, Science and Physics? Leaders must seize this opportunity to guarantee a free, quality, primary *and* secondary education for every child. Some will say this is impractical, or too expensive, or too hard. Or maybe even impossible. But it is time the world thinks bigger.

She repeats the phrase, "The world can no longer accept." The audience applauds here.

25 Dear sisters and brothers, the so-called world of adults may understand it, but we children don't. Why is it that countries which we call "strong" are so powerful in creating wars but are so weak in bringing peace? Why is it that giving guns is so easy but giving books is so hard? Why is it that making tanks is so easy, but building schools is so hard? We are living in the modern age and we believe that nothing is impossible. We have reached the moon 45 years ago and maybe we'll soon land on Mars. Then, in this 21st century, we must be able to give every child quality education.

She repeats the phrase "Why is it." More applause here.

26 Dear sisters and brothers, dear fellow children, we must work . . . not wait. Not just the politicians and the world leaders, we all need to contribute. Me. You. We. It is our duty. Let us become the first generation to decide to be the last that sees empty classrooms, lost childhoods, and wasted potentials. Let this be the last time that a girl or a boy spends their childhood in a factory. Let this be the last time that a girl is forced into early child marriage. Let this be the last time that a child loses life in war. Let this be the last time that we see a child out of school. Let this end with us. Let's begin this ending . . . together . . . today . . . right here, right now. Let's begin this ending now.

She repeats the phrase "Let us become the first generation to decide to be the last," and ends her conclusion with a powerful exhortation.

27 Thank you so much.

MAKING THE GRADE

For more resources to help you understand and apply the information in this chapter, visit the *Understanding Human Communication* website at www.oup.com/us/adleruhc.

OBJECTIVE 11.1 **Describe the importance of topic, purpose, and thesis in effective speech preparation.**

- Choose a topic that is right for you, your audience, and the situation.
- Formulating a clear purpose statement serves to keep you focused while preparing your speech.
- A straightforward thesis statement helps the audience understand your intent.
  - > Why is a carefully worded purpose statement essential for speech success?
  - > Why do you believe your topic has the potential to be effective for you as a speaker?
  - > How would you word your thesis statement for your next speech?

OBJECTIVE 11.2 **Analyze both the audience and occasion in any speaking situation.**

- When analyzing your audience, you should consider the audience's purpose, demographics, attitudes, beliefs, and values.
- When analyzing the occasion, you should consider the time (and date) when your speech will take place, the location, and audience expectations given the occasion.
  - > What are the most important aspects of analyzing your audience?
  - > What are the most important aspects of analyzing the occasion for your next speech?
  - > How will you adapt your next speech to both your audience and occasion?

OBJECTIVE 11.3 **Gather information on your chosen topic from a variety of sources.**

- When researching a speech, students usually think first of online searches.
- Also consider searching the collections or databases at a library, interviewing, making personal observations, and conducting survey research.
  - > Why are multiple forms of research important in speech preparation?

- > How do you analyze the reliability of information you find online?
- > What is the most important piece of research you need for your next speech?

OBJECTIVE 11.4 **Assess and manage debilitative speaking apprehension.**

- Sources of debilitative communication apprehension often include irrational thinking, such as a belief in one or more fallacies.
- To help overcome communication apprehension, remember that nervousness is natural, and use it to your advantage.
- Other methods of overcoming communication apprehension involve being rational, receiver oriented, positive, and prepared.
  - > What is the relationship between self-talk and communication apprehension?
  - > Which types of self-defeating thoughts create the greatest challenge for you?
  - > What forms of rational self-talk can you use to overcome self-defeating thoughts?

OBJECTIVE 11.5 **Make effective choices in the delivery of your speech.**

- Choose an effective type of delivery, or combinations of extemporaneous, impromptu, manuscript, and memorized speeches.
- Practice your speech thoroughly.
- Consider both visual and auditory aspects of delivery.
  - > What are the primary differences among the main types of speeches?
  - > In your last speech, why did you make the delivery choices that you did? Were they as effective as they could have been?
  - > What visual and auditory aspects of delivery will be most important in the next speech you present.

KEY TERMS

addition p. 314

articulation p. 314

attitude p. 302

audience analysis p. 301

belief p. 302

ACTIVITIES

1. **Formulating Purpose Statements** Write a specific purpose statement for each of the following speeches:

   a. An after-dinner speech at an awards banquet in which you will honor a team who has a winning, but not championship, record. (You pick the team. For example: "After listening to my speech, my audience members will appreciate the individual sacrifices made by the members of the chess team.")

   b. A classroom speech in which you explain how to do something. (Again, you choose the topic: "After listening to my speech, my audience members will know at least three ways to maximize their comfort and convenience on an economy class flight.")

   c. A campaign speech in which you support the candidate of your choice. (For example: "After listening to my speech, my audience members will consider voting for Alexandra Rodman in order to clean up student government.")

   Answer the following questions about each of the purpose statements you make up: Is it result oriented? Is it precise? Is it attainable?

2. **Formulating Thesis Statements** Turn each of the following purpose statements into a statement that expresses a possible thesis. For example, if you had a purpose statement such as this:

   > After listening to my speech, my audience will recognize the primary advantages and disadvantages of home teeth bleaching.

   you might turn it into a thesis statement such as this:

   > Home bleaching your teeth can significantly improve your appearance, but watch out for injury to the gums and damaged teeth.

   a. At the end of my speech, the audience members will be willing to sign my petition supporting the local needle exchange program for drug addicts.

   b. After listening to my speech, the audience members will be able to list five disadvantages of tattoos.

   c. During my speech on the trials and tribulations of writing a research paper, the audience members will show their interest by paying attention and their amusement by occasionally laughing.

3. **Communication Apprehension: A Personal Analysis** To analyze your own reaction to communication apprehension, think back to your last public speech, and rate yourself on how rational, receiver oriented, positive, and prepared you were. How did these attributes affect your anxiety level?

Organization and Support

12

LEARNING OBJECTIVES

12.1

Describe different types of speech outlines and their functions.

12.2

Construct an effective speech outline using the organizing principles described in this chapter.

12.3

Choose an appropriate organizational pattern for your speech.

12.4

Develop a compelling introduction and conclusion, and use transitions at key points in your speech.

12.5

Choose supporting material that will help make your points clear, interesting, memorable, and convincing.

FOR SOME PEOPLE, competing on a speech team can be a truly transformative experience. Just ask Andrew Neylon.

Neylon didn't have an easy time as a child. He grew up in an abusive household, and because his parents moved around a lot, he was constantly "the new kid" at school. In addition, he was visually impaired and color blind.

But after one of his high school teachers asked him to write a speech about his color blindness, he ended up joining the school's speech team. He continued competing at Ball State University, where he eventually reached 11 national finals and won 4 national championships. In his senior year of college, he was named overall best speaker in the country by the National Forensics Association.

With regard to the impact of public speaking, Neylon says, "In my life, and in my speech career, I didn't necessarily have a lot of things handed to me. I'm grateful that speech remains a place where individuals can craft their own legacies and find fulfillment, growth, and acknowledgment where other communities often leave those same people voiceless."[1]

His two biggest lessons as a public speaker have been to find strong supporting material and to organize that information in a way that makes sense to a specific audience. As you'll see in the sample speech at the end of this chapter, Neylon gives special attention to both of those factors while preparing a speech.

As Andrew Neylon learned in his study of public speaking, *knowing* what you are talking about and *communicating* that knowledge aren't the same thing. It's frustrating to realize you aren't expressing your thoughts clearly, and it's equally unpleasant to be unable to follow what a speaker is saying because the material is too jumbled. In the following pages, you will learn methods of organizing and supporting your thoughts effectively.

Structuring Your Speech

As discussed in Chapter 2, people tend to arrange their perceptions in some meaningful way in order to make sense of the world. Being clear to your audience, however, isn't the only benefit of good organization; structuring a message effectively will help you refine your own ideas and construct more persuasive messages.

A good speech is like a good building: Both grow from a careful plan. Chapter 11 showed you how to begin this planning by formulating a purpose, analyzing your audience, and conducting research. You apply that information to the structure of the speech through outlining. Like any other plan, a speech outline is the framework on which your message is built. It contains your main ideas and shows how they relate to one another and to your thesis. Virtually every speech outline ought to follow the basic structure outlined in Figure 12-1.

This **basic speech structure** demonstrates the old aphorism for speakers: "Tell what you're going to say, say it, and then tell what you said." Although this structure sounds redundant, the research on listening cited in Chapter 5 demonstrates that receivers forget much of what they hear. The clear, repetitive nature of the basic speech structure reduces the potential for memory loss, because audiences have a tendency to listen more carefully during the beginning and ending of a speech.[2] Your outline will reflect this basic speech structure.

Outlines come in all shapes and sizes, but the three types that are most important to us here are working outlines, formal outlines, and speaking notes.

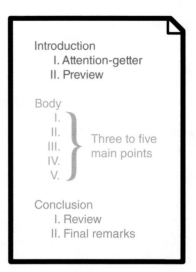

FIGURE 12-1 Basic Speech Structure

Your Working Outline

A **working outline** is a construction tool used to map out your speech. The working outline will probably follow the basic speech structure, but only in rough form. It is for your eyes only, and you'll probably create several drafts as you refine your ideas. As your ideas solidify, your outline will change accordingly, becoming more polished as you go along.

Your Formal Outline

A **formal outline**, such as the ones shown on pages 326 and 342, uses a consistent format and set of symbols to identify the structure of ideas.

A formal outline serves several purposes. In simplified form, it can be displayed as a visual aid or distributed as a handout. It can also serve as a record of a speech that was delivered; many organizations send outlines to members who miss meetings at which presentations were given. Finally, in speech classes, instructors often use speech outlines to analyze student speeches. When one is used for that purpose, it is usually a full-sentence outline and includes the purpose, thesis, and topic or title. Most instructors also require a bibliography of sources at the end of the outline. The bibliography should include full research citations, the correct form for which can be found in any style guide, such as *The Craft of Research*, by Wayne Booth et al.[3] There are at least six standard bibliographic styles. Whichever style you use, you should be consistent in form and remember the two primary functions of a bibliographic citation: to demonstrate the credibility of your source and to enable the readers—in this case, your professor or your fellow students—to find the source if they want to check its accuracy or explore your topic in more detail.

Another person should be able to understand the basic ideas included in your speech by reading the formal outline. In fact, that's one test of the effectiveness of your outline. See if the outline on pages 341–342 passes this test for you.

Your Speaking Notes

Like your working outline, your speaking notes are for your use only, so the format is up to you. Many teachers suggest that speaking notes should be in the form of a brief keyword outline, with just enough information listed to jog your memory but not enough to get lost in.

basic speech structure The division of a speech into introduction, body, and conclusion.

working outline A constantly changing organizational aid used in planning a speech.

formal outline A consistent format and set of symbols used to identify the structure of ideas.

Speech Outline

The following is a basic outline for the sample speech at the end of this chapter. It shows how ideas are divided and arranged with standard symbols. An expanded outline and bibliography for this speech can be found on pages 341–342.

The Amazing Eyeborg

INTRODUCTION

I. Attention-getter

II. Preview main points

BODY

I. The eyeborg opens up a new world for the color blind.
   A. The eye functions via rods and cones.
   B. The eyeborg functions by changing color to sound.

II. The eyeborg has amazing applications.
   A. The eyeborg has amazing applications for the color blind.
   B. The eyeborg has amazing applications for all human beings.

III. The eyeborg also has amazing cultural implications.
   A. The eyeborg augments a physical sense.
   B. The eyeborg suggests that other physical senses can be augmented.
   C. The eyeborg allows humans to become cyborgs.

Conclusion

I. Review

II. Final Remarks

Many teachers suggest that you fit your notes on one side of one 3- by 5-inch note card. Other teachers recommend that you also have your introduction and conclusion on note cards, and still others recommend that your longer quotations be written out on note cards. Andrew Neylon's notes for his speech on color blindness (see the sample outlines, above and on pages 341–342) might look like the ones in Figure 12-2.

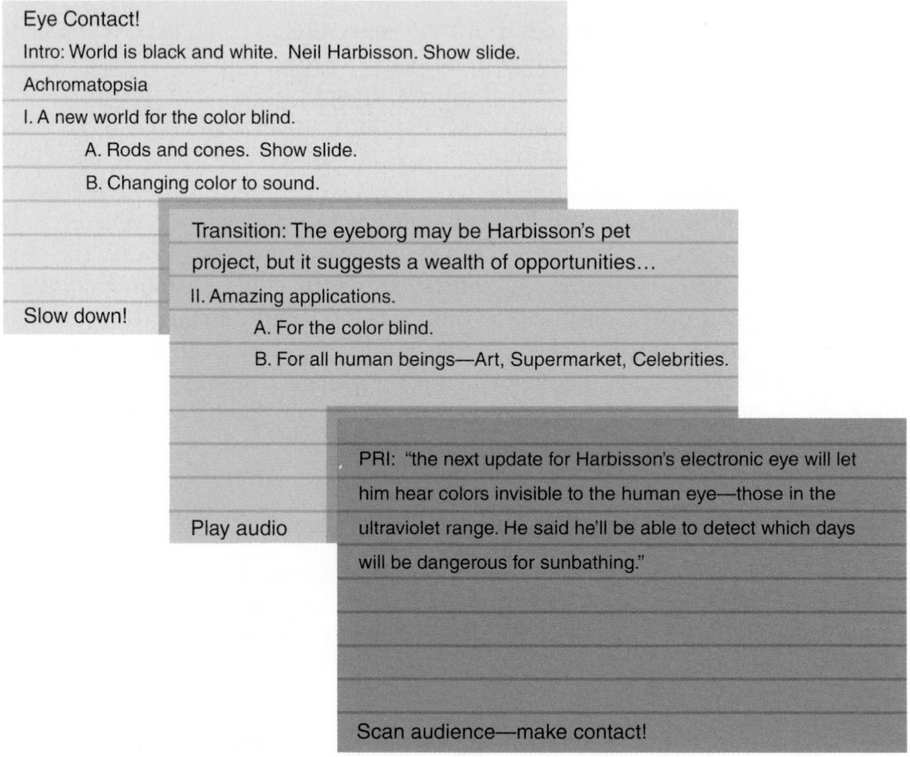

> Eye Contact!
> Intro: World is black and white. Neil Harbisson. Show slide.
> Achromatopsia
> I. A new world for the color blind.
>   A. Rods and cones. Show slide.
>   B. Changing color to sound.
>
> Slow down!

> Transition: The eyeborg may be Harbisson's pet project, but it suggests a wealth of opportunities…
> II. Amazing applications.
>   A. For the color blind.
>   B. For all human beings—Art, Supermarket, Celebrities.
>
> Play audio

> PRI: "the next update for Harbisson's electronic eye will let him hear colors invisible to the human eye—those in the ultraviolet range. He said he'll be able to detect which days will be dangerous for sunbathing."
>
> Scan audience—make contact!

FIGURE 12-2 Speaking Notes These note cards are based on the outlines above and on pages 341–342. Speaking notes are unique to the speaker; yours could be completely different for the same speech.

Principles of Outlining

Over the years, a series of rules or principles for the construction of outlines has evolved. These rules are based on the use of the standard symbols and format discussed next.

Standard Symbols

A speech outline generally uses the following symbols:

I. Main point (roman numeral)
   A. Subpoint (capital letter)
      1. Sub-subpoint (standard number)
         a. Sub-subsubpoint (lowercase letter)

In the examples in this chapter, the major divisions of the speech—introduction, body, and conclusion—are not given in symbols. They are listed by name, and the roman numerals for their main points begin anew in each division. An alternative form is to list these major divisions with roman numerals, main points with capital letters, and so on.

Standard Format

In the sample outlines in this chapter, notice that each symbol is indented a number of spaces from the symbol above it. Besides keeping the outline neat, the indentation of different-order ideas is actually the key to the technique of outlining; it enables you to coordinate and order ideas in the form in which they are most comprehensible to the human mind. If the standard format is used in your working outline, it will help you create a well-organized speech. If it is used in speaking notes, it will help you remember everything you want to say.

Proper outline form is based on a few rules and guidelines, the first of which is the rule of division.

The Rule of Division

In formal outlines, main points and subpoints always represent a division of a whole. Because it is impossible to divide something into fewer than two parts, you always have at least two main points for every topic. Then, if your main points are divided, you will always have at least two subpoints, and so on. Thus, the rule for formal outlines is as follows: Never a "I" without a "II," never an "A" without a "B," and so on.

Three to five is considered to be the ideal number of main points. It is also considered best to divide those main points into three to five subpoints, when necessary and possible. Notice how Andrew Neylon divided the body of his topic as shown in the sample outline on page 326.

The Rule of Parallel Wording

Your main points should be worded in a similar, or "parallel," manner. For example, if you are developing a speech against capital punishment, your main points might look like this:

I. Capital punishment is not effective: It is not a deterrent to crime.
II. Capital punishment is not constitutional: It does not comply with the Eighth Amendment.
III. Capital punishment is not civilized: It does not allow for a reverence for life.

Whenever possible, subpoints should also be worded in a parallel manner. For your points to be worded in a parallel manner, they should each contain one, and only one, idea. (After all, they can't really be parallel if one is longer or contains more

Main Points and Subpoints

To get an idea of your ability to distinguish main points from subpoints, set the "timer" function on your mobile phone and see how long it takes you to fit the following concepts for a speech entitled "The College Application Process" into outline form:

CONCEPTS	RECOMMENDED OUTLINE FORM
Participation in extracurricular activities	I.
Visit and evaluate college websites	A.
Prepare application materials	B.
Career ambitions	II.
Choose desired college	A.
Letters of recommendation	B.
Write personal statement	C.
Visit and evaluate college campuses	III.
Choose interesting topic	A.
Test scores	B.
Include important personal details	1.
Volunteer work	2.
Transcripts	3.

You can score yourself as follows:

A minute or less: Congratulations, organization comes naturally to you.

61–90 seconds: You have typical skills in this area.

More than 90 seconds: Give yourself extra time while building your speech outline.

ideas than the others.) This will enable you to completely develop one idea before moving on to another one in your speech. If you were discussing cures for indigestion, your topic might be divided incorrectly if your main points looked like this:

I. "Preventive cures" help you before eating.
II. "Participation cures" help you during and after eating.

You might actually have three ideas there and thus three main points:

I. Prevention cures (before eating)
II. Participation cures (during eating)
III. Postparticipation cures (after eating)

Organizing Your Outline into a Logical Pattern

An outline should reflect a logical order for your points. You might arrange them from newest to oldest, largest to smallest, best to worst, or in a number of other ways that follow. The organizing pattern you choose ought to be the one that best develops your thesis.

Time Patterns

Arrangement according to **time patterns**, or chronology, is one of the most common patterns of organization. The period of time could be anything from centuries to seconds. In a speech on airline food, a time pattern might look like this:

I. Early airline food: a gourmet treat
II. The middle period: institutional food at 30,000 feet
III. Today's airline food: the passenger starves

Arranging points according to the steps that make up a process is another form of time patterning. The topic "Recording a Hit Song" might use this type of patterning:

I. Record the demo.
II. Post a YouTube video.
III. Get a recording company to listen and view.

Time patterns are also the basis of **climax patterns**, which are used to create suspense. For example, if you wanted to create suspense in a speech about military intervention, you could chronologically trace the steps that eventually led us into Afghanistan or Iraq in such a way that you build up your audience's curiosity. If you told of these steps through the eyes of a soldier who entered military service right before one of those wars, you would be building suspense as your audience wonders what will become of that soldier.

The climax pattern can also be reversed. When it is, it is called *anticlimactic* organization. If you started your military intervention speech by telling the audience that you were going to explain why a specific soldier was killed in a specific war, and then you went on to explain the things that caused that soldier to become involved in that war, you would be using anticlimactic organization. This pattern is helpful when you have an essentially uninterested audience, and you need to build interest early in your speech to get the audience to listen to the rest of it.

One of the greatest speakers of the 20th century, Dr. Martin Luther King Jr., knew well how to construct a speech and how to conclude with passion.

What best practices can you use to boost the impact when organizing your speeches?

Space Patterns

Speech organization by physical area is known as a **space pattern**. The area could be stated in terms of continents or centimeters or anything in between. If you were discussing the Great Lakes, for example, you could arrange them from west to east:

I. Superior
II. Michigan
III. Huron
IV. Erie
V. Ontario

Topic Patterns

A topical arrangement or **topic pattern** is based on types or categories. These categories could be either well known or original; both have their advantages. For example, a division of college students according to well-known categories might look like this:

I. Freshmen
II. Sophomores
III. Juniors
IV. Seniors

time pattern An organizing plan for a speech based on chronology.

climax pattern An organizing plan for a speech that builds ideas to the point of maximum interest or tension.

space pattern An organizing plan in a speech that arranges points according to their physical location.

topic pattern An organizing plan for a speech that arranges points according to logical types or categories.

Well-known categories are advantageous because audiences quickly understand them. But familiarity also has its disadvantages. One disadvantage is the "Oh, this again" syndrome. If the members of an audience feel they have nothing new to learn about the components of your topic, they might not listen to you. To avoid this, you could invent original categories that freshen up your topic by suggesting an original analysis. For example, original categories for "college students" might look like this:

I. Grinds: Students who go to every class and read every assignment before it is due.
II. Renaissance students: Students who find a satisfying balance of scholarly and social pursuits.
III. Burnouts: Students who have a difficult time finding the classroom, let alone doing the work.

Sometimes topics are arranged in the order that will be easiest for your audience to remember. To return to our Great Lakes example, the names of the lakes could be arranged so their first letters spell the word "HOMES." Words used in this way are known as *mnemonics*. Carol Koehler, a professor of communication and medicine, uses the mnemonic "CARE" to describe the characteristics of a caring doctor:

C stands for *concentrate*. Physicians should pay attention with their eyes and ears . . .

A stands for *acknowledge*. Show them that you are listening . . .

R stands for *response*. Clarify issues by asking questions, providing periodic recaps . . .

E stands for *exercise emotional control*. When your "hot buttons" are pushed . . .[4]

One of the greatest speakers of the 20th century, the Reverend Dr. Martin Luther King Jr., knew well how to construct a speech and how to conclude with passion. What best practices can you use to boost the impact when organizing your speeches?

Problem-Solution Patterns

Describing what's wrong and proposing a way to make things better is known as a **problem-solution pattern**. It is usually (but not always) divisible into two distinct parts, as in this example:

I. The Problem: Addiction (which could then be broken down into addiction to cigarettes, alcohol, prescribed drugs, and street drugs)
II. The Solution: A national addiction institute (which would study the root causes of addiction in the same way that the National Cancer Institute studies the root causes of cancer)

We will discuss this pattern in more detail in Chapter 14.

Cause-Effect Patterns

Cause-effect patterns are similar to problem-solution patterns in that they are basically two-part patterns: First you discuss something that happened, and then you discuss its effects.

A variation of this pattern reverses the order and presents the effects first and then the causes. Persuasive speeches often have effect-cause or cause-effect as the first two main points. Elizabeth Hallum, a student at Arizona State University, organized the first two points of a speech on "workplace revenge"[5] like this:

I. The effects of the problem
   A. Lost productivity
   B. Costs of sabotage

problem-solution pattern An organizing pattern for a speech that describes an unsatisfactory state of affairs and then proposes a plan to remedy the problem.

cause-effect pattern An organizing plan for a speech that demonstrates how one or more events result in another event or events.

Nontraditional Patterns of Organization

In addition to the traditional patterns usually taught in public speaking classes, researchers are looking at the use of less linear forms.

One of these is the wave pattern, in which the speaker uses repetitions and variations of themes and ideas. The major points of the speech come at the crest of the wave. The speaker follows these with a variety of examples leading up to another crest, where she repeats the theme or makes another major point.

Perhaps the most famous speech that illustrates this pattern is the Reverend Dr. Martin Luther King Jr.'s "I Have a Dream." King used this memorable line as the crest of a wave that he followed with examples of what he saw in his dream; then he repeated the line. He ended with a "peak" conclusion that emerged from the final wave in the speech—repetition and variation on the phrase "Let freedom ring."

An excerpt from Sojourner Truth's "Ain't I a Woman?" speech also illustrates this pattern:

> That man over there says that women need to be helped into carriages, and lifted over ditches, and to have the best place everywhere. Nobody ever helps me into carriages, or over mud-puddles, or gives me any best place!
>
> And ain't I a woman?
>
> Look at me! Look at my arm! I have ploughed and planted, and gathered into barns, and no man could head me!
>
> And ain't I a woman?
>
> I could work as much and eat as much as a man—when I could get it—and bear the lash as well!
>
> And ain't I a woman?
>
> I have borne thirteen children and seen them most all sold off to slavery, and when I cried out with my mother's grief, none but Jesus heard me!
>
> And ain't I a woman?

Jaffe, C. (2007). *Public speaking: Concepts and skills for a diverse society* (5th ed.). Boston: Wadsworth, © 2007. Reprinted with permission of Wadsworth, an imprint of the Wadsworth Group, a division of Thomson Learning.

Why is diversity important in the messages you send and receive?
Why is message organization such an important factor?

II. The causes of the problem
   A. Employees feeling alienated
   B. Employers' light treatment of incidents of revenge

The third main point in this type of persuasive speech is often "solutions," and the fourth main point is often "the desired audience behavior." Hallum's final points were as follows:

III. Solutions: Support the National Employee Rights Institute.
IV. Desired Audience Response: Log on to www.disgruntled.com.

Cause-effect and problem-solution patterns are often combined in various arrangements. One extension of the problem-solution organizational pattern is Monroe's Motivated Sequence.

Monroe's Motivated Sequence

The Motivated Sequence was proposed by a scholar named Alan Monroe in the 1930s.[6] In this persuasive pattern, the problem is broken down into an attention step and a need step, and the solution is broken down into a satisfaction step, a visualization step, and an action step. In a speech on "random acts of kindness,"[7] the Motivated Sequence might break down like this:

I. The attention step draws attention to your subject. ("Just the other day Ron saved George's life with a small, random, seemingly unimportant act of kindness.")

ASK YOURSELF

Think of an idea that you could present to your class. Which pattern of organization would be most effective for that presentation?

II. The need step establishes the problem. ("Millions of Americans suffer from depression, a life-threatening disease.")

III. The satisfaction step proposes a solution. ("One random act of kindness can lift a person from depression.")

IV. The visualization step describes the results of the solution. ("Imagine yourself having that kind of effect on another person.")

V. The action step is a direct appeal for the audience to do something. ("Try a random act of kindness today!")

Chapter 14 has more to say about the organization of persuasive speeches.

Beginnings, Endings, and Transitions

The introduction and conclusion of a speech are vitally important, although they usually will occupy less than 20% of your speaking time. Listeners form their impression of a speaker early, and they remember what they hear last; it is, therefore, vital to make those few moments at the beginning and end of your speech work to your advantage. It is also essential that you connect sections within your speech using effective transitions.

The Introduction

introduction (of a speech) The first structural unit of a speech, in which the speaker captures the audience's attention and previews the main points to be covered.

A speech **introduction** has four functions: to capture the audience's attention, preview the main points, set the mood and tone of the speech, and demonstrate the importance of the topic.

Capturing Attention   There are several ways to capture an audience's attention. The checklist on page 333 summarizes some of them.

Previewing Main Points   After you capture the attention of the audience, an effective introduction will almost always state the speaker's thesis and give the listeners an idea of the upcoming main points. Katharine Graham, the former publisher of the *Washington Post*, addressed a group of businessmen and their wives in this way:

I am delighted to be here. It is a privilege to address you. And I am especially glad the rules have been bent for tonight, allowing so many of you to bring along your husbands. I think it's nice for them to get out once in a while and see how the other half lives. Gentlemen, we welcome you.

Actually, I have other reasons for appreciating this chance to talk with you tonight. It gives me an opportunity to address some current questions about the press and its responsibilities—whom we are responsible to, what we are responsible for, and generally how responsible our performance has been.[8]

Thus, Graham previewed her main points:

1. To explain who the press is responsible to
2. To explain what the press is responsible for
3. To explain how responsible the press has been

Sometimes your preview of main points will be even more straightforward:

In the *Anchorman* comedies, newscaster Ron Burgundy (Will Ferrell) reads the same opening and closing statements every night. Even if you haven't seen the films, it's easy to imagine the effect of his approach.

How can you make your delivery most authentic?

"I have three points to discuss: They are _____ ,
_____ , and _____ .

Sometimes you will not want to refer directly to your main points in your introduction. Your reasons for not doing so might be based on a plan calling for suspense, humorous effect, or stalling for time to win over a hostile audience. In that case, you might preview only your thesis:

"I am going to say a few words about _____ ."

"Did you ever wonder about _____ ?"

"_____ is one of the most important issues facing us today."

Setting the Mood and Tone of Your Speech   Notice, in the example just given, how Katharine Graham began her speech by joking with her audience. She was a powerful woman speaking before an all-male organization; the only women in the audience were the members' wives. That is why Ms. Graham felt it necessary to put her audience members at ease by joking with them about women's traditional role in society. By beginning in this manner, she assured the men that she would not berate them for the sexist bylaws of their organization. She also showed them that she was going to approach her topic with wit and intelligence. Thus, she set the mood and tone for her entire speech. Imagine how different that mood and tone would have been if she had begun this way:

> Before I start today, I would just like to say that I would never have accepted your invitation to speak here had I known that your organization does not accept women as members. Just where do you Cro-Magnons get off, excluding more than half the human race from your little club?

Demonstrating the Importance of Your Topic to Your Audience
Your audience members will listen to you more carefully if your speech relates to them as individuals. Based on your audience analysis, you should state directly *why* your topic is of importance to your audience members. This importance should be related as closely as possible to their specific needs at that specific time. For example, Emily Meyer, a student at Gustavus Adolphus College, presented a speech about trucking safety. She established the importance of her topic this way:

> Each weekend and every day, all of us, our families and friends, are confronted by a danger we know nothing about. It happens 11 times a day. Last year, it killed 800 times more people than terrorist attacks in the U.S. Despite all this, people continue to believe this isn't terrifying. NBC News of July 30, 2014 proclaims, "In any other industry, thousands of deaths would generate a national outcry." But we pass by trucking accidents every day without batting an eye.[9]

Establishing Credibility   One final consideration for your introduction is to establish your credibility to speak on your topic. One way to do this is to be *well prepared*. Another is to *appear confident* as soon as you face your audience. A third technique is to *tell your audience about*

CHECKLIST ✅
Capturing Audience Attention

☐ **Refer to the audience.** This technique is especially effective if it is complimentary: "Julio's speech about how animals communicate was so interesting that I decided to explore a related topic."

☐ **Refer to the occasion.** "Given our assignment, it seems appropriate to talk about something you might have wondered."

☐ **Refer to the relationship between the audience and the subject.** "It's fair to say that all of us here believe it's important to care for our environment. What you'll learn today will make you care about that environment in a whole new way."

☐ **Refer to something familiar to the audience.** "Most of us have talked to our pets. Today, you'll learn that there are other conversational partners around the house."

☐ **Cite a startling fact or opinion.** "New scientific evidence suggests that plants appreciate human company, kind words, and classical music."

☐ **Ask a question.** "Have you ever wondered why some people have a green thumb, whereas others couldn't make a weed grow?"

☐ **Tell an anecdote.** "The other day, while taking a walk near campus, I saw a man talking quite animatedly to a sunflower."

☐ **Use a quotation.** "The naturalist Max Thornton recently said, 'Psychobiology has proven that plants can communicate. Now humans need to learn how to listen to them.'"

☐ **Tell an (appropriate) joke.** "We once worried about people who talked to plants, but that's no longer the case. Now we only worry if the plants talk back."

conclusion (of a speech) The final part of a speech, in which the main points are reviewed and final remarks are made to motivate the audience or help listeners remember key ideas.

transition A phrase that connects ideas in a speech by showing how one relates to the other.

CHECKLIST ✓
Effective Conclusions

You can make your final remarks most effective by avoiding the following mistakes:

☐ **Don't end abruptly.** Make sure that your conclusion accomplishes everything it is supposed to accomplish. Develop it fully. You might want to use signposts such as, "Finally . . . ," "In conclusion . . . ," or "To sum up what I've been talking about here . . ." to let your audience know that you have reached the conclusion of the speech.

☐ **Don't ramble, either.** Prepare a definite conclusion, and never, ever end by mumbling something like, "Well, I guess that's about all I wanted to say. . . ."

☐ **Don't introduce new points.** The worst kind of rambling is, "Oh, yes, and something I forgot to mention is. . . ."

☐ **Don't apologize.** Don't make statements such as "I'm sorry I didn't have more time to research this subject." You will only highlight the possible weaknesses of your speech, which may have been far more apparent to you than to your audience. It's best to end strong. You can use any of the attention-getters suggested for the introduction to make the conclusion memorable, or you can revisit your attention-getting introduction.

your personal experience with the topic, in order to establish why it is important to you. Andrew Neylon, in the sample speech found at the end of this chapter, uses all three of these techniques.

The Conclusion

Like the introduction, the **conclusion** is an especially important part of your speech. It has three essential functions: to restate the thesis, to review your main points, and to provide a memorable final remark.

You can review your thesis either by repeating it or by paraphrasing it. Or you might devise a striking summary statement for your conclusion to help your audience remember your thesis. Grant Anderson, a student at Minnesota State University, gave a speech against the policy of rejecting blood donations from homosexuals. He ended his conclusion with this statement: "The gay community still has a whole host of issues to contend with, but together all of us can all take a step forward by recognizing this unjust and discriminatory measure. So stand up and raise whatever arm they poke you in to draw blood and say 'Blood is Blood' no matter who you are."[10] Grant's statement was concise but memorable.

Your main points can also be reviewed artistically. For example, first look back at that example introduction by Katharine Graham, and then read her conclusion to that speech:

> So instead of seeking flat and absolute answers to the kinds of problems I have discussed tonight, what we should be trying to foster is respect for one another's conception of where duty lies, and understanding of the real worlds in which we try to do our best. And we should be hoping for the energy and sense to keep on arguing and questioning, because there is no better sign that our society is healthy and strong.

Let's take a closer look at how and why this conclusion was effective. Graham posed three questions in her introduction. She dealt with those questions in her speech and reminded her audience, in her conclusion, that she had answered the questions.

PREVIEW (FROM INTRODUCTION OF SPEECH)	REVIEW (FROM CONCLUSION)
1. To whom is the press responsible?	1. To its own conception of where its duty lies
2. What is the press responsible for?	2. For doing its best in the "real world"
3. How responsible has the press been?	3. It has done its best

Transitions

Transitions keep your message moving forward. They perform the following functions:

> They tell how the introduction relates to the body of the speech.
> They tell how one main point relates to the next main point.
> They tell how your subpoints relate to the points they are part of.
> They tell how your supporting points relate to the points they support.

To be effective, transitions should refer to the previous point and to the upcoming point, showing how they relate to each other and to the thesis. They usually sound something like this:

"Like [previous point], another important consideration in [topic] is [upcoming point]."

"But _____ isn't the only thing we have to worry about. _____ is even more potentially dangerous."

"Yes, the problem is obvious. But what are the solutions? Well, one possible solution is . . ."

Sometimes a transition includes an internal review (a restatement of preceding points), an internal preview (a look ahead to upcoming points), or both:

"So far we've discussed _____ , _____ , and _____ . Our next points are _____ , _____ , and _____ ."

It isn't always necessary to provide a transition between every set of points. You have to choose when one is necessary for your given audience to follow the progression of your ideas. You can find several examples of transitions in the sample speech at the end of this chapter.

Supporting Material

It is important to organize ideas clearly and logically. But clarity and logic by themselves won't guarantee that you'll interest, enlighten, or persuade others; these results call for the use of supporting materials. These materials—the facts and information that back up and prove your ideas and opinions—are the flesh that fills out the skeleton of your speech.

Functions of Supporting Material

There are four functions of supporting material.

To Clarify   As explained in Chapter 4, people of different backgrounds tend to attach different meanings to words. Supporting material can help you overcome this potential source of confusion by helping you clarify key terms and ideas. For example, when Jacoby Cochran, a student at Bradley University in Illinois, spoke on the dangers of "special administrative measures," or SAMs, he needed to clarify what he meant by his key term. He used supporting material in this way:

> AlterNet explains that SAMs are measures including limited communication, solitary confinement, the withholding of evidence and enhanced interrogation, enacted against individuals convicted or suspected of mob or terrorist ties. In the most extreme circumstances, such as keeping a known mob boss from running an organization from inside prison, SAMs have proven beneficial, but their recent expansions have elevated them from a rarely used, extreme holding measure to a legal basis for torture, detentions without charge or trial, and the shredding of constitutional rights.[11]

To Prove   A second function of support is to be used as evidence, to prove the truth of what you are saying. If you were giving a speech on what is known as the immigration crisis, you might want to point out that concerns about immigration are nothing new. The following could be used to prove that point:

> A prominent American once said about immigrants, "Few of their children in the country learn English. . . . The signs in our streets have inscriptions in

Organizing Business Presentations

When top business executives plan an important speech, they often call in a communication consultant to help organize their remarks. Even though they are experts, executives are so close to the topic of their message that they may have difficulty arranging their ideas so others will understand or be motivated by them.

Consultants stress how important organization and message structure are in giving presentations. Seminar leader and corporate trainer T. Stephen Eggleston sums up the basic approach: "Any presentation . . . regardless of complexity . . . should consist of the same four basic parts: an opening, body, summary and closing."[12]

Ethel Cook, a Massachusetts consultant, is very specific about how much time should be spent on each section of a speech. "In timing your presentation," she says, "an ideal breakdown would be:

Opening—10 to 20 percent
Body—65 to 75 percent
Closing—10 to 20 percent."[13]

Business coach Vadim Kotelnikov gives his clients a step-by-step procedure to organize their ideas within the body

of a presentation. "List all the points you plan to cover," he advises. "Group them in sections and put your list of sections in the order that best achieves your objectives. Begin with the most important topics."[14]

Toastmasters International, an organization that runs training programs for business professionals, suggests alternative organizational patterns:

> To organize your ideas into an effective proposal, use an approach developed in the field of journalism—the "inverted pyramid." In the "inverted pyramid" format, the most important information is given in the first few paragraphs. As you present the pitch, the information becomes less and less crucial. This way, your presentation can be cut short, yet remain effective.[15]

While each consultant may offer specific tips, all agree that clear organization is essential when a business speaker wants his or her ideas to be understood and appreciated.

Imagine a business presentation you might have to make in your future career. Why would organization be important in such a presentation? See www.oup.com/us/adleruhc for more on work-related communication.

both languages. . . . Unless the stream of their importation could be turned they will soon so outnumber us that all the advantages we have will not be able to preserve our language, and even our government will become precarious." This sentiment did not emerge from the rancorous debate over the immigration bill defeated not long ago in the Senate. It was not the lament of some . . . candidate intent on wooing bedrock conservative votes. Guess again. Voicing this grievance was Benjamin Franklin. And the language so vexing to him was the German spoken by new arrivals to Pennsylvania in the 1750s, a wave of immigrants whom Franklin viewed as the "most stupid of their nation."[16]

To Make Interesting A third function of support is to make an idea interesting or to catch your audience's attention. For example, when Nathan Dunn, a student at Oklahoma City College, spoke about how special education students are sometimes treated, he started with this concrete example:

> For most high school seniors, their biggest problems revolve around getting into college, getting a job, or finding a date to the prom. Unfortunately, Andre McCollins of Canton, Massachusetts, is not most high school seniors. In October 2002, Andre was an 18-year-old student at the Judge Rotenberg Educational Center, a residential school for students with developmental disabilities less than 20 miles from where we are right now. One morning, Andre did not

respond to a staff member asking him to take off his coat. In response, staff members tied him to a board face down, for 7 hours, with no breaks for food, water, or bathroom use. More horrifying, though, was that staff members electrically shocked Andre 31 times while he was tied down. The school's classroom cameras captured the whole day on a video the school fought to keep secret for years.[17]

To Make Memorable   A final function of supporting materials, related to the preceding one, is to make a point memorable. We have already mentioned the importance of "memorable" statements in a speech conclusion; use of supporting material in the introduction and body of the speech provides another way to help your audience retain important information. When Chris Griesinger of Eastern Michigan University spoke about the importance of pain management, he wanted his audience to remember the severity of the problem, so he used the following as supporting material:

> Every year the National Committee on Treatment of Intractable Pain receives letters from people sharing their stories of loved ones who died in pain. One letter reads, "I lost my mother to cancer. Her pain was so horrid that she lost her mind and ate her bottom lip completely off from clenching her top teeth so tightly. My 13-year-old sister and I watched this for 6 weeks."[18]

Types of Supporting Material

As you may have noted, each function of support could be fulfilled by several different types of material. Let's take a look at these different types of supporting material.

Definitions   It's a good idea to give your audience members definitions of your key terms, especially if those terms are unfamiliar to them or are being used in an unusual way. A good definition is simple and concise. When Erin Schoch, a student at the University of Florida, gave a speech on revenge porn, she needed to define two key terms, *sexting* and *revenge porn*:

> Advancing technology has millions turning to "sexting," or sending provocative images or text messages through a digital means. A study in the journal *Cyberpsychology* reports that 46% of college students have sent a "sext" with a picture. After one's relationship ends, these pictures can turn into "revenge porn," which *USA Today* defines as the posting of someone's sexually explicit images without their consent.[19]

Political satirists like Samantha Bee use extensive supporting material when they poke fun at political and cultural figures. Their quotes, examples, and video clips make their points more interesting, memorable, and persuasive.

How can you enhance the effectiveness of your remarks with compelling supporting material?

factual example A true, specific case that is used to demonstrate a general idea.

hypothetical example An example that asks an audience to imagine an object or event.

statistic Numbers arranged or organized to show how a fact or principle is true for a large percentage of cases.

Examples An example is a specific case that is used to demonstrate a general idea. Examples can be either factual or hypothetical, personal or borrowed. In Erin Schoch's speech, she used the following **factual example** (i.e., a true, specific case to support an argument):

> When Holly Jacobs was 23, she was in a long-distance relationship with Ryan. She loved and trusted him, and at one point during the three years they were together, she sent him nude pictures. After the couple mutually split, Jacobs began dating someone new and posted a picture with her new boyfriend on Facebook. Within hours, strangers emailed saying her nude pictures were on "revenge porn" websites. Jacobs' naked photos were published on over 200 websites with her full name, personal email address, the university she attended, and even a link to where she worked. Horrified at this breach of privacy, Jacobs went to the Miami police. They refused to take any action, however, stating that "because you are over 18 and you consented, technically the photos are his property and he can do whatever he wants with them."[20]

Hypothetical examples, which ask audience members to imagine something, can often be more powerful than factual examples because they create active participation in the thought. Stephanie Wideman of the University of West Florida used a hypothetical example to start off her speech on oil prices:

> The year is 2025. One day you are asked not to come into work, not because of a holiday, but instead because there is not enough energy available to power your office. You see, it is not that the power is out, but that they are out of power.[21]

Statistics Data organized to show that a fact or principle is true for a large percentage of cases are known as **statistics**. These are actually collections of examples, which is why they are often more effective as proof than are isolated examples. Here's the way a newspaper columnist used statistics to demonstrate a point about gun violence:

> I had coffee the other day with Marian Wright Edelman, president of the Children's Defense Fund, and she mentioned that since the murders of Robert Kennedy and the Rev. Martin Luther King Jr. in 1968, well over a million Americans have been killed by firearms in the United States. That's more than the combined U.S. combat deaths in all the wars in all of American history. "We're losing eight children and teenagers a day to gun violence," she said.[22]

Because statistics can be powerful support, you have to follow certain rules when using them. You should make sure that the statistics make sense and that they come from a credible source. You should also cite the source of the statistic when you use it. A final rule is based on effectiveness rather than ethics. You should reduce the statistic to a concrete image if possible. For example, $1 billion in $100 bills. Using concrete images such as this will make your statistics more than "just numbers" when you use them. For example, one observer expressed the idea of Bill Gates's wealth this way:

> Examine Bill Gates' wealth compared to yours: Consider the average American of reasonable but modest wealth. Perhaps he has a net worth of $100,000. Mr. Gates' worth is 400,000 times larger. Which means that if something costs $100,000 to him, to Bill it's as though it costs 25 cents. So for example, you might think a new Lamborghini Diablo would cost $250,000, but in Bill Gates dollars that's 63 cents.[23]

? ASK YOURSELF

Recall a recent occasion in which you tried to change someone's mind. What were your arguments? Which forms of support did you use to back them up?

Analogies/Comparison-Contrast   We use **analogies**, or comparisons, all the time, often in the form of figures of speech, such as similes and metaphors. A simile is a direct comparison that usually uses *like* or *as*, whereas a metaphor is an implied comparison that does not use *like* or *as*. So if you said that the rush of refugees from a war-torn country was "like a tidal wave," you would be using a simile. If you used the expression "a tidal wave of refugees," you would be using a metaphor.

Analogies are extended metaphors. They can be used to compare or contrast an unknown concept with a known one. For example, here's how one writer made her point against separate Academy Awards for men and women:

> Many hours into the Academy Awards ceremony this Sunday, the Oscar for best actor will go to Morgan Freeman, Jeff Bridges, George Clooney, Colin Firth, or Jeremy Renner. Suppose, however, that the Academy of Motion Picture Arts and Sciences presented separate honors for best white actor and best non-white actor, and that Mr. Freeman was prohibited from competing against the likes of Mr. Clooney and Mr. Bridges. Surely, the Academy would be derided as intolerant and out of touch; public outcry would swiftly ensure that Oscar nominations never again fell along racial lines.
>
> Why, then, is it considered acceptable to segregate nominations by sex, offering different Oscars for best actor and best actress?[24]

Anecdotes   An **anecdote** is a brief story with a point, often (but not always) based on personal experience. (The word *anecdote* comes from the Greek, meaning "unpublished item.") Alyssa Gieseck, a student at the University of Akron, used the following anecdote in her speech about the problems some Deaf people encounter with police:

> Jonathan Meister, a deaf man, was retrieving his personal belongings from a friend's home when police arrived, responding to a report of suspicious behavior. Trying to sign to the police officers, Meister was seen as a threat, which is when the officers decided to handcuff him. Handcuffing a deaf person is equivalent to putting duct tape over a hearing person's mouth. Meister initially pulled away to sign that he was deaf, but one police officer pushed him up against a fence, kneed him twice in the abdomen, put him in a choke hold, and tasered him. A second officer repeatedly punched him in the face, tasering him again. But this wasn't the end of the assault. He was then shoved to the ground, kicked, elbowed, tasered for a third time, and put in a second chokehold, which left him unconscious.[25]

Quotations/Testimonies   Using a familiar, artistically stated quotation will enable you to take advantage of someone else's memorable wording. For example, if you were giving a speech on personal integrity, you might quote Mark Twain, who said, "Always do right. This will gratify some people, and astonish the rest." A quotation like that fits Alexander Pope's definition of "true wit": "What was often thought, but ne'er so well expressed."

You can also use quotations as **testimony**, to prove a point by using the support of someone who is more authoritative or experienced on the subject than you are. When Julia Boyle, a student at Northern Illinois University, wanted to prove that spyware stalking was a serious problem, she used testimony this way:

> Michella Cash, advocate for the Women's Service network noted, "Spyware technology is the new form of domestic violence abuse that enables perpetrators to exert round the clock control over their victims."[26]

analogy An extended comparison that can be used as supporting material in a speech.

anecdote A brief, personal story used to illustrate or support a point in a speech.

testimony Supporting material that proves or illustrates a point by citing an authoritative source.

narration The presentation of speech supporting material as a story with a beginning, middle, and end.

Sometimes testimony can be paraphrased. For example, when one business executive was talking on the subject of diversity, he used a conversation he had with Jesse Jackson Sr., an African American leader, as testimony:

> At one point in our conversation, Jesse talked about the stages of advancement toward a society where diversity is fully valued. He said the first stage was emancipation—the end of slavery. The second stage was the right to vote and the third stage was the political power to actively participate in government—to be part of city hall, the Governor's office and Capitol Hill. Jesse was clearly focused, though, on the fourth stage—which he described as the ability to participate fully in the prosperity that this nation enjoys. In other words, economic power.[27]

Styles of Support: Narration and Citation

Most of the forms of support discussed in the preceding section could be presented in either of two ways: through narration or through citation. **Narration** involves telling a story with your information. You put it in the form of a small drama, with a beginning, middle, and end. For example, Evan McCarley of the University of Mississippi narrated the following example in his speech on the importance of drug courts:

> Oakland contractor Josef Corbin has a lot to be proud of. Last year his firm, Corbin Building Inc., posted revenue of over 3 million dollars after funding dozens of urban restoration projects. His company was ranked as one of the 800 fastest-growing companies in the country, all due to what his friends call his motivation for success. Unfortunately, Corbin used this motivation to rob and steal on the streets of San Francisco to support a heroin and cocaine habit.

UNDERSTANDING COMMUNICATION TECHNOLOGY

Plagiarism in a Digital Age

Some experts believe that social media are redefining how students understand the concept of authorship and originality. After all, the Internet is the home of file sharing that allows us to download music, movies, and TV programs without payment. Google and Wikipedia are our main portals to random free information, also. It all seems to belong to us, residing on our computer as it does. Information wants to be free.

According to one expert on the topic, "Now we have a whole generation of students who've grown up with information that just seems to be hanging out there in cyberspace and doesn't seem to have an author. It's possible to believe this information is just out there for anyone to take."[28] Other experts beg to differ. They say students are fully aware of what plagiarism is, online or off, and they know it's cheating. It's just that it's so easy to copy and paste online material, and students like to save time wherever they can.

Public speaking instructors are on the front lines of those fighting plagiarism, because it's so important for successful student speakers to speak from the heart, in their own words and with their own voice. Plus, citing research enhances credibility. Plagiarism in public speaking isn't just cheating, it's ineffective.

The general rule for the digital age is as follows: Thou shalt not cut and paste into a final draft—not for a paper, and not for a speech. Cutting and pasting is fine for research, but everything that's cut and pasted should be placed in a separate "research" file, complete with a full citation for the website you found it in. Then switch to your "draft" file to put everything in your own words, and go back to the research file to find the attribution information when you need to cite facts and ideas that you got from those sources.

Have you ever had a problem with online plagiarism? What was the outcome?

But when he was charged with possession, Josef was given the option to participate in a state drug court, a program targeted at those recently charged with drug use, possession, or distribution. The drug court offers offenders free drug treatment, therapy, employment, education, and weekly meetings with a judge, parole officer and other accused drug offenders.[29]

Citation, unlike narration, is a simple statement of the facts. Citation is shorter and more precise than narration, in the sense that the source is carefully stated. Citation will always include such phrases as, "According to the July 25, 2016, edition of *Time* magazine," or, "As Mr. Smith made clear in an interview last April 24." Evan McCarley cited statistics later in his speech on drug courts:

> Fortunately, Corbin's story, as reported in the May 30th *San Francisco Chronicle*, is not unique, since there are currently over 300 drug courts operating in 21 states, turning first-time and repeat offenders into successful citizens with a 70% success rate.[30]

citation A brief statement of supporting material in a speech.

Some forms of support, such as anecdotes, are inherently more likely to be expressed as narration. Statistics, on the other hand, are nearly always cited rather than narrated. However, when you are using examples, quotation/testimony, definitions, and analogies, you often have a choice.

Sample Speech

The following sample speech was presented by Andrew Neylon, whose story began this chapter. When Andrew was a student at Ball State University, he presented this speech at the National Forensics Association National Tournament.

A full outline for Andrew's speech might appear like the one below. This is an expansion of the outline shown on page 326, which now includes subpoints and transitions. Numbers in parentheses correspond to the numbered paragraphs of the speech.

Speech Outline
The Amazing Eyeborg
INTRODUCTION

I. Attention-getter: "For me, the world is still a lot like a black and white television." (1)
II. Thesis statement: The eyeborg, which allows color-blind people to hear color, has amazing implications for human evolution. (2–4)
III. Preview main points: To truly see the eyeborg for what it is, we'll first take a look at the eye itself, then look up some applications, and finally some implications for the body. (5)

BODY

I. The eyeborg opens up a new world for the color blind. (6–7)
    A. The eye is made up of rods and cones. (7)
        1. Rods affect distance and acuity vision.
        2. Cones make up the ability to see color.
            a. Cones are divided into three sets—red, blue, and green.
            b. Cones combine to form roughly 10 million shades.
            c. The color blind have only one set of cones. (8)

   B. The eyeborg functions by changing color to sound. (9)
      1. The device detects the color frequency of the item.
      2. It turns that color frequency it into a sound frequency.
      3. It passes the information to a chip installed in the user's head.

Transition: The eyeborg may be Harbisson's pet project, but it suggests a wealth of opportunities . . .

Preview of second subpoint (10)

  II. The eyeborg has amazing applications. (10)
    A. The eyeborg has amazing applications for the color blind (11–12)
    B. The eyeborg has amazing applications for all human beings.
      1. The eyeborg can enable you to listen to the colors of artwork. (12)
      2. The eyeborg can enable you to listen to the colors of a supermarket. (12)
      3. The eyeborg can enable you to perceive celebrities in a whole new way. (13)

Transition (14)

  II. The eyeborg also has amazing implications. (14)
    A. The eyeborg augments a physical sense. (15)
    B. The eyeborg suggests that other physical senses can be augmented. (16)
    C. The eyeborg allows humans to become cyborgs. (17)

Transition (18)

CONCLUSION
  I. Review: (19)
  II. Final Remarks: (20)

Andrew collected his supporting material mostly from online sources, as seen in the following bibliography:

Bibliography

Achromatopsia.info. (n.d.) Colorblindness and achromatopsia. Retrieved from http://www.achromatopsia.info/color-blindness/.

Harbisson, N. (2012, June). I listen to color. TEDGlobal. Retrieved from https://www.ted.com/talks/neil_harbisson_i_listen_to_color.

Lee, J. 8. (2012, July 2). A surgical implant for seeing colors through sound. *New York Times*. Retrieved from http://bits.blogs.nytimes.com/2012/07/02/a-surgical-implant-for-seeing-colors-through-sound/.

Mills, C. (2012, July 10). Colourblind artist surgically implants an eyeborg to see through sound. *Gizmodo*. Retrieved from http://www.gizmodo.co.uk/2012/07/colourblind-artist-surgically-implants-an-eyeborg-tosee-through-sound/.

Newitz, A. (2013, December 2). The first person in the world to become a government-recognized cyborg. *Gizmodo*. Retrieved from http://io9.gizmodo.com/the-first-person-in-the-world-to-become-a-government-re-1474975237.

Swift, E. (2014, February 18). The sound of colour. *Haft2*. Retrieved from http://www.haft2.com/2014/02/18/the-sound-of-colour/.

As you read Andrew's speech, notice how he carefully chooses various types of supporting material to back up what he has to say.

The Amazing Eyeborg

1   "For me, the world is still a lot like a black and white television."

    Andrew begins with a quotation that captures his audience's attention.

2   So noted Northern Irish visual artist and activist Neil Harbisson in a recent speech at TEDGlobal in Edinburgh. Harbisson has achromatopsia, an eye condition characterized by little color vision and low visual acuity. For many of you, the mere mention of color blindness will flood your mind with questions. Can Harbisson tell your hair color? What does he see when he looks at the sky? Does he need a cane? Most importantly—who dresses him?

    He defines a key term . . .

    . . . and offers some examples in the form of intriguing questions.

3   I know these questions all too well. Like Harbisson, I have achromatopsia. I'm legally blind and completely color blind. For people like Harbisson and myself, the world can seem like a strange, scary place—that's why Harbisson did something about it.

    He shows the personal importance of this topic and enhances his credibility by citing personal experience.

4   In his TED Talk, Harbisson unveiled the eyeborg, a small sensor attached to his head coupled with a computer chip implanted in his neck, to translate color into sound. Each color is given a unique tonal frequency, developed through eight years of collaborative research between Harbisson and his colleagues. The device allows Harbisson to interact with his world in a unique way.

    He defines a second key term . . .

    . . . and shows a photograph of Harbisson wearing the eyeborg.

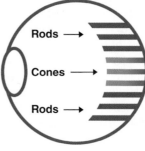

5   Harbisson is more than just a groundbreaker, and, as a recent issue of Gizmodo notes—he's "paving the way for greater acceptance of cybernetics in our culture," realizing two hundred years of scientific research in his quest to push the boundaries of the body. To truly see the eyeborg for what it is, we'll first take a look at the eye itself, then look up some applications, before seeing eye to eye with some implications for the body.

    He cites a striking quotation . . .

    . . . and previews his main points.

6   To begin, let's take a closer look at the eye itself: First, a brief breakdown of the eye and color blindness, and second, how the eyeborg functions and addresses those concerns.

    He previews his first main point and its subpoints.

7   To fully understand the need for the eyeborg, we need to start by explaining a bit about how the eye works. For some of you this may seem rudimentary, but experience has taught me that the concept of color blindness is a toughie for most seeing folks. Essentially, your eye is made up of rods and cones. Rods affect distance and acuity vision, while cones make up the ability to see color. Cones are divided into three sets—red, blue, and green—which combine to form the roughly 10 million shades perceived by the human eye.

    He begins to develop his first main point, provides a concise description of what he is talking about . . .

    . . . and shows a diagram to illustrate.

Rods →

Cones →

Rods →

8   Those with all three sets of functioning cones are called tri-chromats—but Harbisson and I are considered mono-chromats. We only have one set of cones to reflect on, and without any other cones to play off, we essentially see things in two dimensions—as a black and white television. This raises a host of problems

    This analogy compares color blindness to black and white television.

large and small. I can't see the colors of street-lights, thus I can't drive. Harbisson can't tell what color his clothes are—hence his donning of lime green pants for his TED Global Talk. More importantly, the color blind find themselves continually negotiating a world designed for the color seeing. Color films aren't shot with gray in mind, restaurant menus aren't high-contrast, and, naggingly, you still have to match for a speech tournament. Clearly, I didn't pick this outfit out . . .

His examples include a humorous one, which supports the mood of his speech.

9 Which brings us to the electronic eye, or eyeborg, as Harbisson calls it. The eyeborg is the result of eight painstaking years of scientific labor. The device, which Harbisson wears on his head, detects the color frequency of the item that is passed in front of it, turns it into a sound frequency and passes the information to a chip installed at the back of Harbisson's head. He then is able to hear the color through bone conduction, or sound waves that are created as they pass through the bones of the skull to the inner ear. Essentially, he's created a robotic eye capable of detecting the colors around him. Listen for a moment to this recording. (Play it) This is the sound of the color blue.

He explains this subpoint with a series of sub-subpoints.

He defines another key term, "bone conduction."
And he plays a brief recording as an illustration.

10 The eyeborg may be Harbisson's pet project, but it suggests a wealth of opportunities for those seeking to experience their world in a new way. First, we can examine its sensory applications, and second we'll consider how the eyeborg could someday change our cultural interactions.

He transitions to his next subpoint . . .
. . . and previews his sub-subpoints.

11 First, the eyeborg allows the color blind to, for the first time, enter the color seeing world. For Harbisson, the device has changed his life. He now feels "attachment to colors," and has slowly learned about the color, saturation, and hues that color your life. Originally the eyeborg had only six frequencies, but now Harbisson is so good at distinguishing them that he hears 360 microtones in an octave. Where there was once just a single sound for green, there are now gradations for lime, forest, kelly, and seafoam. And, according to a recent program on Public Radio International, "the next update for Harbisson's electronic eye will let him hear colors invisible to the human eye—those in the ultraviolet range. He said he'll be able to detect which days will be dangerous for sunbathing."

He begins to develop his next subpoint . . .
. . . provides a quotation from one of his research sources . . .
. . . provides examples . . .

. . . and finishes off with a longer quotation.

12 This sensory application dramatically changes the everyday experiences of the color blind. Visiting an art museum, oftentimes a bit of a bore for the color blind, yields new excitement for the user who can listen to a Picasso. Harbisson also enjoys walking in supermarkets, which he likens to being in a nightclub, due to all of the different audio sensations. As he puts it, the world is full of different melodies.

He provides examples to help develop this idea.

13 So the eyeborg is at the intersection between the senses and culture, because of its ability to transform single-sense experiences into multi-faceted ones. For example, Harbisson suggests getting kids to eat vegetables by assembling food according to a choice of song. Additionally, continued use of the sensor leads to an association between the tonal quality of the eyeborg and the facial features of the people one encounters. For example, Nicole Kidman and Prince Charles have similar sounding eyes—an entirely new way to perceive celebrity. Because the eyeborg gives sound to color, it also gives color to sound. To Harbisson, listening to Mozart is yellow—Justin Bieber, naturally, is pink. Here are Dr. Martin Luther King's "I Have a Dream" and Hitler's "Mein Kampf" transformed into paintings.

Examples help illustrate each of his points.

14 The eyeborg is an experimental technology, and likely a ways from mass production. Still, its implications for our understanding of our senses, and most importantly—the millennial definition of the body—are staggering.

Transition to next point.

15 Harbisson's device reframes the idea of disability adaptation entirely. One of the most intriguing elements of the eyeborg is its very nature. It doesn't directly

Development of next point.

substitute an experience—that is, it isn't intended (like, say, a prosthetic arm) to directly replace a sensory experience. Harbisson didn't opt to fix his own disability—he augmented it. He's found a way outside of conventional linear thinking to allow us to engage in the world without feeling fixed.

> The comparison to a prosthetic arm helps him explain the difference between fixing a disability and augmenting it.

16 Second, it's not just for the color blind, either. What began as an ingenious way to spruce up Harbisson's wardrobe could become much more. One of the most exciting statements Harbisson made was his own acknowledgment that "the day we start developing applications for our bodies instead of our phones will be an exciting one." These adaptations could become standard. Our senses need not remain disparate.

> He begins his next subpoint . . .

> . . . and uses a quotation as support.

17 Finally, Harbisson is a cyborg. He explained to *BBC* magazine that "It's not the union between the eyeborg and my head that converts me into a cyborg but the union between the software and my brain." This suggests a profound shift in our view of the body. While often relegated to *Star Trek*, the day of the cyborg may be rapidly approaching.

> Next subpoint.
> Quotation again.

18 We have grappled with the idea of body modification since Mary Shelley penned *Frankenstein* in 1818—but this is no science fiction tale. As humans continue to develop, we will increasingly have to grapple with the relationship between man and technology. Harbisson successfully argued with British authorities that he should be allowed to wear the device in his passport picture, on the basis that it is part of his body. This is an early example of the legal questions to come. We may be fast approaching the day of electronic kinship—finally singing Whitman's Body Electric.

> Here he provides an analogy to science fiction . . .

> . . . an example . . .

> . . . a transition to his conclusion, and a reference to Walt Whitman's 1855 poem, "I Sing the Body Electric."

19 We have seen today just how amazing the eyeborg is. We began by examining the way the eye works, and how the eyeborg opens up a host of possibilities for the color blind. We then looked at some applications to see how the eyeborg represents the first bloom in our changing perception of the body's ability to adapt, interact with, and master the world.

> He begins his conclusion with a review of main points.

20 To someone who can't see color, the beauty of the world can sometimes seem hazy and adrift. But Harbisson will be the first to tell you that his foray into the color-seeing world has opened up the eyes of so many. Equally as important, it raises fundamental questions about who we want to become in the technology boom fast approaching. By understanding the eyeborg itself, its applications and implications, we can see the eyeborg for what it is: A giant step on the road to full sensory discovery. And if I can dress myself in the hotel room—well, we all win.

> He finishes his conclusion with a memorable statement.

MAKING THE GRADE

For more resources to help you understand and apply the information in this chapter, visit the *Understanding Human Communication* website at www.oup.com/us/adleruhc.com.

Objective 12.1 Describe different types of speech outlines and their functions.

- A working outline is used to map out the structure of your speech, in rough form.

- A formal outline uses a standard format and a consistent set of symbols.

- Speaking notes are used to jog your memory while giving a speech. Like a working outline, they are for your eyes only.

  > Explain the primary differences among working outlines, formal outlines, and speaking notes.

  > Which type of outline would be most important for your next speech?

  > What style of speaking notes would you use for your next speech?

Objective 12.2   Construct an effective speech outline using the organizing principles described in this chapter.

- Principles for the effective construction of outlines are based on the use of standard symbols and a standard format.

- The rule of division requires at least two divisions of every point or subpoint.

- The rule of parallel wording requires that points at each level of division be worded in a similar manner whenever possible.

  > What are the standard symbols used in a formal outline?

  > Which principle of outlining is least intuitive for you? Which one is most intuitive?

  > How would you divide your next speech into main points and subpoints?

Objective 12.3   Choose an appropriate organizational pattern for your speech.

- The organization of ideas should follow a pattern, such as time, space, topic, problem-solution, cause-effect, or Monroe's motivated sequence.

- Each pattern of organization has its own advantages in helping your audience understand and remember what you have to say.

  > Why is it important to organize your ideas according to a pattern?

  > When should you use each of the six patterns of organization discussed in this chapter?

  > Which pattern of organization would be best for your next speech?

Objective 12.4   Develop a compelling introduction and conclusion, and use effective transitions at key points in your speech.

- The main idea of the speech is established in the introduction, developed in the body, and reviewed in the conclusion.

- The introduction will also gain the audience's attention, preview the main points, set the mood and tone of the speech, and demonstrate the importance of the topic to the audience.

- The conclusion will review your main points and supply the audience with a memory aid in the form of compelling final remarks.

- Effective transitions keep your message moving forward and demonstrate how your ideas are related.

  > Why are the beginning and end of a speech so important? Why are transitions?

> What would be the most important function of the introduction of your next speech?

> What idea would you like to leave your audience with in your next speech?

Objective 12.5   Choose supporting material that will help make your points clear, interesting, memorable, and convincing.

- Supporting materials are the facts and information you use to back up what you say.

- Types of support include *definitions, examples, statistics, analogies, anecdotes, quotations,* and *testimony.*

- Support may be narrated or cited.

  > Give examples of five forms of support that could be used for a speech on financing a college education.

  > Which form of support do you find to be most effective for most speeches?

  > What are the main forms of support that you would use for your next speech?

KEY TERMS

analogy p. 339

anecdote p. 339

basic speech structure p. 325

cause-effect pattern p. 330

citation p. 341

climax pattern p. 329

conclusion (of a speech) p. 334

factual example p. 338

formal outline p. 325

hypothetical example p. 338

introduction (of a speech) p. 332

narration p. 340

problem-solution pattern p. 330

space pattern p. 329

statistic p. 338

testimony p. 339

time pattern p. 329

topic pattern p. 329

transition p. 334

working outline p. 325

ACTIVITIES

1. **Dividing Ideas** For practice in the principle of division, divide each of the following into three to five subcategories:

   a. Clothing

   b. Academic studies

   c. Crime

   d. Health care

   e. Fun

   f. Charities

2. **Organizational Effectiveness** Take any written statement at least three paragraphs long that you consider effective. This statement might be an editorial in your local newspaper, a short magazine article, or even a section of one of your textbooks. Outline this statement according to the rules discussed here. Was the statement well organized? Did its organization contribute to its effectiveness?

3. **The Functions of Support** For practice in recognizing the functions of support, identify three instances of support in each of the speeches at the end of Chapters 13 and 14. Explain the function of each instance of support. (Keep in mind that any instance of support *could* perform more than one function.)

Informative Speaking

LEARNING OBJECTIVES

13.1

Distinguish among the main types of informative speaking.

13.2

Describe the differences between informative and persuasive speaking.

13.3

Increase your informative effectiveness by using the techniques discussed in this chapter.

13.4

Use visual aids appropriately and effectively.

JENNI CHANG AND LISA DAZOLS met in San Francisco. Jenni is a business manager for eBay, and Lisa is a licensed clinical social worker who works in HIV care. When they became a couple, they realized that they both wanted to follow a life of adventure. This led them to travel extensively, and as they did so they realized that their travels might have a larger meaning, both to themselves and to the world. They decided to document their travels on video, in spite of the fact that neither of them had any experience with filmmaking or journalism. They bought a camera and a book on how to make a documentary, and set out on a 15-country sojourn to document what they called "stories of hope."

Their travels led them through Asia, Africa, and South America; they traveled more than 50,000 miles in 342 days, and conducted 55 interviews. Their mission was to interview "Supergays," the people who were leading the movement for gay, lesbian, and transgender equality in the developing world. They titled the resulting film *Out & Around* and produced it in partnership with the It Gets Better Project, which was created "to show young LGBT people the levels of happiness, potential, and positivity their lives will reach—if they can just get through their teen years."[1]

The film was broadcast on the Logo channel in 2015 and has been seen in film festivals around the world. In 2015 they also chronicled their adventure in a TED Talk, which has been viewed more than a million times. That speech is reproduced at the end of this chapter.

Jenni and Lisa are now happily, legally married and living back home in San Francisco. In March 2016, Jenni gave birth to their daughter, Charlie Grace Dazols.

Jenni Chang and Lisa Dazols had a wealth of experience, information, and ideas to transmit to a specific audience. One of the things making that difficult is the huge amount of information competing for our attention these days. Some people call this the age of information; others call it the age of information glut, data smog, and clutter. There are, in fact, a hundred names for it, but they all deal with the same idea: There is just too much information. And the amount of information is increasing exponentially. One information expert estimates that every 2 days now we create as much information as we did from the dawn of civilization up until 2003.[2]

Social scientists tell us that the information glut leads to information overload (see Chapter 10), a form of psychological stress that occurs when people become confused and have trouble sorting through all the information available to them.[3] Some experts use another term, **information anxiety**, for the same phenomenon. To check on how information overload affects you personally, do the self-assessment on p. 351.

Are You Overloaded with Information?

Problems in informative speaking are often the result of information overload—on the speaker's part and the audience's. For each statement to the right, select "often," "sometimes," or "seldom" to assess your level of information overload.

1. I forget information I need to know.

OFTEN   SOMETIMES   SELDOM

2. I have difficulty concentrating on important tasks.

OFTEN   SOMETIMES   SELDOM

3. When I go online, I feel anxious about the work that I don't have time to do.

OFTEN   SOMETIMES   SELDOM

4. I have email messages sitting in my inbox that are more than 2 weeks old.

OFTEN   SOMETIMES   SELDOM

5. I constantly check my online services because I am afraid that if I don't, I will never catch up.

OFTEN   SOMETIMES   SELDOM

6. I find myself easily distracted by things that allow me to avoid work I need to do.

OFTEN   SOMETIMES   SELDOM

7. I feel fatigued by the amount of information I encounter.

OFTEN   SOMETIMES   SELDOM

8. I delay making decisions because of too many choices.

OFTEN   SOMETIMES   SELDOM

9. I make wrong decisions because of too many choices.

OFTEN   SOMETIMES   SELDOM

10. I spend too much time seeking information that is *nice to know* rather than information that I *need to know*.

OFTEN   SOMETIMES   SELDOM

Scoring: Give yourself 3 points for each "often," 2 points for each "sometimes," and 1 point for each "seldom." If your score is:

10–15: Information overload is not a big problem for you. However, it's probably still a significant problem for at least some members of your audience, so try to follow the guidelines for informative speaking outlined in this chapter.

16–24: You have a normal level of information overload. The guidelines in this chapter will help you be a more effective speaker.

25–30: You have a high level of information overload. Along with observing the guidelines in this chapter, you might also want to search online for guidelines to help you overcome this problem.

As Chang and Dazols discovered when planning their speech, the informative speaker's responsibility is not just to provide new information. Informative speaking, when it's done effectively, seeks to relieve information overload by turning information into knowledge for an audience. Information is the raw materials, the sometimes-contradictory facts and competing claims that <u>rain down on</u> public consciousness. **Knowledge** is what you get when you are able to make sense of and use those raw materials. Effective public speakers filter, organize, and illustrate information in order to reach small audiences with messages tailored for them, in an environment in which they can see if the audience is "getting it." If they aren't, the speaker can adjust the message and work with the audience until they do.

Informative speaking goes on all around you: in your professors' lectures or in a mechanic's explanation of how to keep your car from breaking down. You engage

information anxiety The psychological stress that occurs when dealing with too much information.

knowledge The understanding acquired by making sense of the raw material of information.

cultural idiom

rain down on: fall like rain, overwhelm

in this type of speaking frequently whether you realize it or not. Sometimes it is formal, as when you give a report in class. At other times, it is more casual, as when you tell a friend how to prepare your favorite dish. The main objective of this chapter is to give you the skills you need to enhance all of your informative speaking.

Types of Informative Speaking

There are several types of informative speaking. The primary types have to do with the content and purpose of the speech.

By Content

Informative speeches are generally categorized according to their content, and include the following types.

Speeches About Objects   This type of informative speech is about anything that is tangible (that is, capable of being seen or touched). Speeches about objects might include an appreciation of the Grand Canyon (or any other natural wonder) or a demonstration of the newest smartphone (or any other product).

Speeches About Processes   A process is any series of actions that leads to a specific result. If you spoke on the process of aging, the process of learning to juggle, or the process of breaking into a social networking business, you would be giving this type of speech.

Speeches About Events   You would be giving this type of informative speech if your topic dealt with anything notable that happened, was happening, or might happen: an upcoming protest against hydraulic fracturing ("fracking"), for example, or the prospects of your favorite baseball team winning the national championship.

Speeches About Concepts   Concepts include intangible ideas, such as beliefs, theories, ideas, and principles. If you gave an informative speech about postmodernism, vegetarianism, or any other "ism," you would be giving this type of speech. Other topics would include everything from New Age religions to theories about extraterrestrial life to rules for making millions of dollars.

By Purpose

We also distinguish among types of informative speeches depending on the speaker's purpose. We ask, "Does the speaker seek to describe, explain, or instruct?"

Descriptions   A speech of **description** is the most straightforward type of informative speech. You might introduce a new product like a wearable computer to a group of customers, or you might describe what a career in nursing would be like. Whatever its topic, a descriptive speech uses details to create a "word picture" of the essential factors that make that thing what it is.

Explanations   **Explanations** clarify ideas and concepts that are already known but not understood by an audience. For example, your audience members might already know that a U.S. national debt exists, but they might be baffled by the reasons why it has become so large. Explanations often deal with the question of *why* or *how*. Why do we have to wait until the age of 21 to drink legally? How did China evolve from an impoverished economy to a world power in a single generation? Why did tuition need to be increased this semester?

Instructions   **Instructions** teach something to the audience in a logical, step-by-step manner. They are the basis of training programs and orientations. They often

description A type of speech that uses details to create a "word picture" of something's essential factors.

explanations Speeches or presentations that clarify ideas and concepts already known but not understood by an audience.

instructions Remarks that teach something to an audience in a logical, step-by-step manner.

deal with the question of *how to*. This type of speech sometimes features a demonstration or a visual aid. Thus, if you were giving instructions on "how to promote your career via social networking sites," you might demonstrate by showing the social media profile of successful people. For instructions on "how to perform CPR," you could use a volunteer or a dummy.

These types of informative speeches aren't mutually exclusive. As you'll see in the sample speech at the end of this chapter, there is considerable overlap, as when you give a speech about objects that has the purpose of explaining them. Still, even this imperfect categorization demonstrates how wide a range of informative topics is available. One final distinction we need to make, however, is the difference between an informative and a persuasive speech topic.

Informative Versus Persuasive Topics

There are many similarities between an informative and a persuasive speech. In an informative speech, for example, you are constantly trying to "persuade" your audience to listen, understand, and remember. In a persuasive speech, you "inform" your audience about your arguments, your evidence, and so on. However, two basic characteristics differentiate an informative topic from a persuasive topic.

An Informative Topic Tends to Be Noncontroversial

In an informative speech, you generally do not present information that your audience is likely to disagree with. Again, this is a matter of degree. For example, you might want to give a purely informative talk on the differences between hospital births and home-based midwife births by simply describing what the practitioners of each method believe and do. By contrast, a talk either boosting or criticizing one method over the other would clearly be persuasive.

The noncontroversial nature of informative speaking does not mean that your speech topic should be uninteresting to your audience; rather, it means that your approach to it should not engender conflict. You could speak about the animal rights movement, for example, by explaining the points of view of both sides in an interesting but objective manner.

The Informative Speaker Does Not
Intend to Change Audience Attitudes

The informative speaker does seek a response (such as attention and interest) from the listener and does try to make the topic important to the audience. But the speaker's primary intent is not to change attitudes or to make the audience members *feel* differently about the topic. For example, an informative speaker might explain how a microwave oven works but will not try to "sell" a specific brand of oven to the audience.

The speaker's intent is best expressed in a specific informative purpose statement, which brings us to the first of our techniques of informative speaking.

Techniques of Informative Speaking

The techniques of informative speaking are based on a number of principles of human communication in general, and public speaking specifically, that we have discussed in earlier chapters. The most important principles to apply to informative speaking include those that help an audience understand and care about your speech. Let's look at how these principles apply to specific techniques.

UNDERSTANDING DIVERSITY

How Culture Affects Information

Cultural background is always a part of informative speaking, although it's not always easy to spot. Sometimes this is because of ethnocentrism, the belief in the inherent superiority of one's own ethnic group or culture. According to communication scholars Larry Samovar and Richard Porter, ethnocentrism is exemplified by what is taught in schools.

Each culture, whether consciously or unconsciously, tends to glorify its historical, scientific, and artistic accomplishments while frequently minimizing the accomplishments of other cultures. In this way, schools in all cultures, whether or not they intend to, teach ethnocentrism. For instance, the next time you look at a world map, notice that the United States is prominently located in the center—unless, of course, you are looking at a Chinese or Russian map. Many students in the United States, if asked to identify the great books of the world, would likely produce a list of books mainly by Western, white, male authors. This attitude of subtle ethnocentrism, or the reinforcing of the values, beliefs, and prejudices of the culture, is not a uniquely American phenomenon. Studying only the Koran in Iranian schools or only the Old Testament in Israeli classrooms is also a quiet form of ethnocentrism.

From Samovar, L., & Porter, R. (2010). *Communication between cultures* (5th ed.). Boston: Wadsworth. © 2010. Reprinted with permission of Wadsworth, an imprint of the Wadsworth Group, a division of Cengage Learning.

How would culture affect the informative topic you have chosen?

Define a Specific Informative Purpose

As Chapter 11 explained, any good speech must be based on a purpose statement that is audience oriented, precise, and attainable. When you are preparing an informative speech, it is especially important to define in advance, for yourself, a clear informative purpose. An **informative purpose statement** will generally be worded to stress audience knowledge, ability, or both:

> After listening to my speech, my audience will be able to recall the three most important questions to ask when shopping for a smartphone.

> After listening to my speech, my audience will be able to identify the four reasons that online memes go viral.

> After listening to my speech, my audience will be able to discuss the pros and cons of using drones in warfare.

Notice that in each of these purpose statements a specific verb such as *to recall, to identify,* or *to discuss* points out what the audience will be able to do after hearing the speech. Other key verbs for informative purpose statements include these:

Accomplish	Choose	Explain	Name	Recognize
Analyze	Contrast	Integrate	Operate	Review
Apply	Describe	List	Perform	Summarize

A clear purpose statement will lead to a clear thesis statement. As you remember from Chapter 11, a thesis statement presents the central idea of your speech. Sometimes your thesis statement for an informative speech will just preview the central idea:

> Today's smartphones have so many features that it is difficult for the uninformed consumer to make a choice.

> Understanding how memes go viral could make you very wealthy someday.

> Soldiers and civilians have different views on the morality of drones.

informative purpose statement
A complete statement of the objective of a speech, worded to stress audience knowledge and/or ability.

At other times, the thesis statement for an informative speech will delineate the main points of that speech:

> When shopping for a smartphone, the informed consumer seeks to balance price, dependability, and user friendliness.
>
> The four basic principles of aerodynamics—lift, thrust, drag, and gravity—can explain why memes go viral.
>
> Drones can save warrior lives but cost the lives of civilians.

Setting a clear informative purpose will help keep you focused as you prepare and present your speech.

Create Information Hunger

An effective informative speech creates **information hunger**: a reason for your audience members to want to listen to and learn from your speech. To do so, you can use the analysis of communication functions discussed in Chapter 1 as a guide. You read there that communication of all types helps us meet our physical needs, identity needs, social needs, and practical needs. In informative speaking, you could tap into your audience members' physical needs by relating your topic to their survival or to the improvement of their living conditions. If you gave a speech on food (eating it, cooking it, or shopping for it), you would be dealing with that basic physical need. In the same way, you could appeal to identity needs by showing your audience members how to be respected—or simply by showing them that you respect them. You could relate to social needs by showing them how your topic could help them be well liked. Finally, you can relate your topic to practical audience needs by telling your audience members how to succeed in their courses, their job search, or their quest for the perfect outfit.

Make It Easy to Listen

Keep in mind the complex nature of listening, discussed in Chapter 5, and make it easy for your audience members to hear, pay attention, understand, and remember. This means first that you should speak clearly and with enough volume to be heard by all your listeners. It also means that as you put your speech together, you should take into consideration techniques that recognize the way human beings process information.

Limit the Amount of Information You Present   Remember that you probably won't have enough time to transmit all your research to your audience in one sitting. It's better to make careful choices about the three to five main ideas you want to get across and then develop those ideas fully. Remember, too much information leads to overload, anxiety, and a lack of attention on the part of your audience.

Use Familiar Information to Increase Understanding of the Unfamiliar
Move your audience members from familiar information (on the basis of your audience analysis) to your newer information. For example, if you are giving a speech about how the stock market works, you could compare the daily activity of a broker with that of a salesperson in a retail store, or you could compare the idea of capital growth (a new concept to some listeners) with interest earned in a savings account (a more familiar concept).

CHECKLIST ✅

Techniques of Informative Speaking

☐ Define a specific informative purpose.

☐ Create information hunger by relating to audience needs.

☐ Make it easy for audience members to listen.
  - Limit the amount of information presented.
  - Use familiar information to introduce unfamiliar information.
  - Start with simple information before moving to more complex ideas.

☐ Use clear, simple language.

☐ Use clear organization and structure.

☐ Support and illustrate your points.
  - Provide interesting, relevant facts and examples, citing your sources.
  - Use visual aids that help make your points clear, interesting, and memorable.

☐ Emphasize important points.
  - Repeat key information in more than one way.
  - Use signposts: words or phrases that highlight what you are about to say.

☐ Generate audience involvement.
  - Personalize the speech.
  - Use audience participation.
  - Use volunteers.
  - Have a question-and-answer period at the end.

---

information hunger Audience desire, created by a speaker, to learn information.

ETHICAL CHALLENGE
The Ethics of Simplicity

Often, persuasive speakers use language that is purposely complicated or obscure in order to keep the audience uninformed about some idea or piece of information. Informative rather than persuasive intent can often help clear the air. Find any sales message (a print ad or television commercial, for example) or political message (a campaign speech, perhaps) and see if you can transform it into an informative speech. What are the differences?

Use Simple Information to Build Up Understanding of Complex Information Just as you move your audience members from the familiar to the unfamiliar, you can move them from the simple to the complex. An average college audience, for example, can understand the complexities of genetic modification if you begin with the concept of inherited characteristics.

Use Clear, Simple Language

Another technique for effective informative speaking is to use clear language, which means using precise, simple wording and avoiding jargon. As you plan your speech, consult online dictionaries such as Dictionary.com to make sure you are selecting precise vocabulary. Remember that picking the right word seldom means using a word that is unfamiliar to your audience; in fact, just the opposite is true. Important ideas do not have to sound complicated. Along with simple, precise vocabulary, you should also strive for direct, short sentence structure. For example, when Warren Buffett, one of the world's most successful investors, wanted to explain the impact of taxes on investing, he didn't use unusual vocabulary or complicated sentences. He explained it like this:

> Suppose that an investor you admire and trust comes to you with an investment idea. "This is a good one," he says enthusiastically. "I'm in it, and I think you should be, too." Would your reply possibly be this? "Well, it all depends on what my tax rate will be on the gain you're saying we're going to make. If the taxes are too high, I would rather leave the money in my savings account, earning a quarter of 1 percent." So let's forget about the rich and ultrarich going on strike and stuffing their ample funds under their mattresses if—gasp—capital gains rates and ordinary income rates are increased. The ultrarich, including me, will forever pursue investment opportunities.[4]

Each idea within that explanation is stated directly, using simple, clear language.

Use a Clear Organization and Structure

Because of the way humans process information (that is, in a limited number of chunks at any one time),[5] organization is extremely important in an informative speech. Rules for structure may be mere suggestions for other types of speeches, but for informative speeches they are ironclad.

Chapter 12 discusses some of these rules:

- Limit your speech to three to five main points.
- Divide, coordinate, and order those main points.
- Use a strong introduction that previews your ideas.
- Use a conclusion that reviews your ideas and makes them memorable.
- Use transitions, internal summaries, and internal previews.

The repetition that is inherent in strong organization will help your audience members understand and remember those points. This will be especially true if you use a well-organized introduction, body, and conclusion.

The Introduction The following principles of organization from Chapter 12 become especially important in the introduction of an informative speech:

1. Establish the importance of your topic to your audience.
2. Preview the thesis, the one central idea you want your audience to remember.
3. Preview your main points.

For example, Kevin Allocca, the trends manager at YouTube, began his TED Talk "Why Videos Go Viral" with the following introduction:

I professionally watch YouTube videos. It's true. So we're going to talk a little bit today about how videos go viral and then why that even matters. Web video has made it so that any of us or any of the creative things that we do can become completely famous in a part of our world's culture. Any one of you could be famous on the Internet by next Saturday. But there are over 48 hours of video uploaded to YouTube every minute. And of that, only a tiny percentage ever goes viral and gets tons of views and becomes a cultural moment. So how does it happen? Three things: tastemakers, communities of participation, and unexpectedness.[6]

The Body   In the body of an informative speech, the following organizational principles take on special importance:

1. Limit your division of main points to three to five subpoints.
2. Use transitions, internal summaries, and internal previews.
3. Order your points in the way that they will be easiest to understand and remember.

Kevin Allocca followed these principles for organizing his speech on why some videos go viral and some do not. He developed his speech with the following three main points:

1. Tastemakers: Tastemakers like Jimmy Kimmel introduce us to new and interesting things and bring them to a larger audience.
2. Communities of participation: A community of people who share this big inside joke start talking about it and doing things with it.
3. Unexpectedness: In a world where more than 2 days of video get uploaded every minute, only those that are truly unique can go viral.

The Conclusion   Organizational principles are also important in the conclusion of an informative speech:

1. Review your main points.
2. Remind your audience members of the importance of your topic to them.
3. Provide your audience with a memory aid.

For example, this is how Kevin Allocca concluded his speech on viral videos:

> Tastemakers, creative participating communities, complete unexpectedness, these are characteristics of a new kind of media and a new kind of culture where anyone has access and the audience defines the popularity. One of the biggest stars in the world right now, Justin Bieber, got his start on YouTube. No one has to green-light your idea. And we all now feel some ownership in our own pop culture. And these are not characteristics of old media, and they're barely true of the media of today, but they will define the entertainment of the future.

Use Supporting Material Effectively

Another technique for effective informative speaking has to do with the supporting material discussed in Chapter 12. All of the purposes of support (to clarify, to prove, to make interesting, to make memorable) are essential to informative speaking. Therefore, you should be careful to support your thesis in every way possible. Notice the way in which Jenni and Lisa use solid supporting material in the sample speech at the end of this chapter. In particular, notice their use of videos, which can grab your audience members' attention and keep them attuned to your topic throughout your speech.

You should also try to include **vocal citations**, or brief explanations of where your supporting material came from. These citations build the credibility of your

ASK YOURSELF

Do you organize speeches in the order ideas occur to you, or do you plan the organization more strategically? How could you organize your speeches more effectively?

vocal citation A simple, concise, spoken statement of the source of your evidence.

explanations and increase audience trust in the accuracy of what you are saying. For example, when Nicole Wilson of Berry College in Georgia gave a speech on the need for new fire hydrants, she used the following vocal citation:

> According to the March 12, 2014 *Huffington Post*, the contemporary fire hydrant was actually designed in 1801 and hasn't changed much in the last 213 years. The main problem with these relics is how easily they break and deteriorate, and hydrants are deteriorating throughout the country. According the *New Jersey Star Ledger* of August 11, 2013, the 2013 New Jersey fire hydrant inspection revealed that one in seven fire hydrants are currently out of order in Newark, with almost 500 that are completely useless to firefighters.[7]

By telling concisely and simply where her information came from, Nicole reassured her audience that her statistics were credible.

Emphasize Important Points

One specific principle of informative speaking is to stress the important points in your speech through repetition and the use of signposts.

Repetition Repetition is one of the age-old rules of learning. Human beings are more likely to comprehend information that is stated more than once. This is especially true in a speaking situation, because, unlike a written paper, your audience members cannot go back to reread something they have missed. If their minds have wandered the first time you say something, they just might pick it up the second time.

Of course, simply repeating something in the same words might bore the audience members who actually are paying attention, so effective speakers learn to say the same thing in more than one way. Kathy Levine, a student at Oregon State University, used this technique in her speech on contaminated dental water:

> The problem of dirty dental water is widespread. In a nationwide *20/20* investigation, the water used in approximately 90% of dental offices is dirtier than the water found in public toilets. This means that 9 out of 10 dental offices are using dirty water on their patients.[8]

Redundancy can be effective when you use it to emphasize important points.[9] It is ineffective only when (1) you are redundant with obvious, trivial, or boring points or (2) you run an important point into the ground. There is no sure rule for making certain you have not overemphasized a point. You just have to use your best judgment to make sure that you have stated the point enough that your audience members get it without repeating it so often that they want to give it back.

Signposts Another way to emphasize important material is by using **signposts**: words or phrases that emphasize the importance of what you are about to say. You can state, simply enough, "What I'm about to say is important," or you can use some variation of that statement: "But listen to this . . . ," or "The most important thing to remember is . . . ," or "The three keys to this situation are . . . ," and so on.

Generate Audience Involvement

The final technique for effective informative speaking is to get your audience involved in your speech. **Audience involvement** is the level of commitment and attention that listeners devote to a speech. Educational psychologists have long known that the best way to teach people something is to have them do it; social psychologists have added to this rule by proving, in many studies, that involvement in a message increases audience comprehension of, and agreement with, that message.

signpost A phrase that emphasizes the importance of upcoming material in a speech.

audience involvement The level of commitment and attention that listeners devote to a speech.

There are many ways to encourage audience involvement in your speech. One way is by following the rules for good delivery by maintaining enthusiasm, energy, eye contact, and so on. Other ways include personalizing your speech, using audience participation, using volunteers, and having a question-and-answer period.

Personalize Your Speech   One way to encourage audience involvement is to give audience members a human being to connect to. In other words, don't be afraid to be yourself and to inject a little of your own personality into the speech. If you happen to be good at storytelling, make a narration part of your speech. If humor is a personal strength, be funny. If you feel passion about your topic, show it. Certainly if you have any experience that relates to your topic, use it.

Kathryn Schulz, author of *Being Wrong* and a self-proclaimed "wrongologist," personalized her TED speech, "Being Wrong," this way:

> So it's 1995, I'm in college, and a friend and I go on a road trip from Providence, Rhode Island, to Portland, Oregon. And you know, we're young and unemployed, so we do the whole thing on back roads through state parks and national forests—basically the longest route we can possibly take. And somewhere in the middle of South Dakota, I turn to my friend and I ask her a question that's been bothering me for 2,000 miles. "What's up with the Chinese character I keep seeing by the side of the road?"

> My friend looks at me totally blankly. There's actually a gentleman in the front row who's doing a perfect imitation of her look. (Laughter) And I'm like, "You know, all the signs we keep seeing with the Chinese character on them." She just stares at me for a few moments, and then she cracks up, because she figures out what I'm talking about. And what I'm talking about is this:

Right: The Famous Chinese Character for Picnic Area.[10]

Another way to personalize your speech is to link it to the experience of audience members . . . maybe even naming one or more.

Use Audience Participation   Having your listeners actually do something during your speech—**audience participation**—is another way to increase their involvement in your message. For example, if you were giving a demonstration on isometric exercises (which don't require too much room for movement), you could have the entire audience stand up and do one or two sample exercises. If you were explaining how to fill out a federal income tax form, you could give each class member a sample form to fill out as you explain it. Outlines and checklists can be used in a similar manner for just about any speech. Here's how one student organization used audience participation to demonstrate the various restrictions that were once placed on voting rights:

> Voting is something that a lot of us may take for granted. Today, the only requirements for voting are that you are a U.S. citizen aged 18 or older who has lived in the same place for at least 30 days and that you have registered. But it hasn't always been that way. Americans have had to struggle for the right to vote. I'd like to illustrate this by asking everyone to please stand.

audience participation Listener activity during a speech; a technique to increase audience involvement.

[Wait, prod class to stand.]

I'm going to ask some questions. If you answer no to any question, please sit down.

Have you resided at the same address for at least 1 year? If not, sit down. Residency requirements of more than 30 days weren't abolished until 1970.

Are you white? If not, sit down. The 15th Amendment gave nonwhites the right to vote in 1870, but many states didn't enforce it until the late 1960s.

Are you male? If not, sit down. The 19th Amendment only gave women the right to vote in 1920.

Do you own a home? If not, sit down. Through the mid-1800s only property owners could vote.

Are you Protestant? If not, sit down. That's right. Religious requirements existed in the early days throughout the country.[11]

Use Volunteers   Some points or actions are more easily demonstrated with one or two volunteers. Selecting volunteers from the audience will increase the psychological involvement of all audience members, because they will tend to identify with the volunteers.

Kathryn Schulz, in her speech on being wrong, subtly enlisted volunteers when she wanted to impress an important point on her audience. She began by addressing a rhetorical question to her entire audience but then directed it to a few individuals in the front row:

So let me ask you guys something—or actually, let me ask you guys something, because you're right here: How does it feel—emotionally—how does it feel to be wrong?

Schulz then listened to the responses and repeated them for the rest of the audience:

Dreadful. Thumbs down. Embarrassing. . . . Thank you, these are great answers, but they're answers to a different question. You guys are answering the question: How does it feel to *realize* you're wrong? When we're wrong about something—not when we realize it, but before that—*it feels like being right.*

Have a Question-and-Answer Period   One way to increase audience involvement that is nearly always appropriate if time allows is to answer questions at the end of your speech. You should encourage your audience to ask questions. Solicit questions and be patient waiting for the first one. Often no one wants to ask the first question. When the questions do start coming, the following suggestions might increase your effectiveness in answering them:

1. Listen to the substance of the question. Don't zero in on irrelevant details; listen for the big picture, the basic, overall question that is being asked. If you are not really sure what the substance of a question is, ask the questioner to paraphrase it. Don't be afraid to let the questioners do their share of the work.

2. Paraphrase confusing or quietly asked questions. Use the active listening skills described in Chapter 5. You can paraphrase the question in just a few words: "If I understand your question, you are asking _____. Is that right?"

Rubes®   By Leigh Rubin

"In order to adequately demonstrate just how many ways there are to skin a cat, I'll need a volunteer from the audience."

Source: By permission of Leigh Rubin and Creators Syndicate, Inc.

cultural idiom
zero in on: focus directly on

3. Avoid defensive reactions to questions. Even if the questioner seems to be calling you a liar or stupid or biased, try to listen to the substance of the question and not to the possible personality attack.

4. Answer the question briefly. Then check the questioner's comprehension of your answer by observing his or her nonverbal response or by asking, "Does that answer your question?"

Using Visual Aids

Visual aids are graphic devices used in a speech to illustrate or support ideas. Although they can be used in any type of speech, they are especially important in informative speeches. For example, they can be extremely useful when you want to show how things look (photos of your trek to Nepal or the effects of malnutrition) or how things work (a demonstration of a new ski binding or a diagram of how seawater is made drinkable). Visual aids can also show how things relate to one another (a graph showing the relationships among gender, education, and income).

Types of Visual Aids

There is a wide variety of types of visual aids. The most common types include the following.

Objects and Models   Sometimes the most effective visual aid is the actual thing you are talking about. This is true when the thing you are talking about is easily portable and simple enough to use during a demonstration before an audience (e.g., a lacrosse racket). **Models** are scaled representations of the object you are discussing and are used when that object is too large (the new campus arts complex) or too small (a DNA molecule) or simply doesn't exist anymore (a *Tyrannosaurus rex*).

Photos, Videos, and Audio Files   Speaking venues, including college classrooms, are often set up with digital projectors connected to online computers, DVD players, and outlets for personal laptops. In these cases, a wealth of photos, video clips, and audio files are available to enhance your speech. If you can project photos, you can use them to help explain geological formations, for example, or underwater habitats. Videos can be used to play a visual **sound bite** (a brief recorded excerpt) from an authoritative source, or to show a process such as plant growth. Using audio files can help you compare musical styles or demonstrate the differences in the sounds of gas and diesel engines.

Photos can stay up as long as you are referring to them, but videos and audio files should be used sparingly, because they allow audience members to receive information passively. You want your audience to actively participate in the presentation. The general rule when using videos and sound files is *Don't let them get in the way of the direct, person-to-person contact that is the primary advantage of public speaking.*

Videos and sound files should be carefully introduced, controlled, and summarized at the end. They should also be very brief. Jenni Chang and Lisa Dazols use a number of video excerpts in their sample speech at the end of this chapter. As you read the transcript of their speech, you will notice that each video is concise. In addition, the speakers include what media professionals call **wrap-arounds** or **intros** and **outros**: careful introductions and closing summaries of each clip. Using these devices enables the speakers to remain in control of the message.

visual aids Graphic devices used in a speech to illustrate or support ideas.

model (in speeches and presentations) A replica of an object being discussed. It is usually used when it would be difficult or impossible to use the actual object.

sound bite A brief recorded excerpt from a longer statement.

wrap-around A brief introduction before a visual aid is presented, accompanied by a brief conclusion afterward.

intro A brief explanation or comment before a visual aid is used.

outro A brief summary or conclusion after a visual aid has been used.

When Sarah Silverman gave a TED Talk called "A New Perspective on the Number 3000," she projected word and number charts to delineate her main points.

When would this type of visual aid be appropriate?

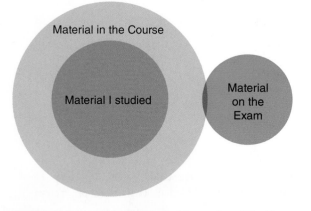

FIGURE 13-2 Venn Diagram: Student Perception of Final Exam

diagram A line drawing that shows the most important components of an object.

word chart A visual aid that lists words or terms in tabular form in order to clarify information.

number chart A visual aid that lists numbers in tabular form in order to clarify information.

pie chart A visual aid that divides a circle into wedges, representing percentages of the whole.

pictogram A visual aid that conveys its meaning through an image of an actual object.

Other types of visual aids are charts and graphs for representing facts and data. They include the following.

Diagrams A **diagram** is any kind of line drawing that shows the most important properties of an object. Diagrams do not try to show everything but just those parts of a thing that the audience most needs to be aware of and understand. Blueprints and architectural plans are common types of diagrams, as are maps and organizational charts. A diagram is most appropriate when you need to simplify a complex object or phenomenon and make it more understandable to the audience. Figure 13-2 shows a humorous depiction of one student's perception of what was covered on a final exam, in the form of a Venn diagram.

Word and Number Charts **Word charts** and **number charts** are visual depictions of key facts or statistics. Your audience will understand and remember these facts and numbers better if you show them than if you just talk about them. Many speakers arrange the main points of their speech, often in outline form, as a word chart. Other speakers list their main statistics. An important guideline for word and number charts is, *Don't read them to your audience; use them to enhance what you are saying and help your audience remember key points.*

Pie Charts Circular graphs cut into with wedges are known as **pie charts**. They are used to show divisions of any whole: where your tax dollars go, the percentage of the population involved in various occupations, and so on. Pie charts are often made up of percentages that add up to 100. Usually, the wedges of the pie are organized from largest to smallest. The pie chart in Figure 13-3 represents one's person's perception of "how princesses spend their time," and Figure 13-4 shows how the U.S. government adapted a pie chart for a new nutrition diagram. Coincidentally, Figure 13-4 is also a **pictogram**, which is a visual aid that conveys its meaning through images of an actual object.

How Princesses Spend Their Time

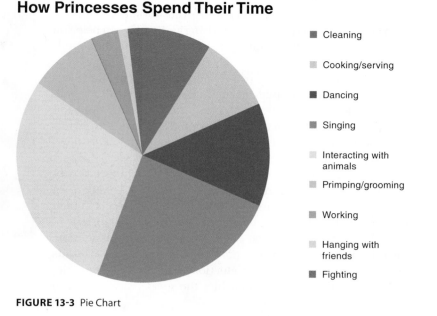

FIGURE 13-3 Pie Chart

FIGURE 13-4 Adaptation of a Pie Chart

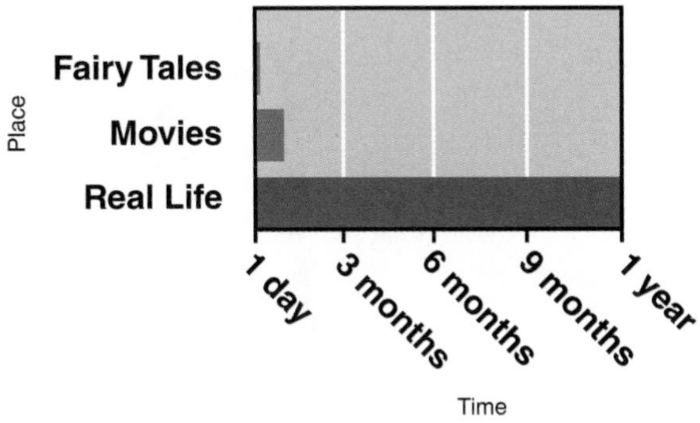

FIGURE 13-5 Bar Chart: Time It Takes to Fall in Love

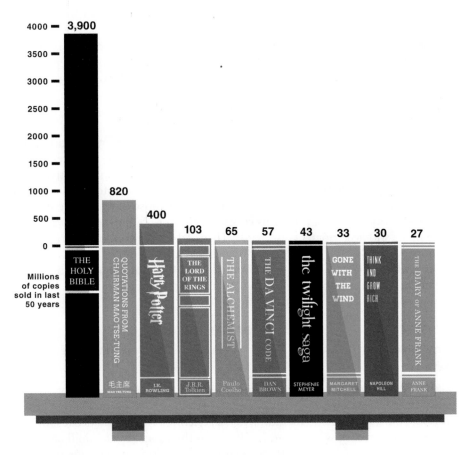

FIGURE 13-6 Column Chart: Most Read Books in the World

Bar and Column Charts   Figure 13-5 is a **bar chart**, a type of chart that compares two or more values by stretching them out in the form of horizontal rectangles. **Column charts**, such as the one shown in Figure 13-6, perform the same function as bar charts but use vertical rectangles.

bar chart A visual aid that compares two or more values by showing them as elongated horizontal rectangles.

column chart A visual aid that compares two or more values by showing them as elongated vertical rectangles.

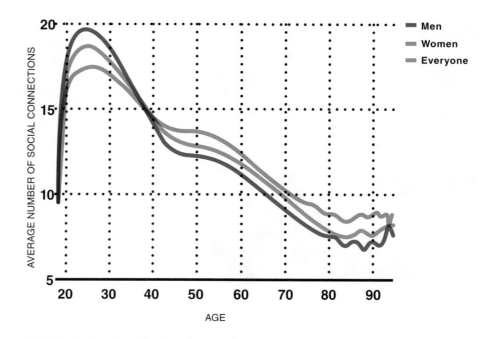

FIGURE 13-7 Line Chart: The Age When People Are the Most Popular

line chart A visual aid consisting of a grid that maps out the direction of a trend by plotting a series of points.

Line Charts A **line chart** maps out the direction of a moving point; it is ideally suited for showing changes over time. The time element is usually placed on the horizontal axis so that the line visually represents the trend over time. Figure 13-7 is a line chart.

Using Presentation Software

Several specialized programs exist just to produce visual aids. Among the most popular of these programs are Microsoft PowerPoint, Apple's Keynote, and Prezi.

In its simplest form, presentation software lets you build an effective slide show out of your basic outline. You can choose color-coordinated backgrounds and consistent formatting that match the tone and purpose of your presentation. Most presentation software programs contain a clip art library that allows you to choose images to accompany your words. They also allow you to import images from outside sources and to build your own charts.

If you would like to learn more about using PowerPoint, Keynote, and Prezi, you can easily find several web-based tutorial programs by typing the name of your preferred program into your favorite search engine.

Alternative Media for Presenting Graphics

When a projector in unavailable, a variety of materials can be used to present visual aids.

Chalkboards, Whiteboards, and Polymer Marking Surfaces The major advantage of these write-as-you-go media is their spontaneity. With them you can create your visual aid as you speak, including items generated from audience responses. Along with the odor of whiteboard markers and the squeaking of chalk, a major disadvantage of these media is the difficulty of preparing visual aids on them in advance, especially if several speeches are scheduled in the same room at the same hour.

Flip Pads and Poster Board Flip pads are like oversized writing tablets attached to a portable easel. Flip pads enable you to combine the spontaneity of the chalkboard (you can write on them as you go) with portability, which enables you to prepare

@WORK

The Pros and Cons of Presentation Software

PowerPoint is by far the most popular form of work presentation today. In fact, as one expert points out, "Today there are great tracts of corporate America where to appear at a meeting without PowerPoint would be unwelcome and vaguely pretentious, like wearing no shoes."[12] Prezi, as an enhanced form of PowerPoint, is subject to many of the same advantages and criticisms.

The Pros   The advantages of PowerPoint are well known. Proponents say that PowerPoint slides can focus the attention of audience members on important information at the appropriate time. The slides also help listeners appreciate the relationship between different pieces of information. By doing so, they make the logical structure of an argument more transparent.

Some experts think the primary advantage of PowerPoint is that it forces otherwise befuddled speakers to organize their thoughts in advance. Most, however, insist that its primary benefit is in providing two channels of information rather than just one. This gives audiences a visual source of information that is a more efficient way to learn than just by listening. One psychology professor puts it this way: "We are visual creatures. Visual things stay put, whereas sounds fade. If you zone out for 30 seconds—and who doesn't?—it is nice to be able to glance up on the screen and see what you missed."[13]

The Cons   For all its popularity, PowerPoint has received some bad press, having been featured in articles with such downbeat titles as "PowerPoint Is Evil"[14] and "Does PowerPoint Make You Stupid?"[15] But a 23-page pamphlet with a less dramatic title, *The Cognitive Style of PowerPoint*,[16] was the statement that truly put the anti-PowerPoint argument on the map, because it was authored by Edward R. Tufte, a well-respected author of several influential books on the effective design of visual aids.

According to Tufte, the use of low-content PowerPoint slides trivializes important information. It encourages oversimplification by asking the presenter to summarize key concepts in as few words as possible—the ever-present bullet points.

Tufte also insists that PowerPoint makes it easier for a speaker to hide lies and logical fallacies.

Perhaps most seriously, opponents of PowerPoint say that it is an enemy of interaction, that it interferes with the spontaneous give-and-take that is so important in effective public speaking. One expert summarized this effect by saying, "Instead of human contact, we are given human display."[17]

The Middle Ground?   PowerPoint proponents say that it is just a tool, one that can be used effectively or ineffectively. They are the first to admit that a poorly done PowerPoint presentation can be boring and ineffective, such as the infamous "triple delivery," in which precisely the same text is seen on the screen, spoken aloud, and printed on the handout in front of you. One proponent insists, "Tufte is correct in that most talks are horrible and most PowerPoint slides are bad—but that's not PowerPoint's fault. Most writing is awful, too, but I don't go railing against pencils or chalk."[18]

PowerPoint proponents say that PowerPoint should not be allowed to overpower a presentation—it should be just one element of a speech, not the whole thing. They point out that even before the advent of the personal computer, some people argued that speeches with visual aids stressed format over content. PowerPoint just makes it extremely easy to stress impressive format over less-than-impressive content, but that's a tendency that the effective speaker recognizes and works against.

After reviewing the pros and cons of PowerPoint, would you say that PowerPoint is a benefit or detriment to effective public speaking? Why?

them in advance. If you plan to use your visuals more than once, you can prepare them in advance on rigid poster board and display them on the same type of easel.

Despite their advantages, flip pads and poster boards are bulky, and preparing professional-looking exhibits on them requires a fair amount of artistic ability.

Handouts   The major advantage of handouts is that audience members can take away the information they contain after your speech. For this reason, handouts are excellent memory and reference aids. The major disadvantage is that they are distracting when handed out during a speech: First, there is the distraction of passing them out, and second, there is the distraction of having them in front of the audience members while you have gone on to something else. It's best, therefore, to pass them out at the end of the speech so audience members can use them as take-aways.

Rules for Using Visual Aids

It's easy to see that each type of visual aid and each medium for its presentation have their own advantages and disadvantages. No matter which type you use, however, there are a few rules to follow.

Simplicity Keep your visual aids simple. Your goal is to clarify, not confuse. Use only key words or phrases, not sentences. The "rule of seven" states that each exhibit you use should contain no more than seven lines of text, each with no more than seven words. Keep all printing horizontal. Omit all nonessential details.

Size Visual aids should be large enough for your entire audience to see them at one time but portable enough for you to get them out of the way when they no longer pertain to the point you are making.

Attractiveness Visual aids should be visually interesting and as neat as possible. If you don't have the necessary artistic or computer skills, try to get help from a friend or at the computer or audiovisual center on your campus.

Appropriateness Visuals must be appropriate to all the components of the speaking situation—you, your audience, and your topic—and they must emphasize the point you are trying to make. Don't make the mistake of using a visual aid that looks good but has only a weak link to the point you want to make—such as showing a map of a city transit system while talking about the condition of the individual cars.

Reliability You must be in control of your visual aid at all times. Test all electronic media (projectors, computers, and so on) in advance, preferably in the room where you will speak. Just to be safe, have nonelectronic backups ready in case of disaster. Be conservative when you choose demonstrations: Wild animals, chemical reactions, and gimmicks meant to shock a crowd can often backfire.

When it comes time for you to use the visual aid, remember one more point: Talk to your audience, not to your visual aid. Some speakers become so wrapped up in their props that they turn their backs on their audience and sacrifice all their eye contact.

Sample Speech

The sample speech for this chapter was presented by Jenni Chang and Lisa Dazols, whose profile began this chapter. This speech was presented at the 2015 TED-Women Conference in Monterey, California. Jenni and Lisa's speech is about an event, and also about a concept. The event was an around-the-world journey they took together, and the concept was LGBT progress in the developing world. It is a speech of description, using a handful of well-chosen examples.

Their informative purpose statement was worded like this:

> After listening to my speech, my audience will understand that courageous individuals outside of the West are fighting for LGBT rights in the developing world, furthering the momentum of global human rights.

Their thesis statement was worded like this:

> Around the world, individuals we like to call "Supergays" are fighting for gay rights.

Throughout their speech, notice how they make it easy for their audience to listen by limiting the number of examples they present, and by making the examples that they do present as interesting as possible. They do this through their use of brief videos with carefully worded intros and outros. Notice also their use of clear, simple language, and the way they generate audience involvement by personalizing their speech.

SAMPLE SPEECH   **Jenni Chang and Lisa Dazols**

This Is What LGBT Life Is Like Around the World

1   Jenni Chang: When I told my parents I was gay, the first thing they said to me was, "We're bringing you back to Taiwan." In their minds, my sexual orientation was America's fault. The West had corrupted me with divergent ideas, and if only my parents had never left Taiwan, this would not have happened to their only daughter. In truth, I wondered if they were right.

2   Of course, there are gay people in Asia, just as there are gay people in every part of the world. But is the idea of living an "out" life, in the "I'm gay, this is my spouse, and we're proud of our lives together" kind of way just a Western idea? If I had grown up in Taiwan, or any place outside of the West, would I have found models of happy, thriving LGBT people?

3   Lisa Dazols: I had similar notions. As an HIV social worker in San Francisco, I had met many gay immigrants. They told me their stories of persecution in their home countries, just for being gay, and the reasons why they escaped to the U.S. I saw how this had beaten them down. After 10 years of doing this kind of work, I needed better stories for myself. I knew the world was far from perfect, but surely not every gay story was tragic.

4   JC: So as a couple, we both had a need to find stories of hope. So we set off on a mission to travel the world and look for the people we finally termed as the "Supergays."

5   These would be the LGBT individuals who were doing something extraordinary in the world. They would be courageous, resilient, and most of all, proud of who they were. They would be the kind of person that I aspire to be. Our plan was to share their stories to the world through film.

6   LD: There was just one problem. We had zero reporting and zero filmmaking experience. We didn't even know where to find the Supergays, so we just had to trust that we'd figure it all out along the way. So we picked 15 countries in Asia, Africa, and South America, countries outside the West that varied in terms of LGBT rights. We bought a camcorder, ordered a book on how to make a documentary, and set off on an around-the-world trip.

7   JC: One of the first countries that we traveled to was Nepal. Despite widespread poverty, a decade-long civil war, and now recently, a devastating earthquake, Nepal has made significant strides in the fight for equality. One of the key figures in the movement is Bhumika Shrestha. A beautiful, vibrant transgendered woman, Bhumika has had to overcome being expelled from school and getting incarcerated because of her gender presentation. But, in 2007, Bhumika and Nepal's LGBT rights organization successfully petitioned the Nepali Supreme Court to protect against LGBT discrimination.

Jenni encourages audience involvement from the beginning by personalizing her speech, and receives a gratifying round of audience laughter.

She defines the important nuance of a key term, *gay*, in familiar words.

The audience laughs at the term *Supergays* . . .

. . . but they understand the meaning and importance of the term when Jenni carefully defines it.

Here Lisa shows a map of the world with their destinations marked, along with photos of Jenni and herself setting up their new video camera.

Video: Street scenes of their arrival in Nepal and their meeting with Bhumika Shrestha.

8 Here's Bhumika: [play video]

(Video) BS: What I'm most proud of? I'm a transgendered person. I'm so proud of my life. On December 21, 2007, the supreme court gave the decision for the Nepal government to give transgender identity cards and same-sex marriage.

9 LD: I can appreciate Bhumika's confidence on a daily basis. Something as simple as using a public restroom can be a huge challenge when you don't fit in to people's strict gender expectations. Traveling throughout Asia, I tended to freak out women in public restrooms. They weren't used to seeing someone like me. I had to come up with a strategy, so that I could just pee in peace.

Photo: Lisa outside a public restroom with very large international symbols for "man" and "woman."

10 So anytime I would enter a restroom, I would thrust out my chest to show my womanly parts, and try to be as non-threatening as possible. Putting out my hands and saying, "Hello," just so that people could hear my feminine voice. This all gets pretty exhausting, but it's just who I am. I can't be anything else.

11 JC: After Nepal, we traveled to India. On one hand, India is a Hindu society, without a tradition of homophobia. On the other hand, it is also a society with a deeply patriarchal system, which rejects anything that threatens the male–female order. When we spoke to activists, they told us that empowerment begins with ensuring proper gender equality, where the women's status is established in society. And in that way, the status of LGBT people can be affirmed as well.

Video: Street scenes of India.

12 LD: There we met Prince Manvendra. He's the world's first openly gay prince. Prince Manvendra came out on the *Oprah Winfrey Show*, very internationally. His parents disowned him and accused him of bringing great shame to the royal family. We sat down with Prince Manvendra and talked to him about why he decided to come out so very publicly.

Here he is:

Photo of Prince Manvendra.

(Video) Prince Manvendra: I felt there was a lot of need to break this stigma and discrimination which is existing in our society. And that instigated me to come out openly and talk about myself. Whether we are gay, we are lesbian, we are transgender, bisexual, or whatever sexual minority we come from, we have to all unite and fight for our rights. Gay rights cannot be won in the courtrooms, but in the hearts and the minds of the people.

13 JC: While getting my hair cut, the woman cutting my hair asked me, "Do you have a husband?" Now, this was a dreaded question that I got asked a lot by locals while traveling. When I explained to her that I was with a woman instead of a man, she was incredulous, and she asked me a lot of questions about my parents' reactions and whether I was sad that I'd never be able to have children.

Here they project a photo of Jenni with salon worker.

14 I told her that there are no limitations to my life and that Lisa and I do plan to have a family some day. Now, this woman was ready to write me off as yet another crazy Westerner. She couldn't imagine that such a phenomenon could happen in her own country.

15 That is, until I showed her the photos of the Supergays that we interviewed in India. She recognized Prince Manvendra from television and soon I had an audience of other hairdressers interested in meeting me.

16 And in that ordinary afternoon, I had the chance to introduce an entire beauty salon to the social changes that were happening in their own country.

17 LD: From India, we traveled to East Africa, a region known for intolerance towards LGBT people. In Kenya, 89 percent of people who come out to their families are disowned. Homosexual acts are a crime and can lead to incarceration.

Video: Street scenes in Kenya, photos of David Kuria.

18 In Kenya, we met the soft-spoken David Kuria. David had a huge mission of wanting to work for the poor and improve his own government. So he decided to run for senate. He became Kenya's first openly gay political candidate. David wanted to run his campaign without denying the reality of who he was. But we were worried for his safety because he started to receive death threats.

Lisa carefully introduces the following video clip, so that the sound bite adds to what she is saying rather than simply repeating it.

(Video) David Kuria: At that point, I was really scared because they were actually asking for me to be killed. And, yeah, there are some people out there who do it and they feel that they are doing a religious obligation.

19 JC: David wasn't ashamed of who he was. Even in the face of threats, he stayed authentic.

Lisa provides an outro for her clip.

20 LD: At the opposite end of the spectrum is Argentina. Argentina's a country where 92 percent of the population identifies as Catholic. Yet, Argentina has LGBT laws that are even more progressive than here in the US. In 2010, Argentina became the first country in Latin America and the 10th in the world to adopt marriage equality. There, we met María Rachid. María was a driving force behind that movement.

Video: street scenes from Argentina.

Photo: María Rachid and her mother. María Rachid (Spanish): I always say that, in reality, the effects of marriage equality are not only for those couples that get married. They are for a lot of people that, even though they may never get married, will be perceived differently by their coworkers, their families and neighbors, from the national state's message of equality. I feel very proud of Argentina because Argentina today is a model of equality. And hopefully soon, the whole world will have the same rights.

(Video) Group of young Chinese people: One, two, three. Welcome gays to Shanghai!

21 JC: When we made the visit to my ancestral lands, I wish I could have shown my parents what we found there. Because here is who we met:

22 A whole community of young, beautiful Chinese LGBT people. Sure, they had their struggles. But they were fighting it out. In Shanghai, I had the chance to speak to a local lesbian group and tell them our story in my broken Mandarin Chinese. In Taipei, each time we got onto the metro, we saw yet another lesbian couple holding hands. And we learned that Asia's largest LGBT pride event happens just blocks away from where my grandparents live. If only my parents knew.

23 LD: By the time we finished our not-so-straight journey around the world, we had traveled 50,000 miles and logged 120 hours of video footage. We traveled to 15 countries and interviewed 50 Supergays. Turns out, it wasn't hard to find them at all.

Statistics.

24 JC: Yes, there are still tragedies that happen on the bumpy road to equality. And let's not forget that 75 countries still criminalize homosexuality today. But there are also stories of hope and courage in every corner of the world. What we ultimately took away from our journey is, equality is not a Western invention.

An essential statistic.

Video: Lisa and Jenni conducting interviews.

25 LD: One of the key factors in this equality movement is momentum, momentum as more and more people embrace their full selves and use whatever opportunities they have to change their part of the world, and momentum as more and more countries find models of equality in one another. When Nepal protected against LGBT discrimination, India pushed harder. When Argentina embraced marriage equality, Uruguay and Brazil followed. When Ireland said yes to equality, the world stopped to notice. When the US Supreme Court makes a statement to the world, that we can all be proud of.

(Applause.)

26 JC: As we reviewed our footage, what we realized is that we were watching a love story. It wasn't a love story that was expected of me, but it is one filled with more freedom, adventure, and love than I could have ever possibly imagined. One year after returning home from our trip, marriage equality came to California. And in the end, we believe, love will win out.

(Video: Lisa and Jenni preparing for their wedding.)

Memorable wording summarizes the take-away from this speech.

(Video: The wedding ceremony.)

Minister: By the power vested in me, by the state of California and by God Almighty, I now pronounce you spouses for life. You may kiss.

(Standing ovation from the audience.)

MAKING THE GRADE

For more resources to help you understand and apply the information in this chapter, visit the *Understanding Human Communication* website at www.oup.com/us/adleruhc.

Objective 13.1 Distinguish among the main types of informative speaking.

- Informative speeches can be classified by the type of content, including speeches about objects, processes, events, and concepts.
- Informative speeches can also be classified according to their purpose, including descriptions, explanations, and instructions.
  > Explain the primary differences among types of informative speeches.
  > Which type of informative speech is most important in your own life?
  > Which type of speech will you use for your next informative speech?

Objective 13.2 Describe the differences between informative and persuasive speaking.

- Two basic characteristics differentiate an informative topic from a persuasive topic.
- In an informative speech, you generally do not present information that your audience is likely to disagree with.
- The speaker's primary intent is not to change attitudes or to make the audience members *feel* differently about the topic.
  > Explain the primary difference between informative and persuasive speaking.
  > How are informative and persuasive speeches similar?
  > Why is it important to distinguish between informative and persuasive speaking?

Objective 13.3 Increase your informative effectiveness by using the techniques discussed in this chapter.

- Decide on a specific purpose statement.
- Create information hunger by tapping into audience needs.

- Make it easy to listen by limiting the amount of information, and by using familiar information, straightforward organization, clear language, and effective supporting material.
- Involve your audience through audience participation, the use of volunteers, and a question-and-answer period.
  - > List at least three techniques for increasing informative effectiveness.
  - > Which is the most important technique in informative speaking?
  - > Which technique do you need to work on for your next speech?

Objective 13.4 Use visual aids appropriately and effectively.

- Visual aids include objects and models, photos, videos, audio files, charts, and graphs.
- Media for the presentation of visual aids include digital projectors with presentation software, chalkboards, whiteboards, flip pads, and handouts.
- Each type of visual aid has its own advantages and disadvantages.
- Keep visual aids simple, large enough for the audience to see, and visually interesting.
  - > List at least three types of visual aids, giving examples of when to use them.
  - > Describe the advantages and disadvantages of the visual aids on your list.
  - > Which types of visual aids will you use for your next informative speech?

KEY TERMS

audience involvement p. 358

audience participation p. 359

bar chart p. 363

column chart p. 363

description p. 352

diagram p. 362

explanations p. 352

information anxiety p. 351

information hunger p. 355

informative purpose statement p. 354

instructions p. 352

intro p. 361

knowledge p. 351

line chart p. 364

model (in speeches and presentations) p. 361

number chart p. 362

outro p. 361

pictogram p. 362

pie chart p. 362

signpost p. 358

sound bite p. 361

visual aids p. 361

vocal citation p. 357

word chart p. 362

wrap-around p. 361

ACTIVITIES

1. **Informative Purpose Statements** For practice in defining informative speech purposes, reword the following statements so that they specifically point out what the audience will be able to do after hearing the speech.

   a. My talk today is about building a wood deck.

   b. My purpose is to tell you about vintage car restoration.

   c. I am going to talk about toilet training.

   d. I'd like to talk to you today about sexist language.

   e. There are six basic types of machines.

   f. The two sides of the brain have different functions.

   g. Do you realize that many of you are sleep deprived?

2. **Effective Repetition** Create a list of three statements, or use the three that follow. Restate each of these ideas in three different ways.

   a. The magazine *Modern Maturity* has a circulation of more than 20 million readers.

   b. Before buying a used car, you should have it checked out by an independent mechanic.

   c. One hundred thousand pounds of dandelions are imported into the United States annually for medical purposes.

3. **Using Clear Language** For practice in using clear language, select an article from any issue of a professional journal in your major field. Using the suggestions in this chapter, rewrite a paragraph from the article so that it will be clear and interesting to a layperson.

4. **Inventing Visual Aids** Take any sample speech. Analyze it for where visual aids might be effective. Describe the visual aids that you think will work best. Compare the visuals you devise with those of your classmates.

Persuasive Speaking

LEARNING OBJECTIVES

14.1

Identify the primary characteristics of persuasion.

14.2

Compare and contrast different types of persuasion.

14.3

Apply the guidelines for persuasive speaking to a speech you will prepare.

14.4

Explain how to best adapt a specific speech to a specific audience.

14.5

Improve your credibility in your next persuasive speech.

KRISTINA MEDERO says she always wanted to change the world. In high school, that led her to work with inner-city kids, teaching them slam poetry. In college, Medero channeled her passion for changing the world into a series of speeches for Western Kentucky University's championship forensics team.

For one of her WKU speech projects, she decided to write about gun violence in America. To her surprise, she couldn't find any objective and reliable research on this massive and obvious problem. This wasn't one of those "I Googled it and couldn't find anything" moments. Medero had done all the in-depth research suggested in Chapter 11 of this book, and she still found mostly conflicting estimates and opinions favoring one side or the other. With a little more digging, she found out what the problem was: a piece of legislation that essentially forbids research into gun violence. As a researcher and a concerned citizen, she was horrified.

This discovery led to one of her most memorable persuasive speeches, titled "Who's Afraid of Gun Facts?" The speech helped her win an All American Forensics award and is reprinted at the end of this chapter as our sample speech.

Medero graduated in 2015 and immediately left for South Africa to volunteer with a nonprofit agency that helps poverty-stricken populations. She returned in 2016 as a Health Program Manager. Medero now says that persuasive speaking has helped her realize that "a healthy community is created by more than just medicine. It takes equal opportunities, unique innovation, and using scientific methods to form lasting solutions."[1] She plans to present a lifetime of persuasive speeches to fulfill that goal.

How persuasion works and how to accomplish it successfully are complex topics. Our understanding of persuasion begins with classical wisdom and extends to the latest psychological research. We begin by looking at what we really mean by the term.

Characteristics of Persuasion

Persuasion is the process of motivating someone, through communication, to change a particular belief, attitude, or behavior. Implicit in this definition are several characteristics of persuasion.

Persuasion Is Not Coercive

Persuasion is not the same thing as coercion. If you put someone in a headlock and said, "Do this, or I'll choke you," you would be acting coercively. Besides being illegal, this approach would be ineffective. As soon as the authorities came and took you away, the person would stop following your demands.

The failure of coercion to achieve lasting results is also apparent in less dramatic circumstances. Children whose parents are coercive often rebel as soon as they can; students who perform from fear of an instructor's threats rarely appreciate the subject matter; and employees who work for abusive and demanding employers are often unproductive and eager to switch jobs as soon as possible. Persuasion, by contrast, makes a listener *want* to think or act differently.

Persuasion Is Usually Incremental

Attitudes do not normally change instantly or dramatically. Persuasion is a process. When it is successful, it generally succeeds over time, in increments, and usually small increments at that. The realistic speaker, therefore, establishes goals and expectations that reflect this characteristic of persuasion.

Communication scientists explain this characteristic of persuasion through **social judgment theory**.[2] This theory tells us that when members of an audience hear a persuasive appeal, they compare it to opinions that they already hold. The preexisting opinion is called an **anchor**, but around this anchor there exist what are called **latitudes of acceptance**, **latitudes of rejection**, and **latitudes of non-commitment**. A diagram of any opinion, therefore, might look something like Figure 14-1.

People who care very strongly about a particular point of view will have a very narrow latitude of noncommitment. People who care less strongly will have a wider latitude of noncommitment. Research suggests that audience members

persuasion The act of motivating a listener, through communication, to change a particular belief, attitude, value, or behavior.

social judgment theory The theory that opinions will change only in small increments and only when the target opinions lie within the receiver's latitudes of acceptance and noncommitment.

anchor The position supported by audience members before a persuasion attempt.

latitude of acceptance In social judgment theory, statements that a receiver would not reject.

latitude of rejection In social judgment theory, statements that a receiver could not possibly accept.

latitude of noncommitment In social judgment theory, statements that a receiver would not care strongly about one way or another.

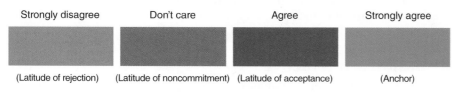

| Strongly disagree | Don't care | Agree | Strongly agree |
| (Latitude of rejection) | (Latitude of noncommitment) | (Latitude of acceptance) | (Anchor) |

FIGURE 14-1 Latitudes of Acceptance, Rejection, and Noncommitment

simply will not respond to appeals that fall within their latitude of rejection. This means that persuasion in the real world takes place in a series of small movements. One persuasive speech may be but a single step in an overall persuasive campaign. The best example of this is the various communications that take place during the months of a political campaign. Candidates watch the opinion polls carefully, adjusting their appeals to the latitudes of acceptance and noncommitment of the uncommitted voters.

Public speakers who heed the principle of social judgment theory tend to seek realistic, if modest, goals in their speeches. For example, if you were hoping to change audience views on the pro-life/pro-choice question, social judgment theory suggests that the first step would be to consider a range of arguments such as this:

Abortion is a sin.

Abortion should be absolutely illegal.

Abortion should be allowed only in cases of rape and incest.

A woman should be required to have her husband's permission to have an abortion.

A girl under the age of 18 should be required to have a parent's permission before she has an abortion.

Abortion should be allowed during the first 3 months of pregnancy.

A girl under the age of 18 should not be required to have a parent's permission before she has an abortion.

A woman should not be required to have her husband's permission to have an abortion.

Abortion is a woman's personal decision.

Abortion should be discouraged but legal.

Abortion should be available anytime to anyone.

Abortion should be considered simply a form of birth control.

You could then arrange these positions on a continuum and estimate how listeners would react to each one. The statement that best represented the listeners' point of view would be their anchor. Other items that might also seem reasonable to them would make up their latitude of acceptance. Opinions that they would reject would make up their latitude of rejection. Those statements that are left would be the listeners' latitude of noncommitment.

Social judgment theory suggests that the best chance of changing audience attitudes would come by presenting an argument based on a position that fell somewhere within the listeners' latitude of noncommitment—even if this wasn't the position that you ultimately wanted them to accept. If you pushed too hard by arguing a position in your audience's latitude of rejection, your appeals would probably backfire, making your audience *more* opposed to you than before.

Persuasion Is Interactive

The transactional model of communication described in Chapter 1 makes it clear that persuasion is not something you do *to* audience members but rather something you do *with* them. This mutual activity is best seen in an argument between two people, in which an openness to opposing arguments is essential to resolution. As one observer has pointed out,

cultural idiom

backfire: produce a result opposite of the one intended

Arguments are not won by shouting down opponents. They are won by changing opponents' minds—something that can happen only if we give opposing arguments a respectful hearing and still persuade their advocates that there is something wrong with those arguments. In the course of this activity, we may well decide that there is something wrong with our own.[3]

Even in public communication, both speaker and audience are active. This might be manifested in the speaker taking an audience survey *before* a speech, a sensitivity to audience reactions *during* a speech, or an open-minded question-and-answer period *after* a speech.

Persuasion Can Be Ethical

Even when they understand the difference between persuasion and coercion, some people are still uncomfortable with the idea of persuasive speaking. They see it as the work of high-pressure hucksters: salespeople with their feet stuck in the door, unscrupulous politicians taking advantage of beleaguered taxpayers, and so on. Indeed, many of the principles we are about to discuss have been used by unethical speakers for unethical purposes, but that is not what all—or even most—persuasion is about. Ethical persuasion plays a necessary and worthwhile role in everyone's life.

It is through ethical persuasion that we influence others' lives in worthwhile ways. The person who says, "I do not want to influence other people," is really saying, "I do not want to get involved with other people," and that is an abandonment of one's responsibilities as a human being. Look at the good you can accomplish through persuasion: You can convince a loved one to give up smoking or to not keep a firearm in the house; you can get members of your community to conserve energy or to join together to refurbish a park; you can persuade an employer to hire you for a job in which your own talents, interests, and abilities will be put to their best use.

Persuasion is considered ethical if it conforms to accepted standards. But what are the standards today? If your plan is selfish and not in the best interest of your audience members, but you are honest about your motives—is that ethical? If your plan is in the best interest of your audience members, yet you lie to them to get them to accept the plan—is that ethical? Philosophers and rhetoricians have argued for centuries over questions like these.

There are many ways to define **ethical persuasion**.[4] For our purpose, we will consider it as *communication in the best interest of the audience that does not depend on false or misleading information to change an audience's attitude or behavior*. The best way to appreciate the value of this simple definition is to consider the many strategies listed in Table 14-1 that do not fit it. For example, faking enthusiasm about a speech topic, plagiarizing material from another source and passing it off as your own, and making up statistics to support your case are clearly unethical.

Besides being wrong on moral grounds, unethical attempts at persuasion have a major practical disadvantage: If your deception is uncovered, your credibility will suffer. If, for example, prospective buyers uncover your attempt to withhold a structural flaw in the condominium you are trying to sell, they will probably suspect that the property has other hidden problems. Likewise, if your speech instructor suspects that you are <u>lifting material</u> from other sources without giving credit, your entire presentation will be suspect. One unethical act can cast doubt on future truthful statements. Thus, for pragmatic as well as moral reasons, honesty really is the best policy. Kristina Medero's speech at the end of this chapter is an example of an honest, ethical speech.

ASK YOURSELF

What does an ethical persuasive speaker need to do to ensure that he or she is sharing a well-founded message?

ethical persuasion Persuasion in an audience's best interest that does not depend on false or misleading information to induce change in that audience.

cultural idiom
lifting material: using another's words or ideas as one's own

ETHICAL CHALLENGE
Analyzing Communication Behaviors

Read Table 14-1 carefully. The behaviors listed there are presented in what some (but certainly not all) communication experts would describe as "most serious to least serious" ethical faults. Do you agree or disagree with the order of this list? Explain your answer and whether you would change the order of any of these behaviors. Are there any other behaviors that you would add to this list?

TABLE 14-1
Unethical Communication Behaviors

1. Committing Plagiarism
  a. Claiming someone else's ideas as your own
  b. Quoting without citing the source

2. Relaying False Information
  a. Deliberate lying
  b. Ignorant misstatement
  c. Deliberate distortion and suppression of material
  d. Fallacious reasoning to misrepresent truth

3. Withholding Information; Suppression
  a. About self (speaker); not disclosing private motives or special interests
  b. About speech purpose
  c. About sources (not revealing sources; plagiarism)
  d. About evidence; omission of certain evidence (card stacking)
  e. About opposing arguments; presenting only one side

4. Appearing to Be What One Is Not; Insincerity
  a. In words, saying what one does not mean or believe
  b. In delivery (for example, feigning enthusiasm)

5. Using Emotional Appeals to Hinder Truth
  a. Using emotional appeals as a substitute or cover-up for lack of sound reasoning and valid evidence
  b. Failing to use balanced appeals

Source: Adapted from Andersen, M. K. (1979). *An analysis of the treatment of ethos in selected speech communication textbooks* (Unpublished dissertation). University of Michigan, Ann Arbor, pp. 244–247.

Categorizing Types of Persuasion

There are several ways to categorize the types of persuasive attempts you will make as a speaker. What kinds of subjects will you focus on? What results will you be seeking? How will you go about getting those results? In the following pages we will look at each of these questions.

By Types of Proposition

Persuasive topics fall into one of three categories, depending on the type of thesis statement (referred to as a "proposition" in persuasion) that you are advancing. The three categories are propositions of fact, propositions of value, and propositions of policy.

Propositions of Fact   Some persuasive messages focus on **propositions of fact**: issues in which there are two or more sides about conflicting information, in which listeners are required to choose the truth for themselves. Here are some examples of questions of fact:

> The National Security Agency was/was not justified in listening in to the phone calls of everyday citizens.

proposition of fact A claim bearing on issue in which there are two or more sides of conflicting factual evidence.

One Pinocchio
Some shading of the facts. Selective telling of the truth. Some omissions and exaggerations, but no outright falsehoods. (You could view this as "mostly true.")

Two Pinocchios
Significant omissions and/or exaggerations. Some factual error may be involved but not necessarily. A politician can create a false, misleading impression by playing with words and using legalistic language that means little to ordinary people. (Similar to "half true.")

Three Pinocchios
Significant factual error and/or obvious contradictions. This gets into the realm of "mostly false." It could include statements that are technically correct (e.g., based on official government data) but that are so taken out of context as to be very misleading. The line between Two and Three can be bit fuzzy, and we do not award half-Pinocchios. So we strive to explain the factors that tipped us toward a Three.

Four Pinocchios
Whoppers.

FIGURE 14-2 Checking Propositions of Fact Fact checking websites evaluate a speaker's propositions of fact. The Washington Post uses a scale that pictures Pinocchio, the fairy tale puppet whose nose grew every time he told a lie.

Windmills are/are not a practical way for the private homeowner to create clean energy.

Bottled water is/is not healthier for you than tap water.

These examples show that many questions of fact can't be settled with a simple "yes" or "no" or with an objective piece of information. Rather, they are open to debate, and answering them requires careful examination and interpretation of evidence, usually collected from a variety of sources. That's why it is possible to debate questions of fact, and that's why these propositions form the basis of persuasive speeches and not informative ones.

proposition of value A claim bearing on an issue involving the worth of some idea, person, or object.

proposition of policy A claim bearing on an issue that involves adopting or rejecting a specific course of action.

convincing A speech goal that aims at changing audience members' beliefs, values, or attitudes.

Propositions of Value   **Propositions of value** go beyond issues of truth or falsity and explore the worth of some idea, person, or object. Propositions of value include the following:

Cheerleaders are/are not just as valuable as the athletes on the field.

The United States is/is not justified in attacking countries that harbor terrorist organizations.

The use of laboratory animals for scientific experiments is/is not cruel and immoral.

In order to deal with most propositions of value, you will have to explore certain propositions of fact. For example, you won't be able to debate whether the experimental use of animals in research is immoral—a proposition of value—until you have dealt with propositions of fact such as how many animals are used in experiments and whether experts believe they actually suffer.

Propositions of Policy   **Propositions of policy** go one step beyond questions of fact or value by recommending a specific course of action (a "policy"). Some questions of policy are these:

The World Bank should/should not create a program of microloans for citizens of impoverished nations.

The Electoral College should/should not be abolished.

Genetic engineering of plants and livestock is/is not an appropriate way to increase the food supply.

Looking at persuasion according to the type of proposition is a convenient way to generate topics for a persuasive speech, because each type of proposition suggests different topics. Selected topics could also be handled differently depending on how they are approached. For example, a campaign speech could be approached as a proposition of fact ("Candidate X has done more for this community than the opponent"), a proposition of value ("Candidate X is a better person than the opponent"), or a proposition of policy ("We should get out and vote for Candidate X"). Remember, however, that a fully developed persuasive speech is likely to contain all three types of propositions. If you were preparing a speech advocating that college athletes should be paid in cash for their talents (a proposition of policy), you might want to first prove that the practice is already widespread (a proposition of fact) and that it is unfair to athletes from other schools (a proposition of value).

By Desired Outcome

We can also categorize persuasion according to two major outcomes: convincing and actuating.

Convincing   When you set out to **convince** an audience, you want to change the way its members think. When we say that convincing an audience changes the way its members think, we do not mean that you have to swing them from one belief or attitude to a completely different one. Sometimes audience members will already think the way you want them to, but they will not be firmly enough committed to that way of thinking. When that is the case, you reinforce, or strengthen, their opinions. For example, if your audience already believed that the federal budget should be balanced but did not consider the idea important, your job would be to reinforce members' current beliefs. Reinforcing is still a type of change, however, because you are causing an audience to adhere more strongly to a belief or attitude. In other cases, a speech to convince will begin to shift attitudes without bringing about a total change of thinking. For example, an effective speech to convince might get a group of skeptics to consider the possibility that bilingual education is/isn't a good idea.

Actuating   When you set about to **actuate** an audience, you want to move its members to a specific behavior. Whereas a speech to convince might move an audience to action, it won't be any specific action that you have recommended. In a speech to actuate, you do recommend that specific action.

There are two types of action you can ask for—adoption or discontinuance. The former asks an audience to engage in a new behavior; the latter asks an audience to stop behaving in an established way. If you gave a speech for a political candidate and then asked for contributions to that candidate's campaign, you would be asking your audience to adopt a new behavior. If you gave a speech against smoking and then asked your audience members to sign a pledge to quit, you would be asking them to discontinue an established behavior.

By Directness of Approach

We can also categorize persuasion according to the directness of approach employed by the speaker.

Direct Persuasion   In **direct persuasion** the speaker will make his or her purpose clear, usually by stating it outright early in the speech. This is the best strategy to use with a friendly audience, especially when you are asking for a response that the audience is likely to give you. Direct persuasion is the kind we hear in most academic situations.

Indirect Persuasion   **Indirect persuasion** disguises or deemphasizes the speaker's persuasive purpose in some way. The question, "Is a season ticket to the symphony worth the money?" (when you intend to prove that it is) is based on indirect persuasion, as is any strategy that does not express the speaker's purpose at the outset.

Indirect persuasion is sometimes easy to spot. A television commercial that shows us attractive young men and women romping in the surf on a beautiful day and then flashes the product name on the screen is pretty indisputably indirect persuasion. Political oratory also is sometimes indirect persuasion, and it can be more difficult to identify as such. A political hopeful ostensibly might be speaking on some great social issue when the real persuasive message is, "Please remember my name, and vote for me in the next election."

In public speaking, indirect persuasion is usually disguised as informative speaking, but this approach isn't necessarily unethical. In fact, it is probably the best approach to use when your audience is hostile to either you or your topic. It is also often necessary to use the indirect approach to get a hearing from listeners who would <u>tune you out</u> if you took a more direct approach. Under such circumstances, you might want to ease into your speech slowly.[5] You might take some time to make your audience feel good about you or the social action you are advocating. If you are speaking in favor of your candidacy for city council, but you are in favor of a tax increase and your audience is not, you might talk for a while about the benefits that a well-financed city council can provide to the community. You might even want to change your desired audience response. Rather than trying to get audience members to rush out to vote for you, you might want them simply to read a policy statement that you have written or become more informed

actuate To move members of an audience toward a specific behavior.

direct persuasion Persuasion that does not try to hide or disguise the speaker's persuasive purpose.

indirect persuasion Persuasion that disguises or deemphasizes the speaker's persuasive goal.

cultural idiom
tune you out: stop listening to you

on a particular issue. The one thing you cannot do in this instance is to begin by saying, "My appearance here today has nothing to do with my candidacy for city council." That would be a false statement. It is more than indirect; it is untrue and therefore unethical.

The test of the ethics of an indirect approach would be whether you would express your persuasive purpose directly if asked to do so. In other words, if someone in the audience stopped you and asked, "Don't you want us to vote for you for city council?," you would admit to it rather than deny your true purpose, if you were ethical.

Creating the Persuasive Message

Persuasive speaking has been defined as "reason-giving discourse." Its principal technique, therefore, involves proposing claims and then backing those claims up with reasons that are true. Preparing an effective persuasive speech isn't easy, but it can be made easier by observing a few simple rules. These include the following: Set a clear, persuasive purpose; structure the message carefully; use solid evidence; and avoid fallacies.

Set a Clear, Persuasive Purpose

Remember that your objective in a persuasive speech is to move the audience to a specific, attainable attitude or behavior. In a speech to convince, the purpose statement will probably stress an attitude:

> After listening to my speech, my audience members will agree that steps should be taken to save whales from extinction.

In a speech to actuate, the purpose statement will stress behavior in the form of a desired audience response. That desired audience response should be as straightforward and clear-cut as possible. For example, in the sample speech at the end of this chapter, Kristina Medero's purpose was to get at least some of her audience members to send an email to their congressional representative.

As Chapter 11 explained, your purpose statement should always be specific, attainable, and worded from the audience's point of view. "The purpose of my speech is to save the whales" is not a purpose statement that has been carefully thought out. Your audience members wouldn't be able to jump into the ocean and save the whales, even if your speech motivated them into a frenzy. They might, however, be able to support a specific piece of legislation.

A clear, specific purpose statement will help you stay on track throughout all the stages of preparation of your persuasive speech. Because the main purpose of your speech is to have an effect on your audience, you have a continual test that you can use for every idea, every piece of evidence, and every organizational structure that you think of using. The question you ask is "Will this help me to get the audience members to think/feel/behave in the manner I have described in my purpose statement?" If the answer is "yes," you forge ahead.

Structure the Message Carefully

A sample structure of the body of a persuasive speech is outlined in Figure 14-3. With this structure, if your objective is to convince, you concentrate on the first two components: establishing the problem and describing the solution. If your objective is to actuate, you add the third component, describing the desired audience reaction.

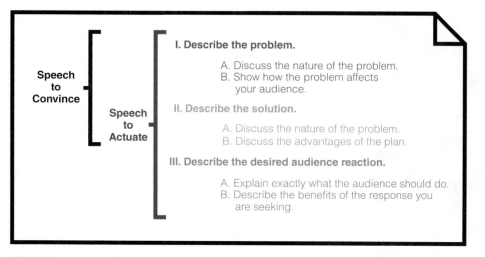

FIGURE 14-3 Sample Structure for a Persuasive Speech

There are, of course, other structures for persuasive speeches. This one can be used as a basic model, however, because it is easily applied to most persuasive topics.

Describe the Problem   To convince an audience that something needs to be changed, you have to show members that a problem exists. After all, if your listeners don't recognize the problem, they won't find your arguments for a solution very important. An effective description of the problem will answer two questions, either directly or indirectly.

What Is the Nature of the Problem?   Your audience members might not recognize that the topic you are discussing is a problem at all, so your first task is to convince them that there is something wrong with the present state of affairs. For example, if your thesis were "This town needs a shelter for homeless families," you would need to show that there are, indeed, homeless families and that the plight of these homeless families is serious.

Your approach to defining your problem will depend largely on your audience analysis, as discussed in Chapter 11. If your prespeech analysis shows that your audience may not feel sympathetic to your topic, you will need to explain why your topic is, indeed, a problem that your audience should recognize. In a speech about the plight of the homeless, you might need to establish that most homeless people are not lazy, able-bodied drifters who choose to panhandle and steal instead of work. You could cite respected authorities, give examples, and maybe even show photographs to demonstrate that some homeless people are hardworking but unlucky parents and innocent children who lack shelter owing to forces beyond their control.

How Does the Problem Affect Your Audience?   It's not enough to prove that a problem exists. Your next challenge is to show your listeners that the problem affects them in some way. This is relatively easy in some cases: the high cost of tuition, the lack of convenient parking near campus, the quality of food in the student center. In other cases, you will need to spell out the impact to your listeners more clearly. Rebecca Yocum, a student at West Chester University of Pennsylvania, presented a speech on problems with the 911 emergency response system. She connected this topic to her audience in the following way:

> On May 15th of last year, a fire broke out at Michelle Dzoch's home in Wilkes-Barre, Pennsylvania. One of Michelle's neighbors called 911 only three minutes after the start of the incident. 911 operators quickly dispatched three fire and

On the reality TV series *Shark Tank*, budding entrepreneurs pitch their ideas to potential investors.

What persuasive strategies distinguish successful contestants from unsuccessful ones? How can you adapt these strategies to your persuasive appeals?

rescue units—to an incorrect address fifteen miles away from the actual fire. Meanwhile, Michelle Dzoch lay dying in the second-floor bathroom of her burning home.

The National Emergency Number Association website estimates that 240 million calls are made to 911 in the United States every year, with the potential for millions of cases of inaccurate dispatches. In order to mitigate this risk for each of us, we need a better system for locating those in need so that our emergency response units can do their jobs with efficiency.[6]

The problem section of a persuasive speech is often broken up into segments discussing the cause and the effect of the problem. (The sample speech at the end of this chapter is an example of this type of organization.)

Describe the Solution   Your next step in persuading your audience members is to convince them that there is an answer to the problem you have just introduced. To describe your solution, you should answer the following two questions.

Will the Solution Work?   A skeptical audience might agree with the desirability of your solution but still not believe that it has a chance of succeeding. In the homeless speech discussed previously, you would need to prove that establishing a shelter can help unlucky families get back on their feet—especially if your audience analysis shows that some listeners might view such a shelter as a way of coddling people who are too lazy to work.

What Advantages Will Result from Your Solution?   You need to describe in specific terms how your solution will lead to the desired changes. This is the step in which you will paint a vivid picture of the benefits of your proposal. In the speech proposing a shelter for homeless families, the benefits you describe would probably include these:

1.   Families will have a safe place to stay, free of the danger of living on the street.
2.   Parents will have the resources that will help them find jobs: an address, telephone, clothes washers, and showers.
3.   The police won't have to apply antivagrancy laws (such as prohibitions against sleeping in cars) to people who aren't the intended target of those laws.
4.   The community (including your listeners) won't need to feel guilty about ignoring the plight of unfortunate citizens.

Describe the Desired Audience Response   When you want to go beyond simply convincing your audience members and impel them to follow your solution, you need to describe exactly what you want them to do. This action step, like the previous ones, should answer two questions.

What Can the Audience Do to Put Your Solution into Action?   Make the behavior you are asking your audience members to adopt as clear and simple as possible for them. If you want them to vote in a referendum, tell them when and where to go to vote and how to go about registering, if necessary (some activists even provide transportation). If you're asking them to support a legislative change, don't expect them to write their congressional representative. *You* write the letter or draft a petition and ask them to sign it. If you're asking for a donation, pass the hat at the conclusion of your speech, or give audience members a stamped, addressed envelope and simple forms that they can return easily.

cultural idioms

get back on their feet: return to a financially stable lifestyle

paint a . . . picture: describe in detail

What Are the Direct Rewards of This Response?   Your solution might be important to society, but your audience members will be more likely to adopt it if you can show that they will get a personal payoff. Show that supporting legislation to reduce acid rain will produce a wide range of benefits, from reduced lung damage to healthier forests to longer life for their car's paint. Explain that saying "no" to a second drink before driving will not only save lives but also help your listeners avoid expensive court costs, keep their insurance rates low, and prevent personal humiliation. Show how helping to establish and staff a homeless shelter can lead to personal feelings of satisfaction and provide an impressive demonstration of community service on a job-seeking résumé.

Adapt the Model Persuasive Structure   Describing the problem and the solution makes up the basic structure for any persuasive speech. However, you don't have to analyze too many successful persuasive speeches to realize that the best of them do far more than this minimum. In one adaptation of the basic model, the speaker will combine the solution with the desired audience response. Another adaptation is known as the Motivated Sequence (discussed in Chapter 12; see pages 331–332).

Use Solid Evidence

All the forms of support discussed in Chapter 12 can be used to back up your persuasive arguments.[7] Your objective here is not just to find supporting material that clarifies your ideas, but rather to find the perfect example, statistic, definition, analogy, anecdote, or testimony to establish the truth of your claim in the mind of this specific audience.

You choose **evidence** that strongly supports your claim, and you should feel free to use **emotional evidence**, which is supporting material that evokes audience feelings such as fear, anger, sympathy, pride, or reverence. Emotional evidence is an ethical fault only when it is used to obscure the truth (see Table 14-1, page 378). It is ethical, however, to use emotion to give impact to a truth.

Whatever type of evidence you use, you should cite your sources carefully. It is important that your audience know that your sources are credible, unbiased, and current. If you are quoting the source of an interview, give a full statement of the source's credentials:

> According to Sean Wilentz, Dayton–Stockton Professor of History, Director of American Studies at Princeton University, and the author of several books on this topic . . .

If the currency of the interview is important, you might add, "I spoke to Professor Wilentz just last week. . . ." If you are quoting an article, give a quick statement of the author's credentials and the full date and title of the magazine:

> According to Professor Sean Wilentz of Princeton University, in an article in the April 21, 2016, *Rolling Stone Magazine* . . .

You do not need to give the title of the article (although you may, if it helps in any way) or the page number. If you are quoting from a book, include a quick statement of the author's credentials:

> According to Professor Sean Wilentz of Princeton University, in his book *The Rise of American Democracy* . . .

You don't need to include the copyright date unless it's important to authenticate the currency of the quotation, and you don't have to mention the publisher or city of publication unless it's relevant to your topic. Generally, if you're unsure about how to cite your sources in a speech, you should err in the direction of too much information rather than too little.

evidence Material used to prove a point, such as testimony, statistics, and examples.

emotional evidence Evidence that arouses the sentiments of an audience.

 ASK YOURSELF

How can you tell if evidence will be effective?

Carefully cited sources are part of a well-reasoned argument. This brings us to our next step in creating a persuasive message.

Avoid Fallacies

fallacy An error in logic.

ad hominem fallacy A fallacious argument that attacks the integrity of a person to weaken his or her position.

reductio ad absurdum fallacy Fallacious reasoning that unfairly attacks an argument by extending it to such extreme lengths that it looks ridiculous.

either–or fallacy Fallacious reasoning that sets up false alternatives, suggesting that if the inferior one must be rejected, then the other must be accepted.

post hoc fallacy Fallacious reasoning that mistakenly assumes that one event causes another because they occur sequentially.

A **fallacy** (from the Latin word meaning "false") is an error in logic. Although the original meaning of the term implied purposeful deception, most logical fallacies are not recognized as such by those who use them. Scholars have devoted lives and volumes to the description of various types of logical fallacies.[8] Here are some of the most common ones to keep in mind when building your persuasive argument:[9]

Attack on the Person Instead of the Argument (*Ad Hominem*)   In an *ad hominem* fallacy the speaker attacks the integrity of a person in order to weaken the argument. At its crudest level, an *ad hominem* argument is easy to detect. "How can you believe that fat slob?" is hardly persuasive. It takes critical thinking to catch more subtle *ad hominem* arguments, however. Consider this one: "All this talk about 'family values' is hypocritical. Take Senator _____ , who made a speech about the 'sanctity of marriage' last year. Now it turns out he was having an affair with his secretary, and his wife is suing him for divorce." Although the senator certainly does seem to be a hypocrite, his behavior doesn't necessarily weaken the merits of family values.

Reduction to the Absurd (*Reductio ad Absurdum*)   A *reductio ad absurdum* fallacy unfairly attacks an argument by extending it to such extreme lengths that it looks ridiculous. "If we allow developers to build homes in one section of this area, soon we will have no open spaces left. Fresh air and wildlife will be a thing of the past." "If we allow the administration to raise tuition this year, soon they will be raising it every year, and before we know it only the wealthiest students will be able to go to school here." This extension of reasoning doesn't make any sense: Developing one area doesn't necessarily mean that other areas have to be developed, and one tuition increase doesn't mean that others will occur. Any of these policies might be unwise or unfair, but the *ad absurdum* reasoning doesn't prove it.

Either–Or   An **either–or fallacy** sets up false alternatives, suggesting that if the inferior one must be rejected, then the other must be accepted. An angry citizen used either–or thinking to support a proposed city ordinance: "Either we outlaw alcohol in city parks, or there will be no way to get rid of drunks." This reasoning overlooks the possibility that there may be other ways to control public drunkenness besides banning all alcoholic beverages. The old saying "America, love it or leave it" provides another example of either–or reasoning. For instance, when an Asian-born college professor pointed out examples of lingering discrimination in the United States, some suggested that if she didn't like her adopted country, she should return to her native home—ignoring that it is possible to admire a country and still envision ways to make it a better place.

False Cause (*Post Hoc Ergo Propter Hoc*)   A *post hoc* fallacy mistakenly assumes that one event causes another because they occur sequentially. An old (and not especially funny) joke illustrates the *post hoc* fallacy. Mac approaches Jack and asks, "Hey, why are you snapping your fingers?" Jack replies, "To keep the elephants away." Mac is incredulous: "What are you talking about? There aren't any elephants within a thousand miles of here." Jack smiles and keeps on snapping: "I know. Works pretty well, doesn't it?"

In real life, *post hoc* fallacies aren't always so easy to detect. For example, one critic of education pointed out that the increase in sexual promiscuity among adolescents began about the same time as prayer in public schools was prohibited by

UNDERSTANDING DIVERSITY

Cultural Differences in Persuasion

Different individuals have a tendency to view persuasion differently, and often these differences are based on cultural background. Even the ability to recognize logical argument is, to a certain extent, culturally determined. Not all cultures use logic in the same way that the European American culture does. The influence of the dominant culture is seen even in the way we talk about argumentation. When we talk about "defending" ideas and "attacking our opponent's position," we are using male-oriented militaristic/aggressive terms. Logic is also based on a trust in objective reality, on information that is verifiable through our senses. As one researcher points out, such a perspective can be culturally influenced:

> Western culture assumes a reality that is materialist and limited to comprehension via the five senses.

African culture assumes a reality that is both material and spiritual viewed as one and the same.[10]

The way logic is viewed also differs between Eastern and Western Hemisphere cultures. As Larry A. Samovar and Richard E. Porter point out:

> Westerners discover truth by active searching and the application of Aristotelian modes of reasoning. On the contrary, many Easterners wait patiently, and if truth is to be known it will make itself apparent.[11]

It is because of cultural differences such as these that speech experts have always recommended a blending of logical and emotional evidence.

---

the courts. A causal link in this case may exist: Decreased emphasis on spirituality could contribute to promiscuity. But it would take evidence to establish a *definite* connection between the two phenomena.

Appeal to Authority (*Argumentum ad Verecundiam*) An ***argumentum ad verecundiam*** **fallacy** involves relying on the testimony of someone who is not an authority in the case being argued. Relying on experts is not a fallacy, of course. A movie star might be just the right person to offer advice on how to seem more glamorous, and a professional athlete could be the best person to comment on what it takes to succeed in organized sports. But an *ad verecundiam* fallacy occurs when the movie star promotes a political candidate or the athlete tells us why we should buy a certain kind of automobile. When considering endorsements and claims, it's smart to ask yourself whether the source is qualified to make them.

Bandwagon Appeal (*Argumentum ad Populum*) An ***argumentum ad populum*** **fallacy** is based on the often dubious notion that, just because many people favor an idea, you should, too. Sometimes, of course, the mass appeal of an idea can be a sign of its merit. If most of your friends have enjoyed a film or a new book, there is probably a good chance that you will, too. But in other cases widespread acceptance of an idea is no guarantee of its validity. In the face of almost universal belief to the contrary, Galileo reasoned accurately that the earth is not the center of the universe, and he suffered for his convictions. The lesson here is simple to comprehend but often difficult to follow: When faced with an idea, don't just <u>follow the crowd</u>. Consider the facts carefully and make up your own mind.

Other Common Fallacies There is a wide range of other common fallacies, as shown in Table 14-2. Often, dogmatic speakers don't even realize they are using faulty logic. Other times, it is purposeful manipulation. How many of these do you recognize from advertising, politics, or everyday arguments? How many other fallacies can you name?

***argumentum ad verecundiam* fallacy** Fallacious reasoning that tries to support a belief by relying on the testimony of someone who is not an authority on the issue being argued.

***argumentum ad populum* fallacy** Fallacious reasoning based on the dubious notion that because many people favor an idea, you should, too.

cultural idiom
follow the crowd: do what the majority does

TABLE 14-2

Other Common Fallacies

Fallacy	Definition	Example
Straw Man	Setting up an argument that was not proposed and then attacking it as if it were the original argument.	"You say we should support animal rights, but many animal rights activists have supported the destruction of research facilities." (The speaker then goes on to argue that the destruction of research facilities is wrong.)
Red Herring	Shifting the focus to a tangential subject, similar to dragging a fish across a trail to distract a bloodhound.	"Bill says that buying a term paper is immoral. But what is morality, anyway?" (The speaker then goes on to discuss this philosophical question.)
Begging the Question	Repeating an argument but never providing support for a point of view.	"I can't believe people eat dog. That's just plain gross. Why? Because it's a dog, of course. How could someone eat a dog?"
Faulty Analogy	Using a comparison suggesting that two things are more alike than they really are.	"If we legalize gay marriage, next we'll legalize marriage between people and their pets."
Hasty Generalization	Reaching an unjustifiable conclusion after making assumptions or misunderstanding statistics.	"You are likely to be shot if you visit NYC." (In fact, fewer people are murdered, per capita, in NYC than in most rural American small towns.)

Adapting to the Audience

It is important to know as much as possible about your audience for a persuasive speech. For one thing, you should appeal to the values of your audience whenever possible, even if they are not *your* strongest values. This advice does not mean you should pretend to believe in something. According to our definition of *ethical persuasion*, pretense is against the rules. It does mean, however, that you have to stress those values that are felt most forcefully by the members of your audience.[12]

In addition, you should analyze your audience carefully to predict the type of response you will get. Sometimes you have to pick out one part of your audience—a **target audience**, the subgroup you must persuade to reach your goal—and aim your speech mostly at those members. Some of your audience members might be so opposed to what you are advocating that you have no hope of reaching them. Still others might already agree with you, so they do not need to be persuaded. A middle portion of your audience members might be undecided or uncommitted, and they would be the most productive target for your appeals.

Of course, you need not ignore that portion of your audience that does not fit your target. For example, if you were giving a speech against smoking, your target might be the smokers in your class. Your main purpose would be to get them to quit, but at the same time, you could convince the nonsmokers not to start and to use their influence to help their smoking friends quit.

All of the methods of audience analysis described in Chapter 11—surveys, observation, interviews, and research—are valuable in collecting information about your audience for a persuasive speech.

Establish Common Ground

It helps to stress as many similarities as possible between yourself and your audience members. This technique helps prove that you understand them—if not, why should they listen to you? Also, if you share a lot of common ground, it

target audience That part of an audience that must be influenced in order to achieve a persuasive goal.

shows you agree on many things. Therefore, it should be easy to settle one disagreement: the one related to the attitude or behavior you would like them to change.

The manager of public affairs for *Playboy* magazine gave a good demonstration of establishing common ground when he reminded a group of Southern Baptists that they shared some important values with him:

> I am sure we are all aware of the seeming incongruity of a representative of *Playboy* magazine speaking to an assemblage of representatives of the Southern Baptist convention. I was intrigued by the invitation when it came last fall, though I was not surprised. I am grateful for your genuine and warm hospitality, and I am flattered (although again not surprised) by the implication that I would have something to say that could have meaning to you people. Both *Playboy* and the Baptists have indeed been considering many of the same issues and ethical problems; and even if we have not arrived at the same conclusions, I am impressed and gratified by your openness and willingness to listen to our views.[13]

Organize According to the Expected Response

It is much easier to get an audience to agree with you if the members have already agreed with you on a previous point. Therefore, you should arrange your points in a persuasive speech so you develop a "yes" response. In effect, you get your audience into the habit of agreeing with you. For example, if you were giving a speech on organ donation, you might begin by asking the audience members if they would like to be able to get a kidney if they needed one. Then you might ask them if they would like to have a major role in curbing tragic and needless dying. The presumed response to both questions is "yes." It is only when you have built a pattern of "yes" responses that you would ask the audience to sign organ donor cards.

An example of a speaker who was careful to organize material according to expected audience response is the late Robert Kennedy. Kennedy, when speaking on civil rights before a group of South Africans who believed in racial discrimination, arranged his ideas so that he spoke first on values that he and his audience shared—values like independence and freedom.[14]

If audience members are already basically in agreement with you, you can organize your material to reinforce their attitudes quickly and then spend most of your time convincing them to take a specific course of action. If, on the other hand, they are hostile to your ideas, you have to spend more time getting the first "yes" out of them.

Neutralize Potential Hostility

One of the trickier problems in audience adaptation occurs when you face an audience hostile to you or your ideas. Hostile audiences are those who have a significant number of members who feel adversely toward you, your topic, or the speech situation. Members of a hostile audience could range from unfriendly to violent. Two guidelines for handling this type of audience are (1) show that you understand their point of view and (2) if possible, use appropriate humor. A good example of a speaker who observed these guidelines was First Lady and literacy activist Barbara Bush when she was invited to speak at the commencement exercises at Wellesley College in 1990. After the invitation was announced,

Celebrities regularly speak out on issues they feel strongly about. Here, Angelina Jolie speaks before the United Nations Security Council on the situation in the Middle East and Syria.

What characteristics of her audience did she have to consider to make her speech effective?

? ASK YOURSELF

Imagine an audience that is as different from you as possible. Assign this audience backgrounds, beliefs, and values that are the opposite of yours. How would you adapt your speech accordingly?

credibility The believability of a speaker or other source of information.

150 graduating seniors at the prestigious women's college signed a petition in protest. They wrote, in part:

> We are outraged by this choice and feel it is important to make ourselves heard immediately. Wellesley teaches us that we will be rewarded on the basis of our own work, not on that of a spouse. To honor Barbara Bush as a commencement speaker is to honor a woman who has gained recognition through the achievements of her husband.[15]

Bush decided to honor her speaking obligation, knowing that these 150 students and others who shared their view would be in the audience of 5,000 people. Bush defused most of this hostility by presenting a speech that stressed that everyone should follow her personal dream and be tolerant of the dreams of others:

> For over 50 years, it was said that the winner of Wellesley's annual hoop race would be the first to get married. Now they say the winner will be the first to become a C.E.O. Both of these stereotypes show too little tolerance. . . . So I offer you today a new legend: the winner of the hoop race will be the first to realize her dream, not society's dream, her own personal dream.[16]

Building Credibility as a Speaker

Credibility refers to the believability of a speaker. Credibility isn't an objective quality; rather, it is a perception in the minds of the audience. In a class such as the one you're taking now, students often wonder how they can build their credibility. After all, the members of the class tend to know one another well by the time the speech assignments roll around. This familiarity illustrates why it's important to earn a good reputation before you speak, through your class comments and the general attitude you've shown.

It is also possible for credibility to change during a speaking event. In fact, researchers speak in terms of initial credibility (what you have when you first get up to speak), derived credibility (what you acquire while speaking), and terminal credibility (what you have after you finish speaking). It is not uncommon for a student with low initial credibility to earn increased credibility while speaking and to finish with much higher terminal credibility.

Without credibility, you won't be able to convince your listeners that your ideas are worth accepting, even if your material is outstanding. On the other hand, if you can develop a high degree of credibility in the eyes of your listeners, they will be likely to open up to ideas they wouldn't otherwise accept. Members of an audience form judgments about the credibility of a speaker based on their perception of many characteristics, the most important of which might be called the "three Cs" of credibility: competence, character, and charisma.[17]

Competence

Competence refers to the speaker's expertise on the topic. Sometimes this competence can come from personal experience that will lead your audience to regard you as an authority on the topic you are discussing. If everyone in the audience knows you've earned big profits in the stock market, they will probably take your investment advice seriously. If you say that you lost 25 pounds from a diet-and-exercise program, most audience members will be likely to respect your opinions on weight loss.

The other way to be seen as competent is to be well prepared for speaking. A speech that is well researched, organized, and presented will greatly increase the audience's perception of the speaker's competence. Your personal credibility will

cultural idiom
roll around: occur, arrive

Persuasive Speech

Use this self-assessment for a persuasive speech you have presented or plan to present.

1. Have you set a clear, persuasive purpose?

I'VE DONE MY BEST. I'VE GOT WORK TO DO. I'VE BARELY STARTED.

2. Is your purpose in the best interest of the audience?

I'VE DONE MY BEST. I'VE GOT WORK TO DO. I'VE BARELY STARTED.

3. Have you structured the message to achieve a "yes" response?

I'VE DONE MY BEST. I'VE GOT WORK TO DO. I'VE BARELY STARTED.

4. Have you used solid evidence for each point?

I'VE DONE MY BEST. I'VE GOT WORK TO DO. I'VE BARELY STARTED.

5. Have you used solid reasoning for each point?

I'VE DONE MY BEST. I'VE GOT WORK TO DO. I'VE BARELY STARTED.

6. Have you adapted to your audience?

I'VE DONE MY BEST. I'VE GOT WORK TO DO. I'VE BARELY STARTED.

7. Have you built your own credibility?

I'VE DONE MY BEST. I'VE GOT WORK TO DO. I'VE BARELY STARTED.

8. Is your information true to the best of your knowledge?

I'VE DONE MY BEST. I'VE GOT WORK TO DO. I'VE BARELY STARTED.

Scoring on this assessment is self-evident: For every area in which you've got work left to do, do it.

---

therefore be enhanced by the credibility of your evidence, including the sources you cite, the examples you choose, the way you present statistics, the quality of your visual aids, and the precision of your language.

Character

Competence is the first component of being believed by an audience. The second is being trusted, which is a matter of character. *Character* involves the audience's perception of at least two ingredients: honesty and impartiality. You should try to find ways to talk about yourself (without boasting, of course) that demonstrate your integrity. You might describe how much time you spent researching the subject or demonstrate your open-mindedness by telling your audience that you changed your mind after your investigation. For example, if you were giving a speech arguing against a proposed tax cut in your community, you might begin this way:

> You might say I'm an expert on the municipal services of this town. As a lifelong resident, I owe a debt to its schools and recreation programs. I've been protected by its police and firefighters and served by its hospitals, roads, and sanitation crews.
>
> I'm also a taxpayer who's on a tight budget. When I first heard about the tax cut that's been proposed, I liked the idea. But then I did some in-depth investigation into the possible effects, not just to my tax bill but to the quality of life of our entire community. I looked into our municipal expenses and into the expenses of similar communities where tax cuts have been mandated by law.

Persuasion Skills in the World of Sales

The skills you develop while learning to prepare persuasive speeches are generalizable to a number of important skills in the world of work. Business consultant George Rodriguez makes it clear that developing a successful sales plan is very much like the planning involved in building a persuasive speech.

"A sales plan is basically your strategic and tactical plan for achieving your marketing objectives," Rodriguez explains. "It is a step-by-step and detailed process that will show how you will acquire new business; and how you will gain more business from your existing customer base."[18]

The process of audience analysis is as important in sales-plan development as it is in persuasive speaking. "The first step is to clearly identify your target markets," Rodriguez says. "Who are more likely to buy your product? The more defined your target market, the better. Your target market can be defined as high-income men ages 30–60 who love to buy the latest electronic gadgets; or mothers with babies 0–12 months old living in urban areas."

And don't forget the guideline that persuasion is interactive. "Prospects are more likely to purchase if you can talk to them about solving their problems," Rodriguez points out.

Rodriguez is far from alone in pointing out the importance of thinking in terms of problems and solutions. Business consultant Barbara Sanfilippo advises her clients to "prepare, prepare, and plan your calls. Today's customers and prospects have very little time to waste. They want solutions. A sales consultant who demonstrates a keen understanding of customers' needs and shows up prepared will earn the business."[19] Sanfilippo suggests reviewing the customer's website and interviewing key people in advance of the meeting.

Sanfilippo also points out the importance of building credibility: "How can you stand out from the pack of sales professionals and consultants all offering similar services?" she asks rhetorically. "Establish Credibility and Differentiate!"

But George Rodriguez probably has the last word in how valuable training in persuasive speaking is to the sales professional. Before you make that first sales call, he says, "You may want to take courses on how to improve your confidence and presentation skills."

Charisma

Charisma is spoken about in the popular press as an almost indefinable, mystical quality. Even the dictionary defines it as "a special quality of leadership that captures the popular imagination and inspires unswerving allegiance and devotion." Luckily, communication scholars favor a more down-to-earth definition. For them, charisma is the audience's perception of two factors: the speaker's enthusiasm and likability. Whatever the definition, history and research have both shown us that audiences are more likely to be persuaded by a charismatic speaker than by a less charismatic one who delivers the same information.

Enthusiasm is sometimes called "dynamism" by communication scholars. Your enthusiasm will mostly be perceived from how you deliver your remarks, not from what you say. The nonverbal parts of your speech will show far better than your words that you believe in what you are saying. Is your voice animated and sincere? Do your gestures reflect your enthusiasm? Do your facial expression and eye contact show you care about your audience?

You can boost your likability by showing that you like and respect your audience. Insincere flattery will probably boomerang, but if you can find a way to give your listeners a genuine compliment, they'll be more receptive to your ideas.

Building your personal credibility through a recognition of the roles of competence, character, and charisma is an important component of your persuasive strategy. When combined with a careful consideration of audience adaptation, persuasive structure, and persuasive purpose, it will enable you to formulate the most effective strategy possible.

cultural idioms
down to earth: practical
boomerang: create a negative effect

Sample Speech

The sample speech for this chapter was presented by Kristina Medero, whose profile began this chapter. Medero presented this speech at the American Forensics Association (AFA) National Individual Events Tournament in 2015.

Her thesis statement could be worded like this:

> Important information about gun violence is being held back by the Dickey Amendment.

Her purpose statement could be worded like this:

> After listening to my speech, some of my audience members will be willing to send an email to their congressional representatives asking for the repeal of the Dickey Amendment.

Medero carefully organized her argument in a problem-causes-solution format, arranging her points for maximum persuasive impact. Her persuasive organization can be seen in the following outline. (Parenthetical numbers refer to paragraphs in the speech.)

INTRODUCTION

   I. Attention-getter (1–2)

  II. Statement of thesis (3)

 III. Preview of main points (4)

BODY

Preview of first main point (5)

   I. The lack of research about gun violence is a serious problem.

     A. The lack of research results in unreliable information about gun violence. (6–7)

     B. The lack of research results in misguided policies. (8–10)

Preview of second main point (10)

  II. The lack of research about gun violence has two primary causes. (11–12)

     A. Political grandstanding: Politicians make speeches about gun violence but then drop the issue when the pain has passed. (11)

     B. Public ignorance: The public is simply unaware of the Dickey Amendment. (12)

Preview of final main point (13)

 III. The lack of research about gun violence can be fixed in two ways: (14–16)

     A. Government Intervention: Send an email asking for the repeal the Dickey Amendment. (14–15)

     B. Information Distribution: Take this leaflet with a link to the Access Denied Report. (16)

CONCLUSION (17)

   I. Review of main points

  II. Restatement of thesis

 III. Final remarks

CHECKLIST ✓

Ethos, Pathos, and Logos

The Greek philosopher Aristotle divided the means of persuasion into three types of appeal: **ethos**, **pathos**, and **logos**. Use the following checklist to make sure you are using all three means in your persuasive speeches.

☐ **Ethos (credibility), or ethical appeal:** Have you established your credibility as a speaker so that your audience believes you to be trustworthy?

☐ **Pathos (emotions):** Have you used emotional appeals effectively to make your case?

☐ **Logos (logic):** Have you made logical arguments to appeal to the audience's sense of reasoning? This was Aristotle's favorite, and the most important form of appeal we have discussed in this chapter.

As you read Medero's speech, notice how she expands on this outline as she develops her argument point by point. Notice also how she uses solid evidence throughout. Because she was preparing this speech for a championship collegiate speech contest, she had to cite more evidence than would be necessary in a classroom speech. The way she cites that evidence, however, is a good model for most persuasive speeches.

Her evidence was culled from the sources listed in the following bibliography. These are just the sources cited in the speech. Medero did extensive reading that is not listed here. Again, the numbers in parentheses at the end of each citation correspond with the paragraph number of the speech in which each source is cited.

Bibliography

APA Reports. (2014). *Gun violence: Prediction, prevention, and policy.* Retrieved from http://www.apa.org/pubs/info/reports/gun-violence-prevention.aspx. (13)

Beckett, L. (2014, May 14). Why don't we know know many people are shot each year in America? *Pro Publica.* Retrieved from https://www.propublica.org/article/why-dont-we-know-how-many-people-are-shot-each-year-in-america. (6)

Fan, M. (2015). Disarming the dangerous: Preventing extraordinary and ordinary violence. *Indiana Law Journal, 90,* 151–179. Retrieved from http://ilj.law.indiana.edu/articles/12-Fan.pdf. (8)

Ferris, S. (2014, October 3). Millions wasted on background checks for guns. *Florida Times Union.* Retrieved from http://jacksonville.com/news/politics/2014-10-03/story/millions-wasted-background-checks-guns. (1)

LaFrance, A. (2015, January 12). America's top killing machine. *The Atlantic.* Retrieved from http://www.theatlantic.com/technology/archive/2015/01/americas-top-killing-machine/384440/. (4)

Mayors Against Illegal Guns. (2013, January 24). Access denied: How the gun lobby is depriving police, policy makers, and the public of the data we need to prevent gun violence. Everytown Research. Retrieved from https://everytownresearch.org/reports/access-denied/. (3, 16)

Murphy, C. (2014, April 22). Still no federal funding for gun violence research. *Colorlines.* Retrieved from http://www.colorlines.com/articles/still-no-federal-funding-gun-violence-research. (11)

Pilkington, E. (2014, June 13). Up to 100 children a year die from accidental shootings, research shows. *The Guardian.* Retrieved from http://www.theguardian.com/world/2014/jun/25/us-accidental-gun-deaths-100-children-yearly. (7)

American College of Emergency Physicians. (2015, February 24). Emergency physicians call for action to prevent firearms injuries. [Press release]. Retrieved from http://newsroom.acep.org/2015-02-24-Emergency-Physicians-Call-for-Action-to-Prevent-Firearms-Injuries. (9)

Rajan, S., cited in: Mental illness is not the biggest reason youth carry guns, study finds. (2014, November 17). *Fox News Health.* Retrieved from http://www.foxnews.com/health/2014/11/17/mental-illness-is-not-biggest-reason-youth-carry-guns-study-finds.html. (12)

Scheiner, E. (2015, February 12). Rep. Blumenauer: "We need more research on the health effects of guns." *Cybercast News Service.* Retrieved from http://www.cnsnews.com/news/article/eric-scheiner/rep-blumenauer-we-need-more-research-health-effects-guns. (15)

Wade, A. (2015, February 4). Apathy toward research, science. *Commonwealth Times*. Retrieved from http://www.commonwealthtimes.org/2015/02/04/apathy-toward-research-science/. (2)

SAMPLE SPEECH | **Kristina Medero**

Who's Afraid of Gun Facts?

1   Called the "most important tool" in preventing gun crime, the National Instant Criminal Background Check System, or NICS, is a database intended to prevent potentially violent individuals from purchasing a firearm. However, according to the October 3, 2014 *Florida Times Union*, the $650 million-a-year program is failing—its ability to identify potential criminals as reliable as a "coin toss."

Medero immediately begins explaining the magnitude of the problem, and also defines a key term.

2   The problem isn't a lack of funding or support, but a lack of information. As *Commonwealth Times* on February 4, 2015 reports, federal public health institutes, where programs like the NICS get their data, cannot receive funding to research gun violence . . . at all. In 1997 Congress passed the Dickey Amendment, blocking federal public health institutes from researching gun violence, on the grounds that any gun research is inherently anti-gun research.

Statement of thesis; a second piece of evidence.

If she wanted to, she could put up a visual aid: "Dickey Amendment: any gun research is inherently anti-gun research."

3   And it's not like other institutions have picked up the slack. As the January 2013 Everytown Research notes, since the Dickey Amendment, peer-reviewed journals have seen a 60% drop in gun violence research. As a result, we are basing policies on dated or, worse, biased research and extrapolating statistics far beyond their intended use.

She explains the effects of the problem, and presents more evidence.

4   It's no surprise programs like NICS are failing. Congress' stranglehold on scientific study isn't ideological, it's a pathological assault on public safety. As a community built on the foundation of dependable research, it is imperative we reverse the Dickey Amendment by discussing the problems, causes, and solutions to its continued existence, especially in a year, as the *Atlantic* from January 2015 reports, marking the first time in decades gun fatalities have surpassed motor vehicle deaths.

Reference to audience—in this case, the AFA. She previews her main points, making it apparent that she will use a problem-causes-solutions organizational pattern. She also introduces another key piece of research.

5   Ironically, the Federal government's open, double peer review system is the only one that assures published studies don't have a political agenda. Its exclusion perpetuates two problems: problematic research and misguided policies.

Transition to first main point and its subpoints.

6   First, the restraints of the Dickey Amendment have resulted in problematic research, drawing on small sample sizes, unorthodox methodologies, and biased intentions. Essentially, research has been done, but not enough to answer basic questions, like how many Americans are shot each year. The answer, according to Pro Publica on May 14th, 2014: We have no clue. While all gun fatalities are reported to the Department of Justice, statistics on nonfatal injuries are extrapolated from only 66 hospitals, or only 1% of the total hospitals, across the entire United States. The best guess is somewhere between 27,000–91,000.

Development of first subpoint: problematic research. More evidence.

7   This vacuum of verifiable evidence also allows for the proliferation of biased research. The *Guardian* on June 13th, 2014 states that both gun control and gun rights advocates used the same data collection, yet one reported 74 school

Second subpoint: biased research. More research cited, along with some quick sarcasm.

shootings within the year, while the other determined there were only 7. Hard to believe they got their information from the same . . . Tweets.

8 Next, as Dr. David Livingston, director of the New Jersey Trauma Center tells the aforementioned Pro Publica, "in the absences of real data, politicians and policy makers do what the hell they want . . . [and] have nobody to call them on it." The 2015 *Indiana Law Journal* reveals that since 2013 over 1500 gun bills have been introduced in state and federal legislatures, many with questionable efficacy. In fact, according to the National Academy of Sciences, our tax dollars fund over 80 programs to teach kids about gun use, yet we have no idea if or how it affects them.

More evidence cited, this time in the form of a quotation.

9 Police, military officers, and doctors feel the impact of the Dickey Amendment daily, with the American College of Physicians estimating on February 24th, 2015 that gun violence costs at least $174 billion a year in health care and policy enforcement. Not only is this limited research driving policies, but it is also impeding those on the front lines trying to combat this issue.

More research.

10 The price of ignorance is not just monetary but includes thousands of preventable deaths and avoidable injuries. Our knowledge is being stifled by two causes: political grandstanding and a vulnerable public.

Transition to second main point, and its subpoints.

11 First, political grandstanding has prevented funding for research that would help us understand gun violence. It is a vicious cycle of ignorance: A tragic shooting occurs and politicians push for change only until the outrage and pain has passed. After Sandy Hook, even staunch republican Representative Jack Kingston loudly proclaimed, we have to "let data lead, rather than our political opinions." But as Colorlines on April 22, 2014 points out, just a year later Kingston declared Obama's efforts to give money to a public health institute for gun-violence research as simply, "Propaganda for gun-grabbing initiatives."

First subpoint, backed up by quotations.

12 Next, most Americans are unaware of the Dickey Amendment. But we know what we see. Sandy Hook, [Trayvon] Martin, Aurora, Michael Brown, and countless other tragedies ignite our fervor for change—new laws, policies, databases. As Fox News on November 17, 2014 reports, "When there is really sensationalized violence, we have a tendency to simplify gun violence." Under these emotional circumstances, lessons learned about quality research are forgotten. Instead we are wasting time and money while delaying meaningful solutions. It is because of those lost innocent lives that we can no longer allow politics to deter our right for knowledge.

Second subpoint.

13 Last year, the American Psychological Association identified gun violence as a public health crisis. The only way to combat this epidemic is to truly allow the data to lead us through two solutions regarding governmental intervention and information distribution.

Transition to final main point and its subpoints.

14 To begin, we must repeal the Dickey Amendment. The aforementioned American College of Physicians reports nearly half a million of their members have contacted their federal and state representatives to lift this suppression of science. Their letter, which I have modified, outlines a plan for conducting research without impeding Second Amendment rights, as vetted by the American Bar Association. If you write your email address and state of residence here, I will forward you the letter as well as your representative's contact information so you can personalize it before sending it off to help break the cycle of ignorance.

Desired audience behavior is made as easy and explicit as possible with a handout.

15 As Representative Earl Blumenauer points out to the February 12, 2015 Cybercast News Service, research has shaped car safety and tobacco policies in the U.S. and has allowed Australia to craft and refine their much touted gun legislation. We can no longer stand for political grandstanding; we must push for our representatives to find reliable strategies.

More evidence.

16 Finally, 10 minutes will not allow me to tackle the vast scope and complexities of this issue, so please take this leaflet with a link to the Access Denied Report. I also have hard copies for the more technology impaired. But until Congress lifts this financial barrier, be wary of studies claiming proofs on gun statistics. As I did for this speech, look at how the research is conducted, the sample size, and who conducted it. Quick fixes have only resulted in a divided country, the inhibition of meaningful solutions, and headlines full of names we'll never forget. We need reliable research. Our vulnerability has left us susceptible to cherry picked half-truths. It's time to get the full picture.

Desired audience behavior is also simplified here—as she offers not just a handout with a link, but the hard copy of the report.

17 After doing a thorough background check on the problems, causes, and solution to our out of control gun control policies, we can hopefully amend this vacuum of knowledge. Twenty years ago Representative Jay Dickey pushed through the amendment restricting research, but after understanding the intrinsic value of research he is now fighting to reverse it. As Dickey notes, "It's like the answer to the question, when's the best time to plant a tree? The best time was 20 years ago. The second best time is right now."

*Review of main points.
Restatement of thesis.*

Ends with a memorable quotation: Rep. Dickey is now in favor of repealing his amendment.

MAKING THE GRADE

For more resources to help you understand and apply the information in this chapter, visit the *Understanding Human Communication* website at www.oup.com/us/adleruhc.

Objective 14.1 Identify the primary characteristics of persuasion.

- Persuasion is the act of moving someone, through communication, toward a belief, attitude, or behavior. Despite a sometimes bad reputation, persuasion can be both worthwhile and ethical.

- Ethical persuasion requires that the speaker be sincere and honest and avoid such behaviors as plagiarism.

- Ethical persuasion must also serve the best interest of the audience, as perceived by the speaker.

  > What are some examples of persuasive speaking in your everyday life?

  > Using the examples you identified earlier, describe the difference between ethical and unethical persuasion.

  > Describe an ethical approach to a persuasive presentation you could deliver.

Objective 14.2 Compare and contrast different types of persuasion.

- Persuasion can be categorized according to the type of proposition (fact, value, or policy).

- Persuasion can be categorized according to the desired outcome (convincing or actuating).

- Persuasion can be categorized according to the type of approach (direct or indirect).

  > For persuasive speeches you could deliver, describe propositions of fact, value, and policy.

  > In your next persuasive speech, do you intend to convince or to actuate your audience? Why have you chosen this goal?

  > In your next persuasive speech, are you planning to use a direct or indirect persuasive approach? Why?

Objective 14.3 Apply the guidelines for persuasive speaking to a speech you will prepare.

A persuasive strategy is put into effect through the use of several strategies. These include

- setting a specific, clear persuasive purpose,
- structuring the message carefully,
- using solid evidence (including emotional evidence),
- using careful reasoning,
- adapting to the audience, and
- building credibility as a speaker.

  > For a speech from your personal experience or that you have watched online (e.g., a TED Talk), identify its purpose, message structure, use of evidence and reasoning, audience adaptation, and enhancement of speaker credibility.

  > Apply the preceding guidelines to a speech you are developing.

Objective 14.4 **Explain how to best adapt a specific speech to a specific audience.**

In adapting to your audience, you should

- establish common ground,
- organize your speech in such a way that you can expect a "yes" response along each step of your persuasive plan, and
- take special care with a hostile audience.
  - > Give examples from speeches you have observed (either in person or online) that illustrate these three strategies. Explain why each strategy contributes to the success of the speech.
  - > Apply these strategies to a speech you are developing.

Objective 14.5 **Improve your credibility in your next persuasive speech.**

- In building credibility, you should keep in mind the audience's perception of your competence, character, and charisma.
  - > For the most effective persuasive speech you can recall, describe how the speaker enhanced his or her competence, character, and charisma.
  - > How can you enhance your perceived competence, character, and charisma in a speech you are developing? Suggest specific improvements.

KEY TERMS

actuate p. 381

ad hominem fallacy p. 386

anchor p. 375

argumentum ad populum fallacy p. 387

argumentum ad verecundiam fallacy p. 387

convincing p. 380

credibility p. 390

direct persuasion p. 381

either–or fallacy p. 386

emotional evidence p. 385

ethical persuasion p. 377

ethos p. 393

evidence p. 385

fallacy p. 386

indirect persuasion p. 381

latitude of acceptance p. 375

latitude of noncommitment p. 375

latitude of rejection p. 375

logos p. 393

pathos p. 393

persuasion p. 375

post hoc fallacy p. 386

proposition of fact p. 378

proposition of policy p. 380

proposition of value p. 380

reductio ad absurdum fallacy p. 386

social judgment theory p. 375

target audience p. 388

ACTIVITIES

1. **Audience Latitudes of Acceptance** To better understand the concept of latitudes of acceptance, rejection, and noncommitment, formulate a list of perspectives on a topic of your choice. This list should contain 8 to 10 statements that represent a variety of attitudes, such as the list pertaining to the pro-life/pro-choice issue on page 376. Arrange this list from your own point of view, from most acceptable to least acceptable. Then circle the single statement that best represents your own point of view. This will be your "anchor." Underline those items that also seem reasonable. These make up your latitude of acceptance on this issue. Then cross out the numbers in front of any items that express opinions that you cannot accept. These make up your latitude of rejection. Those statements that are left would be your latitude of noncommitment. Do you agree that someone seeking to persuade you on this issue would do best by advancing propositions that fall within this latitude of noncommitment?

2. **Personal Persuasion** When was the last time you changed your attitude about something after discussing it with someone? In your opinion, was this persuasion interactive? Not coercive? Incremental? Ethical? Explain your answer.

3. **Propositions of Fact, Value, and Policy** Which of the following are propositions of fact, propositions of value, and propositions of policy?

   a. "Three Strikes" laws that put felons away for life after their third conviction are/are not fair.

   b. Elder care should/should not be the responsibility of the government.

   c. The mercury in dental fillings is/is not healthy for the dental patient.

   d. Congressional pay raises should/should not be delayed until an election has intervened.

   e. Third-party candidates strengthen/weaken American democracy.

   f. National medical insurance should/should not be provided to all citizens of the United States.

   g. Elderly people who are wealthy do/do not receive too many Social Security benefits.

h. Tobacco advertising should/should not be banned from all media.

i. Domestic violence is/is not on the rise.

j. Pit bulls are/are not dangerous animals.

4. **Structuring Persuasive Speeches** For practice in structuring persuasive speeches, choose one of the following topics, and provide a full-sentence outline that conforms to the outline in Figure 14-3, page 383.

   a. It should/should not be more difficult to purchase a handgun.

   b. Public relations messages that appear in news reports should/should not be labeled as advertising.

   c. Newspaper recycling is/is not important for the environment.

   d. Police should/should not be required to carry nonlethal weapons only.

   e. Parole should/should not be abolished.

   f. The capital of the United States should/should not be moved to a more central location.

   g. We should/should not ban capital punishment.

   h. Bilingual education should/should not be offered in all schools in which students speak English as a second language.

5. **Find the Fallacy** Test your ability to detect shaky reasoning by identifying which fallacy is exhibited in each of the following statements.

   - *Ad hominem*
   - *Ad absurdum*
   - Either–or
   - *Post hoc*
   - *Ad verecundiam*
   - *Ad populum*

   a. Some companies claim to be in favor of protecting the environment, but you can't trust them. Businesses exist to make a profit, and the cost of saving the earth is just another expense to be cut.

   b. Take it from me, imported cars are much better than domestics. I used to buy only American, but the cars made here are all junk.

   c. Rap music ought to be boycotted. After all, the number of assaults on police officers went up right after rap became popular.

   d. Carpooling to cut down on the parking problem is a stupid idea. Look around—nobody carpools!

   e. I know that staying in the sun can cause cancer, but if I start worrying about every environmental risk I'll have to stay inside a bomb shelter breathing filtered air, never drive a car or ride my bike, and I won't be able to eat anything.

   f. The biblical account of creation is just another fairy tale. You can't seriously consider the arguments of those Bible-thumping, know-nothing fundamentalists, can you?

6. **The Credibility of Persuaders** Identify someone who tries to persuade you via public speaking or mass communication. This person might be a politician, a teacher, a member of the clergy, a coach, a boss, or anyone else. Analyze this person's credibility in terms of the three dimensions discussed in the chapter. Which dimension is most important in terms of this person's effectiveness?

Notes

CHAPTER 1

1. Siriwardane, V. (2010, June 20). Zappos CEO adds happiness to corporate culture. *The Star-Ledger* at NJ.com. Retrieved from http://www.nj.com/business/index.ssf/2010/06/zappos_ceo_adds_happiness_to_c.html. Quote appears in paragraph 13.

2. Average hours per day spent on socializing and communicating by the U.S. population from 2009 to 2014. (2015). *Statistica.* Retrieved from http://www.statista.com/statistics/189527/daily-time-spent-on-socializing-and-communicating-in-the-us-since-2009/.

3. 100 best companies to work for. (2015). *Fortune.* Retrieved from http://fortune.com/best-companies/zappos-com-86/.

4. Gergen, K. (1991). *The saturated self: Dilemmas of identity in contemporary life.* New York: Basic Books, p. 158.

5. Mottet, T. P., & Richmond V. P. (2001). Student nonverbal communication and its influence on teachers and teaching: A review of literature. In J. L. Chesebro & J. C. McCroskey (Eds.), *Communication for teachers* (pp. 47–61). Needham Heights, MA: Allyn & Bacon.

6. Shannon, C. E., & Weaver, W. (1949). *The mathematical theory of communication.* Urbana: University of Illinois Press.

7. Boaz, K., Epley, N., Carter, T. & Swanson, A. (2011). *Journal of Experimental Social Psychology, 47,* 269–273.

8. For an in-depth look at this topic, see Cunningham, S. B. (2012). Intrapersonal communication: A review and critique. In S. Deetz (Ed.), *Communication yearbook* 15 (pp. 597–620). Newbury Park, CA: Sage.

9. Hsieh, T. (2010). *Delivering happiness: A path to profits, passion, and purpose.* New York: Grand Central Publishing. Quotes appear on page 53.

10. Burns, M. E., & Pearson, J. C. (2011). An exploration of family communication environment, everyday talk, and family satisfaction. *Communication Studies, 62*(2), 171–185.

11. Chui, M., Manyika, J., Bughin, J., Dobbs, R., Roxburgh, C., Sarrazin, H., & Westergren, M. (2012). *The social economy: Unlocking value and productivity through social technologies.* Retrieved from http://www.mckinsey.com/insights/high_tech_telecoms_internet/the_social_economy; Project Management Institute. (2013). The high cost of low performance: The essential role of communications. Retrieved from http://www.pmi.org/~/media/PDF/Business-Solutions/The-High-Cost-Low-Performance-The-Essential-Role-of-Communications.ashx; Silverman, R. E. (2012, February 14). Where's the boss? Trapped in a meeting. *Wall Street Journal.* Retrieved from http://www.wsj.com/articles/SB10001424052970204642604577215013504567548.

12. Gill, B. (2013, June). E-mail: Not dead, evolving. *Harvard Business Review, 91*(6), 32–33.

13. Lucas, K., & Rawlins, J. D. (2015). The competency pivot: Introducing a revised approach to the business communication curriculum. *Business and Professional Communication Quarterly, 78*(2), 167–193.

14. Goo, S. K. (2015, February 19). *The skills Americans say kids need to succeed in life.* Pew Research Center. Retrieved from http://www.pewresearch.org/fact-tank/2015/02/19/skills-for-success/?utm_source=Pew+Research+Center&utm_campaign=ad0be41c05-Internet_newsletter_022015&utm_medium=email&utm_term=0_3e953b9b70-ad0be41c05-399444569.

15. Deming, D. J. (2015). "The growing importance of social skills in the labor market." National Bureau of Economic Research working paper 21473. Retrieved from http://www.nber.org/papers/w21473.

16. Autor, D. H. (2014). Polany's Paradox and the shape of employment growth. Federal Reserve Bank of Kansas City. Retrieved from http://economics.mit.edu/files/9835.

17. National Association of Colleges and Employers. (2012, October 24). *The skills and qualities employers want in their class of 2013 recruits.* Retrieved from http://www.naceweb.org/s10242012/skills-abilities-qualities-new-hires/.

18. Anderson, C., & Gantz, J. F. (2013). *Skills requirements for tomorrow's best jobs: Helping educators provide students with skills and tools they need.* Retrieved from http://news.microsoft.com/download/presskits/education/docs/IDC_101513.pdf.

19. Hollender, J. (2013, March 14). Lessons we can all learn from Zappos CEO Tony Hsieh. *The Guardian.* Retrieved from http://www.theguardian.com/sustainable-business/zappos-ceo-tony-hsieh.

20. Jenks, I. (2009). *Living on the future edge.* Presentation handout, 21st Century Fluency Project. Kelowna, BC, Canada: The Info Savvy Group.

21. United Nations Cyberschoolbus. (n.d.). Retrieved from http://www.un.org/Pubs/CyberSchoolBus/aboutus.html.

22. Aristotle. (1991). *On rhetoric: A theory of civic discourse* (George A. Kennedy, Trans.). New York: Oxford University Press.

23. See, for example, Peters, J. D., Durham, J., & Simonson, P. (1997). *Mass communication and American social thought: Key texts: 1919–1968.* Lanham, MD: Rowman and Littlefield.

24. Heath, R. L., & Bryant, J. (2000). *Human communication theory and research.* Mahwah, NJ: Erlbaum.

25. Heath, R. L., & Bryant, J. (2000). *Human communication theory and research.* Mahwah, NJ: Erlbaum.

26. Hsieh, T. (2010). *Delivering happiness: A path to profits, passion, and purpose.* New York: Grand Central Publishing. Quote appears on p. 143.

27. O'Sullivan, P. B. (2009, May 25). *Masspersonal communication: Rethinking the mass-interpersonal divide.* Paper presented at the annual meeting of the International Communication Association, New York.

28. Lenhart, A. (2016, August 6). Chapter 4: Social media and friendships. In *Teens, technology and friendships.* Pew Research Center Report. Retrieved from http://www.pewinternet.org/2015/08/06/chapter-4-social-media-and-friendships/.

29. Bayer, J. B., Ellison, N. B., Schoenbeck, S. Y. & Falk, E. B. (2016). Sharing the small moments: Ephemeral social interaction on Snapchat. *Information, Communication & Society 19*(7), 956–977.

30. Next Gen 2015: YouTube top 30 influencers. (2015, November 4). *The Hollywood Reporter.* Retrieved from http://www.hollywoodreporter.com/lists/next-gen-2015-youtubes-top-836437/item/adriene-mishler-youtube-next-gen-836480.

31. Johnson, S. (2009, June 5). How Twitter will change the way we live. *Time.* Retrieved from http://www.time.com/time/magazine/article/0,9171,1902818,00.html.

32. Surinder, K. S., & Cooper, R. B. (2003). Exploring the core concepts of media richness theory: The impact of cue multiplicity and feedback immediacy on decision quality. *Journal of Management Information Systems, 20,* 263–299.

33. Severin, W. J., & Tankard, J. W. (1997). *Communication theories: Origins, methods, and uses in the mass media* (4th ed.). New York: Longman, pp. 197–214.

34. Ruggiero, T. E. (2000). Uses and gratifications theory in the 21st century. *Mass Communication & Society, 3,* 3–37. For a somewhat different categorization of uses and gratifications, see Joinson, A. N. (2008, April 5–10). "Looking at," "looking up" or "keeping up with" people? Motives and uses of Facebook. In *Proceedings of the 26th annual SIGCHI Conference on Human Factors in Computing Systems* (Florence, Italy) (pp. 1027–1036). New York: ACM. See also Flanagan, A. J. (2005). IM online: Instant messaging use among college students. *Communication Research Reports, 22,* 173–187.

35. Jaremka, L. M., Andridge, R. R., Fagundes, C. P., Alfano, C. M., Povoski, S. P., Lipari, A. M., . . . Kiecolt-Glaser, J. K. (2014). Pain, depression, and fatigue: Loneliness as a longitudinal risk factor. *Health Psychology, 33*(9), 948–957.

36. Newall, N. G., Chipperfield, J. G., Bailis, D. S., & Stewart, T. L. (2013). Consequences of loneliness on physical activity and mortality in older adults and the power of positive emotions. *Health Psychology, 32*(8), 921–924.

37. Cacioppo, S., Capitanio, J. P., & Cacioppo, J. T. (2014). Toward a neurology of loneliness. *Psychological Bulletin, 140*(6), 1464–1504.

38. Stewart, J. (2004). *Bridges, not walls: A book about interpersonal communication* (9th ed.). New York: McGraw-Hill, p. 11.

39. Shattuck, R. (1980). *The forbidden experiment: The story of the wild boy of Aveyron.* New York: Farrar, Straus & Giroux, p. 37.

40. For a fascinating account of Genie's story, see Rymer, R. (1993). *Genie: An abused child's flight from silence.* New York: HarperCollins. Linguist Susan Curtiss (1977) provides a more specialized account of the case in her book *Genie: A psycholinguistic study of a modern-day "wild child."* San Diego, CA: Academic Press.

41. Rubin, R. B., Perse, E. M., & Barbato, C. A. (1988). Conceptualization and measurement of interpersonal communication motives. *Human Communication Research, 14,* 602–628.

42. Robinson, B. (2015, July 19). Tech CEO worth $820 million lives with his pet alpaca in a tiny Airstream caravan at Las Vegas trailer park called "Llamapolis." *Daily Mail.* Retrieved from http://www.dailymail.co.uk/news/article-3167130/Zappos-CEO-Tony-Hsieh-cushy-net-worth-820-million-opts-live-trailer-park-instead-mansion.html.

43. Goldschmidt, W. (1990). *The human career: The self in the symbolic world.* Cambridge, MA: Basil Blackmun.

44. National Association of Colleges and Employers. (2014). *Job outlook 2015.* Bethlehem, PA: Author. Retrieved from https://www.umuc.edu/upload/NACE-Job-Outlook-2015.pdf.

45. Arroyo, A., & Segrin, C. (2011). The relationship between self- and other-perceptions of communication competence and friendship quality. *Communication Studies, 62*(5), 547–562.

46. Määttä, K., & Uusiautti, S. (2013). Silence is not golden: Review of studies of couple interaction. *Communication Studies, 64*(1), 33–48.

47. Docan-Morgan, T., Manusov, V., & Harvey, J. (2013). When a small thing means so much: Nonverbal cues as turning points in relationships. *Interpersona, 7*(1), 110–124. Quote appears on p. 118.

48. See Wiemann, J. M., Takai, J., Ota, H., & Wiemann, M. (1997). A relational model of communication competence. In B. Kovacic (Ed.), *Emerging theories of human communication* (pp. 25–44). Albany: SUNY. These goals, and the strategies used to achieve them, needn't be conscious. See Fitzsimons, G. M., & Bargh, J. A. (2003). Thinking of you: Nonconscious pursuit of interpersonal goals associated with relationship partners. *Journal of Personality and Social Psychology, 84,* 148–164.

49. Burleson, B. R., & Samter, W. (1994). A social skills approach to relationship maintenance. In D. Canary & L. Stafford (Eds.), *Communication and relationship maintenance.* San Diego, CA: Academic Press, p. 12.

50. Guerrero, L. K. (2014). Jealousy and relational satisfaction: Actor effects, partner effects, and the mediating role of destructive communicative responses to jealousy. *Western Journal of Communication, 78*(5), 586–611.

51. See, for example, Heisel, A. D., McCroskey, J. C., & Richmond, V. P. (1999). Testing theoretical relationships and nonrelationships of genetically-based predictors: Getting started with communibiology. *Communication Research Reports, 16,* 1–9; and McCroskey, J. C., & Beatty, K. J. (2000). The communibiological perspective: Implications for communication in instruction. *Communication Education, 49,* 1–6.

52. Hunter, K. M., Westwick, J. N., & Haleta, L. L. (2014). Assessing success: The impacts of a fundamentals of speech course on decreasing public speaking anxiety. *Communication Education, 63*(2), 124–135.

53. Gearhart, C. C., Denham, J. P., & Bodie, G. D. (2014). Listening as a goal-directed activity. *Western Journal of Communication, 78*(5), 668–684.

54. Wang, S., Hu, Q., & Dong, B. (2015). Managing personal networks: An examination of how high self-monitors achieve better job performance. *Journal of Vocational Behavior, 91,* 180–188.

55. Jones, K. (2014). At-risk students and communication skill deficiencies: A preliminary study. *Journal of Education and Human Development, 3,* 1–8.

56. Weigel, D. J., Brown, C., & O'Riordan, C. K. (2011). Everyday expressions of commitment and relational uncertainty as predictors of relationship quality and stability over time. *Communication Reports, 24*(1), 38–50.

57. Griffith, E. (2015, January 22). How to quit social media (and why you should). *PC Magazine.* Retrieved from http://www.pcmag.com/article2/0,2817,2475453,00.asp. Quote appears in paragraph 5.

58. Roberts, J. A., Luc Honore Petnji, Y., & Manolis, C. (2014). The invisible addiction: Cell-phone activities and addiction among male and female college students. *Journal of Behavioral Addictions, 3*(4), 254–265.

59. U.S. Bureau of Labor Statistics. (2015). American time use survey: Time use on an average weekday of full-time university and college students. Retrieved from http://www.bls.gov/tus/charts/students.htm.

60. Mullen, C. (n.d.). 2 reasons you should limit your social media time. Elevate. Design your best life [Blog post]. Retrieved from http://chrismullen.org/2-reasons-you-should-limit-your-social-media-time-2/. Quotes appear in paragraph 3.

61. Moose A. (2015, May 28). 7 reasons why you need to take a break from social media. *Elite Daily*. Retrieved from http://elitedaily.com/life/limit-time-on-social-media/1036660/. Quote appears in paragraph 5.

62. Grumstrup, C. (2015, June 3). 7 reasons to limit your social media that will lead to a happier you. DOSE [Blog post]. Retrieved from http://www.dose.com/theworld/21439/7-Reasons-To-Limit-Your-Social-Media-That-Will-Lead-To-A-Happier-You. Quote appears in paragraph 3.

63. Moose A. (2015, May 28). 7 reasons why you need to take a break from social media. *Elite Daily*. Retrieved from http://elitedaily.com/life/limit-time-on-social-media/1036660/. Quote appears in paragraph 5.

64. Griffith, E. (2015, January 22.) How to quit social media (and why you should). *PC Magazine*. Retrieved from http://www.pcmag.com/article2/0,2817,2475453,00.asp. Quote appears in paragraph 2.

65. Sher, L. (2012, March 7). Miss Seattle insists she doesn't hate Seattle after Twitter rant. *ABC News*. Retrieved from http://abcnews.go.com/blogs/headlines/2012/03/miss-seattle-insists-she-doesnt-hate-seattle-after-twitter-rant.

66. Lenhart, A. (2009, December 15). Teens and sexting. *Pew Research Center: Internet, Science & Tech*. Retrieved from http://www.pewinternet.org/Reports/2009/Teens-and-Sexting.aspx.

67. *A thin line: MTV-AP digital abuse study* (executive summary). (2009). Retrieved from http://www.athinline.org/MTV-AP_Digital_Abuse_Study_Executive_Summary.pdf.

68. Meyer, E. J. (2009, December 16). "Sexting" and suicide. *Psychology Today Online*. Retrieved from http://www.psychologytoday.com/blog/gender-and-schooling/200912/sexting-and-suicide.

69. Bauerlein, M. (2009, September 4). Why Gen-Y Johnny can't read nonverbal cues. *Wall Street Journal*. Retrieved from http://www.wsj.com/articles/SB10001424052970203863204574348493483201758.

70. Watts, S. A. (2007). Evaluative feedback: Perspectives on media effects. *Journal of Computer-Mediated Communication, 12*(2), 384–411. Retrieved from http://jcmc.indiana.edu/vol12/issue2/watts.html. See also Turnage, A. K. (2007). Email flaming behaviors and organizational conflict. *Journal of Computer-Mediated Communication, 13*(1), 43–59. Retrieved from http://jcmc.indiana.edu/vol13/issue1/turnage.html.

71. LeBlanc, J. C. (2012, October 20). *Cyberbullying and suicide: A retrospective analysis of 22 cases*. Paper presented at the American Academy of Pediatrics National Conference, New Orleans. Retrieved from https://aap.confex.com/aap/2012/webprogram/Paper18782.html.

72. National Crime Prevention Council. (2007, February 28). *Teens and cyberbullying: Executive summary of a report on research conducted for National Crime Prevention Council (NCPC)*. Retrieved from http://www.ncpc.org/resources/files/pdf/bullying/Teens%20and%20Cyberbullying%20Research%20Study.pdf.

73. Caplan, S. E. (2005). A social skill account of problematic Internet use. *Journal of Communication, 55*, 721–736; Schiffrin, H., Edelman, A., Falkenstein, M., & Stewart, C. (2010). Associations among computer-mediated communication, relationships, and well-being. *Cyberpsychology, Behavior, and Social Networking, 13*, 1–14; Morrison, C. M., & Gore, H. (2010). The relationship between excessive Internet use and depression: A questionnaire-based study of 1,319 young people and adults. *Psychopathology, 43*, 121–126.

74. Caplan, 2005.

75. Ko, C., Yen, J., Chen, C., Chen, S., & Yen, C. (2005). Proposed diagnostic criteria of Internet addiction for adolescents. *The Journal of Nervous and Mental Disease, 11*, 728–733.

76. See, for example, Lenhart, A., Madden, M., Smith, A., & MacGill, A. (2007, December 19). Teens creating content. *Pew Internet & American Life Project*. Retrieved from http://www.pewinternet.org/Reports/2007/Teens-and-Social-Media/3-Teens-creating-content/18-Videos-are-not-restricted-as-often-as-photos.aspx?r=1.

77. Strayer, D. L., Drews, F. A., Crouch, D. J., & Johnston, W. A. (2005). Why do cell phone conversations interfere with driving? In W. R. Walker & D. Herrmann (Eds.), *Cognitive technology: Transforming thought and society* (pp. 51–68). Jefferson, NC: McFarland.

78. U.S. Department of Transportation, National Highway Safety Administration. (2010, September). Traffic safety facts. Retrieved from http://www-nrd.nhtsa.dot.gov/Pubs/811379.pdf .

79. Strayer, D. L., & Drew, F. A. (2004, Winter). Profiles in driver distraction: Effects of cell phone conversations on younger and older drivers. *Human Factors, 46*, 640–649.

80. Smith, J. L., Ickes, W., & Hodges, S. (Eds.). (2010). *Managing interpersonal sensitivity: Knowing when—and when not—to understand others.* Hauppauge, NY: Nova Science.

81. Pearce, W. B., & Pearce, K. A. (2000). Extending the theory of the coordinated management of meaning (CMM) through a community dialogue process. *Communication Theory, 10*, 405–423. See also Griffin, E. M. (2003). *A first look at communication theory* (5th ed.). New York: McGraw-Hill, pp. 66–81.

82. Meerloo, J. A. M. (1952). *Conversation and communication.* Madison, CT: International Universities Press, p. 91.

83. McCroskey, J. C., & Wheeless, L. R. (1976). *Introduction to human communication.* Boston: Allyn & Bacon, p. 5.

84. Leonardi, P. M., Treem, J. W., & Jackson, M. H. (2010). The connectivity paradox: Using technology to both decrease and increase perceptions of distance in distributed work arrangements. *Journal of Applied Communication Research, 38*, 85–105.

85. Taylor, B. (2009, July 23). Amazon and Zappos: A savvy deal. *Harvard Business Review*. Retrieved from https://hbr.org/2009/07/a-savvy-deal-from-amazon-to-za. Quote appears in paragraph 12.

CHAPTER 2

1. Bissinger, B. (2015, July). Caitlyn Jenner talks about her mother's reaction and transgender fans. *Vanity Fair*. Retrieved from http://www.vanityfair.com/hollywood/2015/06/caitlyn-jenner-photos-interview-buzz-bissinger. Quote appears in the final paragraph.

2. Bruce Jenner—The interview. (2015, April). *20/20*. ABC. Retrieved from http://abc.go.com/shows/2020/listing/2015-04-24-bruce-jenner-the-interview.

3. Bruce Jenner—The interview. (2015, April). *20/20*. ABC. Retrieved from http://abc.go.com/shows/2020/listing/2015-04-24-bruce-jenner-the-interview.

4. Cvencek, D., Greenwald, A.G., & Meltzoff, A.N. (2015). Implicit measures for preschool children confirm self-esteem's role in maintaining a balanced identity. *Journal of Experimental Social Psychology, 62*, 50–57.

5. Saunders, J. (2012). The role of self-esteem in the misinformation effect. *Memory, 20*(2), 90–99.

6. Peterson, J. L., & DeHart, T. (2013). Regulating connection: Implicit self-esteem predicts positive non-verbal behavior during romantic relationship-threat. *Journal of Experimental Social Psychology, 1*, 99.

7. Ferris, D. L., Lian, H., Brown, D. J., & Morrison, R. (2015). Ostracism, self-esteem, and job performance: When do we self-verify and when do we self-enhance? *Academy of Management Journal, 58*(1), 279–297.

8. Zeigler-Hill, V., Besser, A., Myers, E. M., Southard, A. C., & Malkin, M. L. (2013). The status-signaling property of self-esteem: The role of self-reported self-esteem and perceived self-esteem in personality judgments. *Journal of Personality, 81*(2), 209–220.

9. Zeigler-Hill, V., Besser, A., Myers, E. M., Southard, A. C., & Malkin, M. L. (2013). The status-signaling property of self-esteem: The role of self-reported self-esteem and perceived self-esteem in personality judgments. *Journal of Personality, 81*(2), 209–220.

10. Baumeister, R. F. (2005). *The cultural animal: Human nature, meaning, and social life.* New York: Oxford University Press; and Baumeister, R. F., Campbell, J. D., Krueger, J. I., & Vohs, K. D. Does high self-esteem cause better performance, interpersonal success, happiness, or healthier lifestyles? *Psychological Science in the Public Interest, 4*, 1–44.

11. Vohs, K. D., & Heatherton, T. F. (2004). Ego threats elicit different social comparison process among high and low self-esteem people: Implications for interpersonal perceptions. *Social Cognition, 22*, 168–191.

12. For more, see Kandler, C., Riemann, R., & Kämpfe, N. (2009). Genetic and environmental mediation between measures of personality and family environment in twins reared together. *Behavioral Genetics, 39*, 24–35; and Caspi, A., Harrington, H., Milne, B., Amell, J. W., Theodore, R. F., & Moffitt, T. E. (2003). Children's behavioral styles at age 3 are linked to their adult personality traits at age 26. *Journal of Personality, 71*, 495–514. doi: 10.1111/1467-6494.7104001.

13. Ashton, M. C., Lee, K., Perugini, M., Szarota, P., de Vries, R. E., Di Blas, L., . . . De Raad, B. (2004, February). A six-factor structure of personality-descriptive adjectives: Solutions from psycholexical studies in seven languages. *Journal of Personality and Social Psychology, 86*(2), 356–366.

14. Vukasović, T., & Bratko, D. (2015). Heritability of personality: A meta-analysis of behavior geneticstudies. *Psychological Bulletin, 141*(4), 769–785.

15. See Gong, P., Zheng, A., Zhang, K., Lei, X., Li, F., Chen, D., . . . Zhang, F. (2010). Association analysis between 12 genetic variants of ten genes and personality traits in a young Chinese Han population. *Journal of Molecular Neuroscience,42*, 120–126; and Heck, A., Lieb, R., Ellgas, A., Pfister, H., Lucae, S., Roeske, D., . . . Ising, M. (2009). Investigation of 17 candidate genes for personality traits confirms effects of the HTR2A gene on novelty seeking. *Genes, Brain and Behavior, 8*, 464–472. doi: 10.1111/j.1601-183X.2009.00494.x.

16. Cole, J. G., & McCroskey, J. C. (2000). Temperament and sociocommunicative orientation. *Communication Research Reports, 17*, 105–114.

17. Dweck, C. (2008). Can personality be changed? The role of beliefs in personality and change. *Current Directions in Psychological Science, 6*, 391–394.

18. Begney, S. (2008, December 1). When DNA is not destiny. *Newsweek, 152*, 14.

19. Bruce Jenner—The interview. (2015, April). *20/20.* ABC. Retrieved from http://abc.go.com/shows/2020/listing/2015-04/24-bruce-jenner-the-interview.

20. National Coalition of Anti-Violence Programs. (2015). *Lesbian, gay, bisexual, transgender, queer, and HIV-affected hate violence in 2014* (2015 release edition). New York: New York City Anti-Violence Project. Retrieved from http://www.avp.org/storage/documents/Reports/2014_HV_Report-Final.pdf.

21. Green, E. (2015, June 4). The real Christian debate on transgender identity. *The Atlantic.* Retrieved from http://www.theatlantic.com/politics/archive/2015/06/the-christian-debate-on-transgender-identity/394796/.

22. Krcmar, M., Giles, S., & Helme, D. (2008). Understanding the process: How mediated and peer norms affect young women's body esteem. *Communication Quarterly, 56*, 111–130.

23. Cho, A., & Lee, J. (2013).Body dissatisfaction levels and gender differences in attentional biases toward idealized bodies. *Body Image, 10*, 95–102.

24. Ata, R. N., Thompson, J. K., & Small, B. J. (2013). Effects of exposure to thin-ideal media images on body dissatisfaction: Testing the inclusion of a disclaimer versus warning label. *Body Image, 10*, 472–480.

25. Zuro, A. (2014). *Measuring up: Social comparisons on Facebook and contributions to self-esteem and mental health* (Master's thesis). Available from University of Michigan Deep Blue database (UMI No. 65059383), http://hdl.handle.net/2027.42/107346.

26. Köber, C., Schmiedek, F., & Habermas, T. (2015). Characterizing lifespan development of three aspects of coherence in life narratives: A cohort-sequential study. *Developmental Psychology, 51*(2), 260–275.

27. Sobo, E. J., & Loustaunau, M. O. (2010). *The cultural context of health, illness, and medicine.* Santa Barbara, CA: Praeger. Adapted from dialogue on p. 86.

28. Samovar, L. A., Porter, R. E., & McDaniel, E. R. (2007). *Communication between cultures* (7th ed.). Boston, MA: Cengage, p. 91.

29. Yamagishi, T., Hashimoto, H., Cook, K. S., Kiyonari, T., Shinada, M., Mifune, N., & . . . Li, Y. (2012). Modesty in self-presentation: A comparison between the USA and Japan. *Asian Journal of Social Psychology, 15*(1), 60–68.

30. Brozovich, F. A., & Heimberg, R. G. (2013). Mental imagery and post-event processing in anticipation of a speech performance among socially anxious individuals. *Behavior Therapy, 44*, 701–716.

31. DiPaola, B. M., Roloff, M. E., & Peters, K. M. (2010). College students' expectations of conflict intensity: A self-fulfilling prophecy. *Communication Quarterly, 58*(1), 59–76.

32. Stinson, D. A., Cameron, J. J., Wood, J. V., Gaucher, D., & Holmes J. G. (2009). Deconstructing the "reign of error": Interpersonal warmth explains the self-fulfilling prophecy of anticipated acceptance. *Personality and Social Psychology, 35*, 1165–1178.

33. Dimberg, U., & Söderkvist, S. (2011). The voluntary facial action technique: A method to test the facial feedback hypothesis. *Journal of Nonverbal Behavior, 35*, 17–33.

34. Holmes, J. G. (2002). Interpersonal expectations as the building blocks of social cognition: An interdependence theory perspective. *Personal Relationships, 9*, 1–26.

35. Rosenthal, R.,& Jacobson, L. (1968). *Pygmalion in the classroom.* New York: Holt, Rinehart and Winston.

36. For a detailed discussion of how self-fulfilling prophecies operate in relationships, see Watzlawick, P. (2005). Self-fulfilling prophecies. In J. O'Brien & P. Kollock (Eds.), *The production of reality* (3rd ed., pp. 382–394). Thousand Oaks, CA: Pine Forge Press.

37. James, W. (1920). *The letters of William James* (H. James, Ed.). Boston, p. 462.

38. Ruby, F. M., Smallwood, J., Engen, H., & Singer, T. (2013). How self-generated thought shapes mood—the relation between mind-wandering and mood depends on the socio-temporal content of thoughts. *PLOS ONE, 8*(10), e77554.

39. Knobloch, L. K., Miller, L. E., Bond, B. J., & Mannone, S. E. (2007). Relational uncertainty and message processing in marriage. *Communication Monographs, 74*, 154–180.

40. Macrae, C. N., & Bodenhausen, G. V. (2001). Social cognition: Categorical person perception. *British Journal of Psychology, 92*, 239–256.

41. Greenwell, T. C., Hancock, M., Simmons, J. M., & Thorn, D. (2015). The effects of gender and social roles on the marketing of combat sport. *Sport Marketing Quarterly, 24*(1), 19–29.

42. Heisler, J., & Crabill, S. (2006). Who are "stinkybug" and "packerfan4"? Email pseudonyms and participants' perceptions of demography, productivity, and personality. *Journal of Computer-Mediated Communication,12*, article 6. Retrieved from http://jcmc.indiana.edu/vol12/issue1/heisler.html.

43. Clark, A. (2000). *A theory of sentience.* New York: Oxford University Press.

44. Weigold, A., Weigold, I., Natera, S., & Russell, E. (2013). The role of face familiarity in judgments of personality and intelligence. *Current Psychology, 32*(3), 242–251.

45. Merolla, A. J. (2008). Communicating forgiveness in friendships and dating relationships. *Communication Studies, 59*, 114–131.

46. Miró, E., Cano, M. C., Espinoza-Fernández, L., & Beula-Casal, G. (2003). Time estimation during prolonged sleep deprivation and its relation to activation measures. *Human Factors, 45*, 148–159.

47. Alaimo, K., Olson, C. M., & Frongillo, E. A. (2001). Food insufficiency and American school-aged children's cognitive, academic, and psychosocial development. *Pediatrics, 108*, 44–53.

48. Koukkari, W. L., & Sothern, R. B. (2006). *Introducing biological rhythms: A primer on the temporal organization of life, with implications for health, society, reproduction and the natural environment.* New York: Springer.

49. Hasler, B. P., & Troxel, W. M. (2010). Couples' nighttime sleep efficiency and concordance: Evidence of bidirectional associations with daytime relationship functioning. *Psychosomatic Medicine, 72*, 794–801.

50. Goldstein, S. (2008). Current literature in ADHD. *Journal of Attention Disorders, 11*, 614–616.

51. Von Briesen, P. D. (2007). Pragmatic language skills of adolescents with ADHD. *DAI, 68*(5-B), 3430.

52. Babinski, D. E., Pelham W. E., Jr., Molina, B. S. G., Gnagy, E. M., Waschbusch, D. A., Yu, J., & Karch, K. M. (2011). Late adolescent and young adult outcomes of girls diagnosed with ADHD in childhood: An exploratory investigation. *Journal of Attention Disorders, 15*, 204–214.

53. National Institute of Mental Health. (2008, April 3). *Attention deficit hyperactivity disorder.* Retrieved from http://www.nimh.nih.gov/health/publications/attention-deficit-hyperactivity-disorder/index.shtml.

54. Hormone Health Network. (2011, February). Hyperprolactinemia diagnosis and treatment: A patient's guide. Retrieved from http://www.hormone.org/~/media/hormone/files/patient-guides/pituitary/pghyperprolactinemia_2014.pdf.

55. Bem, S. L. (1974). The measurement of psychological androgyny. *Journal of Consulting and Clinical Psychology, 42*, 155–162.

56. Versalle, A., & McDowell, E. E. (2004–2005). The attitudes of men and women concerning gender differences in grief. *Omega: Journal of Death and Dying, 50*, 53–67.

57. Versalle, A., & McDowell, E. E. (2004–2005). The attitudes of men and women concerning gender differences in grief. *Omega: Journal of Death and Dying, 50*, 53–67.

58. Shechory Bitton, M., & Ben Shaul, D. (2013). Perceptions and attitudes to sexual harassment: An examination of sex differences and the sex composition of the harasser-target dyad. *Journal of Applied Social Psychology, 43*(10), 2136–2145.

59. Solomon, D. H., & Williams, M. L. M. (1997). Perceptions of social-sexual communication at work: The effects of message, situation, and observer characteristics on judgments of sexual harassment. *Journal of Applied Communication Research, 25*, 197–216.

60. Zimbardo, P. G. (1971). *The psychological power and pathology of imprisonment.* Statement prepared for the U.S. House of Representatives Committee on the Judiciary, Subcommittee No. 3, Robert Kastemeyer, Chairman. Unpublished manuscript, Stanford University. See also Zimbardo, P. G. (1977). *Shyness: What it is, what to do about it.* Reading, MA: Addison-Wesley.

61. Swami, V., & Furnham, A. (2008). *The psychology of physical attraction.* New York: Routledge/Taylor & Francis.

62. Gonzaga, G. G., Haselton, M. G., Smurda J., Davies, M., & Poore, J. C. (2008). Love, desire, and the suppression of thoughts of romantic alternatives. *Evolution and Human Behavior, 29*, 119–126.

63. Shaw, C. L. M. (1997). Personal narrative: Revealing self and reflecting other. *Human Communication Research, 24*, 302–319.

64. Kellas, J. K., Baxter, L., LeClair-Underberg, C., Thatcher, M., Routsong, T., Normand, E. L., & Braithwaite, D. O. (2014). Telling the story of stepfamily beginnings: The relationship between young-adult stepchildren's stepfamily origin stories and their satisfaction with the stepfamily. *Journal of Family Communication, 14*(2), 149–166. Quotes appear on p. 160.

65. Martz, J. M., Verette, J., Arriaga, X. B., Slovik, L. F., Cox, C. L., & Rusbult, C. E. (1998). Positive illusion in close relationships. *Personal Relationships, 5*, 159–181.

66. Pearson, J. C. (2000). Positive distortion: "The most beautiful woman in the world." In K. M. Galvin & P. J. Cooper (Eds.), *Making connections: Readings in relational communication* (2nd ed., pp. 184–190). Los Angeles, CA: Roxbury.

67. Willis, J., & Todorov, A. (2006). First impressions: Making up your mind after a 100-ms exposure to a face. *Psychological Science, 17*, 592–598.

68. Nelson, T. D. (2005). Ageism: Prejudice against our featured future self. *Journal of Social Issues, 61*, 207–221.

69. Zenmore, S. E., Fiske, S. T., & Kim, H. J. (2000). Gender stereotypes and the dynamics of social interaction. In T. Eckes & H. M. Trautner (Eds.), *The developmental social psychology of gender* (pp. 207–241). Mahwah, NJ: Erlbaum.

70. Maner, J. K., & Miller, S. L. (2013). Adaptive attentional attunement: Perceptions of danger and attention to outgroup men. *Social Cognition, 31*(6), 733–744.

71. Burgess, M. R., Dill, K. E., Stermer, S., Burgess, S. R., & Brown, B. P. (2011). Playing with prejudice: The prevalence and consequences of racial stereotypes in video games. *Media Psychology, 14*(3), 289–311. doi:10.1080/15213269.2011.596467.

72. Ramasubramanian, S. (2011). Television exposure, model minority portrayals, and Asian-American stereotypes: An exploratory study. *Journal of Intercultural Communication, 26*, 4.

73. Block, C. J., Aumann, K., & Chelin, A. (2012). Assessing stereotypes of black and white managers: A diagnostic approach. *Journal of Applied Social Psychology.* Advance online publication retrieved from http://onlinelibrary.wiley.com/doi/10.1111/j.1559-1816.2012.01014.x/abstract.

74. For a review of these perceptual biases, see Hamachek, D. (1992). *Encounters with the self* (3rd ed.). Fort Worth, TX: Harcourt Brace Jovanovich. See also Bradbury, T. N., & Fincham, F. D. (1990). Attributions in marriage: Review and critique. *Psychological Bulletin, 107*, 3–33. For information on the self-serving bias, see Shepperd, J., Malone, W., & Sweeny, K. (2008). Exploring causes of the self-serving bias. *Social and Personality Psychology Compass, 2/2*, 895–908.

75. Easton, S. S., & Bommelje, R. K. (2011). Interpersonal communication consequences of email non-response. *Florida Communication Journal, 39*(2), 45–63.

76. Black, S. L., & Johnson, A. F. (2012). Employers' use of social networking sites in the selection process. *Journal of Social Media in Society, 1*(1), 7–28.

77. Marek, C. I., Wanzer, M. B., & Knapp, J. L. (2004). An exploratory investigation of the relationship between roommates' first impressions and subsequent communication patterns. *Communication Research Reports, 21*, 210–220.

78. Henningsen, D., Henningsen, M., McWorthy, E., McWorthy, C., & McWorthy, L. (2011). Exploring the effects of sex and mode of presentation in perceptions of dating goals in video-dating. *Journal of Communication, 61*(4), 641–658. doi:10.1111/j.1460-2466.2011.01564.x.

79. See, for example, Walther, J. B., DeAndrea, D. C., & Tong, S. T. (2009, November). *Computer-mediated communication versus vocal communication in the amelioration of stereotypes: A replication with three theoretical models.* Paper presented at the annual meeting of the National Communication Association, Chicago, IL.

80. Okdie, B. M., Guadgno, R. E., Bemien, F. J., Geers, A. L., & Mclarney-Vesotski, A. R. (2011). Getting to know you: Face-to-face versus online interactions. *Computers in Human Behavior, 27*, 153–159.

81. Pempek, T. A., Yermolayeva, Y. A., & Calvert, S. L. (2009). College students' social networking experiences on Facebook. *Journal of Applied Developmental Psychology, 30*, 227–238.

82. Bad grammar is bad for business. (2014, September 19). Hallam. Retrieved from https://www.hallaminternet.com/2014/bad-grammar-bad-business/.

83. Stiff, J. B., Dillard, J. P., Somera, L., Kim, H., & Sleight, C. (1988). Empathy, communication, and prosocial behavior. *Communication Monographs, 55*, 198–213.

84. Quann, V., & Wien, C. A. (2006, July). The visible empathy of infants and toddlers. *Young Children.* Retrieved from https://www.naeyc.org/files/yc/file/200607/Quann709BTJ.pdf.

85. Walter, H. (2012). Social cognitive neuroscience of empathy: Concepts, circuits, and genes. *Emotion Review, 4*, 9–17.

86. Miklikowka, M., Duriez, M., & Soenens, B. (2011). Family roots of empathy-related characteristics: The role of perceived maternal and paternal needs support in adolescence. *Developmental Psychology, 47*, 1342–1352.

87. Spivak, A. L., & Farran, D. C. (2012). First-grade teacher behaviors and children's prosocial actions in classrooms. *Early Education & Development, 23*(5), 623.

88. Van der Graaff, J., Branje, S., De Wied, M., Hawk, S., Van Lier, P., & Meeus, W. (2014). Perspective taking and empathic concern in adolescence: Gender differences in developmental changes. *Developmental Psychology, 50*(3), 881–888.

89. Aradhye, C., Vonk, J., & Arida, D. (2015). Adults' responsiveness to children's facial expressions. *Journal of Experimental Child Psychology, 135*, 56–71.

90. Goffman, E. (1971). *The presentation of self in everyday life.* Garden City, NY: Doubleday; and Goffman, E. (1971). *Relations in public.* New York: Basic Books.

91. Cupach, W. R., & Metts, S. (1994). *Facework.* Thousand Oaks, CA: Sage. See also Brown, P., & Levinson, S. C. (1987). *Politeness: Some universals in language usage.* Cambridge: Cambridge University Press.

92. Sharkey, W. F., Park, H. S., & Kim, R. K. (2004). Intentional self embarrassment. *Communication Studies, 55*, 379–399.

93. Urciuoli, B. (2009). The political topography of Spanish and English: The view from a New York Puerto Rican neighborhood. *American Ethnologist, 10*, 295–310.

94. Benet-Martínez, V., Leu, J., Lee, F., & Morris, M. (2002). Negotiating biculturalism: Cultural frame switching in biculturals with oppositional versus compatible cultural identities. *Journal of Cross-Cultural Psychology, 33*, 492–516.

95. Toomey, A., Dorjee, T., & Ting-Toomey, S. (2013). Bicultural identity negotiation, conflicts, and intergroup communication strategies. *Journal of Intercultural Communication Research, 42*(2), 112–134. Quote appears on p. 120.

96. Centorrino, S., Djemai, E., Hopfensitz, A., Milinski, M., & Seabright, P. (2015). Honest signaling in trust interactions: Smiles rated as genuine induce trust and signal higher earning opportunities. *Evolution and Human Behavior, 36*, 8–16.

97. Newell, C. B. (2012). Your face looks funny: The role of emotion on perceived attractiveness of face images [Supplement]. *Perception, 41*, 106.

98. Snyder, M. (1979). Self-monitoring processes. In L. Berkowitz (Ed.), *Advances in experimental social psychology* (pp. 85–128). New York: Academic Press; and M. Snyder. (1983, March). The many me's of the self-monitor. *Psychology Today*, 34f.

99. Fleming, P., & Sturdy, A. (2009). "Just be yourself!": Towards neo-normative control in organisations? *Employee Relations, 31*, 569–583.

100. Ragins, B. R. (2008). Disclosure disconnects: Antecedents and consequences of disclosing invisible stigmas across life domains. *Academy of Management Review, 33*, 194–215.

101. Ragins, B. R., Singh, R., & Cornwell, J. M. (2007). Making the invisible visible: Fear and disclosure of sexual orientation at work. *Journal of Applied Psychology, 92*, 1103–1118.

102. Pachankis, J. E. (2007). The psychological implications of concealing a stigma: A cognitive-affective-behavioral model. *Psychological Bulletin, 133*, 328–345.

103. Sezer, O., Gino, F., & Norton, M. I. (2015, April 15). *Humblebragging: A distinct—and ineffective—self-presentation strategy.* Harvard Busines School Marketing Unit Working Paper No. 15-080. Retrieved from http://papers.ssrn.com/sol3/papers.cfm?abstract_id=2597626.

104. Sezer, O., Gino, F., & Norton, M. I. (2015, April 15). *Humblebragging: A distinct—and ineffective—self-presentation strategy.* Harvard Busines School Marketing Unit Working Paper No. 15-080. Retrieved from http://papers.ssrn.com/sol3/papers.cfm?abstract_id=2597626. Quote appears on p. 16.

105. Siibak, A. (2009). Constructing the self through the photo selection: Visual impression management on social networking websites. *Cyberpsychology: Journal of Psychosocial Research on Cyberspace, 3*, article 1. Retrieved from http://www.cyberpsychology.eu/view.php?cisloclanku=2009061501& article=1.

106. Hancock, J. T., & Durham, P. J. (2001). Impression formation in computer-mediated communication revisited: An analysis of the breadth and intensity of impressions. *Communication Research, 28*, 325–347.

107. Suler, J. R. (2002). Identity management in cyberspace. *Journal of Applied Psychoanalytic Studies, 4*, 455–459.

108. Gibbs, J. L., Ellison, N. B., & Heino, R. D. (2006). Self-presentation in online personals: The role of anticipated future interaction, self-disclosure, and perceived success in Internet dating. *Communication Research, 33*, 1–26.

109. Wilcox, K., & Stephen, A. T. (2013). Are close friends the enemy? Online social networks, self-esteem, and self-control. *Journal of Consumer Research, 40*, 90–103.

110. Sunstrum, K. (2014). How social media affect our self-perception. *World of Psychology.* Retrieved from http://psychcentral.com/blog/archives/2014/03/14/how-social-media-affects-our-self-perception/. Quote appears in paragraph 8.

111. Kim, J. W., & Chock, T. M. (2015). Body image 2.0: Associations between social grooming on Facebook and body image concerns. *Computers in Human Behavior, 48*, 331–339.

112. Gummow, J. (2014, March 7). 7 telltale signs social media is killing your self-esteem. *Alternet.* Retrieved from http://www.alternet.org/personal-health/7-telltale-signs-social-media-killing-your-self-esteem.

113. Garber, M. (2014, July 29). Kapitalism, with Kim Kardashian. *The Atlantic.* Retrieved from http://www.theatlantic.com/entertainment/archive/2014/07/lessons-in-capitalism-from-kim-kardashian/375252/. Quote appears in paragraph 8.

114. Toma, C., Hancock, J., & Ellison, N. (2008). Separating fact from fiction: An examination of deceptive self-presentation in on-line dating profiles. *Personality and Social Psychology Bulletin, 34,* 1023–1036.

115. Kyung, E. K. (2015, September 9). Caitlyn Jenner tells Matt Lauer about her "new" life as a woman in TODAY exclusive. *TODAY.* Retrieved from http://www.today.com/popculture/caitlyn-jenner-tells-matt-lauer-about-life-woman-today-show-t42981. Quote appears in paragraph 5.

CHAPTER 3

1. Wang, J. (2012, September 21). China sends more students abroad than any other country. *The Epoch Times.* Retrieved from http://www.theepochtimes.com/n3/1481649-china-sends-more-students-abroad-than-any-other-country/.

2. Kroeber, A. L., & Kluckholn, C. (1952). *Culture: A critical review of concepts and definitions.* Harvard University, Peabody Museum of American Archeology and Ethnology Papers 47.

3. Samovar, L. A., & Porter, R. E. (2007). *Communication between cultures* (6th ed.). Belmont, CA: Wadsworth, quote on p. 395.

4. Buzzanell, P. M. (1999). Tensions and burdens in employment interviewing processes: Perspectives of non-dominant group members. *Journal of Business Communication, 36,* 143–162.

5. Ferguson, G. M., & Cramer, P. (2007). Self-esteem among Jamaican children: Exploring the impact of skin color and rural/urban residence. *Journal of Applied Developmental Psychology, 28,* 345–359.

6. Golash-Boza, T., & Darity, W. (2008). Latino racial choices: The effects of skin colour and discrimination on Latinos' and Latinas' racial self-identifications. *Ethnic & Racial Studies, 31,* 899–934.

7. Brown, H. K., Ouellette-Kuntz, H., Lysaght, R., & Burge, P. (2011). Students' behavioural intentions towards peers with disability. *Journal of Applied Research in Intellectual Disabilities, 24,* 322–332. doi:10.1111/j.1468-3148.2010.00616.x.

8. Binder, J., Brown, R., Zagefka, H., Funke, F., Kessler, T., Mummendey, A., . . . Leyens, J.-F. (2009). Does contact reduce prejudice or does prejudice reduce contact? A longitudinal test of the contact hypothesis among majority and minority groups in three European countries. *Journal of Personality & Social Psychology, 96*(4), 843–856.

9. Bryan, C. (2009, August 8). Michelle Obama's dark skin inspired women of color all over the world. *Examiner.com.* Retrieved from http://www.examiner.com/article/michelle-obama-s-dark-skin-inspires-women-of-color-all-over-the-world.

10. Duggan, M., & Brenner, J. (2013, February 14). The demographics of social media users—2012. *Pew Internet & American Life Project.* Retrieved from http://www.pewinternet.org/Reports/2013/Social-media-users.aspx.

11. Collier, M. J. (1996). Communication competence problematics in ethnic relationships. *Communication Monographs, 63,* 314–336. See also Kline, S., Horton, B. W., & Zhang, S. (2008). How we think, feel and express love: A cross-cultural comparison between American and East Asian culture. *International Journal of Intercultural Relations, 32,* 200–214.

12. Tajfel, H., & Turner, J. C. (1986). The social identity theory of inter-group behavior. In S. Worchel & L. W. Austin (Eds.), *Psychology of intergroup relations* (pp. 7–24). Chicago: Nelson-Hall.

13. Triandis, H. C. (1995). *Individualism and collectivism.* Boulder, CO: Westview.

14. Servaes, J. (1989). Cultural identity and modes of communication. In J. A. Anderson (Ed.), *Communication yearbook 12* (pp. 383–416). Newbury Park, CA: Sage.

15. Samovar, L. A., & Porter, R. E. (2004). *Communication between cultures* (5th ed.). Belmont, CA: Wadsworth.

16. Gudykunst, W. B. (1993). *Communication in Japan and the United States.* Albany: State University of New York Press.

17. Cai, D. A., & Fink, E. L. (2002). Conflict style differences between individualists and collectivists. *Communication Monographs, 69,* 67–87.

18. See Triandis, H. C., Bontempo, R., Villareal, M., Asai, M., & Lucca, N. (1988). Individualism and collectivism: Cross-cultural perspectives of self-ingroup relationships. *Journal of Personality and Social Psychology, 54,* 323–338.

19. See, for example, Moss, G., Kubacki, K., Hersh, M., & Gunn, R. (2007). Knowledge management in higher education: A comparison of individualistic and collectivist cultures. *European Journal of Education, 42,* 377–394.

20. Merkin., R. S. (2009). Cross-cultural communication patterns—Korean and American communication. *Journal of Intercultural Communication, 20,* 5.

21. Wu, S., & Keysar, B. (2007). Cultural effects on perspective taking. *Psychological Science, 18,* 600–606.

22. Takano, Y., & Sogun, S. (2008). Are Japanese more collectivistic than Americans? Examining conformity in in-groups and the reference-group effect. *Journal of Cross-Cultural Psychology, 39*(3), 237–250.

23. Ting-Toomey, S. (1988). A face-negotiation theory. In Y. Kim & W. Gudykunst (Eds.), *Theory in interpersonal communication* (pp. 213–238). Newbury Park, CA: Sage.

24. Hall, E. T. (1959). *Beyond culture.* New York: Doubleday.

25. Chen, Y.-S., Chen, C.-Y. D., & Chang, M.-H. (2011). American and Chinese complaints: Strategy use from a cross-cultural perspective. *Intercultural Pragmatics, 8,* 253–275.

26. Leets, L. (1993). Explaining perceptions of racist speech. *Communication Research, 28,* 676–706; and Leets, L. (1993). Disentangling perceptions of subtle racist speech: A cultural perspective. *Journal of Language and Social Psychology, 22,* 1–24.

27. Hofstede, G. (2001). *Culture's consequences: Comparing values, behaviors, institutions, and organizations across nations* (2nd ed.). Thousand Oaks, CA: Sage.

28. Adapted from Hofstede, G. (2011). Dimensionalizing cultures: The Hofstede Model in context. *Online Readings in Psychology and Culture, 2*(1). Retrieved from http://scholarworks.gvsu.edu/cgi/viewcontent.cgi?article=1014&context=orpc.

29. Hofstede, G. (2001). *Culture's consequences: Comparing values, behaviors, institutions, and organizations across nations* (2nd ed.). Thousand Oaks, CA: Sage.

30. Adapted from Hofstede, G. (2011). Dimensionalizing cultures: The Hofstede Model in context. *Online Readings in Psychology and Culture, 2*(1). Retrieved from http://scholarworks.gvsu.edu/cgi/viewcontent.cgi?article=1014&context=orpc.

31. Cohen, A. (2007). One nation, many cultures: A cross-cultural study of the relationship between personal cultural values and commitment in the workplace to in-role performance and organizational citizenship behavior. *Cross-Cultural Research: The Journal of Comparative Social Science, 41,* 273–300.

32. Dailey, R. M., Giles, H., & Jansma, L. L. (2005). Language attitudes in an Anglo-Hispanic context: The role of the linguistic landscape. *Language & Communication, 25*(1), 27–38.

33. Barker, G. G. (2016). Cross-cultural perspectives on intercultural communication competence. *Journal of Intercultural Communication Research, 45,* 13–30.

34. Basso, K. (2012). "To give up on words": Silence in Western Apache culture. In I. Monogahn, J. E. Goodman, & J. M. Robinson (Eds.), *A cultural approach to interpersonal communication: Essential readings* (2nd ed., pp. 73–83). Malden, MA: Blackwell, quote on p. 84.

35. Hofstede, G. (2001). *Culture's consequences: Comparing values, behaviors, institutions, and organizations across nations* (2nd ed.). Thousand Oaks, CA: Sage.

36. What about Taiwan? (n.d.). *The Hofstede Centre*. Retrieved from http://geert-hofstede.com/taiwan.html.

37. What about Taiwan? (n.d.).

38. Ayoun, B., Palakurthi, R., & Moreo, P. (2010). Cultural influences on strategic behavior of hotel executives: Masculinity and femininity. *International Journal of Hospitality & Tourism Administration, 11*, 1–21.

39. Castle, S. (April 30, 2016). "Muslim's Labour Party Candidacy Shapes London Mayoral Race." *New York Times*. http://www.nytimes.com/2016/05/01/world/europe/muslims-labour-candidacy-shapes-london-mayoral-race.html.

40. Ten things everyone should know about race. (2003). *Race—The power of an illusion*. California Newsreel, Public Broadcasting System. Retrieved from http://www.pbs.org/race/000_About/002_04-background-01-x.htm.

41. Interview with Jonathan Marks. (2003). Background readings for *Race—The power of an illusion*. California Newsreel, Public Broadcasting System. Retrieved from http://www.pbs.org/race/000_About/002_04-background-01-08.htm.

42. Samovar, L. A., Porter, R. E., McDaniel, E. R. (2013). *Communication between cultures* (8th ed.). Boston, MA: Wadsworth.

43. Bowleg, L. (2008). When black + lesbian + woman ≠ black lesbian woman: The methodological challenges of qualitative and quantitative intersectionality research. *Sex Roles, 59*(5/6), 312–325. Quote appears on p. 312.

44. DeFrancisco, V. P., & Palczewski, C. H. (2014). *Gender in communication: A critical introduction*. Thousand Oaks, CA: Sage, p. 9.

45. Orbe, M., Allen, B. J., & Flores, L. A. (Eds.). (2006). *The same and different: Acknowledging the diversity within and between cultural groups*. Washington, DC: National Communication Association (International and Intercultural Communication Annual, XXIX).

46. Saulny, S. (2011, October 12). In strangers' glances at family, tensions linger. *The New York Times*. Retrieved from http://www.nytimes.com/2011/10/13/us/for-mixed-family-old-racial-tensions-remain-part-of-life.html?pagewanted=1&_r=0&ref=raceremixed.

47. Bonam, C. M., & Shih, M. (2009). Exploring multiracial individual's comfort with intimate interracial relationships. *Journal of Social Issues, 65*, 87–103.

48. For a summary of research on this subject, see Bradac, J. J. (1990). Language attitudes and impression formation. In H. Giles & W. P. Robinson (Eds.), *The handbook of language and social psychology* (pp. 387–413). Chichester, England: Wiley. See also Ng, S. H., & Bradac, J. J. (1993). *Power in language: Verbal communication and social influence*. Newbury Park, CA: Sage.

49. Bailey, R. W. (2003). Ideologies, attitudes, and perceptions. *American Speech, 88*, 115–143.

50. Frumkin, L. (2007). Influences of accent and ethnic background on perceptions of eyewitness testimony. *Psychology, Crime & Law, 13*, 317–331.

51. Gluszek, A., & Dovidio, J. F. (2010). Perceptions of bias, communication difficulties, and belonging in the United States. *Journal of Language & Social Psychology, 29*, 224–234.

52. Tannen, D. (2005). *Conversational style: Analyzing talk among friends* (Rev. ed.). New York: Oxford University Press.

53. Tannen, D. (2012, October 18). Would you please let me finish. . . . *New York Times*, p. A33.

54. Birdwhistell, R. L. (1970). *Kinesics and context*. Philadelphia: University of Philadelphia Press, pp. 30–31.

55. Andersen, P., Lustig, M., & Anderson, J. (1987). *Changes in latitude, changes in attitude: The relationship between climate, latitude,*

and interpersonal communication predispositions. Paper presented at the annual convention of the Speech Communication Association, Boston; Andersen, P., Lustig, M., & Andersen, J. (1988). *Regional patterns of communication in the United States: Empirical tests*. Paper presented at the annual convention of the Speech Communication Association, New Orleans.

56. What is LGBTQ? (n.d.). Iknowmine.org., sponsored by Alaska Native Tribal Health Consortium, Community Health Services. Retrieved from http://www.iknowmine.org/for-youth/what-is-glbt.

57. Federal Bureau of Investigation. (2012, December 10). Hate crimes accounting: Annual report released. Retrieved from http://www.fbi.gov/news/stories/2012/december/annual-hate-crimes-report-released/annual-hate-crimes-report-released.

58. All of the statements by Anderson Cooper in this paragraph are from Sullivan, A. (2012, July 2). Anderson Cooper: "The fact is, I'm gay." *The Dish*. Retrieved from http://dish.andrewsullivan.com/2012/07/02/anderson-cooper-the-fact-is-im-gay/.

59. Potter, J. E. (2002). Do ask, do tell. *Annals of Internal Medicine, 137*(5), 341–343, quote on p. 342.

60. Hancox, L. (n.d.). Top 8 tips for coming out as trans. *Ditch the Label*. Retrieved from http://www.ditchthelabel.org/8-tips-for-coming-out-as-trans/. Quote appears in Tip 1.

61. Thurman, S. (2003). *The only grammar book you'll ever need*. Avon, MA: Adams Media.

62. American Dialect Society. (2016, January 8). 2015 word of the year is singular "they." Retrieved from http://www.americandialect.org/2015-word-of-the-year-is-singular-they.

63. GLAAD. (2015, May). Tips for allies of transgender people. Retrieved from http://www.glaad.org/transgender/allies.

64. Russell, G. M., & Bohan, J. S. (2005, December). The gay generational gap: Communicating across the LGBT generational divide. *Institute for Gay and Lesbian Strategic Studies, 8*(1), 1–8, quote on p. 3.

65. It Gets Better Project. (2013). About the It Gets Better Project. Retrieved from http://www.itgetsbetter.org/pages/about-it-gets-better-project/.

66. Dan Savage: It gets better. (2013, January 14). *Take part*. Retrieved from http://www.itgetsbetter.org/.

67. Hussein, Y. (2015, December 3). Are you afraid to be Muslim in America? *Huffington Religion*. Retrieved from http://www.huffingtonpost.com/yasmin-hussein/are-you-afraid-to-be-muslim-in-america_b_8710826.html. Quote appears in paragraph 8.

68. Milevsky, A., Shifra Niman, D., Raab, A., & Gross, R. (2011). A phenomenological examination of dating attitudes in Ultra-Orthodox Jewish emerging adult women. *Mental Health, Religion & Culture, 14*, 311–322. doi:10.1080/13674670903585105.

69. Pew Forum on Religion & Public Life. (2008, June). *U.S. religious landscape survey. Religious beliefs and practices: Diverse and politically relevant*. Retrieved from http://religions.pewforum.org/pdf/report2-religious-landscape-study-full.pdf.

70. Bartkowski, J. P., Xiaohe, X., & Fondren, K. M. (2011). Faith, family, and teen dating: Examining the effects of personal and household religiosity on adolescent romantic relationships. *Review of Religious Research, 52*, 248–265.

71. Reiter, M. J., & Gee, C. B. (2008). Open communication and partner support in intercultural and interfaith romantic relationships: A relational maintenance approach. *Journal of Social & Personal Relationships, 25*, 539–559. doi:10.1177/0265407508090872.

72. Colaner, C. (2009). Exploring the communication of evangelical families: The association between evangelical gender role ideology and family communication patterns. *Communication Studies, 60*, 97–113. doi:10.1080/10510970902834833.

73. Pew Forum on Religion & Public Life (2008, June).

74. Stone, K. G. (1995, February 19). Disability act everyone's responsibility in America. *Albuquerque Journal*, p. H3.

75. Solomon, A. (2012). *Far from the tree: Parents, children, and the search for identity*. New York: Scribner, pp. 68–69.

76. Braithwaite, D. O., & Labrecque, D. (1994). Responding to the Americans with Disabilities Act: Contributions of interpersonal communication research and training. *Journal of Applied Communication Research, 22*, 285–294. See also Braithwaite, D. O. (1991). "Just how much did that wheelchair cost?": Management of privacy boundaries by persons with disabilities. *Western Journal of Speech Communication, 55*, 254–275; and Colvert, A. L., & Smith, J. (2000). What is reasonable? Workplace communication and people who are disabled. In D. O. Braithwaite & T. L. Thompson (Eds.), *Handbook of communication and people with disabilities: Research and application* (pp. 116–130). Mahwah, NJ: Erlbaum.

77. Fitch, V. (1985). The psychological tasks of old age. *Naropa Institute Journal of Psychology, 3*, 90–106.

78. Gergen, K. J., & Gergen, M. M. (2000). The new aging: Self construction and social values. In K. W. Schae & J. Hendricks (Eds.), *The societal impact of the aging process* (pp. 281–306). New York: Springer.

79. Bailey, T. A. (2010). Ageism and media discourse: Newspaper framing of middle age. *Florida Communication Journal, 38*, 43–56.

80. Frijters, P., & Beatoon, T. (2012). The mystery of the U-shaped relationship between happiness and age. *Journal of Economic Behavior & Organization, 82*, 525–542.

81. Giles, H., Ballard, D., & McCann, R. M. (2002). Perceptions of intergenerational communication across cultures: An Italian case. *Perceptual and Motor Skills, 95*, 583–591.

82. Ryan, E. B., & Butler, R. N. (1996). Communication, aging, and health: Toward understanding health provider relationships with older clients. *Health Communication, 8*, 191–197.

83. Harwood, J. (2007). *Understanding communication and aging: Developing knowledge and awareness*. Newbury Park, CA: Sage, p. 79.

84. Kroger, J., Martinussen, M., & Marcia, J. E. (2010). Identity status change during adolescence and young adulthood: A meta-analysis. *Journal of Adolescence, 33*, 683–698.

85. Galanaki, E. P. (2012). The imaginary audience and the personal fable: A test of Elkind's theory of adolescent egocentrism. *Psychology, 3*, 457–466.

86. Myers, K. K., & Sadaghiani, K. (2010). Millennials in the workplace: A communication perspective on Millennials' organizational relationships and performance. *Journal of Business and Psychology*, 225–238. doi:10.1007/s10869-010-9173.7.

87. Lucas, K. (2011). The working class promise: A communicative account of mobility-based ambivalences. *Communication Monographs, 78*, 347–369.

88. Stuber, J. M. (2006). Talk of class. *Journal of Contemporary Ethnography, 35*, 285–318, quote on p. 306.

89. Kim, Y. K., & Sax, L. J. (2009). Student–faculty interaction in research universities: Differences by student gender, race, social class, and first-generation status. *Research in Higher Education, 50*, 437–459. doi:10.1007/s11163.009-9127-x.

90. Kaufman, P. (2003). Learning to not labor: How working-class individuals construct middle-class identities. *Sociological Quarterly, 44*, 481–504.

91. Lubrano, A. (2004). *Limbo: Blue-collar roots, white-collar dreams*. Hoboken, NJ: Wiley; and Lucas, K. (2011). The working class promise: A communicative account of mobility-based ambivalences. *Communication Monographs, 78*, 347–369. doi:10.1080/03. 637751.2011.589461. For a case study on social class mobility, see Lucas, K. (2010). Moving up: The challenges of communicating a new social class identity. In D. O. Braithwaite & J. T. Wood (Eds.), *Casing interpersonal communication: Case studies in personal and social relationships* (pp. 17–24). Dubuque, IA: Kendall-Hunt.

92. Orbe, M. P., & Groscurth, C. R. (2004). A co-cultural theoretical analysis of communicating on campus and at home: Exploring the negotiation strategies of first generation college (FGC) students. *Qualitative Research Reports in Communication, 5*, 41–47.

93. Orbe & Groscurth (2004), p. 45.

94. Lopez, J., Perez, J., III, & Cortinas, G., Jr. (2011, Fall). The gang life: A culture of its own. *Journal of Border Educational Research, 9*, https://journals.tdl.org/jber/index.php/jber/article/viewFile/7062/6327.

95. National Youth Violence Prevention Resource Center. (2007, December 20). Gangs fact sheet. Retrieved from http://www.helpinggangyouth.com/statistics.html.

96. Young, S. (2014, March 29). Nordstrom & customer service. *Sam Young: Acts of Leadership* [Blog post]. Retrieved from http://www.samyoung.co.nz/2014/03/nordstrom-customer-service_29.html.

97. Hartnell, C. A., Ou, A., & Kinicki, A. (2011). Organizational culture and organizational effectiveness: A meta-analytic investigation of the competing values framework's theoretical suppositions. *Journal of Applied Psychology, 96*(4), 677–694.

98. Arasaratnam, L. A. (2006). Further testing of a new model of intercultural communication competence. *Communication Research Reports, 23*, 93–99.

99. Pettigrew, T. F., & Tropp, L. R. (2000). Does intergroup contact reduce prejudice? Recent meta-analytic findings. In S. Oskamp (Ed.), *Reducing prejudice and discrimination: Social psychological perspectives* (pp. 93–114). Mahwah, NJ: Erlbaum.

100. Pettigrew, T. F., & Tropp, L. R. (2006, May). A meta-analytic test of intergroup contact theory. *Journal of Personality and Social Psychology, 90*, 751–783.

101. Kassing, J. W. (1997). Development of the Intercultural Willingness to Communicate Scale. *Communication Research Reports, 14*, 399–407.

102. Broockman, D., & Kalla, J. (2016). Durably reducing transphobia: A field experiment on door-to-door canvassing. *Science, 352*(6282), 220–224.

103. Amichai-Hamburger, Y., & McKenna, K. Y. A. (2006). The contact hypothesis reconsidered: Interacting via the Internet. *Journal of Computer-Mediated Communication, 11*(3), 825–843. Retrieved from http://onlinelibrary.wiley.com/doi/10.1111/j.1083-6101.2006.00037.x/abstract.

104. Iyer, P. (1990). *The lady and the monk: Four seasons in Kyoto*. New York: Vintage, pp. 129–130.

105. Ibid., pp. 220–221.

106. Steves, R. (1996, May–September). Culture shock. *Europe Through the Back Door Newsletter, 50*, 9.

107. Pew Research Center, U.S. Politics & Policy. (2016, March 31). Campaign exposes fissures over issues, values and how life has changed in the U.S. Retrieved from http://www.people-press.org/2016/03/31/campaign-exposes-fissures-over-issues-values-and-how-life-has-changed-in-the-u-s/.

108. Lah, K. (2011, March 23). Plastic surgery boom as Asians seek "western" look. *CNN*. Retrieved from http://www.cnn.com/2011/WORLD/asiapcf/05/19/korea.beauty/.

109. Chun, D. M. (2011). Developing intercultural communicative competence through online exchanges. *CALICO Journal, 28*(2), 392–419.

110. Kim, M. S., Hunter, J. E., Miyahara, A., Horvath, A. M., Bresnahan, M., & Yoon, H. (1996). Individual- vs. culture-level dimensions of individualism and collectivism: Effects on preferred conversational styles. *Communication Monographs, 63*, 28–49.

111. Gudykunst, W. B., & Nishida, T. (2001). Anxiety, uncertainty, and perceived effectiveness of communication across relationships and cultures. *Journal of Intercultural Relations, 25*, 55–71.

112. Berger, C. R. (1979). Beyond initial interactions: Uncertainty, understanding, and the development of interpersonal relationships. In H. Giles & R. St. Clair (Eds.), *Language and social psychology* (pp. 122–144). Oxford: Blackwell.

113. Carrell, L. J. (1997). Diversity in the communication curriculum: Impact on student empathy. *Communication Education, 46,* 234–244.

114. Oberg, K. (1960). Cultural shock: Adjustment to new cultural environments. *Practical Anthropology, 7,* 177–182.

115. Obert (1960).

116. Bruhwiler, B. (2012, November 12). Culture shock! [Blog post]. Retrieved from http://www.joburgexpat.com/2012/11/culture-shock.html.

117. Chang, L. C.-N. (2011). My culture shock experience. *ETC: A Review of General Semantics, 68*(4), 403–405.

118. Kim, Y. Y. (2008). Intercultural personhood: Globalization and a way of being. *International Journal of Intercultural Relations, 32,* 359–368.

119. Kim, Y. Y. (2005). Adapting to a new culture: An integrative communication theory. In W. B. Gudykunst (Ed.), *Theorizing about intercultural communication* (pp. 375–400). Thousand Oaks, CA: Sage.

120. See Kim (2008).

121. Gender Neutral Pronoun Blog. (2010, January 24). The need for a gender-neutral pronoun. Retrieved from https://genderneutral pronounwordpress.com/tag/xe/.

CHAPTER 4

1. Nagel, D. (2015). Interview: Scott H. Young's Year Without English project. *The Messofanti Guild.* Retrieved from http://www.mezzoguild.com/scott-h-young-year-without-english/. Quote appears in video interview.

2. Scott H. Young: Language hacks for everyday life (episode 353). *Art of Charm.* Retrieved from http://theartofcharm.com/podcast-episodes/scott-h-young-language-hacks-everyday-life-episode-353/. Quotes are part of a video interview.

3. Wilson, R. (2013, December 2).What dialect do you speak? A map of American English. *Washington Post.* Retrieved from https://www.washingtonpost.com/blogs/govbeat/wp/2013/12/02/what-dialect-to-do-you-speak-a-map-of-american-english/.

4. Sacks, O. (1989). *Seeing voices: A journey into the world of the deaf.* Berkeley: University of California Press, p. 17.

5. Adapted from O'Brien, J., & Kollock, P. (2001). *The production of reality* (3rd ed., p. 66). Thousand Oaks, CA: Pine Forge Press.

6. Ogden, C. K., & Richards, I. A. (1923). *The meaning of meaning.* New York: Harcourt Brace, p. 11.

7. Gaudin, S. (2011, March 25). OMG! Text shorthand makes the Oxford English Dictionary. *Computerworld.* Retrieved from http://www.computerworld.com/s/article/9215079/OMG_Text_shorthand_makes_the_dictionary.

8. W. B. Pearce & V. Cronen. (1980). *Communication, action, and meaning.* New York: Praeger. See also J. K. Barge. (2004). Articulating CMM as a practical theory. *Human Systems: The Journal of Systemic Consultation and Management, 15,* 193–204, and E. M. Griffin. (2006). *A first look at communication theory* (6th ed.). New York: McGraw-Hill.

9. Croom, A. M. (2013). How to do things with slurs: Studies in the way of derogatory words. *Language & Communication, 33*(3), 177–204.

10. McLeod, L. (2011). *Swearing in the 'tradie' environment as a tool for solidarity* (Vol. 4, pp. 1–10).Griffith Working Papers in Pragmatics and Intercultural Communication.

11. Genesis 2:19. This biblical reference was noted by Mader, D. C. (1992, May). *The politically correct textbook: Trends in publishers'* guidelines for the representation of marginalized groups. Paper presented at the annual convention of the Eastern Communication Association, Portland, ME.

12. Laham, S. M., Koval, P., & Alter, A. L. (2012). The name-pronunciation effect: Why people like Mr. Smith more than Mr. Colquhoun. *Journal of Experimental Social Psychology, 48*(3), 752–756.

13. Borget, J. (2012, November 9). Biracial names for biracial babies. *Mom Stories.* Retrieved from http://blogs.babycenter.com/mom_stories/biracial-baby-names-110912/.

14. Derous, E., Ryan, A. M., & Nguyen, H. D. (2012). Multiple categorization in resume screening: Examining effects on hiring discrimination against Arab applicants in field and lab settings. *Journal of Organizational Behavior, 33*(4), 544–570.

15. Bertrand, M., & Mullainathan, S. (2004). Are Emily and Greg more employable than Lakisha and Jamal? A field experiment on labor market discrimination. *The American Economic Review, 4,* 991–1013.

16. No names, no bias? (2015, October 31). *The Economist.* Retrieved from http://www.economist.com/news/business/21677214-anonymising-job-applications-eliminate-discrimination-not-easy-no-names-no-bias.

17. Cotton, J. L., O'Neill, B. S., & Griffin, A. (2008). The "name game": Affective and hiring reactions to first names. *Journal of Managerial Psychology, 23,* 18–39.

18. Brunning, J. L., Polinko, N. K., Zerbst, J. I., & Buckingham, J. T. (2000). The effect on expected job success of the connotative meanings of names and nicknames. *Journal of Social Psychology, 140,* 197–201.

19. Coffey, B., & McLaughlin, P. A. (2009). Do masculine names help female lawyers become judges? Evidence from South Carolina. *American Law and Economics Review, 11,* 112–133.

20. Naftulin, D. H., Ware, J. E., Jr., & Donnelly, F. A. (1973, July). The Doctor Fox Lecture: A paradigm of educational seduction. *Journal of Medical Education, 48,* 630–635. See also Cory, C. T. (Ed.). (1980, May). Bafflegab pays. *Psychology Today,13,* 12; and Marsh, H. W., & Ware, J. E., Jr. (1982). Effects of expressiveness, content coverage, and incentive on multidimensional student rating scales: New interpretations of the "Dr. Fox" effect. *Journal of Educational Psychology, 74,* 126–134.

21. Segrest Purkiss, S. L., Perrewé, P. L., Gillespie, T. L., Mayes, B. T., & Ferris, G. R. (2006). Implicit sources of bias in employment interview judgments and decisions. *Organizational Behavior and Human Decision Processes, 101*(2), 152–167.

22. Hosoda, M., Nguyen, L. T., & Stone-Romero, E. F. (2012). The effect of Hispanic accents on employment decisions. *Journal of Managerial Psychology, 27*(4), 347–364; Hosoda, M., & Stone-Romero, E. (2010). The effects of foreign accents on employment-related decisions. *Journal of Managerial Psychology, 25*(2), 113–132.

23. For a summary of scholarship supporting the notion of linguistic determinism, see Boroditsky, L. (2010, July 23). Lost in translation. *Wall Street Journal Online.* Retrieved from http://www.wsj.com/articles/SB10001424052748703467304575383131592767868.

24. Vervecken, D., & Hannover, B. (2015). Yes I can! Effects of gender fair job descriptions on children's perceptions of job status, job difficulty, and vocational self-efficacy. *Social Psychology, 46*(2), 76–92.

25. Lee, J. K. (2015). "Chairperson" or "chairman"?—A study of Chinese EFL teachers' gender inclusivity. *Australian Review of Applied Linguistics, 38*(1), 24–49.

26. Here in Finland, the language is completely gender-neutral—they don't have any gender specific pronouns like "he" or "she." (2015, May 2). *OMG Facts.* Retrieved from http://www.omgfacts.com/health/14822/Here-in-Finland-the-language-is-completely-gender-neutral-they-don-t-have-any-gender-specific-pronouns-like-he-or-she. Quotes appear in paragraph 2.

27. Prewitt-Freilino, J. L., Caswell, T. A., & Laakso, E. K. (2012). The gendering of language: A comparison of gender equality in countries with gendered, natural gender, and genderless languages. *Sex Roles, 66*(3/4), 268–281.

28. Dalton, D. (2015, March 18). 28 beautiful words the English language should steal. BuzzFeed. Retrieved from http://www.buzzfeed.com/danieldalton/ever-embasan#.abBQB66ep; DeMain, B., Sweetland Edwards, H., &Oltuski, R. 38 wonderful foreign words we could use in English.Retrieved from http://mentalfloss.com/article/50698/38-wonderful-foreign-words-we-could-use-english.

29. Granadillo, E. D., & Mendez, M. F.(2016). Pathological joking or witzelsucht revisited. *Journal of Neuropsychiatry & Clinical Neurosciences.* Advance online publication. Retrieved from http://www.ncbi.nlm.nih.gov/pubmed/26900737.

30. For a summary of scholarship supporting the notion of linguistic determinism, see Boroditsky, L. (2010, July 23). Lost in translation. *Wall Street Journal Online.* Retrieved from http://www.wsj.com/articles/SB10001424052748703467304575383131592767868.

31. Whorf, B. (1956). The relation of habitual thought and behavior to language. In J. B. Carroll (Ed.), *Language, thought, and reality* (pp. 134–159). Cambridge, MA: MIT Press. See also Hoijer, H. (1994). The Sapir-Whorf hypothesis. In Larry A. Samovar & Richard E. Porter (Eds.), *Intercultural communication: A reader* (7th ed., pp. 194–200). Belmont, CA: Wadsworth.

32. Davidoff, J., Goldstein, J., Tharp, I., Wakui, E., & Fagot, J. (2012). Perceptual and categorical judgements of colour similarity. *Journal of Cognitive Psychology, 24*(7), 871–892.

33. Pullum, G. K. (1991). *The great Eskimo vocabulary hoax and other irreverent essays on the study of language.* Chicago: University of Chicago Press.

34. For a discussion of racist language, see Bosmajian, H. A. (1983). *The language of oppression.* Lanham, MD: University Press of America.

35. Mader, D. C. (1992, May).*The politically correct textbook: Trends in publishers' guidelines for the representation of marginalized groups.* Paper presented at the annual convention of the Eastern Communication Association, Portland, ME. See pp. 5 and 9.

36. Kirkland, S. L., Greenberg, J., & Pysczynski, T. (1987). Further evidence of the deleterious effects of overheard derogatory ethnic labels: Derogation beyond the target. *Personality and Social Psychology Bulletin, 12,* 216–227.

37. Erickson, B., Lind, E. A., Johnson, B. C., & O'Barr, W. M. (1978). Speech style and impression formation in a court setting: The effects of "powerful" and "powerless" speech. *Journal of Experimental Social Psychology, 14,* 266–279.

38. Parton, S., Siltanen, S. A., Hosman, L. A., & Langenderfer, J. (2002). Employment interview outcomes and speech style effects. *Journal of Language and Social Psychology, 21,* 144–161.

39. Reid, S. A., Keerie, N., & Palomares, N. A. (2003). Language, gender salience, and social influence. *Journal of Language and Social Psychology, 22,* 210–233.

40. Andrew. (2011). Manners in Spanish—The basics of being polite in Spanish-speaking cultures. How to learn Spanish online: Resources, tips, tricks, and techniques. Retrieved from http://howlearnspanish.com/2011/01/manners-in-spanish/.

41. Guenzi, P., & Georges, L. (2010). Interpersonal trust in commercial relationships: Antecedents and consequences of customer trust in the salesperson. *European Journal of Marketing, 44,* 114–138.

42. Young, S. H. (2014). Looking back a year (almost) without English [Blog post]. *Scott H. Young.* Retrieved from https://www.scotthyoung.com/blog/2014/09/01/tywe-review/. Quote appears in paragraph 20.

43. Speer, R. B., Giles, H., & Denes, A. (2013). Investigating stepparent-stepchild interactions: The role of communication accommodation. *Journal of Family Communication, 13*(3), 218–241.

44. Xiaosui, X. (2014). Constructing common ground for cross-cultural communication. *China Media Research, 10*(4), 1–9.

45. Giles, H., Coupland, J., & Coupland, N. (Eds.). (1991). *Contexts of accommodation: Developments in applied sociolinguistics.* Cambridge: Cambridge University Press.

46. Cassell, J., & Tversky, D. (2005). The language of online intercultural community formation. *Journal of Computer-Mediated Communication, 10,* Article 2.

47. Baruch, Y. & Jenkins, S. (2007). Swearing at work and permissive leadership culture: When anti-social becomes social and incivility is acceptable. *Leadership & Organization Development Journal, 28*(6), 492–507.

48. Maass, A., Salvi, D., Arcuri, L., & Semin, G. R. (1989). Language use in intergroup context. *Journal of Personality and Social Psychology, 57,* 981–993.

49. Weiner, M., & Mehrabian, A. (1968). *A language within language.* New York: Appleton-Century-Crofts.

50. Kubanyu, E. S., Richard, D. C., Bower, G. B., & Muraoka, M. Y. (1992). Impact of assertive and accusatory communication of distress and anger: A verbal component analysis. *Aggressive Behavior, 18,* 337–347.

51. Scott, T. L. (2000, November 27). Teens before their time. *Time,* p. 22.

52. Motley M. T., & Reeder, H. M. (1995). Unwanted escalation of sexual intimacy: Male and female perceptions of connotations and relational consequences of resistance messages. *Communication Monographs, 62,* 356–382.

53. How many is a couple? A few? Several? [Forum post]. (2012, November 1). *The Escapist.* Retrieved from http://www.escapistmagazine.com/forums/read/18.392824-How-many-is-a-couple-A-few-Several.

54. Labov, W. (1992). Social and language boundaries among adolescents. *American Speech, 4,* 339–366.

55. Hayakawa, S. I. (1964). *Language in thought and action.* New York: Harcourt Brace.

56. Twitter killing English, says actor. (2011, October 29). *Herald Sun.* Retrieved from http://www.heraldsun.com.au/entertainment/twitter-killing-english-says-actor/story-e6frf96x-1226179939640.

57. Henry, J. (2013, January 20). Art of essay-writing damaged by Twitter and Facebook, Cambridge don warns. *The Telegraph.* Retrieved from http://www.telegraph.co.uk/technology/social-media/9813109/Art-of-essay-writing-damaged-by-Twitter-and-Facebook-Cambridge-don-warns.html.

58. Evans, N., & Levinson, S. C. (2009). The myth of language universals: Language diversity and its importance for cognitive science. *Behavioral and Brain Sciences, 32,* 429–492.

59. Knapp, A. (2011, October 31). No, Twitter isn't ruining the English language. *Forbes.* Retrieved from http://www.forbes.com/sites/alexknapp/2011/10/31/no-twitter-isnt-ruining-the-english-language/.

60. Lang, M. (2016, January 8). How lifting limits might change Twitter's character. *San Francisco Chronicle.* Retrieved from http://www.sfchronicle.com/business/article/How-lifting-limits-might-change-Twitter-s-6746507.php.

61. OMG! The impact of social media on the English language [Blog post]. (2012, August 28). *iMedia.* Retrieved from http://blogs.imediaconnection.com/blog/2012/08/28/omg-the-impact-of-social-media-on-the-english-language/.

62. Mabillard, A. (2000). Words Shakespeare invented. *Shakespeare Online.* Retrieved from http://www.shakespeare-online.com/biography/wordsinvented.html.

63. Alberts, J. K. (1988). An analysis of couples' conversational complaints. *Communication Monographs, 55,* 184–197.

64. Streisand, B. (1992). Crystal Award speech delivered at the Crystal Awards, Women in Film luncheon.

65. Morrison, B. (2000). What you won't hear the pilot say. *USA Today,* p. A1.

66. Eisenberg, E. M. (Ed.). (2007). *Strategic ambiguities: Essays on communication, organization and identity.* Thousand Oaks, CA: Sage.

67. Hudson, P. (2015, January 16). "I don't understand women"—well read on the for the full explanation. *Mirror.* Retrieved from http://www.mirror.co.uk/lifestyle/dating/i-dont-understand-women—4993587.

68. Q&A: 8 things we don't understand about men. (n.d.). *Wewomen.* Retrieved from http://www.wewomen.com/understanding-men/what-women-don-t-understand-about-men-questions-and-answers-d30896x64063.html.

69. Mehl, M. R., Vazire, S., Ramírez-Esparza, N., Slatcher, R. B., & Pennebaker, J. W. (2007, July). Are women really more talkative than men? *Science, 317,* 82.

70. Sehulster, J. R. (2006). Things we talk about, how frequently, and to whom: Frequency of topics in everyday conversation as a function of gender, age, and marital status. *The American Journal of Psychology, 119,* 407–432.

71. Sehulster, J. R. (2006). Things we talk about, how frequently, and to whom: Frequency of topics in everyday conversation as a function of gender, age, and marital status. *The American Journal of Psychology, 119,* 407–432.

72. Kapidzic, S., & Herring, S. C. (2011). Gender, communication, and self-presentation in teen chatrooms revisited: Have patterns changed? *Journal of Computer-Mediated Communication, 17*(1), 39–59.

73. Cohen, M. M. (2016). It's not you, it's me . . . no, actually it's you: Perceptions of what makes a first date successful or not. *Sexuality & Culture, 20*(1), 173–191.

74. Cohen, 2016.

75. Wood, J. T. (2001). *Gendered lives: Communication, gender, and culture* (4th ed.). Belmont, CA: Wadsworth, p. 141.

76. Fox, A. B., Bukatko, D., Hallahan, M., & Crawford, M. (2007). The medium makes a difference: Gender similarities and differences in instant messaging. *Journal of Language and Social Psychology, 26,* 389–397.

77. Pfafman, T. M., & McEwan, B. (2014). Polite women at work: Negotiating professional identity through strategic assertiveness. *Women's Studies in Communication, 37*(2), 202–219.

78. Booth-Butterfield, M. M., Wanzer, M. B., Weil, N., & Krezmien, E. (2014). Communication of humor during bereavement: Intrapersonal and interpersonal emotion management strategies. *Communication Quarterly, 62*(4), 436–454.

79. Menchhofer, T. O. (2015, April). Planting the seed of emotional literacy: Engaging men and boys in creating change. *The Vermont Connection, 24,* article 4. Retrieved from http://scholarworks.uvm.edu/cgi/viewcontent.cgi?article=1197&context=tvc.

80. Schoenfeld, E. A., Bredow, C. A., & Huston, T. L. (2012). Do men and women show love differently in marriage? *Personality & Social Psychology Bulletin, 11,* 1396–1409.

81. Hancock, A., B., Stutts, H. W., & Bass, A. (2015). Perceptions of gender and femininity based on language: Implications for transgender communication therapy. *Language & Speech, 58*(3), 315–333.

82. Jones, A. C., & Josephs, R. A. (2006). Interspecies hormonal interactions between man and the domestic dog (*Canis familiaris*). *Hormones and Behavior, 50*(3), 393–400.

83. Pennebaker, J. W., Groom, C. J., Loew, D., & Dabbs, J. M. (2004). Testosterone as a social inhibitor: Two case studies of the effect of testosterone treatment on language. *Journal of Abnormal Psychology, 113*(1), 172.

84. Chen, C. P., Cheng, D. Z., Luo, Y.-J. (2011). Estrogen impacts on emotion: Psychological, neuroscience and endocrine studies. *Science China Life, 41*(11). Retrieved from http://www.eurekalert.org/pub_releases/2012-01/sicp-tio010912.php.

85. Premenstrual syndrome (PMS) fact sheet.How common is PMS? (2010). U.S. Department of Health and Human Services. Retrieved from http://womenshealth.gov/publications/our-publications/fact-sheet/premenstrual-syndrome.cfm#e.

86. Mulac, A., Giles, H., Bradac, J. J., & Palomares, N. A. (2013). The gender-linked language effect: An empirical test of a general process model. *Language Sciences, 38,* 22–31.

87. Hancock, A. B., & Rubin, B. A. (2015). Influence of communication partner's gender on language. *Journal of Language & Social Psychology, 34*(1), 46–64.

88. Fandrich, A. M., & Beck, S. J. (2012). Powerless language in health media: The influence of biological sex and magazine type on health language. *Communication Studies, 63*(1), 36–53.

89. Nagel, D. (2015). Interview: Scott H. Young's Year Without English project. *The Messofanti Guild.* Retrieved from http://www.mezzoguild.com/scott-h-young-year-without-english/ Quote appears in video interview.

CHAPTER 5

1. A brave new world: A chat with *YFS* magazine's Erica Nicole. (2015, February 25). *The City Influencer.* Retrieved from http://thecityinfluencer.com/a-brave-new-world-a-chat-with-yfs-magazines-erica-nicole/. Quote appears in paragraph 1.

2. Nicole, E. (2013, November 6). How to use social media to make real-world connections. *Women 2.0.* Retrieved from http://women2.com/2013/11/06/use-social-media-make-real-world-connections/. Quote appears in Tip 3, paragraph 5.

3. Crompton, S. (2015, February 4). The value of listening to your employees. *YFS.* Retrieved from http://yfsmagazine.com/2015/02/04/the-value-of-listening-to-your-employees/. Quote appears in Tip 2.

4. Suzuno, M. (2014, January 21). 5 things recruiters wish you knew about career fairs. *After College.* Retrieved from http://blog.after-college.com/5-things-recruiters-wish-knew-career-fairs/. Quotes in appear in Tip 2 and in the title, respectively.

5. Covey, S. (1989). *The 7 habits of highly effective people.* New York: Simon & Schuster.

6. Kalargyrou, V., & Woods, R. H. (2011). Wanted: Training competencies for the twenty-first century. *International Journal of Contemporary Hospitality Management, 23*(3), 361–376.

7. Kalargyrou, V., & Woods, R. H. (2011). Wanted: Training competencies for the twenty-first century. *International Journal of Contemporary Hospitality Management, 23*(3), 361–376.

8. Davis, J., Foley, A., Crigger, N., & Brannigan, M. C. (2008). Healthcare and listening: A relationship for caring. *International Journal of Listening, 22*(2), 168–175.

9. Pryor, S., Malshe, A., & Paradise, K. (2013). Salesperson listening in the extended sales relationship: an exploration of cognitive, affective, and temporal dimensions. *Journal of Personal Selling & Sales Management, 33*(2), 185–196.

10. Brockner, J., & Ames, D. (2010, December 1). Not just holding forth: The effect of listening on leadership effectiveness. *Social Science Electronic Publishing.* Retrieved from http://papers.ssrn.com/sol3/papers.cfm?abstract_id=1916263.

11. Ames, D., Maissen, L. B., & Brockner, J. (2012). The role of listening in interpersonal influence. *Journal of Research in Personality, 46,* 345–349.

12. Gordon, P., James Allan, C., Nathaniel, B., Derek, J. K., & Jonathan, A. F. (2015). On the reception and detection of pseudo-profound bullshit. *Judgment and Decision Making, 10*(6), 549–563. The preceding quote appears in the title, the subsequent one in the abstract.

13. Brooks, A. W., Gino, F., & Schweitzer, M. E. (2015). Smart people ask for (my) advice: Seeking advice boosts perceptions of competence. *Management Science, 61*(6), 1421–1435. Quote appears on p. 1421.

14. Bodie, G. D., Vickery, A. J., & Gearhart, C. C. (2013). The nature of supportive listening, I: Exploring the relation between supportive listeners and supportive people. *International Journal of Listening, 27*, 39–49.

15. Fletcher, G. O., Kerr, P. G., Li, N. P., & Valentine, K. A. (2014). Predicting romantic interest and decisions in the very early stages of mate selection: Standards, accuracy, and sex differences. *Personality & Social Psychology Bulletin, 4*, 540–550.

16. Advisor Louise PhD. (2012, October 3.) What women want from men: A good listener. Ingenio Advisor Blogs. Retrieved from http://www.ingenio.com/CommunityServer/UserBlogPosts/Advisor_Louise_PhD/What-Women-Want-from-Men--A-Good-Listener/630187.aspx. Quote appears in paragraph 3.

17. Jalongo, M. (2010). Listening in early childhood: An interdisciplinary review of the literature. *International Journal of Listening, 24*, 1–18.

18. Horowitz, S. (2012, November 11). The science and art of listening. *New York Times*, p. SR10.

19. Powers, W. G., & Witt, P. L. (2008). Expanding the theoretical framework of communication fidelity. *Communication Quarterly, 56*, 247–267; Fitch-Hauser, M., Powers, W. G., O'Brien, K., & Hanson, S. (2007). Extending the conceptualization of listening fidelity. *International Journal of Listening, 21*, 81–91; Powers, W. G., & Bodie, G. D. (2003). Listening fidelity: Seeking congruence between cognitions of the listener and the sender. *International Journal of Listening, 17*, 19–31.

20. Kim, Y. G. (2016). Direct and mediated effects of language and cognitive skills on comprehension of oral narrative texts (listening comprehension) for children. *Journal of Experimental Child Psychology, 141*, 101–120.

21. Crompton, S. (2015, February 4). The value of listening to your employees. *YFS*. Retrieved from http://yfsmagazine.com/2015/02/04/the-value-of-listening-to-your-employees/. Quotes appears in Tip 1 and paragraph 1, respectively.

22. Fontana, P. C., Cohen, S. D., & Wolvin, A. D. (2015). Understanding listening competency: A systematic review of research scales. *International Journal of Listening, 29*(3), 148–176.

23. Thomas, T. L., & Levine, T. R. (1994). Disentangling listening and verbal recall: Related but separate constructs? *Human Communication Research, 21*, 103–127.

24. Cowan, N., & AuBuchon, A. M. (2008). Short-term memory loss over time without retroactive stimulus interference. *Psychonomic Bulletin and Review, 15*, 230–235.

25. Brownell, J. (1990). Perceptions of effective listeners: A management study. *Journal of Business Communication, 27*, 401–415.

26. rurounikenji. (2012, July 8). Girlfriend literally never listens to me [Message board post]. The Student Room. Retrieved from http://www.thestudentroom.co.uk/showthread.php?t=2075372. Quote appears in first paragraph.

27. Chapman, S. G. (2012). *The five keys to mindful communication: Using deep listening and mindful speech to strengthen relationships, heal conflicts, and accomplish your goals.* Boulder, CO: Shambhala Publications.

28. Ting-Toomey, S., & Chung, L. C. (2011). *Understanding intercultural communication* (2nd ed.). New York: Oxford University Press.

29. Rautalinko, E., Lisper, H., & Ekehammar, B. (2007). Reflective listening in counseling: Effects of training time and evaluator social skills. *American Journal of Psychotherapy, 61*, 191–209.

30. Dean, M., & Street, J. L. (2014). Review: A 3-stage model of patient-centered communication for addressing cancer patients' emotional distress. *Patient Education and Counseling, 94*, 143–148.

31. Shafir, R. Z. (2003). *The Zen of listening: Mindful communication in the age of distraction.* Wheaton, IL: Quest Books.

32. Chapman, S. G. (2012). *The five keys to mindful communication: Using deep listening and mindful speech to strengthen relationships, heal conflicts, and accomplish your goals.* Boulder, CO: Shambhala Publications.

33. Hansen, J. (2007). *24/7: How cell phones and the Internet change the way we live, work, and play.* New York: Praeger. See also Turner, J. W., & Reinsch, N. L. (2007). The business communicator as presence allocator: Multicommunicating, equivocality, and status at work. *Journal of Business Communication, 44*, 36–58.

34. Sarampalis, A., Kalluri, S., Edwards, B., & Hafter, E. (2009). Objective measures of listening effort: Effects of background noise and noise reduction. *Journal of Speech, Language & Hearing Research, 52*, 1230–1240.

35. Drullman, R., & Smoorenburg, G. F. (1997). Audio-visual perception of compressed speech by profoundly hearing-impaired subjects. *Audiology, 36*, 165–177.

36. Info stupidity. (2005, April 30). *New Scientist, 186*, 6–7.

37. Lin, L. (2009, September 15). Breadth-biased versus focused cognitive control in media multitasking behaviors. *Proceedings of the National Academy of Sciences, 106*, 15521–15522. Retrieved from http://www.pnas.org/content/106/37/15521.full.pdf.

38. Ophir, E., Nass, C., & Wagner, A. (2009). Cognitive control in media multitaskers. *Proceedings of the National Academy of Sciences, 106*, 15583–15587.

39. Hearing problems can stress relationships. (2008, May 1). *Audiology Online.* Retrieved from http://www.audiologyonline.com/releases/listen-to-this-hearing-problems-3780.

40. Agrawal, Y., Platz, E. A., & Niparko, J. K. (2008). Prevalance of hearing loss and differences by demographic characteristics among US adults. Data from the National Health and Nutrition Examination Survey, 1999–2004. *Journal of the American Medical Association, 168*, 1522.

41. National Institute on Deafness and Other Communication Disorders. (2008, August). *Quick Statistics.* Bethesda, MD: U.S. Department of Health and Human Services.

42. Imhof, M. (2003). The social construction of the listener: Listening behaviors across situations, perceived listener status, and cultures. *Communication Research Reports, 20*, 357–366.

43. Zohoori, A. (2013). A cross-cultural comparison of the HURIER Listening Profile among Iranian and U.S. students. *International Journal of Listening, 27*, 50–60.

44. Imhof, M. (2003). The social construction of the listener: Listening behaviors across situations, perceived listener status, and cultures. *Communication Research Reports, 20*, 357–366.

45. Halvorson, H. G. (2010, August 17). Stop being so defensive! A simple way to learn to take criticism gracefully [Blog post]. *Psychology Today.* Retrieved from https://www.psychologytoday.com/blog/the-science-success/201008/stop-being-so-defensive.

46. Valdes, A. (2012, June 19). 8 tips to help you stop being defensive. *Mamiverse.* Retrieved from http://mamiverse.com/8-tips-to-help-you-stop-being-defensive-13577/.

47. Vangelisti, A. L., Knapp, M. L., & Daly, J. A. (1990). Conversational narcissism. *Communication Monographs, 57*, 251–274.

48. Kline, N. (1999). *Time to think: Listening to ignite the human mind.* London: Ward Lock, p. 21.

49. Derber, C. (2000). *The pursuit of attention: Power and ego in everyday life* (2nd ed.). New York: Oxford University Press.

50. Wilson Mizner quotes. (n.d.). Brainy Quote. Retrieved from http://www.brainyquote.com/quotes/authors/w/wilson_mizner.html.

51. Weger, H., Jr., Castle, G. R., & Emmett, M. C. (2010). Active listening in peer interviews: The influence of message paraphrasing on perceptions of listening skill. *International Journal of Listening, 24,* 34–49.

52. Villaume, W. A., & Bodie, G. D. (2007). Discovering the listener within us: The impact of trait-like personality variables and communicator styles on preferences for listening style. *International Journal of Listening, 21,* 102–123.

53. Gearhart, C. G., & Bodie, G. D. (2011). Active-empathic listening as a general social skill: Evidence from bivariate and canonical correlations. *Communication Reports, 24,* 86–98. doi:10.1080/08934215.2011.610731.

54. Paraschos, S. (2013). Unconventional doctoring: A medical student's reflections on total suffering. *Journal of Palliative Medicine, 16,* 325.

55. Huerta-Wong, J. E., & Schoech, R. (2010). Experiential learning and learning environments: The case of active listening skills. *Journal of Social Work Education, 46,* 85–101.

56. Luedtke, K. (1987, January 7). What good is free speech if no one listens? *Los Angeles Times.* Retrieved from http://articles.latimes.com/1987-01-07/local/me-2347_1_free-speech.

57. Adapted from Infante, D. A. (1988). *Arguing constructively.* Prospect Heights, IL: Waveland, pp. 71–75.

58. Fallacies. (n.d.). Internet encyclopedia of philosophy. Retrieved from http://www.iep.utm.edu/fallacy/.

59. Gearhart, C. C., Denham, J. P., & Bodie, G. D. (2013, November). *Listening is a goal-directed activity.* Paper presented at the annual meeting of the National Communication Association, Washington, DC.

60. Bodie, G. D., Vickery, A. J., & Gearhart, C. C. (2013). The nature of supportive listening, I: Exploring the relation between supportive listeners and supportive people. *International Journal of Listening, 27,* 39–49.

61. Sarah Q. (n.d.). When I needed a friend. *TeenInk.* Retrieved from http://www.teenink.com/nonfiction/educator_of_the_year/article/101090/When-I-Needed-a-Friend/. Quotes appear in paragraphs 11 and 10, respectively.

62. Chia, H. L. (2009). Exploring facets of a social network to explicate the status of social support and its effects on stress. *Social Behavior & Personality: An International Journal, 37*(5), 701–710. See also Segrin, C., & Domschke, T. (2011). Social support, loneliness, recuperative processes, and their direct and indirect effects on health. *Health Communication, 26,* 221–232.

63. Giles, L. C., Glonek, G. F., Luszcz, M. A., & Andrews, G. R. (2005, July). Effect of social networks on 10-year survival in very old Australians: The Australian longitudinal study of aging. *Journal of Epidemiology & Community Health, 59*(7), 574–579.

64. Robinson, J. D., & Tian, Y. (2009). Cancer patients and the provision of informational social support. *Health Communication, 24,* 381–390.

65. Sarah Q. (n.d.). When I needed a friend. *TeenInk.* Retrieved from http://www.teenink.com/nonfiction/educator_of_the_year/article/101090/When-I-Needed-a-Friend/. Quotes appear in the comments section.

66. Segrin, C., & Domschke, T. (2011). Social support, loneliness, recuperative processes, and their direct and indirect effects on health. *Health Communication, 26,* 221–232.

67. Tanis, M. (2007). Online support groups. In A. Joinson, K. McKenna, T. Postmes, & U. Reips (Eds.), *The Oxford handbook of Internet psychology* (pp. 137–152). Oxford: Oxford University Press.

68. Humorous example of social media monitoring: Sydney University [Blog post]. (2009, September 7). Retrieved from http://www.altimetergroup.com/2009/09/humorous-example-of-social-media-monitoring-sydney-university.html.

69. Caruso, R. (2011). A real example of effective social media monitoring and engagement. *Bundle Post.* Retrieved from http://bundlepost.wordpress.com/2011/11/07/a-real-example-of-effective-social-media-monitoring-and-engagement/.

70. Petrocelli, T. (2012, December 20). One rule with social media and social networking: Don't be creepy [Blog post]. Retrieved from http://www.esg-global.com/blogs/one-rule-with-social-media-and-social-networking-dont-be-creepy/.

71. Tannen, D. (2010). He said, she said. *Scientific American Mind, 21*(2), 55–59.

72. Tannen, D. (2010). He said, she said. *Scientific American Mind, 21*(2), 55-59. Quote appears in paragraph 11.

73. Weaver, J. B., & Kirtley, M. D. (1995). Listening styles and empathy. *Southern Communication Journal, 60,* 131–140.

74. Goldsmith, D. (2000). Soliciting advice: The role of sequential placement in mitigating face threat. *Communication Monographs, 67,* 1–19.

75. MacGeorge, E. L., Feng, B., & Thompson, E. R. (2008). "Good" and "bad" advice: How to advise more effectively. In M. T. Motley (Ed.), *Studies in applied interpersonal communication* (pp. 145–164). Thousand Oaks, CA: Sage.

76. Hample, D. (2006). Anti-comforting messages. In K. M. Galvin & P. J. Cooper (Eds.), *Making connections: Readings in relational communication* (4th ed., pp. 222–227). Los Angeles, CA: Roxbury.

77. 10 things to say (and not to say) to someone with depression. (n.d.). *Health.* Retrieved from http://www.drbalternatives.com/articles/cc2.html.

78. Stewart, M., Letourneau, N., Masuda, J. R., Anderson, S., Cicutto, L., McGhan, S., & Watt, S. (2012). Support needs and preferences of young adolescents with asthma and allergies: "Just no one really seems to understand." *Journal of Pediatric Nursing, 27*(5), 479–490.

79. DRB Alternatives, Inc. (n.d.). 10 things to say (and not to say) to someone with depression. Retrieved from http://www.drbalternatives.com/articles/cc2.html.

80. Helping adults, children cope with grief. (2001, September 13). *Washington Post.*

81. Guo, J., & Turan, B. (2016). Preferences for social support during social evaluation in men: The role of worry about a relationship partner's negative evaluation. *Journal of Social Psychology, 156*(1), 122–129.

82. Olson, R. (2014). A time-sovereignty approach to understanding carers of cancer patients' experiences and support preferences. *European Journal of Cancer Care, 23*(2), 239–248.

83. Burleson, B. (2008). What counts as effective emotional support?" In M. T. Motley (Ed.), *Studies in Applied Interpersonal Communication* (pp. 207–227). Thousand Oaks, CA: Sage.

84. Young, R. W., & Cates, C. M. (2004). Emotional and directive listening in peer mentoring. *International Journal of Listening, 18,* 21–33.

85. Svokos, A. Bill Nye tells Rutgers grads: We are "much more alike than different." (2015, May 20). *Huffpost College.* Retrieved from http://www.huffingtonpost.com/2015/05/20/bill-nye-rutgers-commencement_n_7338214.html. Quote appears in the third to last paragraph.

86. Gottman, J. M. (1999). *The marriage clinic: A scientifically-based marital therapy.* New York: Norton, p. 10.

87. Lewis, T., & Manusov, V. (2009). Listening to another's distress in everyday relationships. *Communication Quarterly, 57,* 282–301.

CHAPTER 6

1. Cuddy, A. (2012, October). Your body language shapes who you are. TED.com. Retrieved from http://www.ted.com/talks/amy_cuddy_your_body_language_shapes_who_you_are/transcript?language=en#t-235180. All quotes in this profile are from this source.

2. For a survey of the issues surrounding the definition of nonverbal communication, see Knapp, M., & Hall, J. A. (2010). *Nonverbal communication in human interaction* (6th ed.). Belmont, CA: Wadsworth, Chapter 1.

3. Keating, C. F. (2006). Why and how the silent self speaks volumes. In V. Manusov & M. L. Patterson (Eds.), *The SAGE handbook of nonverbal communication* (pp. 321–340). Thousand Oaks, CA: Sage.

4. For a discussion of intentionality, see Knapp & Hall (2010), pp. 9–12.

5. Palmer, M. T., & Simmons, K. B. (1995). Communicating intentions through nonverbal behaviors: Conscious and nonconscious encoding of liking. *Human Communication Research, 22,* 128–160.

6. Tracy, J. L., & Matsumoto, D. (2008, August 19). The spontaneous expression of pride and shame: Evidence for biologically innate nonverbal displays. *Proceedings from the National Academy of Science, 105*(33), 11655–11660.

7. Dennis, A. R., Kinney, S. T., & Hung, Y. T. (1999). Gender differences in the effects of media richness. *Small Group Research, 30,* 405–437.

8. See Smith, S. W. (1994). Perceptual processing of nonverbal relational messages. In D. E. Hewes (Ed.), *The cognitive bases of interpersonal communication* (pp. 87–110). Hillsdale, NJ: Erlbaum.

9. Farris, C., Treat, T. A., Viken, R. J., & McFall, R. M. (2008). Perceptual mechanisms that characterize gender differences in decoding women's sexual intent. *Psychological Science, 19*(4), 348–354. See also Lim, G. Y., & Roloff, M. E. (1999). Attributing sexual consent. *Journal of Applied Communication Research, 27,* 1–23.

10. Safeway clerks object to "service with a smile." (1998, September 2). *San Francisco Chronicle.*

11. Druckmann, D., Rozelle, R. M., & Baxter, J. C. (1982). *Nonverbal communication: Survey, theory, and research.* Newbury Park, CA: Sage.

12. Knapp, M., Hall, J., & Horgan, T. (2013). *Nonverbal communication in human interaction.* Boston, MA: Cengage Learning, pp. 278–293.

13. Nowicki, S., & Duke, M. (2013). Accuracy in interpreting nonverbal cues. *Nonverbal Communication, 2,* 441.

14. Lieberman, M. D., & Rosenthal, R. (2001). Why introverts can't always tell who likes them: Multitasking and nonverbal decoding. *Journal of Personality and Social Psychology, 80*(2), 294.

15. Rosip, J. C., & Hall, J. A. (2004). Knowledge of nonverbal cues, gender, and nonverbal decoding accuracy. *Journal of Nonverbal Behavior, 28*(4), 267–286; Hall, J. A. (1979). Gender, gender roles, and nonverbal communication skills. In R. Rosenthal (Ed.), *Skill in nonverbal communication: Individual differences* (pp. 32–67). Cambridge, MA: Oelgeschlager, Gunn, and Hain.

16. Research supporting these claims is cited in Burgoon, J. K., & Hoobler, G. D. (2002). Nonverbal signals. In M. L. Knapp & J. A. Daly (Eds.), *Handbook of interpersonal communication* (3rd ed., pp. 240–299). Thousand Oaks, CA: Sage.

17. Jones, S. E., & LeBaron, C. D. (2002). Research on the relationship between verbal and nonverbal communication: Emerging interactions. *Journal of Communication, 52,* 499–521.

18. Rourke, B. P. (1989). *Nonverbal learning disabilities: The syndrome and the model.* New York: Guilford Press.

19. Fudge, E. S. (n.d.). Nonverbal learning disorder syndrome? Retrieved from http://www.nldontheweb.org/fudge.htm.

20. Ekman, P., & Friesen, W. (1975). *Unmasking the face.* New York: Prentice Hall.

21. Birdwhistell, R. (1970). *Kinesics and context.* Philadelphia: University of Pennsylvania Press, Chapter 9.

22. Hall, E. (1969). *The hidden dimension.* Garden City, NY: Anchor Books.

23. Kelly, D. J., Liu, S., Rodger, H., Miellet, S., Ge, L., & Caldara, R. (2011). Developing cultural differences in face processing. *Developmental Science, 14*(5), 1176–1184.

24. Yuki, M., Maddux, W. W., & Masuda, T. (2007). Are the windows to the soul the same in the East and West? Cultural differences in using the eyes and mouth as cues to recognize emotions in Japan and the United States. *Journal of Experimental Social Psychology, 43,* 303–311.

25. Rubin, D. L. (1986). "Nobody play by the rules he know": Ethnic interference in classroom questioning events. In Y. Y. Kim (Ed.), *International and intercultural communication yearbook* (pp. 158–177). Beverly Hills, CA: Sage.

26. Linneman, T. J. (2013). Gender in *Jeopardy!* Intonation variation on a television game show. *Gender & Society, 27,* 82–105; Wolk, L., Abdelli-Beruh, N. B., & Slavin, D. (2012). Habitual use of vocal fry in young adult female speakers. *Journal of Voice, 26,* 111–116.

27. Anderson, R. C., Klofstad, C. A., Mayew, W. J., & Venkatachalam, M. (2014). Vocal fry may undermine the success of young women in the labor market. *PLoS ONE, 9,* e97506.

28. Yuasa, I. P. (2010). Creaky voice: A new feminine voice quality for young urban-oriented upwardly mobile American women? *American Speech, 85,* 315–337.

29. Warnecke, A. M., Masters, R. D., & Kempter, G. (1992). The roots of nationalism: Nonverbal behavior and xenophobia. *Ethnology and Sociobiology, 13,* 267–282.

30. Weitz, S. (Ed.). (1974). *Nonverbal communication: Readings with commentary.* New York: Oxford University Press.

31. Booth-Butterfield, M., & Jordan, F. (1988). *"Act like us": Communication adaptation among racially homogeneous and heterogeneous groups.* Paper presented at the Speech Communication Association meeting, New Orleans.

32. Hall, J. A. (2006). Women and men's nonverbal communication. In V. Manusov & M. L. Patterson (Eds.), *The SAGE handbook of nonverbal communication* (pp. 201–218). Thousand Oaks, CA: Sage.

33. Mayo, C., & Henley, N. M. (Eds.). (2012). *Gender and nonverbal behavior.* Springer Science & Business Media.

34. Knöfler, T., & Imhof, M. (2007). Does sexual orientation have an impact on nonverbal behavior in interpersonal communication? *Journal of Nonverbal Behavior, 31,* 189–204.

35. Hall, J. A., Carter, J. D., & Horgan, T. G. (2001). Status roles and recall of nonverbal cues. *Journal of Nonverbal Behavior, 25,* 79–100.

36. Cross, E. S., & Franz, E. A. (2003, March 30–April 1). *Talking hands: Observation of bimanual gestures as a facilitative working memory mechanism.* Paper presented at the 10th annual meeting of the Cognitive Neuroscience Society, New York.

37. See Hall (1969).

38. Kleinke, C. R. (1977). Compliance to requests made by gazing and touching experimenters in field settings. *Journal of Experimental Social Psychology, 13,* 218–233.

39. Argyle, M. F., Alkema, F., & Gilmour, R. (1971). The communication of friendly and hostile attitudes: Verbal and nonverbal signals. *European Journal of Social Psychology, 1,* 385–402.

40. Buller, D. B., & Burgoon, J. K. (1994). Deception: Strategic and nonstrategic communication. In J. Daly & J. M. Wiemann (Eds.), *Interpersonal communication* (pp. 191–223). Hillsdale, NJ: Erlbaum.

41. Burgoon, J. K., Buller, D. B., Guerrero, L. K., & Feldman, C. M. (1994). Interpersonal deception: VI. Effects on preinteractional and international factors on deceiver and observer perceptions of deception success. *Communication Studies, 45*, 263–280; and Burgoon, J. K., Buller, D. B., & Guerrero, L. K. (1995). Interpersonal deception: IX. Effects of social skill and nonverbal communication on deception success and detection accuracy. *Journal of Language and Social Psychology, 14*, 289–311.

42. Riggio, R. G., & Freeman, H. S. (1983). Individual differences and cues to deception. *Journal of Personality and Social Psychology, 45*, 899–915.

43. Vrij, A. (2006). Nonverbal communication and deception. In V. Manusov & M. L. Patterson (Eds.), *The SAGE handbook of nonverbal communication* (pp. 341–359). Thousand Oaks, CA: Sage.

44. DePaulo, B. M., Lindsay, J. J., Malone, B. E., Muhlenbruck, L., Charlton, K., & Cooper, H. (2003). Cues to deception. *Psychological Bulletin, 129*, 74–118; and Vrij, A., Edward, K., Roberts, K. P., & Bull, R. (2000). Detecting deceit via analysis of verbal and nonverbal behavior. *Journal of Nonverbal Behavior, 24*, 239–263.

45. Dunbar, N. E., Ramirez, A., Jr., & Burgoon, J. K. (2003). The effects of participation on the ability to judge deceit. *Communication Reports, 16*, 23–33.

46. Vrig, A., Akehurst, L., Soukara, S., & Bull, R. (2004). Detecting deceit via analyses of verbal and nonverbal behavior in children and adults. *Human Communication Research, 30*, 8–41.

47. Millar, M. G., & Millar, K. U. (1998). The effects of suspicion on the recall of cues to make veracity judgments. *Communication Reports, 11*, 57–64.

48. McCornack, S. A., & Parks, M. R. (1990). What women know that men don't: Sex differences in determining the truth behind deceptive messages. *Journal of Social and Personal Relationships, 7*, 107–118.

49. McCornack, S. A., & Levine, T. R. (1990). When lovers become leery: The relationship between suspicion and accuracy in detecting deception. *Communication Monographs, 7*, 219–230.

50. Burgoon, J. K., & Levine, T. R. (2010). Advances in deception detection. In S. W. Smith & S. R. Wilson (Eds.), *New directions in interpersonal communication research* (pp. 201–220). Thousand Oaks, CA: Sage.

51. Guerrero, L. K., & Floyd, K. (2006). *Nonverbal communication in close relationships.* Mahwah, NJ: Erlbaum.

52. Levine, T. (2009). To catch a liar. *Communication Currents, 4*, 1–2.

53. Maurer, R. E., & Tindall, J. H. (1983). Effect of postural congruence on client's perception of counselor empathy. *Journal of Counseling Psychology, 30*, 158–163. See also Hustmyre, C., & Dixit, J. (2009, January 1). Marked for mayhem. *Psychology Today.* Retrieved from https://www.psychologytoday.com/articles/200901/marked-mayhem.

54. Ray, G., & Floyd, K. (2006). Nonverbal expressions of liking and disliking in initial interaction: Encoding and decoding perspectives. *Southern Communication Journal, 71*(1), 45–65.

55. Myers, M. B., Templer, D., & Brown, R. (1984). Coping ability of women who become victims of rape. *Journal of Consulting and Clinical Psychology, 52*, 73–78. See also Hustmyre & Dixit (2009).

56. Carney, D. R., Cuddy, A. C., & Yap, A. J. (2010). Power posing: Brief nonverbal displays affect neuroendocrine levels and risk tolerance. *Psychological Science, 21*, 1363–1368.

57. Cuddy, A. C., Wilmuth, C. A., Yap, A. J., & Carney, D. R. (2015). Preparatory power posing affects nonverbal presence and job interview performance. *Journal of Applied Psychology, 100*, 1286–1295.

58. Waters, H. (2013, December 13). Fake it 'til you become it: Amy Cuddy's power poses, visualized. *TED Blog.* Retrieved from http://blog.ted.com/fake-it-til-you-become-it-amy-cuddys-power-poses-visualized/. Quote appears in title.

59. Iverson, J. M. (1999). How to get to the cafeteria: Gesture and speech in blind and sighted children's spatial descriptions. *Developmental Psychology, 35*, 1132–1142.

60. Ekman, P. (2009). *Telling lies: Clues to deceit in the marketplace, politics, and marriage.* New York: Norton, pp. 109–110.

61. Donaghy, W., & Dooley, B. F. (1994). Head movement, gender, and deceptive communication. *Communication Reports, 7*, 67–75.

62. Ekman, P. (2004). Emotional and conversational nonverbal signals. In J. M. Larrazabal & L. A. Perez (Eds.), *Language, knowledge, and representation* (pp. 39–50). The Netherlands: Springer.

63. Musicus, A., Tal, A., & Wansink, B. (2014). Eyes in the aisles: Why is Cap'n Crunch looking down at my child? *Environment and Behavior, 47*(7), 715–733.

64. Murphy, K. (2014, May 18). Psst. Look Over Here . . . *New York Times*, SR6.

65. Farroni, T., Csibra, G., Simion, F., & Johnson, M. H. (2002). Eye contact detection in humans from birth. *Proceedings of the National Academy of Sciences of the United States of America, 99*(14), 9602–9605. doi:10.1073/pnas.152159999.

66. Elsabbagh, M., Mercure, E., Hudry, K., Chandler, S., Pasco, G., Charman, T., . . . Johnson, M. H. (2012). Infant neural sensitivity to dynamic eye gaze is associated with later emerging autism. *Current Biology, 22*(4), 338–342. doi:10.1016/j.cub.2011.12.056.

67. Dadds, M. R., Allen, J. A., Oliver, B. R., Faulkner, N., Legge, K., Moul, K., . . . Scott, S. (2012). Love, eye contact and the developmental origins of empathy v. psychopathy. *The British Journal of Psychiatry, 200*(3), 191–196; doi:10.1192/bjp.bp.110.085720.

68. Murphy, N. A. (2007). Appearing smart: The impression management of intelligence, person perception accuracy, and behavior in social interaction. *Personality and Social Psychology Bulletin, 33*(3), 325–339.

69. Sutton, R., & Rafaeli, A. (1988). Untangling the relationship between displayed emotions and organizational sales: The case of convenience stores. *Academy of Management Journal, 31*, 461–487. Cited examples appear on p. 463.

70. Matsumoto, D. (2006). Culture and nonverbal behavior. In V. Manusov & M. L. Patterson (Eds.), *The SAGE handbook of nonverbal communication* (pp. 219–235). Thousand Oaks, CA: Sage.

71. Knapp, M., Hall, J., & Horgan, T. (2013). *Nonverbal communication in human interaction.* Boston, MA: Cengage Learning, pp. 59–88.

72. Dzhelyova, M. P. (2013). *Face evaluation: Perceptual and neurophysiological responses to pro-social attributions* (Doctoral dissertation, University of St Andrews). Retrieved from https://research-repository.st-andrews.ac.uk/handle/10023/3514.

73. Starkweather, J. A. (1961). Vocal communication of personality and human feeling. *Journal of Communication, 11*(2), 63–72; and Scherer, K. R., Koiwunaki, J., & Rosenthal, R. (1972). Minimal cues in the vocal communication of affect: Judging emotions from content-masked speech. *Journal of Psycholinguistic Speech, 1*(3), 269–285. See also Cox, F. S., & Olney, C. (1985). *Vocalic communication of relational messages.* Paper presented at the annual meeting of the Speech Communication Association, Denver.

74. Burns, K. L., & Beier, E. G. (1973). Significance of vocal and visual channels for the decoding of emotional meaning. *Journal of Communication, 23*, 118–130. See also Hegstrom, T. G. (1979). Message impact: What percentage is nonverbal? *Western Journal of Speech Communication, 43*, 134–143; and McMahan, E. M.

(1976). Nonverbal communication as a function of attribution in impression formation. *Communication Monographs, 43,* 287–294.

75. Mehrabian, A., & Weiner, M. (1967). Decoding of inconsistent communications. *Journal of Personality and Social Psychology, 6,* 109–114.

76. Buller, D., & Aune, K. (1992). The effects of speech rate similarity on compliance: Application of communication accommodation theory. *Western Journal of Communication, 56,* 37–53. See also Buller, D., LePoire, B. A., Aune, K., & Eloy, S. V. (1992). Social perceptions as mediators of the effect of speech rate similarity on compliance. *Human Communication Research, 19,* 286–311; and Francis, J., & Wales, R. (1994). Speech a la mode: Prosodic cues, message interpretation, and impression formation. *Journal of Language and Social Psychology, 13,* 34–44.

77. Kimble, C. E., & Seidel, S. D. (1991). Vocal signs of confidence. *Journal of Nonverbal Behavior, 15,* 99–105.

78. Tusing, K. J., & Dillard, J. P. (2000). The sounds of dominance: Vocal precursors of perceived dominance during interpersonal influence. *Human Communication Research, 26,* 148–171.

79. Zuckerman, M., & Driver, R. E. (1989). What sounds beautiful is good: The vocal attractiveness stereotype. *Journal of Nonverbal Behavior, 13,* 67–82.

80. Hosoda, M., & Stone-Romero, E. (2010). The effects of foreign accents on employment-related decisions. *Journal of Managerial Psychology, 25,* 113–132.

81. For a summary, see Knapp, M. L., & Hall, J. A. (1992). *Nonverbal communication in human interaction* (3rd ed.). New York: Holt, Rinehart and Winston, pp. 93–132. See also Hensley, W. (1992). Why does the best looking person in the room always seem to be surrounded by admirers? *Psychological Reports, 70,* 457–469.

82. Bennett, J. (2010, July 19). The beauty advantage: How looks affect your work, your career, your life. *Newsweek.* Retrieved from http://www.newsweek.com/2010/07/19/the-beauty-advantage.html.

83. Guerrero, L. K., & Hecht, M. L. (2008). *The nonverbal communication reader: Classic and contemporary readings* (3rd ed.). Long Grove, IL: Waveland Press.

84. Persico, N., Postlewaite, A., & Silverman, D. (2004). The effect of adolescent experience of labor market outcomes: The case of height. *Journal of Political Economy, 112,* 1019–1053.

85. Furnham, A. (2014, April 22). Lookism at work. *Psychology Today;* Gordon, R., Crosnoe, R., & Wang, X. (2013). Physical attractiveness and the accumulation of social and human capital in adolescence and young adulthood. *Monographs of the Society for Research in Child Development, 78,* 1–137.

86. Agthe, M., Sporrle, M., & Maner, J. K. (2011). Does being attractive always help? Positive and negative effects of attractiveness on social decision making. *Personality and Social Psychology Bulletin, 37,* 1042–1054.

87. Frevert, T. K., & Walker, L. S. (2014). Physical attractiveness and social status. *Sociology Compass, 8,* 313–323.

88. Abdala, K. F., Knapp, M. L., & Theune, K. E. (2002). Interaction appearance theory: Changing perceptions of physical attractiveness through social interaction. *Communication Theory, 12,* 8–40.

89. Bickman, L. (1974). The social power of a uniform. *Journal of Applied Social Psychology, 4,* 47–61.

90. Lawrence, S. G., & Watson, M. (1991). Getting others to help: The effectiveness of professional uniforms in charitable fund raising. *Journal of Applied Communication Research, 19,* 170–185.

91. Rehman, S. U., Nietert, P. J., Cope, D. W., & Kilpatrick, A. O. (2005). What to wear today? Effect of doctor's attire on the trust and confidence of patients. *The American Journal of Medicine, 118,* 1279–1286.

92. Bickman, L. (1974, April). Social roles and uniforms: Clothes make the person. *Psychology Today, 7,* 48–51.

93. Temple, L. E., & Loewen, K. R. (1993). Perceptions of power: First impressions of a woman wearing a jacket. *Perceptual and Motor Skills, 76,* 339–348.

94. Hoult, T. F. (1954). Experimental measurement of clothing as a factor in some social ratings of selected American men. *American Sociological Review, 19,* 326–327.

95. Hart, S., Field, T., Hernandez-Reif, M., & Lundy, B. (1998). Preschoolers' cognitive performance improves following massage. *Early Child Development and Care, 143,* 59–64. For more about the role of touch in relationships, see Keltner, D. (2009). *Born to be good: The science of a meaningful life.* New York: Norton, pp. 173–198.

96. Field, T. (2010). Touch for socioemotional and physical well-being: A review. *Developmental Review, 30*(4), 367–383.

97. Montagu, A. (1972). *Touching: The human significance of the skin.* New York: Harper & Row, p. 93.

98. Feldman, R. (2011). Maternal touch and the developing infant. In M. Hertenstein & S. Weiss (Eds.), *Handbook of touch* (pp. 373–407). New York: Springer.

99. Camps, J., Tuteleers, C., Stouten, J., & Nelissen, J. (2013). A situational touch: How touch affects people's decision behavior. *Social Influence, 8*(4), 237–250.

100. Willis, F. N., & Hamm, H. K. (1980). The use of interpersonal touch in securing compliance. *Journal of Nonverbal Behavior, 5,* 49–55.

101. Jacob, C., & Guéguen, N. (2014). The effect of compliments on customers' compliance with a food server's suggestion. *International Journal of Hospitality Management, 40,* 59–61.

102. Guéguen, N., & Vion, M. (2009). The effect of a practitioner's touch on a patient's medication compliance. *Psychology, Health and Medicine, 14,* 689–694.

103. Segrin, C. (1993). The effects of nonverbal behavior on outcomes of compliance gaining attempts. *Communication Studies, 11,* 169–187.

104. Hornik, J. (1992). Effects of physical contact on customers' shopping time and behavior. *Marketing Letters, 3,* 49–55.

105. Smith, D. E., Gier, J. A., & Willis, F. N. (1982). Interpersonal touch and compliance with a marketing request. *Basic and Applied Social Psychology, 3,* 35–38. See also Soars, B. (2009). Driving sales through shoppers' sense of sound, sight, smell and touch. *International Journal of Retail & Distribution Management, 37*(3), 286–298.

106. Field, T., Lasko, D., Mundy, P., Henteleff, T., Kabat, S., Talpins, S., & Dowling, M. (1997). Brief report: Autistic children's attentiveness and responsivity improve after touch therapy. *Journal of Autism and Developmental Disorders, 27,* 333–338.

107. Kraus, M. W., Huang, C., & Keltner, D. (2010). Tactile communication, cooperation, and performance: An ethological study of the NBA. *Emotion, 10,* 745–749.

108. Chan, Y. K. (1999). Density, crowding, and factors intervening in their relationship: Evidence from a hyper-dense metropolis. *Social-Indicators-Research, 48,* 103–124.

109. See Hall (1969), pp. 113–130.

110. LeFebvre, L., & Allen, M. (2014). Teacher immediacy and student learning: An examination of lecture/laboratory and self-contained course sections. *Journal of the Scholarship of Teaching and Learning, 14*(2), 29–45.

111. Wouda, J. C., & van de Wiel, H. B. (2013). Education in patient–physician communication: How to improve effectiveness? *Patient Education and Counseling, 90*(1), 46–53.

112. Mehrabian, A. (1976). *Public places and personal spaces: The psychology of work, play, and living environments.* New York: Basic Books, p. 69.

113. Sadalla, E. (1987). Identity and symbolism in housing. *Environment and Behavior, 19,* 569–587.

114. Maslow, A. H., & Mintz, N. L. (1956). Effects of esthetic surroundings. *Journal of Psychology, 41,* 247–254.

115. Sommer, R. (1969). *Personal space: The behavioral basis of design.* Englewood Cliffs, NJ: Prentice-Hall, p. 78. See also McPherson, M., Smith-Lovin, L., & Brashears, M. E. (2006). Social isolation in America: Changes in core discussion networks over two decades. *American Sociological Review, 71*(3), 353–375.

116. Sommer, p. 35.

117. Bruneau, T. J. (2012). Chronemics: Time-binding and the construction of personal time. *ETC: A Review of General Semantics, 69*(1), 72.

118. Ballard, D. I., & Seibold, D. R. (2000). Time orientation and temporal variation across work groups: Implications for group and organizational communication. *Western Journal of Communication, 64,* 218–242.

119. Levine, R. (1997). *A geography of time: The temporal misadventures of a social psychologist.* New York: Basic Books.

120. See, for example, Hill, O. W., Block, R. A., & Buggie, S. E. (2000). Culture and beliefs about time: Comparisons among black Americans, black Africans, and white Americans. *Journal of Psychology, 134,* 443–457.

121. Levine, R., & Wolff, E. (1985, March). Social time: The heartbeat of culture. *Psychology Today, 19,* 28–35. See also Levine, R. (1987, April). Waiting is a power game. *Psychology Today, 21,* 24–33.

122. Burgoon, J. K., Buller, D. B., & Woodall, W. G. (1996). *Nonverbal communication.* New York: McGraw-Hill, p. 148. See also White, L. T., Valk, R., & Dialmy, A. (2011). What is the meaning of "on time"? The sociocultural nature of punctuality. *Journal of Cross-Cultural Psychology, 42*(3), 482–493.

123. Carlson, E. N. (2013). Overcoming barriers to self-knowledge: Mindfulness as a path to seeing yourself as you really are. *Perspectives on Psychological Science, 8,* 173–186.

CHAPTER 7

1. Eidell, L. (2015, May 28). *Bachelor* and *Bachelorette* couples: The complete list. *Glamour.* Retrieved from http://www. glamour.com/entertainment/blogs/obsessed/2015/05/ bachelor-bachelorette-couples-history.

2. Boardman, M. (2015, February 7). Sean Lowe, Catherine Giudici explain why so few *Bachelor* couples get married. *US Weekly.* Retrieved from http://www.usmagazine.com/entertainment/ news/sean-lowe-catherine-giudici-explain-why-so-few-bachelor-couples-wed-201572. Quote appears in paragraph 6.

3. Melas, C. (2011, June 29). Emily Maynard: "Why I wouldn't marry Brad Womack! "*Hollywood Life.* Retrieved from http://hollywoodlife.com/2011/06/29/ emily-maynard-brad-womack-break-up-people-magazine.

4. Away with the bling! *The Bachelor* winner Courtney Robertson removes engagement ring after splitting from fiancé Ben Glajnik. (2012, October 8). *Daily Mail.* Retrieved from http://www. dailymail.co.uk/tvshowbiz/article-2214821/Bachelor-winner-Courtney-Robertson-heartbroken-break-up.html.

5. Stewart, J. (2012). *Bridges not walls: A book about interpersonal communication* (11th ed.). Boston: McGraw-Hill. Quote appears on p. 36.

6. Patton, B. R., & Giffin, K. (1974). *Interpersonal communication: Basic text and readings.* New York: Harper & Row.

7. Ledbetter, A. M. (2014). The past and future of technology in interpersonal communication theory and research. *Communication Studies, 65*(4), 456–459.

8. Stewart, J. (2012). *Bridges not walls: A book about interpersonal communication* (11th ed.). Boston: McGraw-Hill. Quote appears on p. 36.

9. Baiocco, R., Laghi, F., Schneider, B.H., Dalessio, M., Amichai-Hamburger, Y., Coplan, R. J., . . . Flament, M. (2011). Daily patterns of communication and contact between Italian early adolescents and their friends. *Cyberpsychology, Behavior, and Social Networking, 14*(7–8), 467–471.

10. Lee, S. J. (2009). Online communication and adolescent social ties: Who benefits more from Internet use? *Journal of Computer-Mediated Communication, 14,* 509–531.

11. Jin, B., & Peña, J.F. (2010). Mobile communication in romantic relationships: Mobile phone use, relational uncertainty, love, commitment, and attachment styles.*Communication Reports, 23,* 39–51.

12. Hammick, J. K., & Lee, M. J. (2014). Do shy people feel less communication apprehension online? The effects of virtual reality on the relationship between personality characteristics and communication outcomes. *Computers in Human Behavior, 33,* 302–310.

13. Tannen, D. (1994, May 16). High tech gender gap. *Newsweek,* pp. 52–53.

14. Wright, K. B. (2012). Emotional support and perceived stress among college students using Facebook.com: An exploration of the relationship between source perceptions and emotional support. *Communication Research Reports, 29,* 175–184.

15. Hales, K. D. (2012). *Multimedia use for relational maintenance in romantic couples.* Paper presented at the annual meeting of the International Communication Association, Phoenix, AZ.

16. Ndasauka, Y., Hou, J., Wang, Y., Yang, L., Yang, Z., Ye, Z., & . . . Zhang, X. (2016). Research report: Excessive use of Twitter among college students in the UK: Validation of the Microblog Excessive Use Scale and relationship to social interaction and loneliness. *Computers in Human Behavior, 55*(Pt. B), 963–971.

17. The phubbing truth. (2013, October 8). Wordability. Retrieved from http://wordability.net/2013/10/08/the-phubbing-truth.

18. Przybylski, A. K., & Weinstein, N. (2013). Can you connect with me now? How the presence of mobile communication technology influences face-to-face conversation quality. *Journal of Social and Personal Relationships, 30,* 237–246.

19. Yao, M. Z., & Zhong, Z. (2014). Loneliness, social contacts and Internet addiction: A cross-lagged panel study. *Computers in Human Behavior, 30,* 164–170.

20. Hunt, E. (2015, November 3). Essena O'Neill quits Instagram claiming social media "is not real life." *The Guardian.* Retrieved from http://www.theguardian.com/media/2015/nov/03/ instagram-star-essena-oneill-quits-2d-life-to-reveal-true-story-behind-images.

21. Hunt, E. (2015, November 3). Essena O'Neill quits Instagram claiming social media "is not real life." *The Guardian.* Retrieved from http://www.theguardian.com/media/2015/nov/03/ instagram-star-essena-oneill-quits-2d-life-to-reveal-true-story-behind-images. Quote appears in paragraph 6.

22. Lemay, E. P., Jr., Clark, M. S., & Greenberg, A. (2010). What is beautiful is good because what is beautiful is desired: Physical attractiveness stereotyping as projection of interpersonal goals. *Personality and Social Psychology Bulletin, 36,* 339–353.

23. Wang, S., Moon, S., Kwon, K., Evans, C. A., & Stefanone, M. A. (2010). Face off: Implications of visual cues on initiating friendship on Facebook. *Computers in Human Behavior, 26,* 226–234.

24. Toma, C. L., & Hancock, J. T. (2010). Looks and lies: The role of physical attractiveness in online dating self-presentation and deception. *Communication Research, 37,* 335–351.

25. Luo, S., & Klohnen, E. (2005). Assortive mating and marital quality in newlyweds: A couple-centered approach. *Journal of Personality and Social Psychology, 88*, 304–326. See also Amodio, D. M., & Showers, C. J. (2005). Similarity breeds liking revisited: The moderating role of commitment. *Journal of Social and Personal Relationships, 22*, 817–836.

26. Mackinnon, S. P., Jordan, C., & Wilson, A. (2011). Birds of a feather sit together: Physical similarity predicts seating distance. *Personality and Social Psychology Bulletin, 37*, 879–892.

27. Heatherington, L., Escudero, V., & Friedlander, M. L. (2005). Couple interaction during problem discussions: Toward an integrative methodology. *Journal of Family Communication, 5*, 191–207.

28. Toma, C., Yzerbyt, V., & Corneille, O. (2012). Reports: Nice or smart? Task relevance of self-characteristics moderates interpersonal projection. *Journal of Experimental Social Psychology, 48*, 335–340.

29. Dindia, K. (2002). Self-disclosure research: Knowledge through meta-analysis. In M. Allen & R. W. Preiss (Eds.), *Interpersonal communication research: Advances through meta-analysis* (pp. 169–185). Mahwah, NJ: Erlbaum.

30. Flora, C. (2004, January/February). Close quarters. *Psychology Today, 37*, 15–16.

31. Haythornthwaite, C., Kazmer, M. M., & Robbins, J. (2000). Community development among distance learners: Temporal and technological dimensions. *Journal of Computer-Mediated Communication, 6*(1), Article 2. Retrieved from http://jcmc.indiana.edu/vol6/issue1/haythornthwaite.html.

32. See, for example, Roloff, M. E. (1981). *Interpersonal communication: The social exchange approach.* Beverly Hills, CA: Sage.

33. Duck, S. W. (2011). Similarity and perceived similarity of personal constructs as influences on friendship choice. *British Journal of Clinical Psychology, 12*, 1–6.

34. Mackinnon, S. P., Jordan, C., & Wilson, A. (2011). Birds of a feather sit together: Physical similarity predicts seating distance. *Personality and Social Psychology Bulletin, 37*, 879–892.

35. Sias, P. M., Drzewiecka, J. A., Meares, M., Bent, R., Konomi, Y., Ortega, M., & White, C. (2008). Intercultural friendship development. *Communication Reports, 21*, 1–13.

36. Alley, T. R., & McCanless, E. R. (2002). *Body shape and muscularity preferences in short-term and long-term relationships.* Paper resented at the Biennial International Conference on Human Ethology, Vienna.

37. Hamachek, D. (1982). *Encounters with others: Interpersonal relationships and you.* New York: Holt, Rinehart and Winston.

38. Singj, R., & Tor, X. L. (2008). The relative effects of competence and likability on interpersonal attraction. *Journal of Social Psychology, 148*(2), 253–255. Quote appears on p. 253.

39. Finkel, E. J., Eastwick, P. W., Karney, B. R., Reis, H. T., & Sprecher, S. (2012). Online dating: A critical analysis from the perspective of psychological science. *Psychological Science in the Public Interest, 13*(1), 3–66.

40. The quote, which came up via a Tumblr search on "friend-zoned," was adapted from Bender, C. (Producer), & Kumble, R. (Director). (2005). *Just Friends* [Motion picture]. Los Angeles, CA: New Line Cinema.

41. Amir, L. (2001, November). Plato's theory of love: Rationality as passion. *Practical Philosophy, 4*(3), 24–32. Retrieved from http://www.society-for-philosophy-in-practice.org/journal/pdf/4-3%2006%20Amir%20-%20Plato%20Love.pdf.

42. Demir, M., & Özdemir, M. (2010). Friendship, need satisfaction and happiness. *Journal of Happiness Studies, 11*, 243–259.

43. Buote, V. M., Pancer, S., Pratt, M. W., Adams, G., Birnie-Lefcovitch, S., Polivy, J., & Wintre, M. (2007). The importance of friends: Friendship and adjustment among 1st-year university students. *Journal of Adolescent Research, 22*, 665–689.

44. Demir, M., Özdemir, M., & Marum, K. (2011). Perceived autonomy support, friendship maintenance, and happiness. *Journal of Psychology, 145*, 537–571.

45. Deci, E., La Guardia, J., Moller, A., Scheiner, M., & Ryan, R. (2006). On the benefits of giving as well as receiving autonomy support: Mutuality in close friendships. *Personality & Social Psychology Bulletin, 32*, 313–327.

46. Hartup, W. W., & Stevens, N. (1997). Friendships and adaptation in the life course. *Psychological Bulletin, 121*, 355–370.

47. Pecchioni, L. (2005). Friendship throughout the life span. In L. Pecchioni, K. B. Wright, & J. F. Nussbaum (Eds.), *Life-span communication* (pp. 97–116). Mahwah, NJ: Erlbaum; Samter, W. (2003). Friendship interaction skills across the life-span. In J. O. Greene & B. R. Burleson (Eds.), *Handbook of communication and social interaction skills* (pp. 637–684). Mahwah, NJ: Erlbaum.

48. Bleske-Rechek, A., Somers, E., Micke, C., Erickson, L., Matteson, L., Stocco, C., . . . Ritchie, L. (2012). Burden or benefit? Attraction in cross-sex friendship. *Journal of Social and Personal Relationships, 29*, 569–596.

49. Hall, J. A. (2011). Sex differences in friendship expectations: A meta-analysis. *Journal of Social and Personal Relationships, 28*, 723–747.

50. Hall (2011).

51. Ward, A. F. (2012, October 23). Men and women can't be "just friends." *Scientific American.* Retrieved from http://www.scientificamerican.com/article.cfm?id=men-and-women-cant-be-just-friends.

52. Rawlins, W. K. (1992). *Friendship matters, communication, dialectics, and the life course.* New York: De Gruyter, p. 105.

53. Kalmijn, M. (2003). Shared friendship networks and the life course: An analysis of survey data on married and cohabiting couples. *Social Networks, 25*, 231–249.

54. Nussbaum, J. F., Pecchioni, L., Baringer, D., & Kundrat, A. (2002). Lifespan communication. In W. B. Gudykunst (Ed.), *Communication yearbook 26* (pp. 366–389). Mahwah, NJ: Erlbaum.

55. Becker, J. H., Johnson, A., Craig, E. A., Gilchrist, E. S., Haigh, M. M., & Lane, L. T. (2009). Friendships are flexible, not fragile: Turning points in geographically close and long-distance friendships. *Journal of Social & Personal Relationships, 26*, 347–369. Quote appears on p. 347.

56. Johnson, A., Haigh, M. M., Craig, E. A., & Becker, J. H. (2009). Relational closeness: Comparing undergraduate college students' geographically close and long-distance friendships. *Personal Relationships, 16*, 631–646.

57. Migliaccio, T. (2009). Men's friendships: Performances of masculinity. *Journal of Men's Studies, 17*, 226–241.

58. Hall, J. A. (2011). Sex differences in friendship expectations: A meta-analysis. *Journal of Social and Personal Relationships, 28*, 723–747.

59. Bello, R. S., Brandau-Brown, F. E., Zhang, S., & Ragsdale, J. (2010). Verbal and nonverbal methods for expressing appreciation in friendships and romantic relationships: A cross-cultural comparison. *International Journal of Intercultural Relations, 34*, 294–302.

60. Hall, J. A. (2011). Sex differences in friendship expectations: A meta-analysis. *Journal of Social and Personal Relationships, 28*, 723–747.

61. Hall (2011).

62. Baiocco, R., Laghi, F., Di Pomponio, I., & Nigito, C. S. (2012). Self-disclosure to the best friend: Friendship quality and internalized sexual stigma in Italian lesbian and gay adolescents. *Journal of Adolescence, 35*, 381–387.

63. Guerrero, L. K., Farinelli, L., & McEwan, B. (2009). Attachment and relational satisfaction: The mediating effect of emotional communication. *Communication Monographs, 76,* 487–514.

64. Tabak, B., McCullough, M., Luna, L., Bono, G., & Berry, J. (2012). Conciliatory gestures facilitate forgiveness and feelings of friendship by making transgressors appear more agreeable. *Journal of Personality, 80,* 503–536.

65. Davis, J. R., & Gold, G. J. (2011). An examination of emotional empathy, attributions of stability, and the link between perceived remorse and forgiveness. *Personality & Individual Differences, 50,* 392–397.

66. Bello, R. S., Brandau-Brown, F. E., Zhang, S., & Ragsdale, J. D. (2010). Verbal and nonverbal methods for expressing appreciation in friendships and romantic relationships: A cross-cultural comparison. *International Journal of Intercultural Relations, 34,* 294–302.

67. van der Horst, M., & Coffe, H. (2012). How friendship network characteristics influence subjective well-being. *Social Indicators Research, 107,* 509–529.

68. Rawlins, W. K., & Holl, M. (1987). The communicative achievement of friendship during adolescence: Predicaments of trust and violation. *Western Journal of Speech Communication, 51,* 345–363.

69. Deci, E., La Guardia, J., Moller, A., Scheiner, M., & Ryan, R. (2006). On the benefits of giving as well as receiving autonomy support: Mutuality in close friendships. *Personality & Social Psychology Bulletin, 32,* 313–327.

70. Hall (2011).

71. Russell, E. M., DelPriore, D. J., Butterfield, M. E., & Hill, S. E. (2013). Friends with benefits, but without the sex: Straight women and gay men exchange trustworthy mating advice. *Evolutionary Psychology, 11*(1), 132–147.

72. Quinn, B. (2011, May 8). Social network users have twice as many friends online as in real life. *The Guardian.* Retrieved from http://www.theguardian.com/media/2011/may/09/social-network-users-friends-online.

73. Whitty, M., & Joinson, A. (2009). *Truth, lies & trust on the Internet.* New York: Routledge.

74. Chan, D. K.-S., & Cheng, G. H.-L. (2004). A comparison of offline and online friendship qualities at different stages of relationship development. *Journal of Social and Personal Relationships, 21,* 305–320.

75. Walther, J. B. (1996). Computer-mediated communication: Impersonal, interpersonal, and hyperpersonal interaction. *Communication Research, 23,* 3–43; Okdie, B. M., Guadagno, R. E., Bernieri, F. J., Geers, A. L., & Mclarney-Vesotski, A. R. (2011). Getting to know you: Face-to-face versus online interactions. *Computers in Human Behavior, 27,* 153–159.

76. Galvin, K. M. (2006). Diversity's impact of defining the family: Discourse-dependence and identity. In R. L. West & L. H. Turner (Eds.), *The family communication sourcebook* (pp. 3–20). Thousand Oaks, CA: Sage.

77. Minow, M. (1998). Redefining families: Who's in and who's out? In K. V. Hansen & A. I. Garey (Eds.), *Families in the U.S.: Kinship and domestic policy* (pp. 7–19). Philadelphia: Temple University Press. (Originally published in the *University of Colorado Law Review,* 1991, *62,* 269–285.)

78. Trace the evolution of this model by exploring McLeod, M. J., & Chaffee, S. H. (1972). The construction of social reality. In J. Tedeschi (Ed.), *The social influence process* (pp. 177–195). Hillsdale, NJ: Erlbaum; Ritchie, L. D., & Fitzpatrick, M. A. (1990), Family communication patterns: Measuring interpersonal perceptions of interpersonal relationships. *Communication Research, 17,* 523–544; and Koerner, A. F., & Fitzpatrick, M. A. (2006). Family communication patterns theory: A social cognitive approach. In D. O. Braithwaite & L. A. Baxter (Eds.), *Engaging theories in family communication: Multiple perspectives* (pp. 50–65). Thousand Oaks, CA: Sage.

79. Edwards, R., Hadfield, L., Lucey, H., & Mauthner, M. (2006). *Sibling identity and relationships: Sisters and brothers.* New York: Routledge. Quote appears on p. 63.

80. Stewart, R. B., Kozak, A. L., Tingley, L. M., Goddard, J. M., Blake, E. M., & Cassel, W. A. (2001). Adult sibling relationships: Validation of a typology. *Personal Relationships, 8,* 299–324.

81. Mansson, D. H. (2012). A qualitative analysis of grandparents' expressions of affection for their young adult grandchildren. *North American Journal of Psychology, 14,* 207–219.

82. Geurts, T., van Tilburg, T., & Poortman, A. (n.d.). The grandparent-grandchild relationship in childhood and adulthood: A matter of continuation? *Personal Relationships, 19,* 267–278.

83. Breheny, M., Stephens, C., & Spilsbury, L. (2013). Involvement without interference: How grandparents negotiate intergenerational expectations in relationships with grandchildren. *Journal of Family Studies, 19,* 174–184.

84. Feiler, B. (2013). *The secrets of happy families: Improve your mornings, rethink family dinner, fight smarter, go out and play, and much more.* New York: HarperCollins.

85. Ahmetoglu, G., Swami, V., & Chamorro-Premuzic, T. (2010). The relationship between dimensions of love, personality, and relationship length. *Archives of Sexual Behavior, 34,* 1181–1190.

86. Malouff, J. M., Schutte, N. S., & Thorsteinsson, E. B. (2013). Trait emotional intelligence and romantic relationship satisfaction: A meta-analysis. *American Journal of Family Therapy, 42,* 53–66.

87. Erickson, E. H. (1963). *Childhood and society* (2nd ed.). New York: Norton.

88. Floyd, K., Hess, J. A., Miczo, L. A., Halone, K. K., Mikkelson, A. C., & Tusing, K. (2005). Human affection exchange: VIII. Further evidence of the benefits of expressed affection. *Communication Quarterly, 53,* 285–303.

89. Bond, B. J. (2009). He posted, she posted: Gender differences in self-disclosure on social network sites. *Rocky Mountain Communication Review, 6*(2), 29–37.

90. MacGeorge, E. L., Graves, A. R., Feng, B., Gillihan, S. J., & Burleson, B. R. (2004). The myth of gender cultures: Similarities outweigh differences in men's and women's provision of and responses to supportive communication. *Sex Roles, 50,* 143–175.

91. Information in this paragraph is from Hall, E., Travis, M., Anderson, S., & Henley, A. (2013). Complaining and Knapp's relationship stages: Gender differences in instrumental complaints. *Florida Communication Journal, 41,* 49–61.

92. Duke, M. P. (2013, March 23). The stories that bind us: What are the twenty questions? *The Blog.* Retrieved from http://www.huffingtonpost.com/marshall-p-duke/the-stories-that-bind-us-_b_2918975.html.

93. Guerrero, L. K., Farinelli, L., & McEwan, B. (2009). Attachment and relational satisfaction: The mediating effect of emotional communication. *Communication Monographs, 76,* 487–514.

94. Young, S. L. (2009). The function of parental communication patterns: Reflection-enhancing and reflection-discouraging approaches. *Communication Quarterly, 57,* 379–394.

95. Information in this paragraph is from Petronio, S. (2010). Communication privacy management theory: What do we know about family privacy regulation? *Journal of Family Theory & Review, 2,* 175–196.

96. Baraldi, C., & Iervese, V. (2010). Dialogic mediation in conflict resolution education. *Conflict Resolution Quarterly, 27,* 423–445.

97. Strom, R. E., & Boster, F. J. (2011). Dropping out of high school: Assessing the relationship between supportive messages from family and educational attainment. *Communication Reports, 24,* 25–37.

98. Feiler, B. (2013). *The secrets of happy families: Improve your mornings, rethink family dinner, fight smarter, go out and play, and much more.* New York: HarperCollins.

99. Elliott, S., & Umberson, O. (2008). The performance of desire: Gender and sexual negotiation in long-term marriages. *Journal of Marriage and Family, 70,* 391–406.

100. Balsam, K. F., Beauchaine, T. P., Rothblum, E. D., & Solomon, S. E. (2008). Three-year follow-up of same-sex couples who had civil unions in Vermont, same-sex couples not in civil unions, and heterosexual married couples. *Developmental Psychology, 44,* 102–116.

101. Gottman, J., Levenson, R., & Swanson, C. (2003). Observing gay, lesbian and heterosexual couples' relationships: Mathematical modeling of conflict interaction. *Journal of Homosexuality, 45,* 65–91.

102. Chapman, G. (2010). *The five love languages: The secret to love that lasts.* Chicago: Northfield Publishing.

103. Frisby, B. N., & Booth-Butterfield, M. (2012). The "how" and "why" of flirtatious communication between marital partners. *Communication Quarterly, 60,* 465–480.

104. Merolla, A. J. (2010). Relational maintenance during military deployment: Perspectives of wives of deployed US soldiers. *Journal of Applied Communication Research, 38*(1), 4–26.

105. Haas, S. M., & Stafford, L. (2005). Maintenance behaviors in same-sex and marital relationships: A matched sample comparison. *Journal of Family Communication, 5,* 43–60.

106. Haas & Stafford (2005).

107. Soin, R. (2011). Romantic gift giving as chore or pleasure: The effects of attachment orientations on gift giving perceptions. *Journal of Business Research, 64,* 113–118.

108. Guéguen, N. (2010). The effect of a woman's incidental tactile contact on men's later behavior. *Social Behavior and Personality: An International Journal, 38,* 257–266.

109. Floyd, K., Boren, J. P., & Hannawa, A. F. (2009). Kissing in marital and cohabiting relationships: Effects of blood lipids, stress, and relationship satisfaction. *Western Journal of Communication, 73,* 113–133.

110. Chapman, G. (1995). *The five love languages: How to express heartfelt commitment to your mate.* Chicago: Northfield Publishing, p. 17. See also Egbert, N., & Polk, D. (2006). Speaking the language of relational maintenance: A validity test of Chapman's (1992) five love languages. *Communication Research Reports, 23*(1), 19–26.

111. Knapp, M. L., & Vangelisti, A. L. (2009). *Interpersonal communication and human relationships* (6th ed.). Boston: Allyn and Bacon.

112. Canary, D. J., & Stafford, L. (Eds.). (1994). *Communication and relational maintenance.* San Diego: Academic Press. See also Lee, J. (1998). Effective maintenance communication in superior-subordinate relationships. *Western Journal of Communication, 62,* 181–208.

113. Burgess, B. (n.d.). What to do if it gets awkward on an Internet date. *Synonym* [blog]. Retrieved from http://classroom.synonym.com/gets-awkward-internet-date-23312.html. Quote appears in paragraph 1.

114. Caughlin, J. P., & Sharabi, L. L. (2013). A communicative interdependence perspective of close relationships: The connections between mediated and unmediated interactions matter. *Journal of Communication, 63*(5), 873–893.

115. Flaa, J. (2013, October 29). I met my spouse online: 9 online dating lessons learned the hard way. *The Blog.* Retrieved from http://www.huffingtonpost.com/

116. Aslay, J. (2012, November 3). You lost me at hello, how to get past the awkward first meeting. *Understand Men Now* [blog]. Retrieved from http://www.jonathonaslay.com/2012/11/03/you-lost-me-at-hello-how-to-get-past-the-awkward-first-meeting/. Quote appears in paragraph 11.

117. Flaa, J. (2013, October 29). I met my spouse online: 9 online dating lessons learned the hard way. *The Blog.* Retrieved from http://www.huffingtonpost.com/jennifer-flaa/9-online-dating-lessons_b_4174334.html. Quote appears in paragraph 13.

118. Wilson, S. R., Kunkel, A. D., Robson, S. J., Olufowote, J. O., & Soliz, J. (2009). Identity implications of relationship (re)definition goals: An analysis of face threats and facework as young adults initiate, intensify, and disengage from romantic relationships. *Journal of Language and Social Psychology, 28,* 32–61.

119. Baxter, L. A. (1987). Symbols of relationship identity in relationship culture. *Journal of Social and Personal Relationships, 4,* 261–280.

120. Fox, J., Warber, K. M., & Makstaller, D. (2013). The role of Facebook in romantic relationship development: An exploration of Knapp's relational stage model. *Journal of Social and Personal Relationships, 30,* 771–794.

121. Dunleavy, K., & Booth-Butterfield, M. (2009). Idiomatic communication in the stages of coming together and falling apart. *Communication Quarterly, 57,* 416–432.

122. Meyers, S. (2014, December 23.) 5 ways to put your date at ease (and alleviate awkward tension). *Fox News Magazine.* Retrieved from http://magazine.foxnews.com/love/5-ways-put-your-date-at-ease-and-alleviate-awkward-tension.

123. Meyers (2014). Quote appears in the last paragraph.

124. Brown, B. (2010). *The gifts of imperfection.* Center City, MN: Hazelden.

125. Knapp, M. L. (1984). *Interpersonal communication and human relationships.* Boston, MA: Allyn & Bacon.

126. Lim, T. S., & Bowers, J. W. (1991). Facework: Solidarity, approbation, and tact. *Human Communication Research,17,* 415–450.

127. Frei, J. R., & Shaver, P. R. (2002). Respect in close relationships: Prototype, definition, self-report assessment, and initial correlates. *Personal Relationships, 9,* 121–139.

128. See Rossiter, C. M., Jr. (1974). Instruction in metacommunication. *Central States Speech Journal, 25,* 36–42; and Wilmot, W. W. (1980). Metacommunication: A reexamination and extension. In D. Nimmo (Ed.), *Communication yearbook 4.* New Brunswick, NJ: Transaction.

129. Tamir, D. I., & Mitchell, J. P. (2012). Disclosing information about the self is intrinsically rewarding. *Proceedings of the National Academy of Science, 109*(21), 8038–8043.

130. Altman, I., & Taylor, D. A. (1973). *Social penetration: The development of interpersonal relationships.* New York: Holt, Rinehart and Winston.

131. Luft, J. (1969). *Of human interaction.* Palo Alto, CA: National Press.

132. Chen, Y., & Nakazawa, M. (2009). Influences of culture on self-disclosure as relationally situated in intercultural and interracial friendships from a social penetration perspective. *Journal of Intercultural Communication Research, 38*(2), 77–98.

133. Friendship: An all or nothing proposition? (2009, September 15). *INTJ Forum.* Retrieved from http://intjforum.com/showthread.php?t=23349. Quote appears in second paragraph.

134. Morgan. (2016, March 28, 10:59 P.M.). Re: Needy friends: A friend indeed? [Blog comment].*The Friendship Blog.* Retrieved from http://www.thefriendshipblog.com/needy-friends-friend-indeed/.

135. See, for example, Baxter, L. A., & Montgomery, B. M. (1998). A guide to dialectical approaches to studying personal relationships. In B. M. Montgomery & L. A. Baxter (Eds.), *Dialectical approaches to studying personal relationships* (pp. 1–16). Mahwah, NJ: Erlbaum; and Ebert, L. A., & Duck, S. W. (1997). Rethinking satisfaction in personal relationships from a dialectical perspective. In R. J. Sternberg & M. Hojjatr (Eds.), *Satisfaction in close relationships* (pp. 190–216). New York: Guilford.

136. Summarized by Baxter, L. A. (1994). A dialogic approach to relationship maintenance. In D. J. Canary & L. Stafford (Eds.), *Communication and relational maintenance* (pp. 233–254). San Diego: Academic Press.

137. Baxter (1994).

138. Tannen, D. (1986).*That's not what I meant: How conversational style makes or breaks relationships.* New York: Ballantine, p. 17.

139. Morris, D. (1971). *Intimate behavior.* New York: Kodansha Globe, pp. 21–29.

140. Adapted from Baxter & Montgomery, *A guide to dialectical approaches*, pp. 1–16.

141. Siffert, A., & Schwarz, B. (2011). Spouses' demand and withdrawal during marital conflict in relation to their subjective well-being. *Journal of Social and Personal Relationships, 28,* 262–277.

142. Serota, K. B., Levine, T. R., & Boster, F. J. (2010). The prevalence of lying in America: Three studies of self-reported lies. *Human Communication Research, 36*(1), 2–25.

143. Harrell, E. (2009, August 19). Why we lie so much. *Time.* Retrieved from http://www.time.com/time/health/article/0,8599,1917215,00.html.

144. McCornack, S. A., & Levine, T. R. (1990). When lies are uncovered: Emotional and relational outcomes of discovered deception. *Communication Monographs, 57,* 119–138.

145. Guthrie, J., & Kunkel, A. (2013). Tell me sweet (and not-so-sweet) little lies: Deception in romantic relationships. *Communication Studies, 64*(2), 141–157.

146. Kaplar, M. E., & Gordon, A. K. (2004). The enigma of altruistic lying: Perspective differences in what motivates and justifies lie telling within romantic relationships. *Personal Relationships, 11,* 489–507.

147. Bryant, E. (2008). Real lies, white lies and gray lies: Towards a typology of deception. *Kaleidoscope: A Graduate Journal of Qualitative Communication Research, 7,* 723–748.

148. Bavelas, J. B. (2009). Equivocation. In H. T. Reis & S. Sprecher (Eds.), *Encyclopedia of human relationships* (Vol. 1, pp. 537–539). Thousand Oaks, CA: Sage. Retrieved from http://web.uvic.ca/psyc/bavelas/2009Equivocation.pdf.

149. Gunderson, P. R., & Ferrari, J. R. (2008). Forgiveness of sexual cheating in romantic relationships: Effects of discovery method, frequency of offense, and presence of apology. *North American Journal of Psychology, 10,* 1–14.

150. Jang, S. A., Smith, S. W., & Levine, T. R. (2002). To stay or to leave? The role of attachment styles in communication patterns and potential termination of romantic relationships following discovery of deception. *Communication Monographs, 69,* 236.

151. Lee, T. (2016, March 17). Jillian Harri's advice for new "Bachelorette" JoJo Fletcher: "It sounds cliché, but be yourself." *Hello!* Retrieved from http://us.hellomagazine.com/health-and-beauty/12016031712860/jillian-harris-bachelorette-advice-jojo-fletcher-motherhood/. Quote appears in headline.

CHAPTER 8

1. Brown, B. (2015).*Rising strong.*New York: Spiegel & Grau. Quotes appear on p. 31.

2. Brown, B. (2015). *Rising strong.*New York: Spiegel & Grau. Quotes appear on p. 31.

3. Brown, B. (2015). *Rising strong.* New York: Spiegel & Grau. Quote appears on p. 32.

4. Sillars, A. L. (2009). Interpersonal conflict. In C. Berger, M. Roloff, & D. R. Roskos-Ewoldsen (Eds.), *Handbook of communication science* (2nd ed., pp. 273–289). Thousand Oaks, CA: Sage.

5. Brown, B. (2015). *Rising strong.*New York: Spiegel & Grau. Quote appears on p. 33.

6. Cissna, K. N. L., &Seiburg, E. (1995).Patterns of interactional confirmation and disconfirmation. In J. Stewart (Ed.), *Bridges not walls: A book about interpersonal communication* (6th ed., pp. 237–246). New York: McGraw-Hill.

7. Brown, B. (2015). *Rising strong.* New York: Spiegel & Grau. Quote appears on p. 32.

8. Cissna, K. N. L., & Seiburg, E. (1995). Patterns of interactional confirmation and disconfirmation. In J. Stewart (Ed.), *Bridges not walls: A book about interpersonal communication* (6th ed., pp. 237–246). New York: McGraw-Hill.

9. Brown, B. (2015). *Rising strong.* New York: Spiegel & Grau. Quote appears on p. 32.

10. De Vries, R. E., Bakker-Pieper, A., & Oostenveld, W. (2010). Leadership = communication? The relations of leaders' communication styles with leadership styles, knowledge sharing and leadership outcomes. *Journal of Business & Psychology, 25,* 367–380.

11. Singh, R., & Simons, J. J. P. (2010). Attitudes and attraction: Optimism and weight as explanations for the similarity-dissimilarity asymmetry. *Social and Personality Psychology Compass, 12,* 1206–1219.

12. Imai, T., & Vangelisti, A. L. (2011). *The influence of plans to marry in dating couples on relationship quality, confirmation, and desire for evaluation.* Presented at the annual meeting of the International Communication Association, Boston, MA.

13. Markman, H. J., Rhoades, G. K., Stanley, S. M., & Ragan, E. P. (2010). The premarital communication roots of marital distress and divorce: The first five years of marriage. *Journal of Family Psychology, 24,* 289–298.

14. Spott, J., Pyle, C., & Punyanunt-Carter, N. (2010). Positive and negative nonverbal behaviors in relationships: A study of relationship satisfaction and longevity. *Human Communication, 13,* 29–41.

15. Burns, M. E., & Pearson, J. C. (2011). An exploration of family communication environment, everyday talk, and family satisfaction. *Communication Studies, 62*(2), 171–185.

16. Ellis, K. (2004). The impact of perceived teacher confirmation on receiver apprehension, motivation, and learning. *Communication Education, 53,* 1–20.

17. Dailey, R. M. (2008). Assessing the contribution of nonverbal behaviors in displays of confirmation during parent-adolescent interactions: An actor-partner interdependence model. *Journal of Family Communication, 8,* 62–91.

18. For a discussion of reactions to disconfirming responses, see Vangelisti, A. L., & Crumley, L. P. (1998). Reactions to messages that hurt: The influence of relational contexts. *Communication Monographs, 64,* 173–196. See also Cortina, L. M., Magley, V. J., Williams, J. H., & Langhout, R. D. (2001). Incivility in the workplace: Incidence and impact. *Journal of Occupational Health Psychology, 6,* 64–80.

19. Turkle, S. (2007, May 7). Can you hear me now? *Forbes.* Retrieved from http://www.forbes.com/free_forbes/2007/0507/176.html.

20. Gottman, J. M., & Levenson, R. W. (2002). A two-factor model for predicting when a couple will divorce: Exploratory analyses using 14-year longitudinal data. *Family Process, 41*(1), 83–96; Gottman, J. M., Coan, J., Carrere, S., & Swanson, C. (1998). Predicting marital happiness and stability from newlywed interactions. *Journal of Marriage and the Family, 60*(1), 5–22. Retrieved from http://www.jstor.org/pss/353438; Carrere, S., Buehlman, K. T., Gottman, J. M., Coan, J. A., & Ruckstuhl, L. (2000). Predicting

marital stability and divorce in newlywed couples. *Journal of Family Psychology, 14*(1), 42–58; Gottman, J. M. (1991). Predicting the longitudinal course of marriages. *Journal of Marital and Family Therapy, 17*(1), 3–7; Gottman, J. M., & Krokoff, L. J. (1989). The relationship between marital interaction and marital satisfaction: A longitudinal view. *Journal of Consulting and Clinical Psychology, 57,* 47–52; Carrere, S., & Gottman, J. M. (1999). Predicting divorce among newlyweds from the first three minutes of a marital conflict discussion.*Family Process, 38*(3), 293–301.

21. Gottman, J. (1994). *Why marriages succeed or fail: And how you can make yours last.* New York: Simon & Schuster.

22. Elium, D. (n.d.). What is the difference between a complaint and a criticism? Retrieved from http://www.donelium.com/complaints.html#.V46d0ZMrKCd.

23. Gottman, J. M. (2009). *The marriage clinic.* New York: Norton.

24. See Wilmot, W. W. (1987). *Dyadic communication.* New York: Random House, pp. 149–158; and Andersson, L. M., & Pearson, C. M. (1999). Tit for tat? The spiraling effect of incivility in the workplace. *Academy of Management Review, 24,* 452–471. See also Olson, L. N., & Braithwaite, D. O. (2004). "If you hit me again, I'll hit you back": Conflict management strategies of individuals experiencing aggression during conflicts. *Communication Studies, 55,* 271–286.

25. Harper, M. S., & Welsh, D. P. (2007). Keeping quiet: Self-silencing and its association with relational and individual functioning among adolescent romantic couples. *Journal of Social & Personal Relationships, 24,* 99–116.

26. Peterson, K. M., & Smith, D. A. (2010). To what does perceived criticism refer? Constructive, destructive, and general criticism. *Journal of Family Psychology, 24,* 97–100.

27. Bates, C. E., & Samp, J. A. (2011). Examining the effects of planning and empathic accuracy on communication in relational and nonrelational conflict interactions. *Communication Studies, 62,* 207–223.

28. Wilmot, W. W., & Hocker, J. L. (2007). *Interpersonal conflict* (7th ed.). New York: McGraw-Hill, pp. 21–22.

29. Wilmot, W. W., & Hocker, J. L. (2007). *Interpersonal conflict* (7th ed.). New York: McGraw-Hill, pp. 23–24.

30. Brown, B. (2015). *Rising strong.* New York: Spiegel & Grau. Quote appears on p. 33.

31. Blake, R. R., & Mouton, J. S. (1964). *The managerial grid.* Houston, TX: Gulf.

32. Kilmann, R. H., & Thomas, K. W. (1977). Developing a forced-choice measure of conflict-handling behavior. The "MODE" instrument. *Educational and Psychological Measurement, 37,* 309–325.

33. Gibb, J. (1961). Defensive communication. *Journal of Communication,11,* 141–148. See also Eadie, W. F. (1982). Defensive communication revisited: A critical examination of Gibb's theory. *Southern Speech Communication Journal, 47,* 163–177.

34. Lannutti, P. J., & Monahan, J. I. (2004). "Not now, maybe later": The influence of relationship type, request persistence, and alcohol consumption on women's refusal strategies. *Communication Studies, 55,* 362–377.

35. McNulty, J. K., & Russell, V. (2010). When "negative" behaviors are positive: A contextual analysis of the long-term effects of problem-solving behaviors on changes in relationship satisfaction. *Journal of Personality & Social Psychology, 98,* 587–604.

36. See Kellermann, K., & Shea, B. C. (1996). Threats, suggestions, hints, and promises: Gaining compliance efficiently and politely. *Communication Quarterly, 44,* 145–465.

37. Curl, T. S., & Drew, P. (2008). Contingency and action: A comparison of two forms of requesting. *Research on Language & Social Interaction, 41,* 129–153.

38. Bach, G. R., & Goldberg, H. (1974). *Creative aggression.* Garden City, NY: Doubleday.

39. Meyer, J. R. (2004). Effect of verbal aggressiveness on the perceived importance of secondary goals in messages.*Communication Studies, 55,* 168–184; New Mexico Commission on the Status of Women. (2002). *Dealing with sexual harassment.* Retrieved from http://www.womenscommission.state.nm.us/Publications/sexhbrochre.pdf.

40. Adapted from Adler, R. B., &Elmhorst, J. M. (2010). *Communicating at work: Principles and practices for business and the professions* (10th ed., pp. 118–119). New York: McGraw-Hill.

41. For information on filing a formal complaint, see http://www.eeoc.gov/laws/types/sexual_harassment.cfm.

42. Information in this paragraph is from Rose, A. J., & Rudolph, K. D. (2006). A review of sex differences in peer relationship processes: Potential trade-offs for the emotional and behavioral development of girls and boys. *Psychological Bulletin, 132,* 98–131.

43. Holmstrom, A. J. (2009). Sex and gender similarities and differences in communication values in same-sex and cross-sex friendships. *Communication Quarterly, 57,* 224–238.

44. Baillargeon, R. H., Zoccolillo, M., Keenan, K., Côté, S., Pérusse, D., Wu, H.-X., . . . Tremblay, R. E. (2007). Gender differences in physical aggression: A prospective population-based survey of children before and after 2 years of age. *Developmental Psychology, 43,* 13–26.

45. Daly, M., & Wilson, M. (1983).*Sex, evolution, and behavior* (2nd ed.). Belmont, CA: Wadsworth.

46. Joseph, R. (2000). The evolution of sex differences in language, sexuality, and visual-spatial skills. *Archives of Sexual Behavior, 29,* 35–66.

47. Root, A., & Rubin, K. H. (2010). Gender and parents' reactions to children's emotion during the preschool years. *New Directions for Child & Adolescent Development, 128,* 51–64.

48. Niederle, M., & Versterlund, L. (2007). Do women shy away from competition? Do men complete too much? *Quarterly Journal of Economics, 122,* 1067–1101.

49. The information in this paragraph is drawn from research summarized by Wood, J. T. (2005). *Gendered lives* (6th ed.). Belmont, CA: Wadsworth.

50. Wood (2005).

51. Kapidzic, S., & Herring, S. C. (2011). Gender, communication, and self-presentation in teen chatrooms revisited: Have patterns changed? *Journal of Computer-Mediated Communication, 17,* 39–59.

52. Carothers, B. J., & Reis, H. T. (2013). Men and women are from Earth: Examining the latent structure of gender. *Journal of Personality and Social Psychology, 104,* 385–407.

53. Tan, R., Overall, N. C., & Taylor, J. K. (2012). Let's talk about us: Attachment, relationship-focused disclosure, and relationship quality. *Personal Relationships, 19,* 521–534.

54. Niemiec, E. (2010, September 27). Emotions and Italians. Retrieved from http://www.lifeinitaly.com/italian/emotions.

55. The following research is summarized in Tannen, D. (1990). *You just don't understand: Women and men in conversation.* New York: William Morrow, p. 160.

56. Modern Family Script VO, Season 4. (n.d.). HypnoseriesTV. Retrieved from https://www.hypnoseries.tv/modern-family/episodes/saison-4/404---the-butler-s-escape/script-vo-404.192.1220/. Quote appears in third to last script block.

57. Hammer, M. R. (2009). Solving problems and resolving conflict using the Intercultural Style Model and Inventory. In M. A. Moodian (Ed.), *Contemporary leadership and intercultural competence* (pp. 219–232). Thousand Oaks, CA: Sage.

58. Ting-Toomey, S., Yee-Jung, K. K., Shapiro, R. B., Garcia, W., & Wright, T. (1994, November). *Ethnic identity salience and conflict styles in four ethnic groups: African Americans, Asian Americans, European Americans, and Latino Americans.* Paper presented at the annual conference of the Speech Communication Association, New Orleans.

59. Hammer, M. R. (2009). Solving problems and resolving conflict using the Intercultural Style Model and Inventory. In M. A. Moodian (Ed.), *Contemporary leadership and intercultural competence* (pp. 219–232). Thousand Oaks, CA: Sage.

60. Kim-Jo, T., Benet-Martinez, V., & Ozer, D. J. (2010). Culture and interpersonal conflict resolution styles: Role of acculturation. *Journal of Cross-Cultural Psychology, 41*, 264–269.

61. Information in this paragraph is from Hammer (2009).

62. Hammer (2009).

63. Hammer (2009).

64. University of Colorado Conflict Research Consortium. (1998). *Shuttle diplomacy/mediated communication.* Boulder, CO: International Online Training Program on Intractable Conflict. Retrieved from http://www.colorado.edu/conflict/peace/treatment/shuttle.htm.

65. Ellis, D. G. (2010). Online deliberative discourse and conflict resolution. *Landscapes of Violence, 1*(1), article 6. Retrieved from http://scholarworks.umass.edu/lov/vol1/iss1/6.

66. Aakhus, M., & Rumsey, E. (2010). Crafting supportive communication online: A communication design analysis of conflict in an online support group. *Journal of Applied Communication Research, 38*, 65–84.

67. Filley, A. C. (1975). *Interpersonal conflict resolution.* Glenview, IL: Scott Foresman, p. 23.

68. For a brief discussion of constructive problem solving, see Gallo, A. (2009, May 12). The right way to fight. *Harvard Business Review.* Retrieved from http://blogs.hbr.org/hmu/2010/05/the-right-way-to-fight.html.

69. Brown, B. (2015). *Rising strong.* New York: Spiegel & Grau. Quote appears on p. 56.

70. Brown (2015). Quote appears on p. 56.

71. Brown (2015). Quote appears on p. 34.

72. Brown (2015). Quote appears on p. 35.

73. Brown (2015). Quote appears on p. 37.

CHAPTER 9

1. Stengel, R. (2009). *Mandela's way: Lessons on life, love, and courage.* New York: Random House.

2. Stengel, R. (2008, July 9). Mandela: His 8 lessons of leadership. *Time.* Retrieved from http://www.time.com/time/magazine/article/0,9171,1821659-1,00.html.

3. Nelson Mandela interview with Morgan Freeman. (2006). Audio transcripts. Retrieved from https://www.nelsonmandela.org/images/uploads/6.Audio_.pdf. Quote appears in first excerpt.

4. Information in this paragraph is from Marquard, L. (1969). *The people and policies of South Africa* (4th ed.). New York: Oxford University Press.

5. Limb, P. (2008). *Nelson Mandela: A biography.* Westport, CT: Greenwood Press.

6. Sampson, A. (1999). *Mandela: The authorized biography.* New York: Knopf.

7. See Stengel (2008).

8. Marby, E. A. (1999). The systems metaphor in group communication. In L. R. Frey (Ed.), *Handbook of group communication theory and research* (pp. 71–91). Thousand Oaks, CA: Sage.

9. Krakauer, J. (1997). *Into thin air.* New York: Anchor, pp. 212–213.

10. Rothwell, J. D. (2004). *In mixed company: Small group communication* (5th ed.). Belmont, CA: Wadsworth, pp. 29–31.

11. Is your team too big? Too small? What's the right number? (2006, June 14). Knowledge@Wharton. Retrieved from Knowledge@Wharton, http://knowledge.wharton.upenn.edu/article.cfm?articleid=1501.

12. Lowry, P., Roberts, T. L., Romano, N. C., Jr., Cheney, P. D., & Hightower, R. T. (2006).The impact of group size and social presence on small-group communication. *Small Group Research, 37*, 631–661.

13. Hackman, J. (1987). The design of work teams. In J. Lorsch (Ed.), *Handbook of organizational behavior* (pp. 315–342). Englewood Cliffs, NJ: Prentice Hall.

14. LaFasto, F., & Carson, C. (2001). When teams work best: 6,000 team members and leaders tell what it takes to succeed. Thousand Oaks, CA: Sage; Larson, C. E., & LaFasto, F. M. J. (1989). *Teamwork: What must go right, what can go wrong.* Thousand Oaks, CA: Sage.

15. Kirschner, F., Paas, F., & Kirschner, P. A. (2010). Superiority of collaborative learning with complex tasks: A research note on alternative affective explanation. *Computers in Human Behavior, 27*, 53–57.

16. Stepper, J. (2015). *Working out loud: For a better career and life.* Farnborough, NH: Ikigai Press.

17. Stepper, J. (n.d.) About me. Working Out Loud. Retrieved from http://workingoutloud.com/about/.

18. Anderson, D. M., & Haddad, C. J. (2005). Gender, voice and learning in online course environments. *Journal of Asynchronous Learning Networks, 9*, 3–14.

19. Schaefer, R. A. B., & Erskine, L. (2012). Virtual team meetings: Reflections on a class exercise exploring technology choice. *Journal of Management Education, 36*, 777–801.

20. Nunamaker, J. F., Jr., Reinig, B. A., & Briggs, R. O. (2009).Principles for effective virtual teamwork. *Communications of the ACM, 52*(4), 113–117.

21. Capdeferro, N., & Romero, M. (2012). Are online learners frustrated with collaborative learning experiences? *International Review of Research in Open & Distance Learning, 13*, 26–44.

22. Alge, B. J., Wiethoff, C., & Klein, H. J. (2003). When does the medium matter? Knowledge-building experiences and opportunities in decision-making teams. *Organizational Behavior and Human Decision Processes, 91*, 26–37. See also Hobman, E. V., Bordia, P., Irmer, B., & Chang, A. (2002). The expression of conflict in computer-mediated and face-to-face groups. *Small Group Research, 33*, 439–465.

23. Knopf, A. S. (1999). Book excerpt. *Mandela: The authorized biography. The New York Times.* Retrieved from https://www.nytimes.com/books/first/s/sampson-mandela.html. Quote appears in paragraph 20.

24. Simms, A., & Nichols, T. (2014). Social loafing: A review of the literature. *Journal of Management Policy & Practice, 15*(1), 58–67.

25. Wagner, R., & Harter, J. K. (2006). When there's a freeloader on your team. Excerpt from *The elements of great managing.* Washington, DC: Gallup Press. Retrieved from http://www.stybelpeabody.com/newsite/pdf/Freeloader_on_Your_Team.pdf.

26. Paknad, D. (n.d.). The 5 dynamics of low performing teams. Don't let freeloaders or fear undermine your team. *Workboard.* Retrieved from http://www.workboard.com/blog/dynamics-of-low-performing-teams.php.

27. Tuckman, B. (1965). Developmental sequence in small groups. *Psychological Bulletin, 63*, 384–399.

28. Gouran, D. S., Hirokawa, R. Y., Julian, K. M., & Leatham, G. B. (1992). The evolution and current status of the functional perspective on communication in decision-making and problem-solving groups. In S. A. Deetz (Ed.), *Communication yearbook 16* (pp. 573–600). Newbury Park, CA: Sage. See also Wittenbaum, G. M., Hollingshead, A. B., Paulus, P. B., Hirokawa, R. Y., Ancona, D. G., Peterson, R. S., . . . Yoon, K. (2004). The functional perspective as a lens for understanding groups. *Small Group Research, 35*, 17–43.

29. Mayer, M. E. (1998). Behaviors leading to more effective decisions in small groups embedded in organizations. *Communication Reports, 11*, 123–132.

30. Bales, R. F., & Strodbeck, P. L. (1951). Phases in group problem solving. *Journal of Abnormal and Social Psychology, 46*, 485–495.

31. Postman, N. (1976). *Crazy talk, stupid talk.* New York: Dell.

32. Collins, J. (2001). *Good to great: Why some companies make the leap . . . and others don't.* New York: HarperCollins.

33. Aristotle. (1958). *Politics.* New York: Oxford University Press, Book 7.

34. Crwys-Williams, J. (Ed.). (2010). *In the words of Nelson Mandela.* New York: Walker. Quote appears on p. 62.

35. Van Wart, M. (2013). Lessons from leadership theory and the contemporary challenges of leaders. *Public Administration Review, 73*(4), 553–565.

36. Van Wart, M. (2013). Lessons from leadership theory and the contemporary challenges of leaders. *Public Administration Review, 73*(4), 553–565.

37. Greenleaf, R. K., & Spears, L. C. (2002). *Servant leadership: A journey into the nature of legitimate power and greatness.* New York: Paulist Press.

38. Alonderiene, R., & Majauskaite, M. (2016). Leadership style and job satisfaction in higher education institutions. *International Journal of Educational Management, 30*(1), 140–164.

39. Bhatti, N., Maitlo, G. M., Shaikh, N., Hashmi, M. A., & Shaikh, F. M. (2012). The impact of autocratic and democratic leadership style on job satisfaction. *International Business Research, 5*(2), 192–201.

40. Skogstad, A., Hetland, J., Glasø, L., & Einarsen, S. (2014). Is avoidant leadership a root cause of subordinate stress? Longitudinal relationships between laissez-faire leadership and role ambiguity. *Work & Stress, 28*(4), 323–341.

41. Asplund, J., & Blacksmith, N. (2012, March 6). Strengths-based goal setting. *Gallup Business Journal.* Retrieved from http://www.gallup.com/businessjournal/152981/strengths-based-goal-setting.aspx.

42. Tischler, L., Giambatista, R., McKeage, R., & McCormick, D. (2016). Servant leadership and its relationships with core self-evaluation and job satisfaction. *The Journal of Values-Based Leadership, 9*(1), 1–20. Retrieved from http://scholar.valpo.edu/cgi/viewcontent.cgi?article=1148&context=jvbl.

43. Bradberry, T. (2015, October 15). 7 things that make great bosses unforgettable. *Forbes.* Retrieved from http://www.forbes.com/sites/travisbradberry/2015/10/15/7-things-that-make-great-bosses-unforgettable/#4dcc029853b7. Quote appears in paragraph 7.

44. Blake, R. R., & McCanse, A. A. (1991). *Leadership dilemmas grid solutions.* Houston, TX: Gulf.

45. Kuhnert, K. W., & Lewis, P. (1987). Transactional and transformational leadership: A constructive/developmental analysis. *Academy of Management Review, 12*, 648–657.

46. Boies, K., Fiset, J., & Gill, H. (2015). Communication and trust are key: Unlocking the relationship between leadership and team performance and creativity. *The Leadership Quarterly, 26*, 1080–1094.

47. Hersey, P., & Blanchard, K. (2001). *Management of organizational behavior: Utilizing human resources* (8th ed.). Upper Saddle River, NJ: Prentice Hall.

48. Amin, M., Tatlah, I. A., & Khan, A. M. (2013). Which leadership style to use? An investigation of conducive and non-conducive leadership style(s) to faculty job satisfaction. *International Research Journal of Art & Humanities, 41*, 229–253.

49. Hersey, P., & Blanchard, K. (2001). *Management of organizational behavior: Utilizing human resources* (8th ed.). Upper Saddle River, NJ: Prentice Hall.

50. Senge, P. M. (2006). *The fifth discipline: The art and practice of the learning organization.* New York: Doubleday/Currency.

51. Weinstein, B. (2007, August 17). How to handle an off-the-wall boss. *CIO.* Retrieved from http://www.cio.com/article/2438178/staff-management/how-to-deal-with-bully-bosses.html. Quotes appear in last two paragraphs on p. 1.

52. Kuhnert, K. W., & Lewis, P. (1987). Transactional and transformational leadership: A constructive/developmental analysis. *Academy of Management Review, 12*, 648–657.

53. Bass, B. M. (1990). From transactional to transformational leadership: Learning to share the vision. *Organizational Dynamics, 3*, 19–31.

54. Pierro, A., Raven, B. H., Amato, C., & Bélanger, J. J. (2013). Bases of social power, leadership styles, and organizational commitment. *International Journal of Psychology, 48*(6), 1122–1134.

55. Mandela, N. (1994). *The long walk to freedom.* London: Little, Brown. Quote appears on p. 617.

56. Mandela, N. (1990, February 11.) Remarks by Nelson Mandela in Cape Town on 11 [sic] February 11, 1990 after his release from Victor Verster. Nelson Mandela Centre of Memory. Retrieved from https://www.nelsonmandela.org/omalley/index.php/site/q/03lv03445/04lv04015/05lv04154/06lv04191.htm. Quote appears in paragraph 3.

57. Hackman, M. Z., & Johnson, C. E. (2004). *Leadership: A communication perspective.* Long Grove, IL: Waveland. See also Anderson, C., & Kilduff, G. J. (2009). Why do dominant personalities attain influence in face-to-face groups? The competence-signaling effects of trait dominance. *Journal of Personality and Social Psychology, 96*, 491–503.

58. See Bormann (1990). For a succinct description of Bormann's findings, see Rothwell (2004), p. 165.

59. Claros Group. (2010). Leaving a job professionally: Wrapping up your current position before moving on. Retrieved from http://www.clarosgroup.com/leavingjob.pdf.

60. Kelley, R. E. (2008). Rethinking followership. In R. E. Riggio, I. Chaleff, & J. Lipman-Blumen (Eds.), *The art of followership: How great followers create great leaders and organizations* (pp. 5–16). San Francisco: Jossey-Bass; Ibid., p. 8.

61. Agho, A. O. (2009). Perspectives of senior-level executives on effective followership and leadership. *Journal of Leadership & Organizational Studies, 16*, 159–166.

62. The long walk of Nelson Mandela. The prisoner. (n.d.) *Frontline.* PBS. Retrieved from http://www.pbs.org/wgbh/pages/frontline/shows/mandela/prison/.

63. Kellerman, B. (2008). *Followership: How followers are creating change and changing leaders.* Boston, MA: Harvard Business Press.

64. Kelley, R. E. (1992). *The power of followership.* New York: Doubleday Business.

65. Ibid., p. 179.

66. The following types of power are based on the categories developed by French, J. R., & Raven, B. (1968). The basis of social power. In D. Cartright & A. Zander (Eds.), *Group dynamics* (pp. 259–269). New York: Harper & Row, p. 265.

67. Rothwell, J. D. (2004). *In mixed company: Small group communication* (5th ed.). Belmont, CA: Wadsworth, pp. 247–282.

68. See Limb (2008).

69. Nelson Mandela: In his own words. (2013, December 6). *The Telegraph.* Retrieved from http://www.telegraph.co.uk/news/worldnews/nelson-mandela/9734032/Nelson-Mandela-in-his-own-words.html. Quote appears in the last paragraph.

CHAPTER 10

1. Automattic. (2015). Press. Retrieved from https://automattic.com/press/.

2. Robles, M. M. (2012). Executive perceptions of the top 10 soft skills needed in today's workplace. *Business Communication Quarterly, 75*, 453–465.

3. Lutgen-Sandvik, P., Riforgiate, S., & Fletcher, C. (2011). Work as a source of positive emotional experiences and the discourses informing positive assessment. *Western Journal of Communication, 75*, 2–27.

4. See, for example, Pavitt, C. (2003). Do interacting groups perform better than aggregates of individuals? *Human Communication Research, 29,* 592–599; Wittenbaum, G. M. (2004). Putting communication into the study of group memory. *Human Communication Research, 29,* 616–623; and Frank, M. G., Feely, T. H., Paolantonio, N., & Servoss, T. J. (2004). Individual and small group accuracy in judging truthful and deceptive communication. *Group Decision and Negotiation, 13,* 45–54.

5. Rae-Dupree, J. (2008, December 7). Innovation is a team sport. *The New York Times.* Retrieved from http://www.nytimes.com/2008/12/07/business/worldbusiness/07iht-innovate.1.18456109.html?_r=0.

6. Aritz, J., & Walker, R. C. (2009). Group composition and communication styles: An analysis of multicultural teams in decision-making meetings. *Journal of Intercultural Communication Research, 38*(2), 99–114.

7. Stier, J., & Kjellin, M. (2010). Communicative challenges in multinational project work: Obstacles and tools for reaching common understandings. *Journal of Intercultural Communication, 24,* 1–12.

8. Tadmor, C. T., Satterstrom, P., Jang, S., & Polzer, J. T. (2012). Beyond individual creativity: The superadditive benefits of multicultural experience for collective creativity in culturally diverse teams. *Journal of Cross-Cultural Psychology, 43,* 384–392.

9. Information drawn from García, M., & Cañado, M. (2011). Multicultural teamwork as a source of experiential learning and intercultural development. *Journal of English Studies, 9,* 145–163; and van Knippenberg, D., van Ginkel, W. P., & Homan, A. C. (2013, July). Diversity mindsets and the performance of diverse teams. *Organizational Behavior and Human Decision Processes, 121,* 183–193.

10. van Knippenberg, van Ginkel, & Homan (2013).

11. The Crommunist Manifesto. (2011, July 12). Diversity makes us smarter [Blog post]. Retrieved from http://freethoughtblogs.com/crommunist/2011/07/12/diversity-makes-us-smarter/.

12. Adler, R. B., & Elmhorst, J. M. (2010). *Communicating at work: Principles and practices for business and the professions* (10th ed.). New York: McGraw-Hill, pp. 278–279.

13. Krista. (2013, September 17). *The year without pants: An interview with author Scott Berkun.* Hot off the Press. Retrieved from https://en.blog.wordpress.com/2013/09/17/scott-berkun-interview/. Quote appears in the third to last paragraph.

14. Hülsheger, U. R., Anderson, N., & Salgado, J. F. (2009). Team-level predictors of innovation at work: A comprehensive meta-analysis spanning three decades of research. *Journal of Applied Psychology, 94,* 1128–1145.

15. Clark, D. (2012, May 23). How to deal with difficult co-workers. *Forbes.* Retrieved from http://www.forbes.com/sites/dorieclark/2012/05/23/how-to-deal-with-difficult-co-workers/#6c21476a191d.

16. Dugan, D. (n.d.). Co-workers from hell: Dealing with difficult colleagues. Salary.com. Retrieved from http://www.salary.com/co-workers-from-hell-dealing-with-difficult-colleagues/.

17. Fisher, B. A. (1970). Decision emergence: Phases in group decision making. *Speech Monographs, 37,* 53–66.

18. Bohm, D. (1996). *On dialogue* (L. Nichol, Ed.). London: Routledge & Kegan Paul.

19. Isaacs, W. (1999). *Dialogue: The art of thinking together.* New York: Currency.

20. Kraut, R. E., Gergle, D., & Fussell, S. R. (2002). The use of visual information in shared visual spaces: Informing the development of virtual co-presence. *Carnegie Mellon University Research Showcase.*

21. Rico, R., Alcover, C., Sánchez-Manzanares, M., & Gil, F. (2009). The joint relationships of communication behaviors and task interdependence on trust building and change in virtual project teams. *Social Science Information, 48,* 229–255.

22. Altschuller, S., & Benbunan-Fich, R. (2010). Trust, performance, and the communication process in ad hoc decision-making virtual teams. *Journal of Computer-Mediated Communication, 16,* 27–47.

23. Mullenweg, M. (2014, April). The CEO of Automattic on holding "auditions" to build a strong team. *Harvard Business Review.* Retrieved from https://hbr.org/2014/04/the-ceo-of-automattic-on-holding-auditions-to-build-a-strong-team. Quote appears in paragraph 2 of the "When 9 to 5 Fails" section.

24. Snow. S. (2014, September 11). How Matt's machine works. *Fast Company.* Retrieved from http://www.fastcompany.com/3035463/how-matts-machine-works. Quotes appear in paragraph 7 of the "A Secret Sauce" section.

25. Berry, G. R. (2011). Enhancing effectiveness on virtual teams. *Journal of Business Communication, 48,* 186–206.

26. Altschuller, S., & Benbunan-Fich, R. (2010). Trust, performance, and the communication process in ad hoc decision-making virtual teams. *Journal of Computer-Mediated Communication, 16,* 27–47.

27. Johnson, G. M. (2006). Synchronous and asynchronous text-based CMC in educational contexts: A review of recent research. *Tech Trends, 50,* 46–53.

28. Lin, C., Standing, C., & Liu, Y. (2008). A model to develop effective virtual teams. *Decision Support Systems, 45,* 1031–1045.

29. Baltes, B. B., Dickson, M. W., Sherman, M. P., Bauer, C. C., & LaGanke, J. S. (2002). Computer-mediated communication and group decision making: A meta-analysis. *Organizational Behavior and Human Decision Processes,* 156–179.

30. Nowak, K., Watt, J. H., & Walther, J. B. (2009). Computer mediated teamwork and the efficiency framework: Exploring the influence of synchrony and cues on media satisfaction and outcome success. *Computers in Human Behavior, 25,* 1108–1119.

31. Steinel, W., Van Kleef, G. A., & Harinck, F. (2008). Are you talking to *me*?! Separating the people from the problem when expressing emotions in negotiation. *Journal of Experimental Social Psychology, 44,* 362–369.

32. Shipman, A. S., & Mumford, M. D. (2011). When confidence is detrimental: Influence of overconfidence on leadership effectiveness. *The Leadership Quarterly, 22,* 649–665.

33. Dewey, J. (1910). *How we think.* New York: Heath.

34. Poole, M. S. (1991). Procedures for managing meetings: Social and technological innovation. In R. A. Swanson & B. O. Knapp (Eds.), *Innovative meeting management* (pp. 53–109). Austin, TX: 3M Meeting Management Institute. See also Poole, M. S., & Holmes, M. E. (1995). Decision development in computer-assisted group decision making. *Human Communication Research, 22,* 90–127.

35. Lewin, K. (1951). *Field theory in social science.* New York: Harper & Row, pp. 30–59.

36. Osborn, A. (1959). *Applied imagination.* New York: Scribner's.

37. Hastle, R. (1983). *Inside the jury.* Cambridge, MA: Harvard University Press.

38. Ming-Yi, W. (2008). Comparing expected leadership styles in Taiwan and the United States: A study of university employees. *China Media Research, 4,* 36–46.

39. Ibid.

40. Dimitratos, P., Petrou, A., Plakoyiannaki, E., & Johnson, J. E. (2011). Strategic decision-making processes in internationalization: Does national culture of the local firm matter? *Journal of World Business, 46,* 194–204.

41. Adapted from Adler, R. B., & Elmhorst, J. M. (2005). *Communicating at work: Principles and practices for business and the professions* (8th ed.). New York: McGraw-Hill, p. 269.

42. Mullenweg, M. (2014, April). The CEO of Automattic on holding "auditions" to build a strong team. *Harvard Business Review.* Retrieved from https://hbr.org/2014/04/the-ceo-of-automattic-on-holding-auditions-to-build-a-strong-team. Quote appears in the last paragraph of the "Tryouts Trump Interviews" section.

43. Mullenweg, M. (2014, April). The CEO of Automattic on holding "auditions" to build a strong team. *Harvard Business Review.* Retrieved from https://hbr.org/2014/04/the-ceo-of-automattic-on-holding-auditions-to-build-a-strong-team. Quote appears in the last paragraph of the "Tryouts Trump Interviews" section.

44. Rothwell, J. D. (2013). *In mixed company* (8th ed.). Boston: Wadsworth-Cengage, pp. 139–142.

45. Carmeli, A., Sheaffer, Z., & Helevi, M. Y. (2009). Does participatory decision-making in top management teams enhance decision effectiveness and firm performance? *Personnel Review, 38,* 696–714.

46. Rains, S. A. (2007). The impact of anonymity on perceptions of source credibility and influence in computer-mediated group communication. *Communication Research, 34,* 100–125.

47. Waller, B. M., Hope, L., Burrowes, N., & Morrison, E. R. (2011). Twelve (not so) angry men: Managing conversational group size increases perceived contribution by decision makers. *Group Processes & Intergroup Relations, 14,* 835–843.

48. Orbach, M., Demko, M., Doyle, J., Waber, B. N., & Pentland, A. (2015). Sensing informal networks in organizations. *American Behavioral Scientist, 59*(4), 508.

49. Janis, I. (1982). *Groupthink: Psychological studies of policy decisions and fiascoes.* Boston: Houghton Mifflin. See also Baron, R. S. (2005). So right it's wrong: Groupthink and the ubiquitous nature of polarized group decision making. In M. P. Zanna (Ed.), *Advances in experimental social psychology* (Vol. 37, pp. 219–253). San Diego, CA: Elsevier Academic Press.

50. Peralta, E., Memmott, M., & Coleman, K. (2012, July 12). Paterno, others slammed in report for failing to protect Sandusky's victims. *National Public Radio.*

51. Bort, J. (2013, November 5). How Automattic grew into a startup worth $1 billion with no email and no office workers. *Business Insider.* Retrieved from http://www.businessinsider.com/automattic-no-email-no-office-workers-2013-11. Quote appears in the last paragraph.

52. Adapted from Rothwell, J. D. (2013). *In mixed company* (8th ed.). Boston: Wadsworth-Cengage, pp. 223–226.

CHAPTER 11

1. McKinnon, J. (2013, January 18). Will Malala's influence stretch to Europe? *Deutsche Welle.* Retrieved from http://www.dw.com/en/will-malalas-influence-stretch-to-europe/a-16532149. Quote appears in paragraph 1.

2. Global Coalition to Protect Education from Attack. (2014). *Education Under Attack 2014.* New York: Author. Retrieved from http://www.protectingeducation.org/sites/default/files/documents/eua_2014_full.pdf.

3. Yousafzai, M. (2013, July 12). The full text: Malala Yousafzai delivers defiant riposte to Taliban militants with speech to the UN General Assembly. *The Independent.* Retrieved from http://www.independent.co.uk/news/world/asia/the-full-text-malala-yousafzai-delivers-defiant-riposte-to-taliban-militants-with-speech-to-the-un-general-assembly-8706606.html.

4. Dwyer, K. K., & Davidson, M. M. (2012, April–June). Is public speaking really more feared than death? *Communication Research Reports, 29,* 99–107. This study found that public speaking was selected more often as a common fear than any other fear, including death. However, when students were asked to select a top fear, students selected death most often.

5. For an example of how demographics have been taken into consideration in great speeches, see Stephens, G. (1997, Fall). Frederick Douglass' multiracial abolitionism: "Antagonistic cooperation" and "redeemable ideals" in the July 5 speech. *Communication Studies, 48,* 175–194. On July 5, 1852, Douglass gave a speech titled "What to the Slave Is the 4th of July," attacking the hypocrisy of Independence Day in a slaveholding republic. It was one of the greatest antislavery speeches ever given, and part of its success stemmed from the way Douglass sought common ground with his multiracial audience.

6. See, for example, Jaschik, S. (2007, June 28). 2 kinds of part-time students. *Inside Higher Education.* Retrieved from https://www.insidehighered.com/news/2007/06/28/parttime.

7. Polisetti, S. (2012). The loss of Native American culture. In L. Schnoor (Ed.), *Winning orations, 2012* (pp. 163–165). Mankato, MN: Interstate Oratorical Association, p. 163. Polisetti was coached by Lee Mayfield and Ken Young.

8. See, for example, Rolfe-Redding, J., Maibach, E. W., Feldman, L., & Leiserowitz, A. (2011, November 7). Republicans and climate change: An audience analysis of predictors for belief and policy preferences (SSRN Scholarly Paper 2026002). Retrieved from http://papers.ssrn.com/sol3/papers.cfm?abstract_id=2026002.

9. Stutman, R. K., & Newell, S. E. (1984, Fall). Beliefs versus values: Silent beliefs in designing a persuasive message. *Western Journal of Speech Communication, 48*(4), 364.

10. Mahoney, B. (2015). They're not a burden: The inhumanity of anti-homeless legislation. *Winning Orations, 2015* (pp. 20–22). Mankato, MN: Interstate Oratorical Association. Brianna was coached by Kellie Roberts.

11. Mahoney (2015).

12. In information science parlance, these are referred to as Boolean terms.

13. Some recent literature specifically refers to public speaking anxiety, or PSA. See, for example, Bodie, G. D. (2010, January). A racing heart, rattling knees, and ruminative thoughts: Defining, explaining, and treating public speaking anxiety. *Communication Education, 59*(1), 70–105.

14. See, for example, Borhis, J., & Allen, M. (1992, January). Meta-analysis of the relationship between communication apprehension and cognitive performance. *Communication Education, 41*(1), 68–76.

15. Daly, J. A., Vangelisti, A. L., & Weber, D. J. (1995, December). Speech anxiety affects how people prepare speeches: A protocol analysis of the preparation process of speakers. *Communication Monographs, 62,* 123–134.

16. Researchers generally agree that communication apprehension has three causes: genetics, social learning, and inadequate skills acquisition. See, for example, Finn, A. N. (2009). Public speaking: What causes some to panic? *Communication Currents, 4*(4), 1–2.

17. See, for example, Sawyer, C. R., & Behnke, R. R. (1997, Summer). Communication apprehension and implicit memories of public speaking state anxiety. *Communication Quarterly, 45*(3), 211–222.

18. Adapted from Ellis, A. (1977). *A new guide to rational living.* North Hollywood, CA: Wilshire Books. G. M. Philips listed a different set of beliefs that he believes contributes to reticence. The beliefs are as follows: (1) an exaggerated sense of self-importance (reticent people tend to see themselves as more important to others than others see them); (2) effective speakers are born, not made; (3) skillful speaking is manipulative; (4) speaking is not that important; (5) I can speak whenever I want to; I just choose not to; (6) it is better to be quiet and let people think you are a fool than prove it by talking (they assume they will be evaluated negatively); and (7) what is wrong with me requires a (quick) cure. See Keaten, J. A., Kelly, L., & Finch, C. (2000). Effectiveness of the Penn State Program in changing beliefs associated with reticence. *Communication Education, 49*(2), 134–145.

19. Behnke, R. R., Sawyer, C. R., & King, P. E. (1987, April). The communication of public speaking anxiety. *Communication Education, 36*, 138–141.

20. Honeycutt, J. M., Choi, C. W., & DeBerry, J. R. (2009, July). Communication apprehension and imagined interactions. *Communication Research Reports, 26*(2), 228–236.

21. Behnke, R. R., & Sawyer, C. R. (1999, April). Milestones of anticipatory public speaking anxiety. *Communication Education, 48*(2), 165.

22. Hinton, J. S., & Kramer, M. W. (1998, April). The impact of self-directed videotape feedback on students' self-reported levels of communication competence and apprehension. *Communication Education, 47*(2), 151–161. Significant increases in competency and decreases in apprehension were found using this method.

23. Research has confirmed that speeches practiced in front of other people tend to be more successful. See, for example, Smith, T. E., & Frymier, A. B. (2006, February). Get "real": Does practicing speeches before an audience improve performance? *Communication Quarterly, 54*, 111–125.

24. See, for example, Rosenfeld, L. R., & Civikly, J. M. (1976). *With words unspoken.* New York: Holt, Rinehart and Winston, p. 62. Also see Chaiken, S. (1979). Communicator physical attractiveness and persuasion. *Journal of Personality and Social Psychology, 37*, 1387–1397.

25. A study demonstrating this stereotype is Street, R. L., Jr., & Brady, R. M. (1982, December). Speech rate acceptance ranges as a function of evaluative domain, listener speech rate, and communication context. *Speech Monographs, 49*, 290–308.

26. See, for example, Mulac, A., & Rudd, M. J. (1977). Effects of selected American regional dialects upon regional audience members. *Communication Monographs, 44*, 184–195. Some research, however, suggests that nonstandard dialects do not have the detrimental effects on listeners that were once believed. See, for example, Johnson, F. L., & Buttny, R. (1982, March). White listeners' responses to "sounding black" and "sounding white": The effect of message content on judgments about language. *Communication Monographs, 49*, 33–39.

27. Smith, V., Siltanen, S. A., & Hosman, L. A. (1998, Fall). The effects of powerful and powerless speech styles and speaker expertise on impression formation and attitude change. *Communication Research Reports, 15*(1), 27–35. In this study, a powerful speech style was defined as one without hedges and hesitations such as *uh* and *anda*.

28. The jurist, Fredrik Heffermehl, was quoted in Walsh, D. (2014, October 10). Two champions of children are given Nobel Peace Prize. *The New York Times.*

29. Quoted in Borger, J., & Imtiaz, S. (2014, October 10). Malala Yousafzai's Nobel Peach http://www.theguardian.com/world/2014/oct/10/malala-yousafzai-nobel-peace-prize-pakistan-reaction. Conservative politicians in Pakistan expressed concern that Malala was being used by those who wanted to promote Western values.

30. In fact, the All Pakistan Private Schools Federation held an "I Am Not Malala" day to protest against the contents of her biography *I Am Malala.* Bacchi, U. (2014, November 10). Pakistani schools hold "I Am Not Malala" day against Nobel Peace Prize winner. *International Business Times.* Retrieved from http://www.ibtimes.co.uk/pakistani-schools-hold-i-am-not-malala-day-against-nobel-peace-prize-winner-1474090.

31. Yousafzai, M. (2015). Nobel lecture. In K. Grandin (Ed.), *The Nobel Prizes 2014.* Sagamore Beach, MA: Science History Publications/USA.

CHAPTER 12

1. Personal correspondence with authors, June 2016.

2. See, for example, Stern, L. (1985). *The structures and strategies of human memory.* Homewood, IL: Dorsey Press. See also Turner, C. (1987, June 15). Organizing information: Principles and practices. *Library Journal, 112*(11), 58.

3. Booth, W. C., Colomb, G. C., & Williams, J. M. (2003). *The craft of research.* Chicago: University of Chicago Press.

4. Koehler, C. (1998, June 15). Mending the body by lending an ear: The healing power of listening. *Vital Speeches of the Day,* 543.

5. Hallum, E. (1998). Untitled. In L. Schnoor (Ed.), *Winning orations, 1998* (pp. 4–6). Mankato, MN: Interstate Oratorical Association, p. 4. Hallum was coached by Clark Olson.

6. Monroe, A. (1935). *Principles and types of speech.* Glenview, IL: Scott, Foresman.

7. Adapted from http://vaughnkohler.com/wp-content/uploads/2013/01/Monroe-Motivated-Sequence-Outline-Handout1.pdf, accessed May 23, 2013.

8. Graham, K. (1976, April 15). The press and its responsibilities. *Vital Speeches of the Day,* 42.

9. Meyer, E. (2015). "Gridlocked." In L. Schnoor (Ed.), *Winning orations, 2015* (pp. 54–56). Mankato, MN: Interstate Oratorical Association. Meyer was coached by Kristofer Kracht and Cadi Kadlecek.

10. Anderson, G. (2009). Don't reject my homoglobin. In L. Schnoor (Ed.), *Winning Orations, 2009* (pp. 33–35). Mankato, MN: Interstate Oratorical Association, p. 33. Anderson was coached by Leah White.

11. Cochran, J. (2012). Untitled. In L. Schnoor (Ed.), *Winning orations, 2012* (pp. 40–42). Mankato, MN: Interstate Oratorical Association, p. 40. Cochran was coached by Dan Smith and Michael Chen.

12. Eggleston, T. S. (n.d.). The key steps to an effective presentation. Retrieved April 4, 2016 from http://seggleston.com/1/writing-and-communications/key-steps.

13. Cook, E. (n.d.). Making business presentations work. Retrieved May 19, 2010 from www.businessknowhow.com/manage/presentation101.htm.

14. Kotelnikov, V. (n.d.). Effective presentations. Retrieved May 19, 2010 from http://www.1000ventures.com/business_guide/crosscuttings/presentations_main.html.

15. Toastmasters International, Inc.'s Communication and Leadership Program Retrieved May 19, 2010 from www.toastmasters.org.

16. Davis, K. C. (2007, July 3). The founding immigrants. *New York Times.* Retrieved from http://www.nytimes.com/2007/07/03/opinion/03davis.html.

17. Dunn, N. (2012). There's nothing special about restraining and secluding students with disabilities. In L. Schnoor (Ed.), *Winning orations, 2012* (pp. 126–128). Mankato, MN: Interstate Oratorical Association, p. 126. Dunn was coached by Jacob Stutzman and Tai Du.

18. Griesinger, C. (2006). Untitled. In L. Schnoor (Ed.), *Winning orations, 2006* (pp. 37–39). Mankato, MN: Interstate Oratorical Association, p. 37. Griesinger was coached by Ray Quiel.

19. Schoch, E. (2014). Revenge porn: An intimate invasion. In L. Schnoor (Ed.), *Winning orations, 2014* (pp. 27–29). Mankato, MN: Interstate Oratorical Association. Schoch was coached by Kellie Roberts.

20. See Schoch (2014).

21. Wideman, S. (2006). Planning for peak oil: Legislation and conservation. In L. Schnoor (Ed.), *Winning orations, 2006* (pp. 7–9). Mankato, MN: Interstate Oratorical Association, p. 7. Wideman was coached by Brendan Kelly.

22. Herbert, B. (2007, April 26). Hooked on violence. *New York Times*. Retrieved from http://www.nytimes.com/2007/04/26/opinion/26herbert.html.

23. Sherriff, D. (1998, April 1). Bill Gates too rich [Online forum comment]. Retrieved from CRTNET discussion group.

24. Elsesser, K. (2010, March 4). And the gender-neutral Oscar goes to . . . *New York Times*. Retrieved from http://www.nytimes.com/2010/03/04/opinion/04elsesser.html.

25. Gieseck, A. (2015). Problem of police brutality with regard to the deaf community. In L. Schnoor (Ed.), *Winning orations, 2015* (pp. 82–84). Mankato, MN: Interstate Oratorical Association, p. 83.

26. Boyle, J. (2015). Spyware stalking. In L. Schnoor (Ed.), *Winning orations, 2015* (pp. 32–34). Mankato, MN: Interstate Oratorical Association, p. 32. Julia was coached by Judy Santacaterina and Lisa Roth.

27. Notebaert, R. C. (1998, November 1). Leveraging diversity: Adding value to the bottom line. *Vital Speeches of the Day*, 47.

28. Teresa Fishman, director of the Center for Academic Integrity at Clemson University, quoted in Gabriel, T. (2010, August 1). Plagiarism lines blur for students in digital age. *New York Times*. Retrieved from http://www.nytimes.com/2010/08/02/education/02cheat.html?pagewanted-all.

29. McCarley, E. (2009). On the importance of drug courts. In L. Schnoor (Ed.), *Winning orations, 2009* (pp. 36–38). Mankato, MN: Interstate Oratorical Association, p. 36.

30. See McCarley (2009).

CHAPTER 13

1. Mission statement from http://www.itgetsbetter.org/pages/about-it-gets-better-project/.

2. Sigler, M. G. (2010, August 4). Eric Schmidt: Every 2 days we create as much information as we did up to 2003. *TechCrunch*. Retrieved from http://techcrunch.com/2010/08/04/schmidt-data/.

3. See, for example, Wurman, R. S. (2000). *Information anxiety 2*. Indianapolis: Que.

4. Buffett, W. E. (2012, November 25). A minimum tax for the wealthy. *New York Times*. Retrieved from http://www.nytimes.com/2012/11/26/opinion/buffett-a-minimum-tax-for-the-wealthy.html.

5. See Fransden, K. D., & Clement, D. A. (1984). The functions of human communication in informing: Communicating and processing information. In C. C. Arnold & J. W. Bowers (Eds.), *Handbook of rhetorical and communication theory* (pp. 338–399). Boston: Allyn and Bacon.

6. Allocca, K. (2001, November). *Why videos go viral*. TEDYouth talk, New York City. Retrieved from http://www.ted.com/talks/kevin_allocca_why_videos_go_viral.html.

7. Wilson, N. (2015). Dousing disaster: The case for new hydrants. In L. Schnoor (Ed.), *Winning orations, 2015* (pp. 26–28). Mankato, MN: Interstate Oratorical Association, p. 26. Nicole was coached by Hope Willoughby and Randy Richardson.

8. Levine, K. (2001). The dentist's dirty little secret. In L. Schnoor (Ed.), *Winning orations, 2001* (pp. 77–79). Mankato, MN: Interstate Oratorical Association, p. 77. Levine was coached by Trischa Goodnow.

9. Cacioppo, J. T., & Petty, R. E. (1979). Effects of message repetition and position on cognitive response, recall, and persuasion. *Journal of Personality and Social Psychology, 37*, 97–109.

10. Schulz, K. (2011, March). *On being wrong*. Speech presented at the TED Conference. Retrieved from http://www.ted.com/talks/kathryn_schulz_on_being_wrong.html.

11. New York Public Interest Research Group, Brooklyn College chapter. (2004). Voter registration project.

12. Parker, I. (2001, May 28). Absolute PowerPoint. *New Yorker*, p. 78.

13. Steven Pinker, a psychology professor at MIT, quoted in Zuckerman, L. (1999, April 17). Words go right to the brain, but can they stir the heart? *New York Times*, p. 9.

14. Tufte, E. (2003, September). PowerPoint is evil: Power corrupts. PowerPoint corrupts absolutely. *Wired*. Retrieved from http://www.wired.com/wired/archive/11.09/ppt2.html.

15. Simons, T. (2004, March). Does PowerPoint make you stupid? *Presentations*, p. 25.

16. Tufte, E. R. (2003). *The cognitive style of PowerPoint*. Cheshire, CT: Graphics Press.

17. See Parker (2001).

18. Don Norman, a design expert cited in Simons (2004), p. 26.

CHAPTER 14

1. Kristina Medero, personal communication with the authors, May 2016.

2. For an explanation of social judgment theory, see Griffin, E. (2012). *A first look at communication theory* (8th ed.). New York: McGraw-Hill.

3. Lasch, C. (1990, Spring). Journalism, publicity and the lost art of argument. *Gannett Center Journal*, 1–11, p. 3.

4. See, for example, Jaska, J. A., & Pritchard, M. S. (1994). *Communication ethics: Methods of analysis* (2nd ed.). Boston: Wadsworth.

5. Some research suggests that audiences may perceive a direct strategy as a threat to their freedom to form their own opinions. This perception hampers persuasion. See Brehm, J. W. (1966). *A theory of psychological reactance*. New York: Academic Press.

6. Yocum, R. (2015). A deadly miscalculation. In L. Schnoor (Ed.), *Winning orations 2015* (pp. 97–99). Mankato, MN: Interstate Oratorical Association. Rebeka was coached by Mark Hickman.

7. For an excellent review of the effects of evidence, see Reinard, J. C. (1988, Fall). The empirical study of persuasive effects of evidence: The status after fifty years of research. *Human Communication Research, 15*(1), 3–59.

8. There are, of course, other classifications of logical fallacies than those presented here. See, for example, Warnick, B., & Inch, E. (1994). *Critical thinking and communication: The use of reason in argument* (2nd ed.). New York: Macmillan, pp. 137–161.

9. Sprague, J., & Stuart, D. (1992). *The speaker's handbook* (3rd ed.). Fort Worth, TX: Harcourt Brace Jovanovich, p. 172.

10. Myers, L. J. (1981). The nature of pluralism and the African American case. *Theory into Practice, 20*, 3–4. Cited in Samovar, L. A., & Porter, R. E. (1995). *Communication between cultures* (2nd ed.). New York: Wadsworth, p. 251.

11. Samovar & Porter (1995), pp. 154–155.

12. For an example of how one politician failed to adapt to his audience's attitudes, see Hostetler, M. J. (1998, Winter). Gov. Al Smith confronts the Catholic question: The rhetorical legacy of the 1928 campaign. *Communication Quarterly, 46*(1), 12–24. Smith was reluctant to discuss religion, attributed bigotry to anyone who brought it up, and was impatient with the whole issue. He lost the election. Many years later, John F. Kennedy dealt with "the Catholic question" more reasonably and won.

13. Mount, A. (1973). Speech before the Southern Baptist Convention. In W. A. Linkugel, R. R. Allen, & R. Johannessen (Eds.), *Contemporary American speeches* (3rd ed., pp. 203–205). Belmont, CA: Wadsworth, p. 204.

14. Rudolf, H. J. (1983, Summer). Robert F. Kennedy at Stellenbosch University. *Communication Quarterly, 31,* 205–211.

15. Preface to Barbara Bush's speech, "Choices and Change," in Peterson, O. (Ed.). (1991). *Representative American speeches, 1990–1991.* New York: H. W. Wilson, p. 162.

16. Bush, B. (1990, June 1). *Choices and change.* Speech presented to the graduating class of Wellesley College in Wellesley, MA, on June 1, 1990. Reprinted in Peterson (1991), p. 166.

17. DeVito, J. A. (1986). *The communication handbook: A dictionary.* New York: Harper & Row, pp. 84–86.

18. Rodriguez, G. (2010). How to develop a winning sales plan. Retrieved from www.powerhomebiz.com/062006/salesplan.htm.

19. Sanfilippo, B. (2010). Winning sales strategies of top performers. Retrieved from www.selfgrowth.com/articles/Sanfilippo2.html.

Glossary

abstraction ladder A range of more- to less-abstract terms describing an event or object. 109

abstract language Language that lacks specificity or does not refer to observable behavior or other sensory data. 110

actuate To move members of an audience toward a specific behavior. 381

addition The articulation error that involves adding extra parts to words. 314

ad hominem **fallacy** A fallacious argument that attacks the integrity of a person to weaken his or her position. 386

advising response Helping response in which the receiver offers suggestions about how the speaker should deal with a problem. 145

affect blend The combination of two or more expressions, each showing a different emotion. 168

affect displays Facial expressions, body movements, and vocal traits that reveal emotional states. 157

affinity The degree to which people like or appreciate one another. As with all relational messages, affinity is usually expressed nonverbally. 198

all-channel network A communication network pattern in which group members are always together and share all information with one another. 252

altruistic lies Deception intended to be unmalicious, or even helpful, to the person to whom it is told. 207

ambushing A style in which the receiver listens carefully to gather information to use in an attack on the speaker. 132

analogy An extended comparison that can be used as supporting material in a speech. 339

analytical listening Listening in which the primary goal is to fully understand the message, prior to any evaluation. 139

analyzing statement A helping style in which the listener offers an interpretation of a speaker's message. 145

anchor The position supported by audience members before a persuasion attempt. 375

androgynous Combining both masculine and feminine traits. 46

anecdote A brief, personal story used to illustrate or support a point in a speech. 339

argumentum ad populum **fallacy** Fallacious reasoning based on the dubious notion that because many people favor an idea, you should, too. 387

argumentum ad verecundiam **fallacy** Fallacious reasoning that tries to support a belief by relying on the testimony of someone who is not an authority on the issue being argued. 387

articulation The process of pronouncing all the necessary parts of a word. 314

assertive communication A style of communicating that directly expresses the sender's needs, thoughts, or feelings, delivered in a way that does not attack the receiver. 225

asynchronous communication Communication that occurs when there's a lag between receiving and responding to messages. 18

attending The process of focusing on certain stimuli from the environment. 127

attitude The predisposition to respond to an idea, person, or thing favorably or unfavorably. 302

attribution The process of attaching meaning. 49

audience analysis A consideration of characteristics, including the type, goals, demographics, beliefs, attitudes, and values of listeners. 301

audience involvement The level of commitment and attention that listeners devote to a speech. 358

audience participation Listener activity during a speech; a technique to increase audience involvement. 359

authoritarian leadership A style in which the designated leader uses coercive and reward power to dictate the group's actions. 257

avoidance spiral A communication spiral in which the parties slowly reduce their dependence on one another, withdraw, and become less invested in the relationship. 222

bar chart A visual aid that compares two or more values by showing them as elongated horizontal rectangles. 363

basic speech structure The division of a speech into introduction, body, and conclusion. 325

behavioral description An account that refers only to observable phenomena. 110

belief An underlying conviction about the truth of an idea, often based on cultural training. 302

brainstorming A method for creatively generating ideas in groups by minimizing criticism and encouraging a large quantity of ideas without regard to their workability or ownership by individual members. 287

breadth (of self-disclosure) The range of topics about which an individual discloses. 201

breakout groups A strategy used when the number of members is too large for effective discussion. Subgroups simultaneously address an issue and then report back to the group at large. 281

cause-effect pattern An organizing plan for a speech that demonstrates how one or more events result in another event or events. 330

chain network A communication network in which information passes sequentially from one member to another. 252

channel The medium through which a message passes from sender to receiver. 7

chronemics The study of how humans use and structure time. 175

citation A brief statement of supporting material in a speech. 341

climax pattern An organizing plan for a speech that builds ideas to the point of maximum interest or tension. 329

coculture The perception of membership in a group that is part of an encompassing culture. 69

coercive power The power to influence others by the threat or imposition of unpleasant consequences. 266

cohesiveness The totality of forces that causes members to feel themselves part of a group and makes them want to remain in that group. 276

collectivistic culture A culture in which members focus on the welfare of the group as a whole, rather than a concern by individuals for their own success. 72

column chart A visual aid that compares two or more values by showing them as elongated vertical rectangles. 363

comforting A response style in which a listener reassures, supports, or distracts the person seeking help. 146

communication The process of creating meaning through symbolic interaction. 5

communication climate The emotional tone of a relationship as it is expressed in the messages that the partners send and receive. 216

communication competence The ability to maintain a relationship on terms acceptable to all parties. 20

compromise An approach to conflict resolution in which both parties attain at least part of what they seek by giving something up. 235

conclusion (of a speech) The final structural unit of a speech, in which the main points are reviewed and final remarks are made to motivate the audience to act or help listeners remember key ideas. 334

confirming messages Actions and words that express respect and value the other person. 218

conflict An expressed struggle between at least two interdependent parties who perceive incompatible goals, scarce rewards, and interference from the other party in achieving their goals. 215

conflict stage When group members openly defend their positions and question those of others. 280

connection power The influence granted by virtue of a member's ability to develop relationships that help the group reach its goal. 266

connotative meanings Informal, implied interpretations for words and phrases that reflect the people, culture, emotions, and situations involved. 97

consensus Agreement among group members about a decision. 289

contempt Verbal and nonverbal messages that ridicule or belittle the other person. 220

content message A message that communicates information about the subject being discussed. 198

control The social need to influence others. 199

convergence Accommodating one's speaking style to another person, usually a person who is desirable or has higher status. 105

conversational narcissists People who focus on themselves and their interests instead of listening to and encouraging others. 133

convincing A speech goal that aims at changing audience members' beliefs, values, or attitudes. 380

coordination Interaction in which participants interact smoothly, with a high degree of satisfaction but without necessarily understanding one another well. 30

counterfeit question A question that is not truly a request for new information. 135

credibility The believability of a speaker or other source of information. 390

critical listening Listening in which the goal is to evaluate the quality or accuracy of the speaker's remarks. 140

criticism A message that is personal, all-encompassing, and accusatory. 220

culture The language, values, beliefs, traditions, and customs people share and learn. 69

cyber relationship An affiliation between people who know each other *only* through the virtual world. 187

database A computerized collection of information that can be searched in a variety of ways to locate information that the user is seeking. 306

debilitative communication apprehension An intense level of anxiety about speaking before an audience, resulting in poor performance. 307

decode To attach meaning to a message. 7

defensive listening A response style in which the receiver perceives a speaker's comments as an attack. 132

defensiveness Protecting oneself by counterattacking the other person. 220

deletion An articulation error that involves leaving off parts of words. 314

democratic leadership A style in which the nominal leader invites the group's participation in decision making. 257

demographics Audience characteristics that can be analyzed statistically, such as age, gender, education, and group membership. 301

denotative meanings Formally recognized definitions for words, as in those found in a dictionary. 97

depth (of self-disclosure) The level of personal information a person reveals on a particular topic. 201

description A type of speech that uses details to create a "word picture" of the essential factors that make that thing what it is. 352

developmental models of relational maintenance Theoretical frameworks based on the idea that communication patterns are different in various stages of interpersonal relationships. 194

diagram A line drawing that shows the most important components of an object. 362

dialect A version of the same language that includes substantially different words and meanings. 97

dialectical model The perspective that people in virtually all interpersonal relationships must deal with equally important, simultaneous, and opposing forces such as connection and autonomy, predictability and novelty, and openness versus privacy. 204

dialogue A process in which people let go of the notion that their ideas are more correct or superior to others' and instead seek to understand an issue from many different perspectives. 282

direct aggression A message that attacks the position and perhaps the dignity of the receiver. 225

direct persuasion Persuasion that does not try to hide or disguise the speaker's persuasive purpose. 381

disconfirming messages Words and actions that express a lack of caring or respect for another person. 219

disfluencies Vocal interruptions such as stammering and use of "uh," "um," and "er." 170

disinhibition The tendency to transmit messages without considering their consequences. 27

divergence A linguistic strategy in which speakers emphasize differences between their communicative style and that of others to create distance. 106

dyad A two-person unit. 11

dyadic communication Two-person communication. 11

dysfunctional roles Individual roles played by group members that inhibit the group's effective operation. 255

either–or fallacy Fallacious reasoning that sets up false alternatives, suggesting that if the inferior one must be rejected, then the other must be accepted. 386

emblems Deliberate nonverbal behaviors with precise meanings, known to virtually all members of a cultural group. 162

emergence stage When a group moves from conflict toward a single solution. 280

emergent leader A member who assumes leadership roles without being appointed by higher-ups. 261

emotional evidence Evidence that arouses the sentiments of an audience. 385

emotive language Language that conveys an attitude rather than simply offering an objective description. 113

empathy The ability to project oneself into another person's point of view, so as to experience the other's thoughts and feelings. 52

encode Put thoughts into symbols, most commonly words. 7

environment Both the physical setting in which communication occurs and the personal perspectives of the parties involved. 8

equivocal words Words that have more than one dictionary definition. 107

equivocation A deliberately vague statement that can be interpreted in more than one way. 114

escalatory spiral A reciprocal pattern of communication in which messages, either confirming or disconfirming, between two or more communicators reinforce one another. 222

ethical persuasion Persuasion in an audience's best interest that does not depend on false or misleading information to induce change in that audience. 377

ethnicity A social construct that refers to the degree to which a person identifies with a particular group, usually on the basis of nationality, culture, religion, or some other unifying perspective. 78

ethnocentrism The attitude that one's own culture is superior to others'. 89

ethos A speaker's credibility or ethical appeal. 393

euphemism A mild or indirect term or expression used in place of a more direct but less pleasant one. 114

evidence Material used to prove a point, such as testimony, statistics, and examples. 385

expert power The ability to influence others by virtue of one's perceived expertise on the subject in question. 265

explanations Speeches or presentations that clarify ideas and concepts already known but not understood by an audience. 252

extemporaneous speech A speech that is planned in advance but presented in a direct, conversational manner. 311

face The socially approved identity that a communicator tries to present. 55

facework Verbal and nonverbal behavior designed to create and maintain a communicator's face and the face of others. 55

facilitative communication apprehension A moderate level of anxiety about speaking before an audience that helps improve the speaker's performance. 307

factual example A true, specific case that is used to demonstrate a general idea. 338

factual statement A statement that can be verified as being true or false. 112

fallacy An error in logic. 386

fallacy of approval The irrational belief that it is vital to win the approval of virtually every person a communicator deals with. 309

fallacy of catastrophic failure The irrational belief that the worst possible outcome will probably occur. 308

fallacy of overgeneralization Irrational beliefs in which (1) conclusions (usually negative) are based on limited evidence or (2) communicators exaggerate their shortcomings. 309

fallacy of perfection The irrational belief that a worthwhile communicator should be able to handle every situation with complete confidence and skill. 308

family A collection of people who share affection and resources and who think of themselves and present themselves as a family. 191

feedback The discernible response of a receiver to a sender's message.   9

flaming Sending angry and/or insulting emails, text messages, and website postings.   27

focus group A procedure used in market research by sponsoring organizations to survey potential users or the public at large regarding a new product or idea.   281

force field analysis A method of problem analysis that identifies the forces contributing to resolution of the problem and the forces that inhibit its resolution.   287

formal outline A consistent format and set of symbols used to identify the structure of ideas.   325

formal role A role assigned to a person by group members or an organization, usually to establish order.   252

forum A discussion format in which audience members are invited to add their comments to those of the official discussants.   281

frame switching Adopting the perspectives of different cultures.   57

gatekeeper Person in a small group through whom communication among other members flows.   252

gender Socially constructed roles, behaviors, activities, and attributes that a society considers appropriate for men and/or women.   46

general purpose One of three basic ways a speaker seeks to affect an audience: to entertain, inform, or persuade.   299

group A small collection of people whose members interact with one another, usually face-to-face, over time in order to reach goals.   246

group goals Goals that a group collectively seeks to accomplish.   249

groupthink A group's collective striving for unanimity that discourages realistic appraisals of alternatives to its chosen decision.   293

haptics The study of touch.   172

hearing The process wherein sound waves strike the eardrum and cause vibrations that are transmitted to the brain.   126

hegemony The dominance of one culture over another.   89

hidden agendas Individual goals that group members are unwilling to reveal.   249

high-context culture A culture that relies heavily on subtle, often nonverbal cues to maintain social harmony.   93

hypothetical example An example that asks an audience to imagine an object or event.   338

identity management Strategies used by communicators to influence the way others view them.   55

illustrators Nonverbal behaviors that accompany and support verbal messages.   162

immediacy The degree of interest and attraction we feel toward and communicate to others. As with all relational messages, immediacy is usually expressed nonverbally.   199

impromptu speech A speech given "off the top of one's head," without preparation.   311

indirect communication Hinting at a message instead of expressing thoughts and feelings directly.   223

indirect persuasion Persuasion that disguises or deemphasizes the speaker's persuasive goal.   381

individual goals Individual motives for joining a group.   249

individualistic culture A culture in which members focus on the value and welfare of individual members, as opposed to a concern for the group as a whole.   72

inferential statement A conclusion arrived at from an interpretation of evidence.   112

informal role A role usually not explicitly recognized by a group that describes functions of group members, rather than their positions. These are sometimes called "functional roles."   252

information anxiety The psychological stress that occurs when dealing with too much information.   350

information hunger Audience desire, created by a speaker, to learn information.   355

information overload The decline in efficiency that occurs when the rate of complexity of material is too great to manage.   350

information underload The decline in efficiency that occurs when there is a shortage of the information necessary to operate effectively.   291

informative purpose statement A complete statement of the objective of a speech, worded to stress audience knowledge and/or ability.   354

in-groups Groups with which we identify.   71

insensitive listening The failure to recognize the thoughts or feelings that are not directly expressed by a speaker, and instead accepting the speaker's words at face value.   133

instructions Remarks that teach something to an audience in a logical, step-by-step manner.   352

insulated listening A style in which the receiver ignores undesirable information.   133

intergroup communication The interaction between members of different cocultures.   69

interpersonal communication Communication in which the parties consider one another as unique individuals rather than as objects. It is characterized by minimal use of stereotyped labels; unique, idiosyncratic social rules; and a high degree of information exchange.   11, 183

interpretation The perceptual process of attaching meaning to stimuli that have previously been selected and organized.   44

intersectionality The idea that people are influenced in unique ways by the complex overlap and interactions of multiple identities.   79

intimacy A state of closeness between two (or sometimes more) people. Intimacy can be manifested in several ways: physically, intellectually, emotionally, and via shared activities.   192

intimate distance One of Hall's four distance zones, ranging from skin contact to 18 inches.   173

intrapersonal communication Communication that occurs within a single person.   10

intro A brief explanation or comment before a visual aid is used.   361

introduction (of a speech) The first structural unit of a speech, in which the speaker captures the audience's attention and previews the main points to be covered. 332

irrational thinking Beliefs that have no basis in reality or logic; one source of debilitative communication apprehension. 308

jargon Specialized vocabulary used as a kind of shorthand by people with common backgrounds and experience. 108

Johari Window A model that describes the relationship between self-disclosure and self-awareness. 201

judging response A reaction in which the receiver evaluates the sender's message either favorably or unfavorably. 144

kinesics The study of body movement, gesture, and posture. 166

knowledge The understanding acquired by making sense of the raw material of information. 351

laissez-faire leadership A style in which the designated leader gives up his or her formal role, transforming the group into a loose collection of individuals. 257

language A collection of symbols, governed by rules and used to convey messages between individuals. 97

latitude of acceptance In social judgment theory, statements that a receiver would not reject. 375

latitude of noncommitment In social judgment theory, statements that a receiver would not care strongly about one way or another. 375

latitude of rejection In social judgment theory, statements that a receiver could not possibly accept. 375

Leadership Grid A two-dimensional model that identifies leadership styles as a combination of concern for people and for the task at hand. 258

legitimate power The ability to influence a group owing to one's position in a group. 264

linear communication model A characterization of communication as a one-way event in which a message flows from sender to receiver. 7

line chart A visual aid consisting of a grid that maps out the direction of a trend by plotting a series of points. 364

linguistic intergroup bias The tendency to label people and behaviors in terms that reflect their in-group or out-group status. 106

linguistic relativism The notion that language influences the way we experience the world. 102

listening The process wherein the brain reconstructs electro-chemical impulses generated by hearing into representations of the original sound and gives them meaning. 126

listening fidelity The degree of congruence between what a listener understands and what the message sender was attempting to communicate. 127

logos A speaker's use of logical arguments to appeal to an audience's sense of reasoning. 393

lose–lose problem solving An approach to conflict resolution in which neither party achieves its goals. 235

low-context culture A culture that uses language primarily to express thoughts, feelings, and ideas as directly as possible. 93

manipulators Movements in which one part of the body grooms, massages, rubs, holds, pinches, picks, or otherwise manipulates another part. 167

manuscript speech A speech that is read word for word from a prepared text. 311

mass communication The transmission of messages to large, usually widespread audiences via broadcast, print, multimedia, and other forms of media, such as recordings and movies. 13

mediated communication Communication sent via a medium other than face-to-face interaction, e.g., telephone, email, and instant messaging. It can be both mass and personal. 7

memorized speech A speech learned and delivered by rote without a written text. 311

message A sender's planned and unplanned words and nonverbal behaviors. 7

metacommunication Messages (usually relational) that refer to other messages; communication about communication. 200

mindful listening Being fully present with people—paying close attention to their gestures, manner, and silences, as well as to what they say. 128

model (in speeches and presentations) A replica of an object being discussed. It is usually used when it would be difficult or impossible to use the actual object. 361

monochronic The use of time that emphasizes punctuality, schedules, and completing one task at a time. 175

narration The presentation of speech supporting material as a story with a beginning, middle, and end. 340

narrative The stories people create and use to make sense of their personal worlds. 48

noise External, physiological, and psychological distractions that interfere with the accurate transmission and reception of a message. 7

nominal group technique A method for including the ideas of all group members in a problem-solving session. 287

nominal leader The person who is identified by title as the leader of a group. 264

nonassertion The inability or unwillingness to express one's thoughts or feelings. 223

nonverbal communication Messages expressed by other than linguistic means. 155

norms Shared values, beliefs, behaviors, and procedures that govern a group's operation. 250

number chart A visual aid that lists numbers in tabular form in order to clarify information. 362

opinion statement A statement based on the speaker's beliefs. 112

organization The perceptual process of organizing stimuli into patterns. 43

organizational communication Communication that occurs among a structured collection of people in order to meet a need or pursue a goal. 11

organizational culture A relatively stable, shared set of rules about how to behave and a set of values about what is important. 86

orientation stage When group members become familiar with one another's positions and tentatively volunteer their own.   280

out-groups Groups of people that we view as different from us.   71

outro A brief summary or conclusion after a visual aid has been used.   361

panel discussion A discussion format in which participants consider a topic more or less conversationally, without formal procedural rules. Panel discussions may be facilitated by a moderator.   281

paralanguage Nonlinguistic means of vocal expression: rate, pitch, tone, and so on.   168

paraphrasing Feedback in which the receiver rewords the speaker's thoughts and feelings.   136

parliamentary procedure A problem-solving method in which specific rules govern the way issues may be discussed and decisions made.   281

participative decision making A process in which people contribute to the decisions that will affect them.   274

passive aggression An indirect expression of aggression, delivered in a way that allows the sender to maintain a facade of kindness.   225

pathos A speaker's use of emotional appeals to persuade an audience.   393

perceived self The person we believe ourselves to be in moments of candor. It may be identical to or different from the presenting and ideal selves.   55

perception checking A three-part method for verifying the accuracy of interpretations, including a description of the sense data, two possible interpretations, and a request for confirmation of the interpretations.   54

personal distance One of Hall's four distance zones, ranging from 18 inches to 4 feet.   173

personality The set of enduring characteristics that define a person's temperament, thought processes, and social behavior.   38

persuasion The act of motivating a listener, through communication, to change a particular belief, attitude, value, or behavior.   375

phonological rules Linguistic rules governing how sounds are combined to form words.   98

phubbing A mixture of the words *phone* and *snubbing*, used to describe episodes in which people pay more attention to their devices than they do to the people around them.   185

pictogram A visual aid that conveys its meaning through an image of an actual object.   362

pie chart A visual aid that divides a circle into wedges, representing percentages of the whole.   362

pitch The highness or lowness of one's voice.   314

polychronic The use of time that emphasizes flexible schedules in which multiple tasks are pursued at the same time.   175

post hoc **fallacy** Fallacious reasoning that mistakenly assumes that one event causes another because they occur sequentially.   386

power The ability to influence others' thoughts and/or actions.   264

power distance The degree to which members of a group are willing to accept a difference in power and status.   76

pragmatic rules Rules that govern how people use language in everyday interaction.   99

prejudice An unfairly biased and intolerant attitude toward others who belong to an out-group.   89

presenting self The image a person presents to others. It may be identical to or different from the perceived and ideal selves. *See also* face.   55

problem census A technique used to equalize participation in groups when the goal is to identify important issues or problems. Members first put ideas on cards, which are then compiled by a leader to generate a comprehensive statement of the issue or problem.   281

problem-solution pattern An organizing pattern for a speech that describes an unsatisfactory state of affairs and then proposes a plan to remedy the problem.   330

procedural norms Norms that describe rules for the group's operation.   250

prompting Using silence and brief statements of encouragement to draw out a speaker.   147

proposition of fact A claim bearing on issue in which there are two or more sides of conflicting factual evidence.   378

proposition of policy A claim bearing on an issue that involves adopting or rejecting a specific course of action.   380

proposition of value A claim bearing on an issue involving the worth of some idea, person, or object.   380

proxemics The study of how people and animals use space.   173

pseudolistening An imitation of true listening.   132

public communication Communication that occurs when a group becomes too large for all members to contribute. It is characterized by an unequal amount of speaking and by limited verbal feedback.   13

public distance One of Hall's four distance zones, extending outward from 12 feet.   173

purpose statement A complete sentence that describes precisely what a speaker wants to accomplish.   299

questioning An approach in which the receiver overtly seeks additional information from the sender.   135

race A construct originally created to explain differences between people whose ancestors originated in different regions of the world—Africa, Asia, Europe, and so on.   78

rate The speed at which a speaker utters words.   313

reappropriation The process by which members of a marginalized group reframe the meaning of a term that has historically been used in a derogatory way.   100

receiver One who notices and attends to a message.   7

reductio ad absurdum **fallacy** Fallacious reasoning that unfairly attacks an argument by extending it to such extreme lengths that it looks ridiculous.   386

referent power The ability to influence others by virtue of the degree to which one is liked or respected.   266

reflected appraisal The influence of others on one's self-concept.   38

reflecting Listening that helps the person speaking hear and think about the words just spoken.   147

reinforcement stage When group members endorse the decision they have made.   280

relational listening A listening style that is driven primarily by the concern to build emotional closeness with the speaker.   137

relational message A message that expresses the social relationship between two or more individuals.   198

relational spiral A reciprocal communication pattern in which each person's message reinforces the other's.   221

relative words Words that gain their meaning by comparison.   108

remembering The act of recalling previously introduced information. Recall drops off in two phases: short term and long term.   127

residual message The part of a message a receiver can recall after short- and long-term memory loss.   127

respect The degree to which we hold others in esteem.   199

responding Providing observable feedback to another person's behavior or speech.   127

reward power The ability to influence others by the granting or promising of desirable consequences.   266

richness A term used to describe the abundance of nonverbal cues that add clarity to a verbal message.   17

roles The patterns of behavior expected of group members.   252

rule An explicit, officially stated guideline that governs group functions and member behavior.   250

salience How much weight we attach to a particular person or phenomenon.   71

script Habitual, reflexive way of behaving.   57

selection The perceptual act of attending to some stimuli in the environment and ignoring others.   43

selective listening A listening style in which the receiver responds only to messages that interest him or her.   132

self-concept The relatively stable set of perceptions each individual holds of himself or herself.   36

self-disclosure The process of deliberately revealing information about oneself that is significant and that would not normally be known by others.   200

self-esteem The part of the self-concept that involves evaluations of self-worth.   37

self-fulfilling prophecy A prediction or expectation of an event that makes the outcome more likely to occur than would otherwise have been the case.   41

self-serving bias The tendency to interpret and explain information in a way that casts the perceiver in the most favorable manner.   50

semantic rules Rules that govern the meaning of language as opposed to its structure.   99

sender The originator of a message.   7

servant leadership A style based on the idea that a leader's job is mostly to recruit outstanding team members and provide the support they need to do a good job.   257

sex A biological category such as male, female, or intersex.   46

significant other A person whose opinion is important enough to affect one's self-concept strongly.   38

signpost A phrase that emphasizes the importance of upcoming material in a speech.   358

sincere question A question posed with the genuine desire to learn from another person.   135

situational leadership A theory that argues that the most effective leadership style varies according to leader–member relations, the nominal leader's power, and the task structure.   257

slang Language used by a group of people whose members belong to a similar coculture or other group.   108

slurring The articulation error that involves overlapping the end of one word with the beginning of the next.   314

small group communication Communication within a group of a size such that every member can participate actively with the other members.   11

social comparison Evaluating ourselves in terms of how we compare with others.   39

social distance One of Hall's four distance zones, ranging from 4 to 12 feet.   173

social exchange theory The idea that we seek out people who can give us rewards that are greater than or equal to the costs we encounter in dealing with them.   187

social judgment theory An explanation of attitude change that posits that opinions will change only in small increments and only when the target opinions lie within the receiver's latitudes of acceptance and noncommitment.   375

social loafing The tendency of some people to do less work as a group member than they would as an individual.   250

social media Digital communication channels used primarily for personal reasons, often to reach small groups of receivers.   16

social norms Group norms that govern the way members relate to one another.   250

social penetration model A theory that describes how intimacy can be achieved via the breadth and depth of self-disclosure.   201

social roles Emotional roles concerned with maintaining smooth personal relationships among group members. Also termed "maintenance functions."   255

sociogram A graphic representation of the interaction patterns in a group.   251

sound bite A brief recorded excerpt from a longer statement.   361

space pattern An organizing plan in a speech that arranges points according to their physical location.   329

specific purpose The precise effect that the speaker wants to have on an audience. It is expressed in the form of a purpose statement.   299

stage hogs People who are more concerned with making their own points than with understanding the speaker.   133

statistic Numbers arranged or organized to show how a fact or principle is true for a large percentage of cases.   338

stereotyping The perceptual process of applying exaggerated beliefs associated with a categorizing system.   49

stonewalling Refusing to engage with the other person.   220

substitution The articulation error that involves replacing part of a word with an incorrect sound.   314

supportive listening The reception approach to use when others seek help for personal dilemmas.   142

survey research Information gathering in which the responses of a population sample are collected to disclose information about the larger group.   306

symbol An arbitrary sign used to represent a thing, person, idea, event, or relationship in ways that make communication possible.   6

sympathy Compassion for another's situation. *See also* empathy.   53

symposium A discussion format in which participants divide the topic in a manner that allows each member to deliver in-depth information without interruption.   281

synchronous communication Communication that occurs in real time.   18

syntactic rules Rules that govern the ways in which symbols can be arranged as opposed to the meanings of those symbols.   98

target audience That part of an audience that must be influenced in order to achieve a persuasive goal.   388

task norms Group norms that govern the way members handle the job at hand.   250

task-oriented listening A listening style that is primarily concerned with accomplishing the task at hand.   134

task roles Roles group members take on in order to help solve a problem.   255

territory Fixed space that an individual assumes some right to occupy.   174

testimony Supporting material that proves or illustrates a point by citing an authoritative source.   339

thesis statement A complete sentence describing the central idea of a speech.   300

time pattern An organizing plan for a speech based on chronology.   329

topic pattern An organizing plan for a speech that arranges points according to logical types or categories.   329

trait theories of leadership A school of thought based on the belief that some people are born to be leaders and others are not.   257

transactional communication model A characterization of communication as the simultaneous sending and receiving of messages in an ongoing, irreversible process.   9

transformational leaders Defined by their devotion to help a team fulfill an important mission.   260

transition A phrase that connects ideas in a speech by showing how one relates to the other.   334

uncertainty avoidance The cultural tendency to seek stability and honor tradition instead of welcoming risk, uncertainty, and change.   75

understanding The act of interpreting a message by following syntactic, semantic, and pragmatic rules.   127

value A deeply rooted belief about a concept's inherent worth.   302

virtual groups People who interact with one another via mediated channels, without meeting face-to-face.   248

visual aids Graphic devices used in a speech to illustrate or support ideas.   361

vocal citation A simple, concise, spoken statement of the source of your evidence.   357

Web 2.0 The Internet's evolution from a one-way medium into a "masspersonal" phenomenon.   16

wheel network A communication network in which a gatekeeper regulates the flow of information from all other members.   252

win–lose problem solving An approach to conflict resolution in which one party reaches his or her goal at the expense of the other.   234

win–win problem solving An approach to conflict resolution in which the parties work together to satisfy all their goals.   236

word chart A visual aid that lists words or terms in tabular form in order to clarify information.   362

working outline A constantly changing organizational aid used in planning a speech.   325

wrap-around A brief introduction before a visual aid is presented, accompanied by a brief conclusion afterward.   361

Credits

PHOTOGRAPHS

Page 2 Meinzahn/iStock; **4** Courtesy of Tony Hsieh; **5** TCJ2020/Shutterstock.com; **6** © imageBROKER / Alamy Stock Photo; **6** ASSOCIATED PRESS; **11** © Cultura Creative (RF) / Alamy Stock Photo; **12** SolisImages/iStock; **13** Randy Miramontez/Shutterstock.com; **15** © Tribune Content Agency, LLC. All Rights Reserved. Reprinted with permission.; **17** © PhotoAlto / Alamy Stock Photo; **19** Masterfile Royalty Free; **21** ASSOCIATED PRESS; **34** Jonathan Knowles/Stone/Getty Images; **36** Chris Pizzello/Invision/AP; **37** ASSOCIATED PRESS; **38** Vianney Le Caer/Invision/AP; **43** Wayne Eardley/Masterfile; **44** pathdoc/Shutterstock.com; **46** Courtesy of Alia Roth; **47** Jason Henthorn/iStock; **49** JoseGirarte/iStock; **50** mrsmuckers/iStock; **55** kikkerdirk/Fotolia; **56** lassedesignen/Fotolia; **60** © charles taylor / Alamy Stock Photo; **62** Kim Kardashian Hollywood; **66** daizuoxin/Shutterstock.com; **68** Courtesy of Robin Luo; **69** HBO/ALLSTAR; **70** The Canadian Press; **74** Rich Fury/Invision/AP; **78** ASSOCIATED PRESS; **80** shylendrahoode/iStock; **82** FILM MOVEMENT/ALLSTAR; **83** ZUMA Press, Inc. / Alamy Stock Photo; **85** Photo: ©ABC; **87** Masterfile Royalty Free; **94** marekuliasz/iStock; **96** With permission of Scott H. Young and Vat Jaiswal; **97** © Photos 12 / Alamy Stock Photo; **99** Choreograph/iStock; **101** vector illustration/Shutterstock.com; **107** © Todd Elert; **113** ASSOCIATED PRESS; **118** © Mark Wiener / Alamy Stock Photo; **122** © UpperCut Images / Alamy Stock Photo; **124** Courtesy of Erica Nicole; **125** © Blend Images / Alamy Stock Photo; **127** Source: avadiamond.com; **134** © PhotoAlto / Alamy Stock Photo; **139** JGI/Jamie Grill/Getty Images; **141** Hill Street Studios / Getty Images; **152** © GraficallyMinded / Alamy Stock Photo; **154** Photo by Fred J. Field; **155** Nicescene/Shutterstock.com; **157** *The Hackensack Record*; **160** A.RICARDO/Shutterstock.com; **168** © Miriam Dörr / Alamy Stock Photo; **170** Scott Green / © IFC / Courtesy: Everett Collection; **170** K2 images/Shutterstock.com; **174** © evan hurd / Alamy Stock Photo; **176** Helga Esteb/Shutterstock.com; **180** © Clio Media Ltd / Alamy Stock Photo; **182** marla dawn studio/Shutterstock.com; **186** Eamonn McCormack/WireImage; **189** franckreporter/iStock; **199** © Image Source Plus / Alamy Stock Photo; **202** EdStock/iStock; **207** Boden/Ledingham/Masterfile; **212** Robert Simon/iStock; **214** Photo: Danny Clark; **215** Kamil Sarna/iStock; **225** CBS/ALLSTAR; **231** Joe Seer/Shutterstock.com; **233** EdStock/iStock; **242** xuanhuongho/Shutterstock.com; **244** ASSOCIATED PRESS; **247** CatonPhoto/Shutterstock.com; **251** MarcelaC/iStock; **251** Izabela Habur/iStock; **253** dotshock/Shutterstock.com; **254** Dean Mitchell/iStock; **255** © photonic 20 / Alamy Stock Photo; **257** Andrew Burton/Getty Images News/Getty Images; **260** Taylor Hill/WireImage/Getty Images; **265** Tonko Oosterink/Shutterstock.com; **270** Photo_Concepts/iStock; **272** Brian Ach/Stringer/Getty Images; **273** Disney XD/Disney ABC Television Group/Getty Images; **275** Yardi Systems, Inc.; **276** OPOLJA/Shutterstock.com; **281** dotshock/Shutterstock.com; **289** CBS Photo Archive/CBS/Getty Images; **296** Glynnis Jones/Shutterstock.com; **298** Associated Press; **299** Chris Pizzello/Invision/AP; **301** © Rob Walls / Alamy Stock Photo; **304** sdominick/iStock; **307** stevecoleimages/iStock; **308** © DOD Photo / Alamy Stock Photo; **311** Troels Graugaard/iStock; **322** © Tetra Images / Alamy Stock Photo; **324** Courtesy of Andrew Nylon; **329** ASSOCIATED PRESS; **332** DREAMWORKS SKG/ALLSTAR; **337** Dimitrios Kambouris/Getty Images; **343** CC BY 2.0 Neil Harbisson; **348** Rawpixel.com/Shutterstock.com; **350** Out & Around Productions, LLC; **361** Kevin Winter/Getty Images; **368** Out & Around Productions, LLC; **369** Out & Around Productions, LLC; **369** Out & Around Productions, LLC; **369** Out & Around Productions, LLC; **372** Mirko Raatz/Fotolia; **374** Courtesy of Kristina Medero; **379** Courtesy of The Washington Post; **384** Adam Taylor/ Disney ABC Television Group/Getty Images; **389** Jemal Countess/Stringer/Getty Images

CARTOONS

Page 5 © The New Yorker Collection 1984 Warren Miller from cartoonbank.com. All Rights Reserved.; **16** Alex Gregory The New Yorker Collection/The Cartoon Bank; **24** CALVIN and HOBBES © 1994 Watterson. Distributed by UNIVERSAL PRESS SYNDICATE. Reprinted with permission. All Rights Reserved.; **29** Mick Stevens The New Yorker Collection/The Cartoon Bank; **42** Edward Frascino The New Yorker Collection/The Cartoon Bank; **47** Edward Koren The New Yorker Collection/The Cartoon Bank; **53** William Steig The New Yorker Collection/The Cartoon Bank; **61** Peter Steiner The New Yorker Collection/The Cartoon Bank; **79** Paul Noth The New Yorker Collection/The Cartoon Bank; **88** © 2011 Malcolm Evans; **103** Drew Dernavich The New Yorker Collection/The Cartoon Bank; **117** Leo Cullum The New Yorker Collection/The Cartoon Bank; **126** © 2003 Zits Partnership Distribtued by King Features Syndicate, Inc.; **130** DILBERT © 2009 Scott Adams. Used by permission of UNIVERSAL UCLICK. All rights reserved.; **133** CALVIN and HOBBES © 1995 Watterson. Reprinted by permission of UNIVERSAL UCLICK. All Rights Reserved.; **136** Courtesy of Ted Goff; **159** DILBERT © 1991 Scott Adams. Used by permission of UNIVERSAL UCLICK. All rights reserved.; **165** DILBERT © 2006 Scott Adams. Used by permission of UNIVERSAL UCLICK. All rights reserved.; **169** Alex Gregory The New Yorker Collection/The Cartoon Bank; **200** Leo Cullum The New Yorker Collection/The Cartoon Bank; **201** Leo Cullum The New Yorker Collection/The Cartoon Bank; **205** ©2006 Zits Partnership Distributed by King Features Syndicate Inc.; **219** Ted Goff, North America Syndicate, 1994; **234** Leo Cullum The New Yorker Collection/The Cartoon Bank; **245** THE FAR SIDE ©1987 FARWORKS, INC. Used by permission. All rights reserved.; **292** Source: © Original Artist. Reproduction rights obtainable from www.CartoonStock.com; **337** DILBERT © 2006 Scott Adams. Used by permission of UNIVERSAL

UCLICK. All rights reserved.; **360** By permission of Leigh Rubin and Creators Syndicate, Inc.

TABLES AND FIGURES

Page 17 Source: Lenhart, A. (2015). Teens, technologies, and friendships. Pew Research Center. http://www.pewinternet.org/2015/08/06/teens-technology-and-friendships/.; **26** Source: Adapted from R. B. Adler, J. M. Elmhorst, & K. Lucas. (2003). Communicating at work: Strategies for success in business and the professions (11th ed.). New York: McGraw-Hill, p. 14.; **27** Source: Raine, L., & Zickuhr, K. (2015, August 26). Americans' views on mobile etiquette. Pew Research Center. Retrieved from http://www.pewinternet.org/files/2015/08/2015-08-26_mobile-etiquette_FINAL.pdf.; **72** Source: Adapted from F. Trompenaars. (2012). Riding the waves of culture, 3rd ed. New York: McGraw-Hill, p. 67.; **73** Source: Adapted by Sandra Sudweeks from H. C. Triandis. (1990). Cross-cultural studies of individualism and collectivism. In J. Berman (Ed.), Nebraska symposium on motivation (pp. 41–133). Lincoln: University of Nebraska Press; and E. T. Hall. (1976). Beyond culture. New York: Doubleday.; **84** Source: J. Harwood. (2007). Understanding communication and aging: Developing knowledge and awareness. Newbury Park, CA: Sage, p. 76.; **131** Four Thought Patterns. Source: A.D. Wolvin and C.G. Coakley, Perspectives on Listening (Norwood, NJ: Ablex, 1993), p. 115.; **155** Source: Adapted from John Stewart, J., & D'Angelo, G. (1980). Together: Communicating interpersonally (2nd ed.). Reading, MA: Addison-Wesley, p. 22. Copyright © 1993 by McGraw-Hill. Reprinted/adapted by permission.; **165** Source: Based on material from Ekman, P. (1981). Mistakes when deceiving. In T. A. Sebok & R. Rosenthal (Eds.), The clever Hans phenomenon: Communication with horses, whales, apes and people (pp. 269–278). New York: New York Academy of Sciences. See also Samhita, L. & Gross, H. (2103). The "Clever Hans Phenomenon" revisited. Communicative & Integrative Biology 6(6), e27122. doi: 10.4161/cib.27122; **207** Source: Adapted from categories originally presented in Camden, C., Motley, M.T., & Wilson, A. White lies in interpersonal communication: A taxonomy and preliminary investigation of social motivations. *Western Journal of Speech Communication, 48*, 315.; **221** Source: Adapted from Hess, J. A. (2002). Distance regulation in personal relationships: The development of a conceptual model and a test of representational validity. Journal of Social and Personal Relationships, 19, 663–683.; **255** Source: "Functional Roles of Group Members" and "Dysfunctional Roles of Group Members," adapted from Wilson, G., & Hanna, M. (1986). Groups in context: Leadership and participation in decision-making groups, pp. 144–146. Reprinted by permission of McGraw-Hill Companies, Inc.; **256** Source: Adapted with permission of The Free Press, a division of Macmillan, Inc., from Bass, B. M. Stodgill's handbook of leadership (Rev. ed.). Copyright © 1974, 1981 by The Free Press.; **258** Source: The Leadership Grid® Figure from Blake, R. R., & McCanse, A. A. Leadership dilemmas–grid solutions. Houston, TX: Gulf, p. 29. Copyright © 1991 by Scientific Methods, Inc. Reproduced by permission of the owners.; **265** Source: Adapted from Rothwell, J. D. (1998). In mixed company: Small group communication (3rd ed.). Fort Worth, TX: Harcourt Brace, pp. 252–272. Reprinted with permission of Wadsworth, an imprint of the Wadsworth Group, a division of Thomson Learning. Fax 800-730-2215.; **278** Source: Adapted from research summarized in Wheelan, S. A., Murphy, D., Tsumaura, E., & Kline, S. F. (1998). Member perceptions of internal group dynamics and productivity. Small Group Research, 29, 371–393.; **287** Source: Adapted from Brilhart, J., & Galanes, G. Adapting problem-solving methods. Effective group discussion (10th ed.), p. 291. Copyright © 2001. Reprinted by permission of McGraw-Hill Companies, Inc.; **362** Source: http://www.kon.org/urc/v13/ewert.html; **363** Adapted from graphjam.com; **363** Source: http://hubpages.com/literature/mostreadbooks; **364** Source: Washington Post.; **378** Source: Adapted from Andersen, M. K. (1979). An analysis of the treatment of ethos in selected speech communication textbooks (Unpublished dissertation). University of Michigan, Ann Arbor, pp. 244–247.

Index

Diversity
  of American dialects, 315
  of audience members, 301
  communicating with the disabled, 83
  in emotional expressiveness, 232
  of gender pronoun choice, 81
  group, 274
  of identity management and coming
      out, 58
  information affected by, 354
  of language and worldview, 103
  of multicultural teams, 275
  nontraditional organization patterns,
      331
  in other-sex friendships, 189, 190
  persuasion differences, 387
  See also Culture
Division, rule of, 327
Dolezal, Rachel, 37
Dominator (role), 255
Dunn, Nathan, 336–337
Dyad(s), 11
Dyadic communication, 11
Dysfunctional roles, 255, 256, 292

Eggleston, T. Stephen, 336
Either-or fallacy, 386
Elaborator/clarifier (role), 253
Ellis, Albert, 308
Ellis, Donald, 233
Email, 7, 10, 29, 106, 156, 198, 252
  identity management in, 61
  percentage of teens using, 17
  pros and cons of, 26
  at work, 12
Emblems, 162
Emergence stage of groups, 280
Emergent leader, 261
Emoticons, 160, 163
Emotional appeals, 141
Emotional dimension
  of empathy, 52–53
  of nonverbal communication, 157
  of perception, 43
Emotional evidence, 385
Emotional expressiveness, 232, 233
Emotional intelligence, 158
Emotive language, 48, 113–114
Empathy, 23–24
  communication climate and, 223
  conflict resolution and, 222
  defined, 52
  dimensions of, 52–53
  intercultural communication and, 87

perception and, 52–54
Encoding, 10, 158
Energizer (role), 253
Environments, 8–9, 174–175
Epley, Nicholas, 169
Equivocal words, 107–108
Equivocation, 114–116, 207
Escalatory spirals, 222
Ethical persuasion, 377, 382, 385, 388
Ethics
  of adapting to a hostile audience, 390
  of adapting to speaking situations,
      302
  of analyzing communication
      behaviors, 378
  of civility when values clash, 91
  of clothing and impression
      management, 171
  of communicating *vs.* not
      communicating, 31
  of conflict management, 227
  of euphemisms and equivocations,
      114
  of free *vs.* hate speech, 102
  of group participation inequality, 292
  of hidden agendas, 250
  of honesty, 40
  of listening, 139
  of lying and evading, 206
  of multiple identities, 63
  of online extracurricular
      relationships, 185
  of simplicity, 356
  of supporting material, 340
Ethnicity, 78–79
Ethnocentrism, 89, 354
Ethos, 393
Etiquette for social media use, 27–29,
      30
Euphemisms, 114
Evaluator/critic (role), 254
Evasive language, 114–116, 206,
      207–208
Events, speeches about, 352
Evidence, 140, 385–386
Examples, 338
Expectations
  of audience, 304
  perception and, 45
Experimenting stage of relationships,
      195–196
Expert opinion, 290
Expert power, 265–266
Explanations, 352

Expressed struggle, 215
Extemporaneous speeches, 311
External noise. *See* Physical noise
Eye contact, 160, 161, 168, 313

Face, 55, 74
Facebook, 14, 16, 40, 62, 163, 184
"Facebook official" relationship status,
      196
Face saving, 104, 114, 144, 207–208,
      224, 232
Face-to-face communication, 7, 9, 14
  computer-mediated communication
      compared with, 28–29, 156
  mediated communication compared
      with, 17–18
  percentage of teens using, 17
  pros and cons of, 26
Facework, 55
Facial expressions, 80, 159, 161, 168,
      313
Facilitative communication
      apprehension, 307
Factiva, 306
Facts
  citing startling, 333
  inferences *vs.*, 112–113
  opinions *vs.*, 112
  propositions of, 378–380
Factual examples, 338
Fallacies, 141, 308–309, 386–387, 388
False cause fallacy, 386–387
Familiarity, 45, 330, 333, 355, 356
Family
  relationships in, 191–193
  rules and norms in, 251
Faulty analogies, 388
Feedback, 9, 13, 127
Feeling expresser (role), 254
Feiler, Bruce, 192
Feiler, Linda, 192
Feminine sex type/role, 46–47
Fiennes, Ralph, 109
Filley, Albert, 235
First-generation college (FCG) students,
      85–87
First impressions, 51, 187
Flaa, Jennifer, 196
Flaming, 27, 233
Flip pads, 364–366
Focus groups, 281
Foggy mirror concept, 62
Follower (role), 254
Followership, 263–267